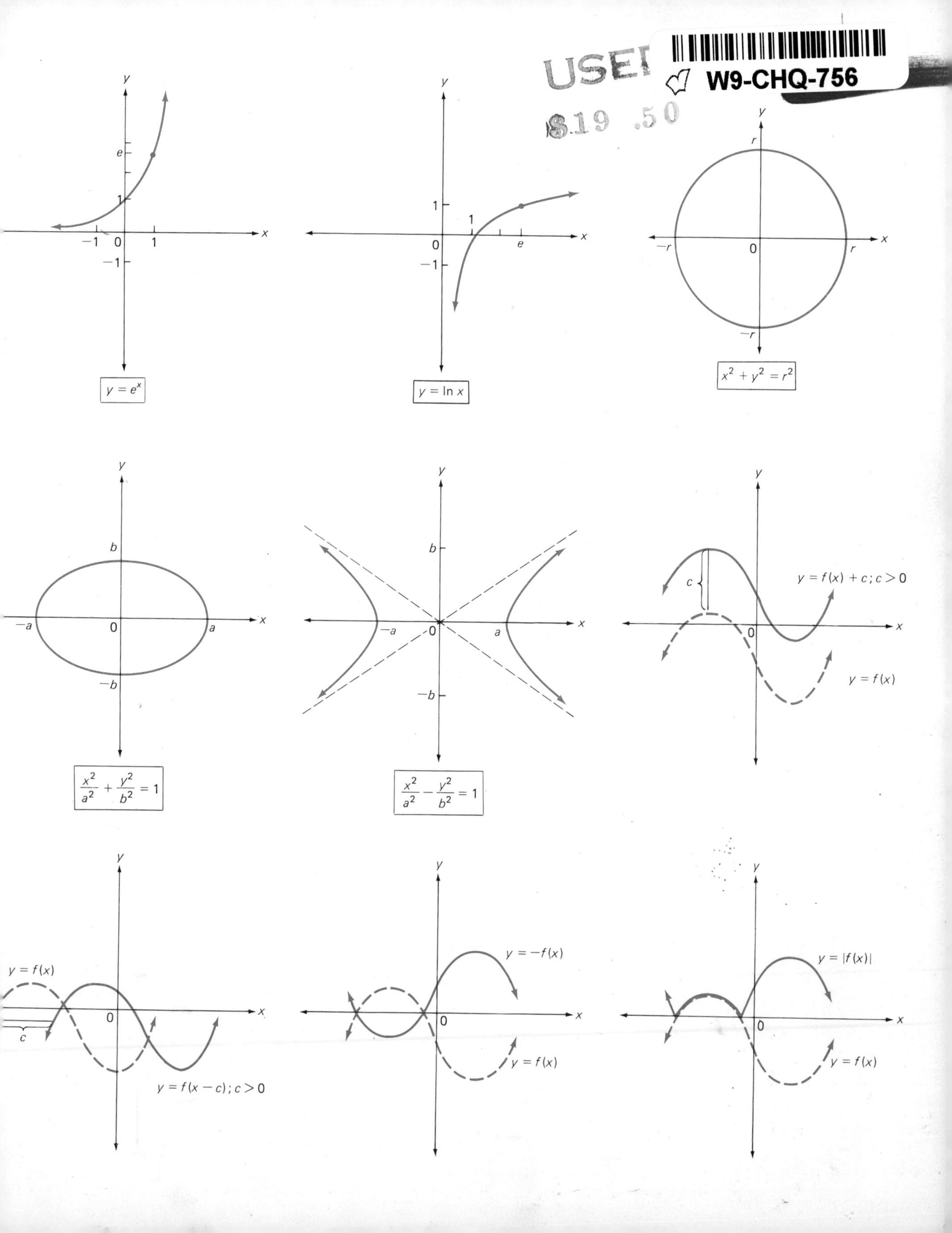
y = e^x
y = ln x
x^2 + y^2 = r^2
x^2/a^2 + y^2/b^2 = 1
x^2/a^2 − y^2/b^2 = 1
y = f(x) + c; c > 0
y = f(x)
y = f(x − c); c > 0
y = −f(x)
y = |f(x)|
USED
$19.50
W9-CHQ-756

COLLEGE ALGEBRA

COLLEGE

PRENTICE-HALL, INC., ENGLEWOOD CLIFFS, NEW JERSEY 07632

ALGEBRA

Max A. Sobel

Montclair State College

Norbert Lerner

State University of New York
at Cortland

Library of Congress Cataloging in Publication Data

Sobel, Max A.
College algebra.

Includes index.
1. Algebra. I. Lerner, Norbert. II. Title.
QA154.2.S62 1983 512.9 82-13335
ISBN 0-13-141796-7

COLLEGE ALGEBRA
Max A. Sobel/Norbert Lerner

Portions of this book were previously published as Chapters 1 through 7 and Chapter 10 of Algebra and Trigonometry: A Pre-Calculus Approach, second edition, by Max A. Sobel and Norbert Lerner.

10 9 8 7 6 5 4 3 2 1

Editorial/production supervision: Kathleen M. Lafferty and Paula Martinac
Interior and cover designs: Walter A. Behnke
Editorial assistant: Susan Pintner
Manufacturing buyer: John B. Hall
Cover illustration: Vitalia Hodgetts

ISBN 0-13-141796-7

PRENTICE-HALL INTERNATIONAL, INC., *London*
PRENTICE-HALL OF AUSTRALIA PTY. LIMITED, *Sydney*
EDITORA PRENTICE-HALL DO BRASIL, LTDA., *Rio de Janeiro*
PRENTICE-HALL CANADA INC., *Toronto*
PRENTICE-HALL OF INDIA PRIVATE LIMITED, *New Delhi*
PRENTICE-HALL OF JAPAN, INC., *Tokyo*
PRENTICE-HALL OF SOUTHEAST ASIA PTE. LTD., *Singapore*
WHITEHALL BOOKS LIMITED, *Wellington, New Zealand*

CONTENTS

2

Fundamentals of algebra

page 35

3

Linear functions and equations

page 77

4

Quadratic functions and equations

page 121

5

Polynomial and rational functions

page 171

6

Radical functions and equations

page 215

7

Exponential and logarithmic functions

page 239

8

Matrices, determinants, and linear systems

page 279

PREFACE

College Algebra has been written to provide the essential concepts and skills of algebra that are needed for further study in mathematics, with special emphasis given to direct preparation for the study of calculus. Thus a major objective of this book is to provide a text that will help students make a more comfortable transition from elementary mathematics to calculus.

There is an extensive review of the fundamentals of algebra in Chapter 2. However, much of the work with solving equations is delayed until the appropriate functions are studied. Thus, for example, quadratic equations are studied in the setting of quadratic functions. This allows for further building on algebraic skills as the course progresses while simultaneously showing how solutions to equations can be viewed as intercepts of related curves.

The text also contains a strong emphasis on graphing throughout. The objective is to have students become familiar with basic graphs and learn how to obtain new curves from these by using translations and reflections.

Improving the students' ability to read mathematics should be a major goal of a pre-calculus course. To assist students in this direction, the exposition here is presented in a relaxed style that avoids unnecessary mathematical jargon without sacrificing mathematical accuracy. Included in this text are the following pedagogical features that are designed to assist the students in using the book for self-study and to reinforce classroom instruction:

TEST YOUR UNDERSTANDING:

These are short sets of exercises (in addition to the regular exercises) that are found within most sections of the text so that students can test knowledge of new material just developed. Answers to these are given at the end of each chapter.

CAUTION ITEMS:

Where appropriate, students are alerted to the typical kind of errors that they should avoid.

ILLUSTRATIVE EXAMPLES AND EXERCISES:

The text contains numerous illustrative examples with detailed solutions. There are approximately 4000 exercises for the students to try, with answers to the odd-numbered problems given at the back of the book.

REVIEW EXERCISES:

Each chapter has a set of review exercises that are keyed directly to the illustrative examples developed in the text. Students can use these as a review of the work of the chapter and compare results with the worked-out solutions that can be found in the body of the chapter.

MARGINAL NOTES:

Marginal notes are a pedagogical feature designed to assist the students' understanding throughout the text.

SAMPLE TEST QUESTIONS:

Sample test questions are given at the end of each chapter, with answers provided in the back of the book.

All too often students are unable to adapt their knowledge of basic mathematics to calculus. In an effort to solve this complex problem of transition, the authors have included exercises and material that can best be described as being *directly supportive* of topics in calculus. Some examples of these *supportive* items follow. Note that the calculus topics themselves are *not* included in the text.

Calculus topic Simplifying derivatives.

Pre-calculus support Procedures needed to simplify derivatives are included using the same algebraic forms as will be encountered in calculus. For example, the students learn to convert $x^{1/3} + \frac{x-8}{3x^{2/3}}$ (the derivative of $x^{1/3}(x-8)$) into the form $\frac{4(x-2)}{3x^{2/3}}$.

The concept of a derivative. *Calculus topic*

Work with difference quotients is introduced early and reinforced throughout as new functions are studied. *Pre-calculus support*

Using the signs of derivatives. *Calculus topic*

Inequalities are considered early in the book. These concepts are applied later to determine the signs of a variety of functions. A convenient tabular format is used throughout that can easily be extended for working with signs of derivatives when the students get to calculus. *Pre-calculus support*

The chain rule for derivatives. *Calculus topic*

In addition to the usual work in forming composites of given functions, special material is included that shows how to reverse this process. For example, the student learns how to view a *given* function, such as $f(x) = \sqrt{(2x - 1)^5}$, as the composition of other functions. Much of the difficulty students have later with the chain rule appears to be related to the inability to do this type of decomposition. *Pre-calculus support*

Applied or verbal problems. *Calculus topic*

A major difficulty that many students have throughout calculus with such problems deals with setting up functions related to practical or geometric situations. Students are introduced to this type of thinking in Section 3.8, and this experience is reinforced later via follow-up exercises as the course progresses. *Pre-calculus support*

COURSE STRUCTURE

The text allows for considerable flexibility in the selection and ordering of topics. The first five chapters should be done in sequence. Thereafter the dependency of the chapters is displayed in the diagram.

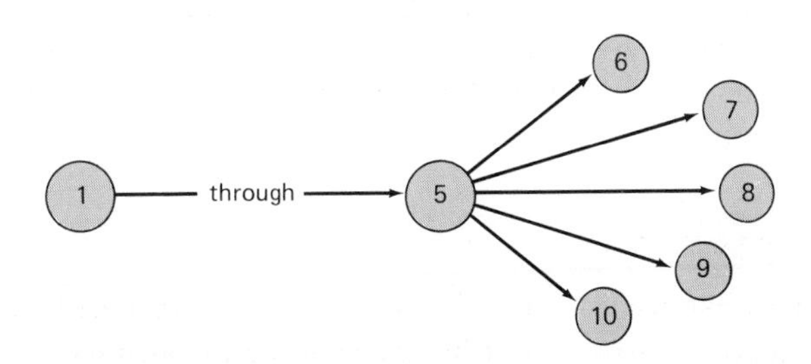

Note: Even though the material in Chapter 8, Matrices, Determinants, and Linear Systems, could be done directly after Chapter 3, Linear Functions and Equations, we have placed it later since the added maturity the students gain in the study of the first five to seven chapters will be helpful in the mastering of Chapter 8.

If the students' background is adequate, much of Chapters 1 and 2 can be assigned as self-study or covered quickly. Also, the first few sections of Chapter 3 can be covered quickly in such classes. This would allow for substantial coverage of much of the remaining chapters.

■ This symbol is used to identify exercises that are more *directly* supportive of topics in calculus than the other exercises. Also, when this symbol appears next to a section heading it means that the section and its exercises fall into this support category. For such sections the exercises are not separately labeled with this symbol.

The subject matter labeled with ■ may be treated as optional. It is not prerequisite to the subsequent developments.

* This symbol is used to identify exercises of a more challenging nature.

There are different points of view at this time as to the role that calculators should play in a mathematics course. The authors feel this is a choice that must be left to the judgment and discretion of the individual instructor. In this text students are advised occasionally to make use of a calculator to enhance the pedagogical development, as well as to complete certain exercises that involve cumbersome computations. However, use of a calculator is neither a prerequisite nor a requirement for the completion of the course.

Additional comments regarding course content are given in the Instructor's Manual, which is available upon adoption.

ACKNOWLEDGMENTS

The preparation of this text was significantly influenced by many individuals. For their detailed reviews, constructive criticisms, and suggestions as we prepared this text we thank the following professors:

Frank Battles, Massachusetts Maritime Academy
C. F. Blakemore, University of New Orleans
Joe W. Fisher, University of Cincinnati
James Hall, University of Wisconsin, Madison
Louise Hasty, Austin Community College
Kay Hudspeth, Pennsylvania State University
Adam J. Hulin, University of New Orleans
Erich Nussbaum, State University of New York, Albany
Paul M. Young, Kansas State University

For their encouragement, support, and assistance we thank the staff at Prentice-Hall, especially Robert Sickles, Kathleen Lafferty, and Paula Martinac.

For his careful proofreading of the manuscript at all stages, as well as for his many valuable suggestions, we thank George Feissner of the State University of New York at Cortland.

Finally, we thank Karin Lerner for her excellent typing of the manuscript as well as the accompanying Instructor's Manual.

The authors sincerely hope that you will find this book teachable and enjoyable, and welcome your comments, criticisms, and suggestions.

Max A. Sobel
Norbert Lerner

SUGGESTIONS FOR THE STUDENT

The most important suggestion we can give regarding your study habits for this mathematics course is that you make every possible effort to keep up to date. Set aside regular time periods for the work in this course and stick to this plan.

During your periods of study, read the text often. Few students do this because they find mathematics very difficult to read, and the frustrations they encounter do not seem to be worth the effort. We ask you to be patient; always try to dig things out on your own in spite of the difficulties you may encounter. Even a modest effort along these lines will prove to be rewarding in the long run. We have tried to make this book readable and have included some special features (listed in the Preface) that are designed to help you during your periods of self-study.

How does one succeed in a mathematics course? Unfortunately, there is no universal prescription guaranteed to work. However, the experience that the authors have had with many students throughout their teaching careers provides some guidelines that seem to help. We suggest that you make an effort to follow these suggestions that have been useful to other students who have taken similar courses in the past:

1 Read the text! We recognize that mathematics is not always easy to read, but we have made every effort to make this book as readable as possible. Don't look upon the textbook merely as a source of exercises. Rather, read each section thoroughly to reinforce classroom instruction.

2 Try to complete the illustrative examples that appear within each section before studying the solutions provided. Mathematics is not "a spectator sport"; study the book with paper and pencil at hand.

3 Attempt each "Test Your Understanding" exercise and check your results with the answers given at the back of each chapter. Re-read the section if you have difficulty with these exercises.

4 Try to complete as many of the exercises as possible at the end of each section. Complete the odd-numbered ones first and check your answers with those given at the back of the book. At times your answers may be in a different form than that given in the book; if so, try to show that the two results are equivalent. Considerable efforts have been made to assure that the answers are correct. However, if you happen to find an occasional incorrect answer, please write to let us know about it.

5 Prior to a test you should make use of the review exercises that appear at the end of each chapter. These are collections of representative examples from within each chapter. You can check your results by referring back to the designated section from which they are taken where you will find the worked-out solution for each one.

6 Each chapter ends with sample test questions that will help tell you whether or not you have understood the work of that chapter. Check your work with the answers that are given at the back of the book.

7 If you find yourself making a careless error in completing a problem, it would be wise to attempt another similar problem. If you make the same mistake again, it is best to go back and review, since a serious misunderstanding of basic concepts may be involved.

8 If convenient, find time to solve problems with a classmate. Such cooperative efforts can be quite beneficial as you attempt to explain ideas to each other.

We hope that you will have a profitable semester studying from this book. Good luck!

Max A. Sobel
Norbert Lerner

COLLEGE ALGEBRA

REAL NUMBERS, EQUATIONS, INEQUALITIES

1

1.1 Real numbers and their properties

Throughout this course we will be concerned with the set of **real numbers**. Here are some examples of such numbers:

$$7 \qquad -5 \qquad \tfrac{2}{3} \qquad \sqrt{13} \qquad 0.25 \qquad 0.333\ldots \qquad \sqrt[3]{17} \qquad \pi$$

We can represent these numbers on a **real number line** and find that each point on the number line may be named by a real number, and each real number is the *coordinate* of a point on the number line. We say that there is a *one-to-one correspondence* between the set of real numbers and the set of points on the number line. Here is a number line with some of its points labeled by real numbers.

WATCH THE MARGINS! We will use these for special notes, explanations, and hints.

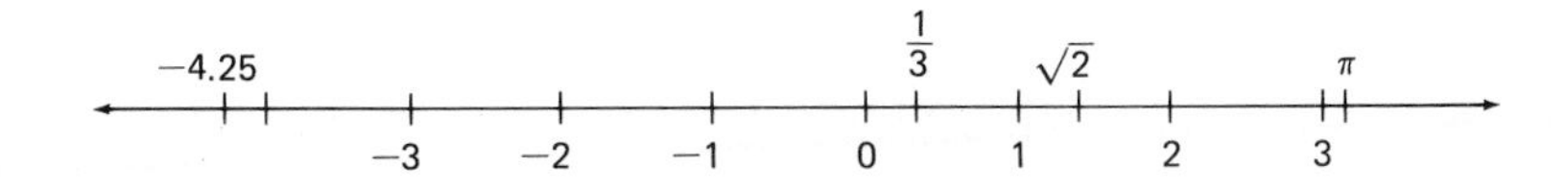

You are undoubtedly familiar with many of the basic properties of the real numbers, although you may not know their specific names. Here is a list of some of these important properties that you may use for reference. In each case letters, called *variables*, are used to represent real numbers.

Closure properties The sum and product of two real numbers is a real number. That is, for any real numbers a and b,

$$a + b \text{ is a real number}$$
$$a \times b \text{ is a real number}$$

The product $a \times b$ may be written in a number of other ways, such as $a \cdot b$, $(a)(b)$, $a(b)$, $(a)b$, or just ab.

Commutative properties The sum and product of two real numbers is not affected by the order in which they are combined. That is, for any real numbers a and b,

$$a + b = b + a$$
$$a \times b = b \times a$$

For example, $12 + 17 = 17 + 12$. Also, $12 \times 17 = 17 \times 12$.

Associative properties The sum and product of three real numbers is the same when the third is combined with the first two or when the first is combined with the last two. That is, for real numbers a, b, and c,

$$(a + b) + c = a + (b + c)$$
$$(a \times b) \times c = a \times (b \times c)$$

For example, $(15 + 23) + 18 = 15 + (23 + 18)$.

Also, $(15 \times 23) \times 18 = 15 \times (23 \times 18)$. The order in which three real numbers are added or multiplied does not affect the result.

Before proceeding with some additional properties of the real numbers, let us pause to be sure that the preceding properties are understood.

TEST YOUR UNDERSTANDING

Throughout this book we shall occasionally pause for you to test your understanding of the ideas just presented. If you have difficulty with these brief sets of exercises, you should reread the material of the section before going ahead. Answers are given at the end of the chapter.

In Exercises 1–6, name the property of real numbers being illustrated.

1 $3 + (\frac{1}{2} + 5) = (3 + \frac{1}{2}) + 5$

2 $3 + (\frac{1}{2} + 5) = (\frac{1}{2} + 5) + 3$

3 $6 + 4(2) = 4(2) + 6$

4 $8\pi = \pi 8$

5 $(17 \cdot 23)59 = (23 \cdot 17)59$

6 $(17 \cdot 23)59 = 17(23 \cdot 59)$

7 Does $3 - 5 = 5 - 3$? Is there a commutative property for subtraction?

8 Give a *counterexample* to show that the set of real numbers is not commutative with respect to division. (That is, use a specific example to show that $a \div b \neq b \div a$.)

9 Does $(8 - 5) - 2 = 8 - (5 - 2)$? Is there an associative property for subtraction?

10 Give a counterexample to show that the set of real numbers is not associative with respect to division.

11 Are $3 + \sqrt{7}$ and $3\sqrt{7}$ real numbers? Explain.

***Counterexamples* are important in mathematics. To disprove a statement, we need only find *one* case where the statement does not hold.**

There is a very important property of the set of real numbers that combines the two operations of addition and multiplication. To introduce this property let us evaluate the expression $5(3 + 9)$ in two different ways.

(a) Add first, then multiply.

$$5(3 + 9) = 5(12) = 60$$

(b) Multiply first, then add.

$$5(3 + 9) = (5)(3) + (5)(9) = 15 + 45 = 60$$

The same result is obtained either way. This example illustrates the use of the **distributive property of multiplication over addition.**

Distributive property

The product of a real number times the sum of two others is the same as the sum of the products of the first number times each of the others. That is, for real numbers a, b, and c,

$$a(b + c) = ab + ac$$
$$(b + c)a = ba + ca$$

Verify, by example, that $a(b - c) = ab - ac$.

EXAMPLE 1 Use the distributive property to find the product 7×61.

Solution Think of 61 as $60 + 1$.

$$7 \times 61 = 7 \times (60 + 1) = (7 \times 60) + (7 \times 1) = 420 + 7 = 427$$

This shows how the distributive property can be used to do mental arithmetic.

Identity properties

The sum of any real number a and zero is the given real number, a.

$$0 + a = a + 0 = a$$

We call zero the *additive identity*.

The product of any real number a and 1 is the given real number, a.

$$1 \cdot a = a \cdot 1 = a$$

We call 1 the *multiplicative identity*.

Inverse properties For each real number a there exists another real number, $-a$, called the negative of a, such that the sum of a and $-a$ is zero.

$$a + (-a) = (-a) + a = 0$$

We also call $-a$ the *additive inverse* or *opposite* of a.

For each real number a different from zero there exists another real number, $\frac{1}{a}$, such that the product of a and $\frac{1}{a}$ is 1.

$$a \cdot \frac{1}{a} = \frac{1}{a} \cdot a = 1 \qquad a \neq 0$$

We call $\frac{1}{a}$ the *multiplicative inverse* or *reciprocal* of a.

EXAMPLE 2 What basic property of the real numbers is illustrated by each of the following?

(a) $6 + (17 + 4) = (17 + 4) + 6$
(b) $\frac{3}{4} + (-\frac{3}{4}) = 0$
(c) $57 \times 1 = 57$
(d) $\frac{2}{3}(12 + 36) = \frac{2}{3}(12) + \frac{2}{3}(36)$
(e) $(-43)\left(\frac{1}{-43}\right) = 1$

Solution **(a)** Commutative, for addition **(b)** inverse, for addition **(c)** identity, for multiplication **(d)** distributive **(e)** inverse, for multiplication

---

Properties of zero The product of a real number a and zero is zero.

$$0 \cdot a = a \cdot 0 = 0$$

This is often referred to as the *multiplication property of zero*.

If the product of two (or more) real numbers is zero, then at least one of the numbers is zero. That is, for real numbers a and b:

If $ab = 0$, then $a = 0$ or $b = 0$, or both $a = 0$ and $b = 0$.

The opposite of the opposite of a real number a is the number a. *Other inverse properties*

$$-(-a) = a$$

The opposite of a sum is the sum of the opposites. That is, for real numbers a and b,

$$-(a + b) = (-a) + (-b)$$

The opposite of a product of two real numbers is the product of one number times the opposite of the other. That is, for real numbers a and b,

$$-(ab) = (-a)b = a(-b)$$

The preceding properties have been described primarily in terms of addition and multiplication. The basic operations of subtraction and division are defined now in terms of addition and multiplication, respectively.

Subtraction

The difference $a - b$ of two real numbers a and b is defined as

$$a - b = a + (-b)$$

For example, $5 - 8 = 5 + (-8)$. Alternatively, we say

$$a - b = c \textit{ if and only if } c + b = a$$

Thus $5 - 8 = -3$ because $-3 + (8) = 5$.

When we say "statement p *if and only if* statement q" it means that if either statement is true, then so is the other. Here we say that $a - b = c$ if and only if $c + b = a$ and thus imply these two results:

"p if and only if q" is a form often used in mathematics.

Whenever $a - b = c$ is true, then so is $c + b = a$.
Whenever $c + b = a$ is true, then so is $a - b = c$.

These statements may also be written in the following briefer "if–then" form.

If $a - b = c$, then $c + b = a$.
If $c + b = a$, then $a - b = c$.

Division

The quotient $a \div b$ of two real numbers a and b is defined as

$$a \div b = a \cdot \frac{1}{b} \qquad b \neq 0$$

For example, $8 \div 2 = 8 \times \frac{1}{2} = 4$. Alternatively, we say

$$a \div b = c \textit{ if and only if } c \times b = a \qquad b \neq 0$$

Thus $8 \div 2 = 4$ because $2 \times 4 = 8$.

The statement $a \div b = c$ *if and only if* $c \times b = a$ means:

If $a \div b = c$, then $c \times b = a$.
If $c \times b = a$, then $a \div b = c$.

This is very important: DIVISION BY ZERO IS NOT POSSIBLE.

Using this definition of division we can see why division by zero is not possible. Suppose we assume that division by zero is possible. Assume, for example, that $2 \div 0 = x$, where x is some real number. Then, by the definition of division, $0 \cdot x = 2$. But $0 \cdot x = 0$, leading to the false statement $2 = 0$. This argument can be duplicated where 2 is replaced by any nonzero number. Can you explain why $0 \div 0$ must also remain undefined? (See Exercise 39.)

EXERCISES 1.1

Name the property illustrated by each of the following.

1 $5 + 7$ is a real number.

2 $8 + \sqrt{7} = \sqrt{7} + 8$

3 $5 + (-5) = 0$

4 $9 + (7 + 6) = (9 + 7) + 6$

5 $(5 \times 7) \times 8 = (7 \times 5) \times 8$

6 $(5 \times 7) \times 8 = 5 \times (7 \times 8)$

7 $\frac{1}{4} + \frac{1}{2} = \frac{1}{2} + \frac{1}{4}$

8 $(4 \times 5) + (4 \times 8) = 4(5 + 8)$

9 $-13 + 0 = -13$

10 $1 \times \frac{1}{9} = \frac{1}{9}$

11 $\frac{1}{2} + (-\frac{1}{2}) = 0$

12 $3 - 7 = 3 + (-7)$

13 $0(\sqrt{2} + \sqrt{3}) = 0$

14 $\sqrt{2} \times \pi$ is a real number.

15 $(3 + 9)(7) = (3)(7) + (9)(7)$

16 $\dfrac{1}{\sqrt{2}} \cdot \sqrt{2} = 1$

Replace the variable n by a real number to make each statement true.

17 $7 + n = 3 + 7$

18 $\sqrt{5} \times 6 = 6 \times n$

19 $(3 + 7) + n = 3 + (7 + 5)$

20 $6 \times (5 \times 4) = (6 \times n) \times 4$

21 $5(8 + n) = (5 \times 8) + (5 \times 7)$

22 $(3 \times 7) + (3 \times n) = 3(7 + 5)$

23 If $5(x - 2) = 0$, then $x = 2$. Explain this statement using an appropriate property of zero.

24 **(a)** Convert $14 + 3 = 17$ into a subtraction statement.

(b) Convert $(-6)12 = -72$ into a division statement.

25 Give a basic property that justifies each of the numbered steps.

$$\begin{aligned} (5 \div 6)(-3) &= -[(5 \div 6)3] && \text{(i)} \\ &= -[(5 \cdot \tfrac{1}{6})3] && \text{(ii)} \\ &= -[5(\tfrac{1}{6} \cdot 3)] && \text{(iii)} \\ &= -[5(\tfrac{1}{2})] \\ &= -\tfrac{5}{2} \end{aligned}$$

26 Give a basic property that justifies each of the numbered steps.

$$\begin{aligned} -(-3)[(-4) - (-9)] &= 3[(-4) - (-9)] && \text{(i)} \\ &= 3[(-4) + (-(-9))] && \text{(ii)} \\ &= 3[(-4) + 9] && \text{(iii)} \\ &= 3(-4) + 3(9) && \text{(iv)} \\ &= -(3 \cdot 4) + 3(9) && \text{(v)} \\ &= 15 \end{aligned}$$

27 Give the basic property that justifies each step.

$$d[(a + b) + c] = d(a + b) + dc \qquad \text{(i)}$$

$$= (da + db) + dc \quad \text{(ii)}$$
$$= da + (db + dc) \quad \text{(iii)}$$
$$= da + (bd + cd) \quad \text{(iv)}$$
$$= (bd + cd) + da \quad \text{(v)}$$
$$= (b + c)d + da \quad \text{(vi)}$$
$$= da + (b + c)d \quad \text{(vii)}$$

28 Give the reasons for the following proof that multiplication distributes over subtraction.

$$a(b - c) = a[b + (-c)] \quad \text{(i)}$$
$$= ab + a(-c) \quad \text{(ii)}$$
$$= ab + [-(ac)] \quad \text{(iii)}$$
$$= ab - ac \quad \text{(iv)}$$

29 Note that $2^4 = 4^2$. Is the operation of raising a number to a power a commutative operation? Justify your answer.

30 Give an example showing that addition does *not* distribute over multiplication; that is, $a + (b \cdot c) \neq (a + b) \cdot (a + c)$.

Give two statements that can be made in place of each of the following "if and only if" statements.

31 $10 - x = 7$ if and only if $7 + x = 10$.

32 $12 \div x = 4$ if and only if $4 \times x = 12$.

33 n is an even integer if and only if n^2 is an even integer.

34 n is an odd integer if and only if n^2 is an odd integer.

35 Triangle ABC is congruent to triangle DEF if and only if the three sides of one triangle are congruent to the three sides of the other triangle.

36 A triangle is a right triangle if and only if the sum of the squares of the two sides is equal to the square of the third side.

Replace each pair of statements by an "if and only if" statement.

37 If n is a multiple of 3, then n^2 is a multiple of 3. If n^2 is a multiple of 3, then n is a multiple of 3.

38 If n is an odd integer, then $n + 1$ is an even integer. If $n + 1$ is an even integer, then n is an odd integer.

***39** Explain why $\frac{0}{0}$ must remain undefined. (*Hint:* If $\frac{0}{0}$ is to be some value, then it must be a unique value.)

***40** Explain why the real numbers have the closure property for both subtraction and division, excluding division by 0.

1.2 Sets of numbers

Within the set of real numbers, there are other collections of numbers that we will have occasion to consider. One such set is the set of **whole numbers**:

$$\{0, 1, 2, 3, 4, 5, \ldots\}$$

This is an example of an **infinite set**; there is no last member. The three dots are used to indicate that this set of numbers continues on without end. By contrast, some of the sets of numbers that we will use will be **finite sets**. The members of a finite set can be listed and counted, and there is an end to this counting. For example, the set of whole numbers that are less than 5 is an example of a finite set:

$$\{0, 1, 2, 3, 4\}$$

The set consisting of the whole numbers greater than 0, {1, 2, 3, 4, 5, . . .}, is often referred to as the set of *natural numbers* or *counting numbers*.

At times a set can be classified as finite even though it has so many elements that no one would really want to list them all. Thus the set of counting numbers from 1 through 1,000,000 is a large set, but nevertheless finite inasmuch as it does eventually have a last member. We adopt the convention of using three dots to indicate that some members of a set are not listed and can write this as

$$\{1, 2, 3, \ldots, 1{,}000{,}000\}$$

How long do you think it would take you to count to 1,000,000 at the rate of one number per second? (Don't use 1 million seconds as your answer!) First guess. Then use a calculator to help find the answer.

*Throughout the text, an asterisk will be used to indicate that an exercise is more difficult than usual or involves some unusual aspect.

Another set of numbers that we will refer to frequently is the set of **integers**:

$$\{\ldots, -3, -2, -1, 0, 1, 2, 3, \ldots\}$$

We can represent the integers on a number line by locating the whole numbers and their opposites. For example, the opposite of 3 is located three units to the left of 0 on the number line and is named -3 (negative three). The opposite of 2 is -2, the opposite of 1 is -1, and the opposite of 0 is 0.

−3 −2 −1 0 1 2 3

Observe that the *positive integers* are located to the right of 0 and that the *negative integers* are located to the left of 0. The number 0 is an integer, but is neither positive nor negative. We see that every integer is the coordinate of some point on the number line. Can every point on the line be named by an integer?

TEST YOUR UNDERSTANDING

The word "list" here means that you are either to show all the members of a set or to indicate the members through the use of three dots.

List the elements in each of the following sets.

1 The set of whole numbers less than 5.

2 The set of whole numbers greater than 10.

3 The set of whole numbers *between* 1 and 10. (*Note:* The word "between" means that 1 and 10 are not included.)

4 The set of integers between -3 and 4.

5 The set of negative integers greater than -5.

6 The set of integers less than 2.

7 Which of the preceding sets are finite?

As you can see, there are many points between the integers. Some of these points can be identified by introducing all the numbers that can be written in the form $\frac{a}{b}$, where a and b are integers, with $b \neq 0$. The set of numbers that can be represented in this way is called the set of **rational numbers**.

A **rational number** is one that can be written in the form $\frac{a}{b}$, where a and b are integers, $b \neq 0$.

Every integer is a rational number because it can be written as the quotient of integers. For example, $5 = \frac{5}{1}$. However, not every rational number is an integer. Thus $\frac{2}{3}$ and $-\frac{3}{4}$ are examples of rational numbers (*fractions*) that are not integers.

Every rational number can be written in decimal form. Sometimes the result will be a terminating decimal, as in the following examples:

$$\frac{3}{4} = 0.75 \qquad \frac{7}{8} = 0.875 \qquad \frac{23}{10} = 2.3$$

Other rational numbers will produce a repeating decimal:

$$\frac{2}{3} = 0.666\ldots \qquad \frac{19}{22} = 0.86363\ldots \qquad \frac{3}{7} = 0.428571428571\ldots$$

Usually, a bar is placed over the set of digits that repeat, so that the preceding illustrations can be written in this way:

$$\frac{2}{3} = 0.\overline{6} \qquad \frac{19}{22} = 0.8\overline{63} \qquad \frac{3}{7} = 0.\overline{428571}$$

Some decimals neither terminate nor repeat. You are probably familiar with the number π from your earlier study of geometry. This number is *not* a rational number; it cannot be expressed as the quotient of two integers. The decimal representation for π goes on endlessly without repetition. In fact, one computer recently computed π as a decimal to 100,000 places. Here are the first 100 places:

A method for converting a repeating decimal into fraction form is considered in Exercises 45 and 46.

$\pi =$ 3.14159 26535 89793 23846 26433 83279 50288 41971 69399
37510 58209 74944 59230 78164 06286 20899 86280 34825
34211 70679 . . .

It is true that every rational number is the **coordinate** of some point on the number line. However, not every point on the number line can be labeled by a rational number. For example, here is a construction that can be used to locate a point on the number line whose coordinate is $\sqrt{2}$, the *square root* of 2.

At the point with coordinate 1 construct a one-unit segment perpendicular to the number line. Connect the endpoint of this segment to the point labeled 0. This becomes the hypotenuse of a right triangle, and by the Pythagorean theorem can be shown to be the square root of 2, written as $\sqrt{2}$. Using a compass, this length can then be transferred to the number line, thus locating a point with coordinate $\sqrt{2}$.

x
1
0
1 $\sqrt{2}$
2

$$x^2 = 1^2 + 1^2$$
$$x^2 = 2$$
$$x = \sqrt{2}$$

It can be proved that $\sqrt{2}$ cannot be expressed as the quotient $\frac{a}{b}$ of two integers, and thus is *not* a rational number. We call such a number an **irrational number**. Some other examples of irrational numbers are $\sqrt{5}$, $\sqrt{12}$, $\sqrt[3]{4}$, and π. The term "irrational" is chosen to indicate that such a number cannot be expressed as the ratio of two integers.

PYTHAGOREAN THEOREM

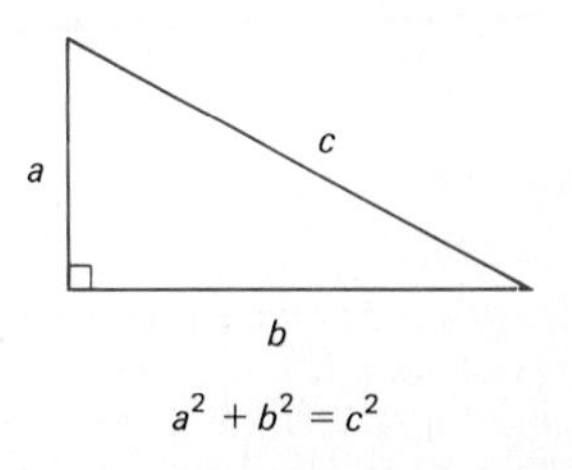

$a^2 + b^2 = c^2$

When the set of rational numbers and the set of irrational numbers, are combined we have the collection of *real numbers* which will suffice for most of our work in this text. Every real number can be represented by a decimal. If the decimal terminates or repeats, the number is a rational number; otherwise, it is an irrational number.

EXAMPLE Classify each number into as many different sets as possible: **(a)** 5 **(b)** $\frac{2}{3}$ **(c)** $\sqrt{7}$ **(d)** -14

Solution
(a) 5 is a whole number, an integer, a rational number, and a real number.
(b) $\frac{2}{3}$ is a rational number and a real number.
(c) $\sqrt{7}$ is an irrational number and a real number.
(d) -14 is an integer, a rational number, and a real number.

EXERCISES 1.2

List the elements in each of the sets described in Exercises 1–10.

1 The set of whole numbers less than 3.
2 The set of whole numbers greater than 100.
3 The set of whole numbers between 2 and 7.
4 The set of integers greater than -3.
5 The set of negative integers greater than -3.
6 The set of positive integers less than 5.
7 The set of integers less than 1.
8 The set of integers that are not whole numbers.
9 The set of whole numbers that are not integers.
10 The set of integers that are also rational numbers.

In Exercises 11–20, answer true or false.

11 Every whole number is an integer.
12 Every integer is a whole number.
13 Every integer is a rational number.
14 Every rational number is an integer.
15 Every real number is a rational number.
16 Every irrational number is a real number.
17 Every point on the number line can be associated with a rational number.
18 Every point on the number line can be associated with a real number.
19 Every rational number is the coordinate of some point on the number line.
20 Every real number is a rational or an irrational number.

Classify each number as being a member of one or more of these sets: (a) whole numbers; (b) integers; (c) rational numbers; (d) irrational numbers; (e) real numbers.

21 -15
22 72
23 $\sqrt{5}$
24 $-\frac{3}{4}$
25 $\frac{16}{2}$
26 0.01
27 $\sqrt{49}$
28 1000
29 $\sqrt{12}$
30 $\sqrt{\frac{9}{16}}$
31 $\frac{-5}{\sqrt{100}}$

32 Draw a number line and locate a point with coordinate $\sqrt{2}$, as shown in this section. At this point construct a perpendicular segment of length 1. Now construct a right triangle with hypotenuse of length $\sqrt{3}$ and use this construction to locate a point on the number line with coordinate $\sqrt{3}$.

*33 Locate a point on the number line with coordinate $\sqrt{5}$.

*34 Locate a point on the number line with coordinate $-\sqrt{2}$.

*35 Take a circular object (like a tin can) and let the diameter be 1 unit. Use this as a unit on a number line. Mark a point on the circular object with a dot and place the circle so that the dot coincides with zero on the number line. Roll the circle to the right and mark the point where the dot coincides with the number line after one revolution. What number corresponds to this position? Why?

*36 From a mathematical point of view, it is common practice to introduce the integers after the counting numbers have been discussed and then develop the rational numbers. Historically, however, the positive rational numbers were developed before the negative integers.

(a) Think of some everyday situations where the negative integers are not usually used but could easily be introduced.

(b) Speculate on the historical necessity for the development of the positive rational numbers.

Use division to convert each fraction into a terminating or repeating decimal.

37 $\frac{4}{5}$ **38** $\frac{257}{100}$

39 $\frac{2}{7}$ **40** $\frac{25}{8}$

41 $\frac{45}{13}$ **42** $\frac{18}{7}$

43 $\frac{13}{900}$ **44** $\frac{279}{55}$

45 Every repeating decimal can be expressed as a rational number in the form a/b. Consider, for example, the decimal 0.727272 . . . and the following process:

Let $n = 0.727272\ldots$ Then $100n = 72.727272\ldots$

$$100n = 72.727272\ldots$$

Subtract: $$n = .727272\ldots$$

$$99n = 72$$

Solve for n: $$n = \frac{72}{99} = \frac{8}{11}$$

Use this method to express each decimal as the quotient of integers and check by division.

(a) 0.454545 . . . ($0.\overline{45}$) (b) 0.373737. . . ($0.\overline{37}$)

(c) 0.234234 . . . ($0.\overline{234}$)

(*Hint:* Let $n = 0.\overline{234}$; multiply by 1000.)

***46** Study the given illustration for $n = 0.2737373\ldots$ and then convert each repeating decimal into a quotient of integers.

Multiply n by 1000: $1000n = 273.\overline{73}$ The decimal point is *behind* the first cycle.

Multiply n by 10: $10n = 2.\overline{73}$ The decimal point is in *front* of the first cycle.

Subtract: $990n = 271$

Divide: $n = \dfrac{271}{990}$

(a) 0.4585858 . . . (b) 3.21444 . . .

(c) $2.0\overline{146}$ (d) $0.00\overline{123}$

1.3 Introduction to equations and problem solving

A statement such as $2x - 3 = 7$ is said to be a **conditional equation**. It is true for some replacements of the variable x, but not true for others. For example, $2x - 3 = 7$ is a true statement for $x = 5$ but is false for $x = 7$. On the other hand, an equation such as $3(x + 2) = 3x + 6$ is called an **identity** because it is true for *all* real numbers x.

To *solve* an equation means to find the real numbers x for which the given equation is true; these are called the **solutions** or **roots** of the given equation. Let us solve the equation $2x - 3 = 7$, showing the important steps.

A flowchart can be used to illustrate an equation.

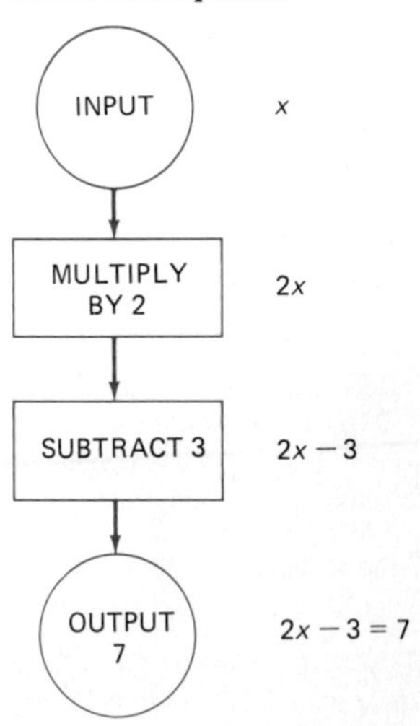

$$2x - 3 = 7 \qquad \text{Add 3 to each side of the equation.}$$

$$(2x - 3) + 3 = 7 + 3$$

$$2x = 10 \qquad \text{Now multiply each side by } \tfrac{1}{2}.$$

$$\tfrac{1}{2}(2x) = \tfrac{1}{2}(10)$$

$$x = 5$$

We can check this solution by substituting 5 for x in the original equation.

$$2(5) - 3 = 10 - 3 = 7$$

In the preceding solution we made use of the following two basic *properties of equality*.

The strength of these two properties is that they produce equivalent equations, equations having the same roots. Thus the addition property converts $2x - 3 = 7$ into the equivalent form $2x = 10$.

ADDITION PROPERTY OF EQUALITY:
For all real numbers a, b, c, if $a = b$, then $a + c = b + c$.

MULTIPLICATION PROPERTY OF EQUALITY:
For all real numbers a, b, c, if $a = b$, then $ac = bc$.

The properties of equality can also be used to solve more complicated equations. The procedure is to *collect all terms with the variable on one side of the equation and all constants on the other side.* Although most of the steps shown can be done mentally, the next two examples will include the essential details that comprise a formal solution.

You should practice showing all steps in the solution of an equation. With experience, however, you will be able to complete many of the details mentally.

EXAMPLE 1 Solve for x: $6x + 9 = 2x + 1$.

Solution

$$6x + 9 = 2x + 1$$

$$6x + 9 + (-9) = 2x + 1 + (-9) \qquad \text{(addition property)}$$

$$6x = 2x - 8$$

$$6x + (-2x) = 2x - 8 + (-2x) \qquad \text{(addition property)}$$

$$4x = -8$$

$$\tfrac{1}{4}(4x) = \tfrac{1}{4}(-8) \qquad \text{(multiplication property)}$$

$$x = -2$$

Check: $6(-2) + 9 = -3$; $2(-2) + 1 = -3$.

EXAMPLE 2 Solve for x: $2(x + 3) = x + 5$.

Solution Here we have an additional step because of the parentheses, which can be eliminated by applying the distributive property. Try to give a reason for each step in the solution that follows.

$$2(x + 3) = x + 5$$

$$2x + 6 = x + 5$$

$$2x + 6 + (-6) = x + 5 + (-6)$$

$$2x = x + (-1)$$

$$2x + (-x) = x + (-1) + (-x)$$

$$x = -1$$

Each of the equations in Examples 1 and 2 is an example of a *linear equation* because the variable x appears only to the first power.

Check this solution. Does $2[(-1) + 3] = (-1) + 5$?

TEST YOUR UNDERSTANDING

Solve each linear equation for x.

1 $x + 3 = 9$	**2** $x - 5 = 12$	**3** $x - 3 = -7$
4 $2x + 5 = x + 11$	**5** $3x - 7 = 2x + 6$	**6** $3(x - 1) = 2x + 7$
7 $3x + 2 = 5$	**8** $5x - 3 = 3x + 1$	**9** $x + 3 = 13 - x$
10 $2(x + 2) = x - 5$	**11** $2x - 7 = 5x + 2$	**12** $4(x + 2) = 3(x - 1)$

The properties of equality can be used to solve equations having more than one variable. The next example shows the use of properties of equality to solve a formula for one of the variables in terms of the others.

EXAMPLE 3 The formula relating degrees Fahrenheit and degrees Celsius is $\frac{5F - 160}{9} = C$. Solve for F in terms of C.

Solution Try to explain each step shown.

$$\frac{5F - 160}{9} = C$$

$$5F - 160 = 9C$$

$$5F = 9C + 160$$

$$F = (\tfrac{1}{5})(9C + 160)$$

$$F = \tfrac{9}{5}C + 32$$

Here is an example of the use of this formula: When $C = 20$, then $F = \frac{9}{5}(20) + 32 = 36 + 32 = 68$. Thus 20° Celsius = 68° Fahrenheit.

Let us now explore the solution of problems that are expressed in words. Our task will be to translate the English sentences of a problem into suitable mathematical language, and develop an equation that we can solve. To illustrate this process let us explore a "think of a number" type of puzzle. The directions to this puzzle are given below at the left. Follow these instructions first with a specific number and verify that your result is 5. Then study the mathematical explanation given at the right as to why the result will always be 5, regardless of the number with which you start.

Instruction	*Example*	*Algebraic representation*
Think of a number.	Let $n = 7$	n
Add 2.	$7 + 2 = 9$	$n + 2$
Multiply by 3.	$3 \times 9 = 27$	$3(n + 2) = 3n + 6$
Add 9.	$27 + 9 = 36$	$(3n + 6) + 9 = 3n + 15$
Multiply by 2.	$2 \times 36 = 72$	$2(3n + 15) = 6n + 30$
Divide by 6.	$72 \div 6 = 12$	$\frac{6n + 30}{6} = \frac{1}{6}(6n + 30) = n + 5$
Subtract the number with which you started.	$12 - 7 = 5$	$(n + 5) - n = 5$
The result is 5.		

Try to make up a similar puzzle of your own and explain why it works.

Solving verbal problems often creates difficulties for many students. To become a good problem solver you need some patience and much practice. Study the solution to the following problem in detail, as well as the examples that follow. Then try to solve as many of those problems as possible given in the exercises.

Problem: The length of a rectangle is 1 centimeter less than twice the width. The perimeter is 28 centimeters. Find the dimensions of the rectangle.

1 *Reread the problem and try to picture the situation given. Make note of all the information stated in the problem.*
The length is one less than twice the width.
The perimeter is 28.

2 *Determine what it is you are asked to find. Introduce a suitable variable, usually to represent the quantity to be found. When appropriate, draw a figure.*
Let x represent the width.
Then $2x - 1$ represents the length.

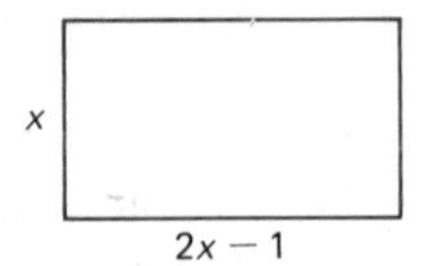

3 *Use the available information to compose an equation that involves the variable.*
The perimeter is the distance around the rectangle. This provides the necessary information to write an equation.

$$x + (2x - 1) + x + (2x - 1) = 28$$

4 *Solve the equation.*

$$x + (2x - 1) + x + (2x - 1) = 28$$
$$6x - 2 = 28$$
$$6x = 30$$
$$x = 5$$

5 *Return to the original problem to see whether the answer obtained makes sense. Does it appear to be a reasonable solution? Have you answered the question posed in the problem?*
The original problem asked for both dimensions. If the width x is 5 centimeters, then the length $2x - 1$ must be 9 centimeters.

6 *Check the solution by direct substitution of the answer into the original statement of the problem.*
As a check, note that the length of the rectangle, 9 centimeters, is 1

centimeter less than twice the width, 5, as given in the problem. Also, the perimeter is 28 centimeters.

7 *Finally, state the solution in terms of the appropriate units of measure.* The dimensions are 5 centimeters by 9 centimeters.

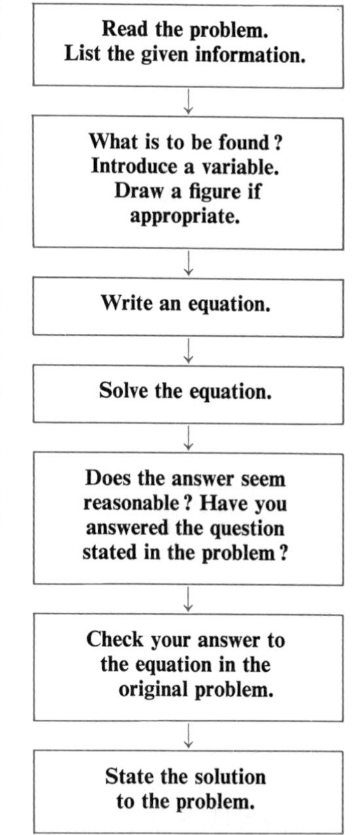

Guidelines for problem solving:
Read the problem. List the given information.
What is to be found? Introduce a variable. Draw a figure if appropriate.
Write an equation.
Solve the equation.
Does the answer seem reasonable? Have you answered the question stated in the problem?
Check your answer to the equation in the original problem.
State the solution to the problem.

EXAMPLE 4 A car leaves a certain town at noon, traveling due east at 40 miles per hour. At 1:00 P.M. a second car leaves the town traveling in the same direction at a rate of 50 miles per hour. In how many hours will the second car overtake the first car?

Solution Problems of motion of this type often prove difficult to students of algebra, but need not be. The basic relationship to remember is that rate multiplied by time equals distance ($r \times t = d$). For example, a car traveling at a rate of 60 miles per hour for 5 hours will travel 60×5 or 300 miles.

Now we need to explore the problem to see what part of the information given will help form an equation. The two cars travel at different rates, and for different amounts of time, but both travel the same distance from the point of departure until they meet. This is the clue: Represent the distance each travels and equate these quantities.

Let us use x to represent the number of hours it will take the second car to overtake the first. Then the first car, having started an hour earlier, travels $x + 1$ hours until they meet. You may find it helpful to summarize this information in tabular form.

	Rate	*Time*	*Distance*
First car	40	$x + 1$	$40(x + 1)$
Second car	50	x	$50x$

Equating the distances, we have an equation that can be solved for x:

$$50x = 40(x + 1)$$
$$50x = 40x + 40$$
$$10x = 40$$
$$x = 4$$

The second car overtakes the first in 4 hours. Does this solution seem reasonable? Let us check the solution. The first car travels 5 hours at 40 miles per hour for a total of 200 miles. The second car travels 4 hours at 50 miles per hour for the same total of 200 miles.
Solution: The second car takes 4 hours to overtake the first.

EXAMPLE 5 David has a total of \$4.10 in nickels, dimes, and quarters. He has twice as many nickels as dimes, and two more quarters than dimes. How many dimes does he have?

Although some of these problems may not seem to be very practical to you, they will help you develop basic skills at problem solving that will be helpful later when you encounter more realistic applications.

Solution Begin by letting x represent the number of dimes. Then from the statement of the problem we can let $2x$ represent the number of nickels, and $x + 2$ the number of quarters.

Next, observe that whenever you have a certain number of a particular coin, then the total value is the number of coins times the value of that coin. For example, if you have 8 dimes, then you have 10×8 or 80¢. If you have 9 nickels, then you have 5×9 or 45¢. For our problem:

Let $10x$ represent the value of the dimes, in cents.
Let $5(2x)$ represent the value of the nickels, in cents.
Let $25(x + 2)$ represent the value of the quarters, in cents.

Since the total value of these coins is \$4.10, or 410¢, we write and solve the following equation:

$$\begin{aligned} 10x + 5(2x) + 25(x + 2) &= 410 \\ 10x + 10x + 25x + 50 &= 410 \\ 45x + 50 &= 410 \\ 45x &= 360 \\ x &= 8 \end{aligned}$$

Always check by returning to the original statement of the problem. If you just check in the constructed equation, and the equation you formed was incorrect, you will not detect the error.

Check: If David has 8 dimes, then he must have $2x$ or 16 nickels, and $x + 2$ or 10 quarters. The total amount is

$$8(10) + 16(5) + 10(25) = 80 + 80 + 250 = 410, \text{ that is, } \$4.10$$

Solution: David has 8 dimes.

Now it's your turn! For Exercises 38–50, use the guidelines suggested earlier in this section, and try to solve as many of the problems as you can. Don't become discouraged; be assured that most mathematics students have trouble with verbal problems. Time and practice will most certainly help you develop your skill at problem solving.

EXERCISES 1.3

Solve for x and check each result.

1 $3x - 2 = 10$

2 $5x + 1 = 21$

3 $-2x + 1 = 9$

4 $-3x - 2 = 10$

5 $-3x - 5 = 7$

6 $3x + 2 = -13$

7 $2x - 1 = -17$

8 $-2x + 3 = -12$

9 $2(x + 1) = 11$

10 $3(x - 2) = 15$

11 $3x + 7 = 2x - 2$

12 $2.5x - 8 = x + 3$

13 $\frac{1}{2}x + 7 = 2x - 3$

14 $\frac{5}{2}x - 5 = 3x + 7$

15 $\frac{4}{3}x - 7 = \frac{1}{3}x + 8$

16 $5x - 1 = 5x + 1$

17 $\frac{3}{5}(x - 5) = x + 1$

18 $5(x + 4) = \frac{5}{2}x - 5$

19 $\frac{7}{2}x + 5 + \frac{1}{2}x = \frac{5}{2}x - 6$

20 $2(x + 3) - x = 2x + 8$

21 $-3(x + 2) + 1 = x - 25$

22 $\frac{4}{3}(x + 8) = \frac{3}{4}(2x + 12)$

23 $1 - 12x = 7(1 - 2x)$

24 $2(3x - 7) - 4x = -2$

25 $x + 2(\frac{1}{6}x + 2) = \frac{6}{5}x + 16$

Solve for the indicated variable.

26 $P = 4s$ for s

27 $P = 2l + 2w$ for w

28 $F = \frac{9}{5}C + 32$ for C

29 $N = 10t + u$ for t

30 $7a - 3b = c$ for b

31 $C = 2\pi r$ for r

32 $2(r - 3s) = 6t$ for s

33 $6 + 4v = w - 1$ for v

Use the formulas in Example 3 to make these conversions.

34 0° Celsius to Fahrenheit

35 10° Celsius to Fahrenheit

36 −4° Fahrenheit to Celsius

37 32° Fahrenheit to Celsius

38 Find a number such that two-thirds of the number increased by one is 13.

39 Find the dimensions of a rectangle whose perimeter is 56 inches if the length is three times the width.

40 Each of the two equal sides of an isosceles triangle is 3 inches longer than the base of the triangle. The perimeter is 21 inches. Find the length of each side.

41 Carlos spent $4.50 on stamps, in denominations of 10¢, 18¢, and 20¢. He bought one-half as many 18¢ stamps as 10¢ stamps, and three more 20¢ stamps than 10¢ stamps. How many of each type did he buy?

42 Maria has $169 in ones, fives, and tens. She has twice as many one-dollar bills as she has five-dollar bills, and five more ten-dollar bills than five-dollar bills. How many of each type bill does she have?

43 Two cars leave a town at the same time and travel in opposite directions. One car travels at the rate of 40 miles per hour, and the other at 60 miles per hour. In how many hours will the two cars be 350 miles apart?

44 Robert goes for a walk at a speed of 3 miles per hour. Two hours later Roger attempts to overtake him by jogging at the rate of 7 miles per hour. How long will it take him to reach Robert?

45 Prove that the measures of the angles of a triangle cannot be represented by consecutive odd integers. (*Hint:* The sum of the measures is 180.)

46 The width of a painting is 4 inches less than the length. The frame that surrounds the painting is 2 inches wide and has an area of 240 square inches. What are the dimensions of the painting? (*Hint:* The total area minus the area of the painting alone is equal to the area of the frame.)

*47 The length of a rectangle is 1 inch less than three times the width. If the length is increased by 6 inches and the width is increased by 5 inches, then the length will be twice the width. Find the dimensions of the rectangle.

*48 The units' digit of a two-digit number is three more than the tens' digit. The number is equal to four times the sum of the digits. Find the number. (*Hint:* We can represent a two-digit number as $10t + u$.)

*49 Find three consecutive odd integers such that their sum is 237.

*50 The length of a rectangle is 1 inch less than twice the width. If the length is increased by 11 inches and the width is increased by 5 inches, then the length will be twice the width. What can you conclude about the data for this problem?

1.4 Properties of order

As you continue your study of mathematics you will find a great deal of attention given to *inequalities*. We begin our discussion of this topic by considering the ordering of the real numbers on the number line. In the following figure we say that *a is less than b* because a lies to the left of b. In symbols, we write $a < b$.

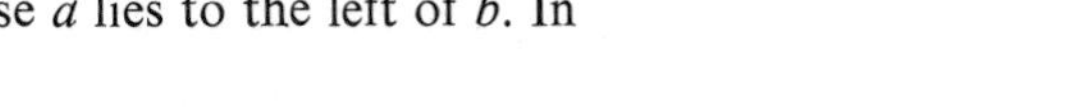

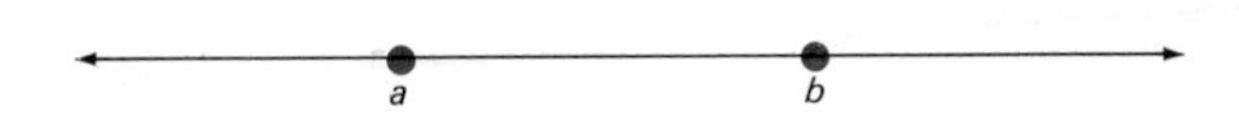

Also note that b lies to the right of a. That is, $b > a$; this is read "*b is greater than a.*" Two inequalities, one using the symbol $<$ and the other $>$, are said to have the *opposite sense*.

The fundamental property behind the study of inequalities is known as the **trichotomy law**. This law states that for any two real numbers a and b exactly one of the following must be true:

$$a < b \qquad a = b \qquad a > b$$

There are a number of important properties of order that are fundamental for later work. We list them in terms of the following rules.

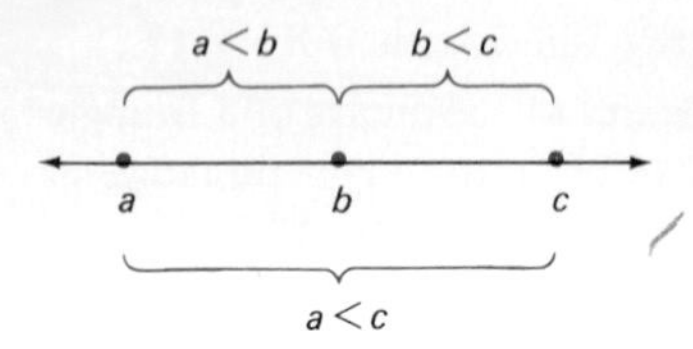

RULE 1. **If $a < b$ and $b < c$, then $a < c$.**
This is known as the **transitive property of order**. Geometrically, it says that if a is to the left of b, and b is to the left of c, then a must be to the left of c on a number line.

RULE 2. **If $a < b$, then $a + c < b + c$ for any real number c.**
Since $5 < 7$, then $5 + 10 < 7 + 10$; that is, $15 < 17$.
Since $5 < 7$, then $5 + (-10) < 7 + (-10)$, or $5 - 10 < 7 - 10$; that is, $-5 < -3$.

CAUTION: Rules 3 and 4 often cause trouble! If you multiply both sides of an inequality by a positive number, the sense of the resulting inequality is the same as the sense of the original inequality. Multiplying by a negative number changes the sense of the inequality.

RULE 3. **If $a < b$ and $c > 0$, then $ac < bc$.**
Since $5 < 10$, then $5(3) < 10(3)$; that is, $15 < 30$.
Since $-4 < 6$, then $-4(\frac{1}{2}) < 6(\frac{1}{2})$; that is, $-2 < 3$.

RULE 4. **If $a < b$ and $c < 0$, then $ac > bc$.**
Since $5 < 10$, then $5(-3) > 10(-3)$; that is, $-15 > -30$.
Since $-4 < 6$, then $-4(-\frac{1}{2}) > 6(-\frac{1}{2})$; that is, $2 > -3$.

RULE 5. **If $a < b$ and $c < d$, then $a + c < b + d$.**
Since $5 < 10$ and $-15 < -4$, then $5 + (-15) < 10 + (-4)$; that is, $-10 < 6$.

RULE 6. **If $0 < a < b$ and $0 < c < d$, then $ac < bd$.**
Since $3 < 7$ and $5 < 9$, then $(3)(5) < (7)(9)$; that is, $15 < 63$.

These rules can also be restated by reversing the sense. For instance, Rule 5 would read as: If $a > b$ and $c > d$, then $a + c > b + d$.

RULE 7. **If $a < b$ and $ab > 0$, then $\frac{1}{a} > \frac{1}{b}$.**
Since $5 < 10$, then $\frac{1}{5} > \frac{1}{10}$.
Since $-3 < -2$, then $-\frac{1}{3} > -\frac{1}{2}$.

We have developed the concept of $a < b$ by saying that a is to the left of b on a number line. It is often useful to state this relationship in terms of this algebraic definition:

> For any two real numbers a and b, $a < b$ (and $b > a$) if and only if $b - a$ is a positive number, that is, if and only if $b - a > 0$.

For example, $2 < 5$ because $5 - 2 = 3$, a positive number; $5 - 2 > 0$. Also, since $-2 - (-7) = 5 > 0$, then $-2 > -7$.

With this definition of order, we can prove the rules discussed earlier. The next example shows how this can be done.

EXAMPLE 1 Prove Rule 1: If $a < b$ and $b < c$, then $a < c$.

Solution We note that $a < b$ and $b < c$ means that $b - a$ and $c - b$ are positive numbers by definition. Since the sum of two positive numbers is also positive, it follows that

$$(b - a) + (c - b) > 0$$

But

$$(b - a) + (c - b) = c - a$$

Hence $c - a$ is positive, and the definition implies that $a < c$.

These additional symbols of inequality are also used:

$a \leq b$ means *a is less than or is equal to b;* that is, $a < b$ or $a = b$.
$a \geq b$ means *a is greater than or is equal to b;* that is, $a > b$ or $a = b$.

We may use these symbols in the rules of order previously listed. Thus Rule 1 may be stated in these forms:

If $a \leq b$ and $b \leq c$, then $a \leq c$.
If $a \geq b$ and $b \geq c$, then $a \geq c$.

The two inequalities $a < b$ and $b < c$ may be written as $a < b < c$. Also, if $a \leq b$ and $b \leq c$, then $a \leq b \leq c$. A similar statement may be made for the $\geq$ relationship.

EXAMPLE 2 Translate each verbal statement into a statement of inequality.
(a) x is to the right of zero.
(b) x is between 2 and 3.
(c) x is greater than or equal to -1.
(d) x is at least 10 but less than 100.

Solution **(a)** $x > 0$ **(b)** $2 < x < 3$ **(c)** $x \geq -1$ **(d)** $10 \leq x < 100$

We can use inequalities to define *bounded intervals* on the number line. This is shown in the next figure, together with specific examples of each.

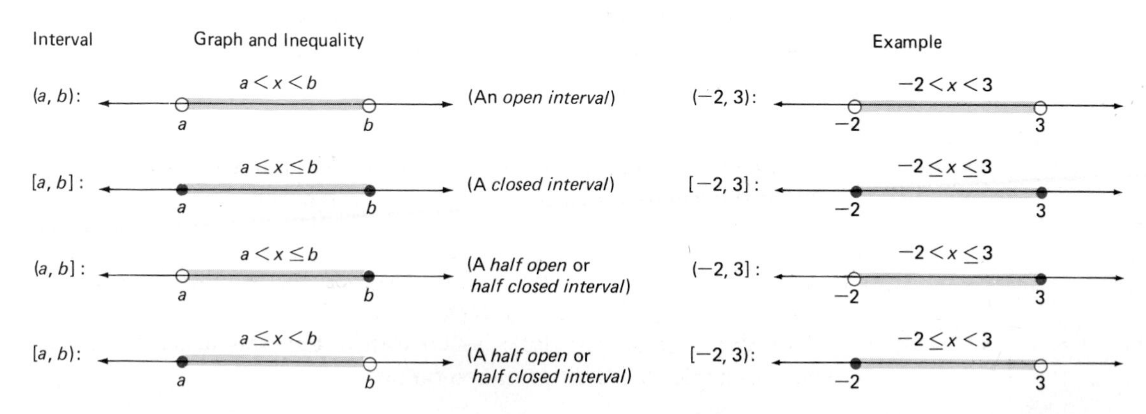

Note that the parenthesis in $(a, b]$ means that the number a is not included in the interval, and that the bracket means that b is included. Also, regardless of the parentheses or brackets, the numbers a and b are boundaries for all of the values x in the interval.

There are also *unbounded intervals*. For example, the set of all $x > 5$ is denoted by $(5, \infty)$. Similarly, $(-\infty, 5]$ represents all $x \leq 5$. The symbols ∞ and $-\infty$ are read "plus infinity" and "minus infinity" but do *not* represent numbers. They are symbolic devices used to indicate that *all* x in a given direction, without end, are included, as in the following figure.

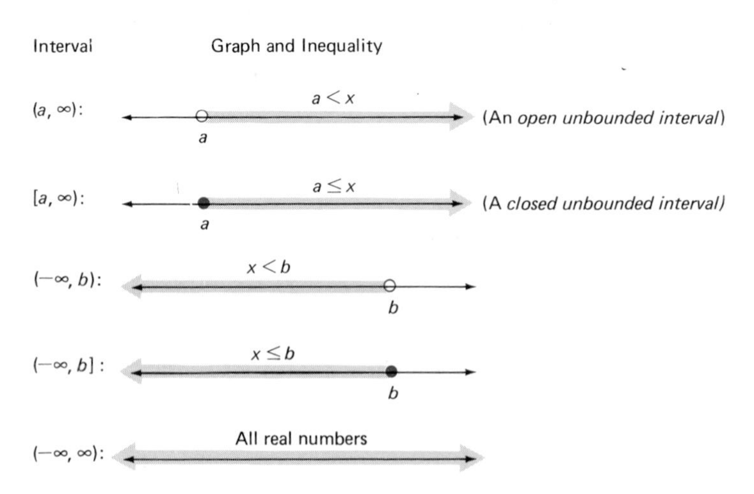

TEST YOUR UNDERSTANDING

Show each of the following intervals as a graph on a number line.

1 $(-4, 1]$ **2** $[0, 3]$ **3** $[-1, 2)$ **4** $(-2, \infty)$

Express each inequality in interval notation.

5 $-1 \leq x \leq 2$ **6** $0 < x < 3$ **7** $x \leq 0$
8 $1 \leq x < 4$ **9** $-3 \leq x$ **10** $-5 < x < -2$

EXERCISES 1.4

In Exercises 1–8, insert the appropriate symbol, $<$, $=$, or $>$, between the given pair of real numbers.

1 -100 ___ 2
2 $\frac{1}{2}$ ___ $\frac{1}{3}$
3 $-\frac{4}{5}$ ___ $-\frac{2}{3}$
4 2.619 ___ 2.621
5 0.7 ___ $\frac{7}{9}$
6 $-\frac{13}{14}$ ___ $-\frac{20}{21}$
7 $\frac{1}{2}(4.02)$ ___ 2.01
8 $\frac{1}{9}$ ___ 0.111

In Exercises 9 and 10, translate each verbal statement into a statement of inequality.

9 **(a)** x is to the left of 1.
(b) x is to the right of $-\frac{2}{3}$.
(c) x is at least as large as 5.
(d) x is more than -10 but less than or equal to 7.

10 **(a)** x is between -1 and 4.

(b) x is no more than 6.
(c) x is to the right of 12.
(d) x is a positive number less than 40.

In Exercises 11–18, classify each statement as true or false. If it is false, give a specific example to explain your answer.

11 If $x > 1$ and $y > 2$, then $x + y > 3$.

12 If $x < 2$, then x is negative.

13 If $x < 5$ and $y < 6$, then $xy < 30$.

14 If $0 < x$, then $-x < 0$.

15 If $x < y < -2$, then $\frac{1}{x} > \frac{1}{y}$.

16 If $0 < x$, then $x < x^2$.

17 If $x \leq y$ and $y < z$, then $x < z$.

18 If $x \leq -5$, then $x - 2 \leq -7$.

Show each of the following intervals as a graph on a number line.

19 $(-3, -1)$ 20 $(-3, -1]$

21 $[-3, -1)$ 22 $[-3, -1]$

23 $[0, 5]$ 24 $(-1, 3)$

25 $(-\infty, 0]$ 26 $[2, \infty)$

Express each inequality in interval notation.

27 $-5 \leq x \leq 2$ 28 $0 < x < 7$

29 $-6 \leq x < 0$ 30 $-2 < x \leq 4$

31 $-10 < x < 10$ 32 $3 \leq x \leq 7$

33 $x < 5$ 34 $x \leq -2$

35 $-2 \leq x$ 36 $2 < x$

37 $x \leq -1$ 38 $x < 3$

39 Complete the following proof of Rule 3: If $a < b$ and $c > 0$, then $ac < bc$.
(a) $a < b$ implies ________, by the definition of $<$.
(b) Since $c > 0$, the product ________ is also positive.
(c) But $(b - a)c =$ ________.
(d) Therefore, ________ is positive.
(e) Thus ________ $<$ ________ by the definition of $<$.

*40 Use Rule 3 to prove that if $0 < a < 1$, then $a^2 < a$.

*41 Use Rule 3 to prove that if $1 < a$, then $a < a^2$.

*42 Prove: If $a < b < 0$ and $c < d < 0$, then $ac > bd$.

1.5 Conditional inequalities

The inequality $3x - 4 > 11$ is true for some values for x, such as 10, and false for others, such as 3. Inequalities that are not true for all allowable values of the variable are called **conditional inequalities**. Solving such an inequality means finding the set of all x for which it is true. Our basic rules of Section 1.4 will be helpful in this work, as in the following example.

$$3x - 4 > 11$$

$$3x - 4 + 4 > 11 + 4 \qquad \text{By Rule 2 we may add 4 to both sides.}$$

$$3x > 15$$

$$\tfrac{1}{3}(3x) > \tfrac{1}{3}(15) \qquad \text{By Rule 3 we may multiply each side by } \tfrac{1}{3}, \text{ or divide each side by 3.}$$

$$x > 5$$

Because we are multiplying (or dividing) by a positive number, we do not change the sense of the inequality.

The answer then consists of all real numbers x such that $x > 5$. This answer may also be displayed graphically on a number line.

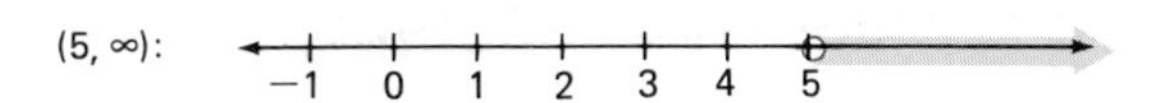

The heavily shaded arrow is used to show that all points in the indicated direction are included. The open circle indicates that 5 is *not* included in the solution. Why not? When the open circle is replaced by a solid dot we have the following graph, which represents the solution of the conditional inequality $3x - 4 \geq 11$. In this case 5 is a solution of the inequality.

$[5, \infty)$: number line from -1 to 5, closed dot at 5, shaded to the right.

EXAMPLE 1 Solve for x and graph: $5(3 - 2x) \geq 10$.

Solution Multiply each side by $\frac{1}{5}$ (or divide by 5).

$$\tfrac{1}{5} \cdot 5(3 - 2x) \geq \tfrac{1}{5}(10) \qquad \text{(Rule 3)}$$
$$3 - 2x \geq 2$$

Add -3 to each side.

$$3 - 2x + (-3) \geq 2 + (-3) \qquad \text{(Rule 2)}$$
$$-2x \geq -1$$

Multiply by $-\frac{1}{2}$ (or divide by -2).

Note the change in the sense of the resulting inequality because we are multiplying (or dividing) by a negative number.

$$-\tfrac{1}{2}(-2x) \leq -\tfrac{1}{2}(-1) \qquad \text{(Rule 4)}$$
$$x \leq \tfrac{1}{2}$$

$(-\infty, \frac{1}{2}]$: number line with -1, 0, $\frac{1}{2}$, 1, 2; closed dot at $\frac{1}{2}$, shaded to the left.

In Example 2, note that $x \neq -4$. Why not?

EXAMPLE 2 Solve: $\dfrac{2}{x + 4} < 0$.

Solution Since the numerator of the fraction is positive, the fraction will be less than zero if and only if the denominator is negative. Thus

$$x + 4 < 0$$
$$x < -4$$

EXAMPLE 3 If $x < 2$, is $x - 5$ positive or negative?

Solution

$$x < 2$$
$$x - 5 < 2 - 5 \qquad \text{subtract 5}$$
$$x - 5 < -3$$

Therefore $x - 5$ is negative.

Example 3 can also be solved by selecting one convenient test value from the interval $(-\infty, 2)$. Thus if $x = 1$, then $x - 5 = 1 - 5 = -4$. From this we can conclude that $x - 5 < 0$ for *all* $x < 2$. You can see that this process makes sense by studying this figure.

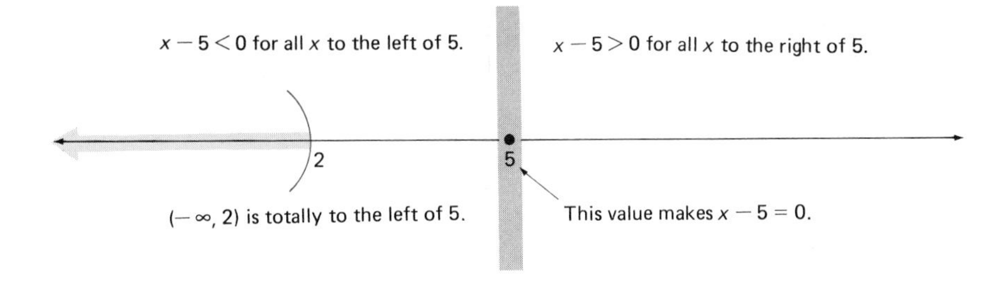

Since $(-\infty, 2)$ is completely to the left of 5, all x in $(-\infty, 2)$ satisfies $x - 5 < 0$. Therefore any specific value of x in that interval may be used as a test value.

TEST YOUR UNDERSTANDING

Solve each inequality for x. Graph the solution for Exercises 1 and 2.

1 $x - 3 < 0$ **2** $x + 5 > 0$ **3** $x + 1 < -3$

4 $x - 2 \geq \frac{1}{2}$ **5** $2x + 7 < 11$ **6** $3x + 1 \leq x - 4$

Solve for x.

7 $-2(x + 6) < 0$ **8** $-2(x + 6) \geq 0$ **9** $-\frac{1}{x} < 0$

10 $\frac{5}{3 - x} < 0$

11 Assume that $x < -1$. Classify as positive or negative: **(a)** $x + 1$; **(b)** $x - 3$.

12 Assume that $4 < x < 5$. Classify as positive or negative: **(a)** $x - 4$; **(b)** $x - 5$.

Let us explore in detail the solution to a more complicated inequality:

$$(x - 2)(x - 5) < 0$$

Since $(x - 2)(x - 5) = 0$ if and only if $x = 2$ or $x = 5$, we need to consider the other possible values of x. First observe that the numbers 2 and 5 determine three intervals.

Recall that $(x - 2)(x - 5) = 0$ if and only if either $x - 2 = 0$ or $x - 5 = 0$. That is, if $ab = 0$, then $a = 0$ or $b = 0$.

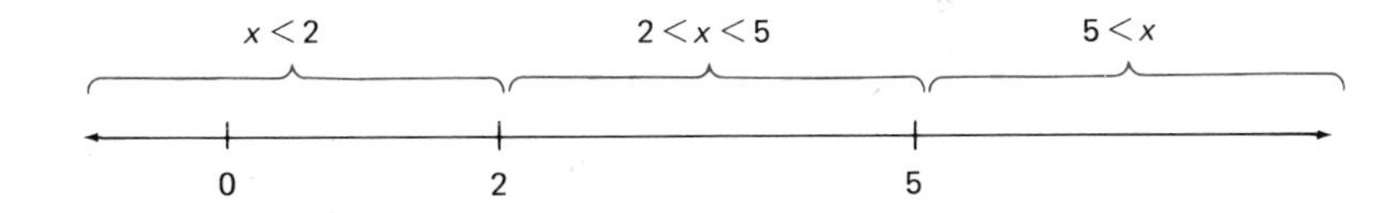

These three intervals are listed left to right at the top of the next table. Now select a value in the first interval $(-\infty, 2)$, such as $x = 0$, and use it as a test value to determine the signs of the factors $x - 2$ and $x - 5$. The minus signs next to the factors $x - 2$ and $x - 5$ under the heading $x < 2$ indicate that these factors are negative for this interval. Furthermore, since the product of two negative factors is positive, a plus sign is written next to $(x - 2)(x - 5)$ for this interval.

Note that 2 and 5 are not within any of the intervals listed at the top of the table.

Interval	$x < 2$	$2 < x < 5$	$5 < x$
Sign of $x - 2$	$-$	$+$	$+$
Sign of $x - 5$	$-$	$-$	$+$
Sign of $(x - 2)(x - 5)$	$+$	$-$	$+$

Using a test value such as $x = 3$ in the interval (2, 5) shows that $x - 2$ is positive, $x - 5$ is negative, and consequently that $(x - 2)(x - 5)$ is negative. This explains the signs in the column under $2 < x < 5$.

The entries for the last column can be checked using $x = 6$ as a specific example. The results in the last row of the table show that $(x - 2)(x - 5) < 0$ only for the interval (2, 5), which is the solution to the given inequality.

EXAMPLE 4 Graph on a number line: $\dfrac{1 - 2x}{x - 3} \leq 0.$

Solution First find the values of x for which the numerator or the denominator is zero.

$$1 - 2x = 0 \qquad x - 3 = 0$$

$$-2x = -1 \qquad x = 3$$

$$x = \tfrac{1}{2}$$

Now locate the three intervals determined by these two points.

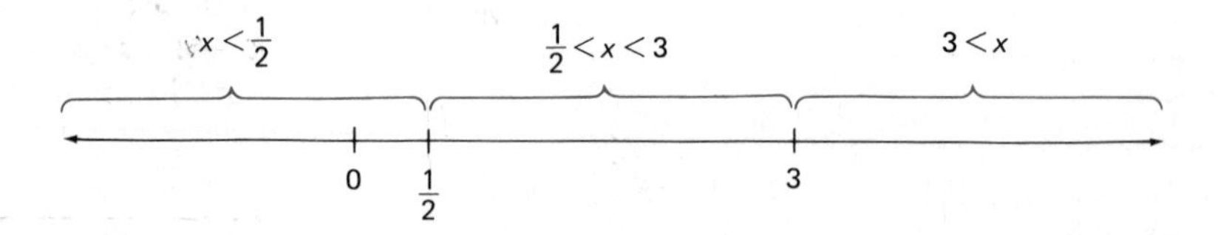

Next, select a specific value to test in each interval and record the results in a table as follows.

Interval	$x < \frac{1}{2}$	$\frac{1}{2} < x < 3$	$3 < x$
Sign of $1 - 2x$	$+$	$-$	$-$
Sign of $x - 3$	$-$	$-$	$+$
Sign of $\dfrac{1 - 2x}{x - 3}$	$-$	$+$	$-$

The table shows that the fraction will be negative when $x < \frac{1}{2}$ or when $x > 3$. Also note that $x = \frac{1}{2}$ satisfies the given inequality but that $x = 3$

is excluded from the solution. Can you explain why this is so? The solution consists of all x such that $x \leq \frac{1}{2}$ or $x > 3$.

This solution cannot be described easily using a single inequality. For example, it makes no sense to write $3 < x < \frac{1}{2}$ since this implies the false result $3 < \frac{1}{2}$.

EXERCISES 1.5

Solve for x and graph.

1 $x + 4 < 0$

2 $2x - 3 \geq 1$

3 $5 - x \leq 0$

4 $5x - 10 > 10$

5 $-3x + 4 \geq 0$

6 $3x - 4 \leq 0$

7 $2x + 7 \leq 5 - 6x$

8 $3x - 2 > x + 5$

9 $-5x + 5 < -3x + 1$

10 $3(x - 1) \geq 2(x - 1)$

11 $2(x + 1) < x - 1$

12 $\frac{1}{3}(x + 4) > 2$

13 $\dfrac{2 - x}{5} \geq 0$

14 $\dfrac{1}{x} < 0$

15 $-\dfrac{2}{x + 1} > 0$

16 Assume that $-2 < x$. Classify as positive or negative: (a) $x + 2$; (b) $x + 4$.

17 Assume that $x < \frac{1}{2}$. Classify as positive or negative: (a) $x - \frac{1}{2}$; (b) $x - 1$.

18 Assume that $-1 < x < 3$. Classify as positive or negative:
(a) $x + 1$ (b) $x + 4$
(c) $x - 3$ (d) $x - 6$

19 Assume that $-\frac{3}{2} < x < -\frac{1}{2}$. Classify as positive or negative:
(a) $x + \frac{3}{2}$ (b) $x + 2$
(c) $x + \frac{1}{2}$ (d) $x - 1$

Solve for x and graph.

20 $3x + 5 \neq 8$ (The symbol $\neq$ is read "is not equal to.")

21 $2x + 1 \ngtr 5$ (The symbol $\ngtr$ is read "is not greater than.")

22 $3x - 2 \nless 1$ (The symbol $\nless$ is read "is not less than.")

23 $12x + 9 \neq 15(x - 2)$

24 $3(x - 1) \ngtr 5(x + 2)$

25 $2(x + 3) \nless 3(x + 1)$

26 $\frac{5}{3}x \nless 2x - 1$

27 $x + 2 \ngtr \frac{3}{4}x$

*28 $\frac{2}{9}(3x + 7) \ngeq 1 - \frac{4}{3}x$

▬ *Construct a table to solve each inequality.*

29 $(x + 1)(x - 1) < 0$

30 $(x + 3)(x + 2) > 0$

31 $(x + 5)(2x - 1) \leq 0$

32 $(2x - 1)(3x + 1) > 0$

33 $\dfrac{x}{x + 1} > 0$

34 $\dfrac{x - 2}{x + 3} < 0$

35 $\dfrac{x + 5}{2 - x} < 0$

36 $\dfrac{1}{x} \geq 2$

37 $(x + 3)(x - 1)(x - 2) \geq 0$

38 $(x + 5)(x + 1)(x - 3) \leq 0$

*39 Here is an alternative (graphic) method for solving conditional inequalities like $(x - 2)(x - 5) < 0$. On the top number line, the bold (dark) half represents those x for which $x - 2$ is positive $(x > 2)$, and the other half gives those x for which $x - 2$ is negative $(x < 2)$. The bottom line has the same information for the factor $x - 5$.

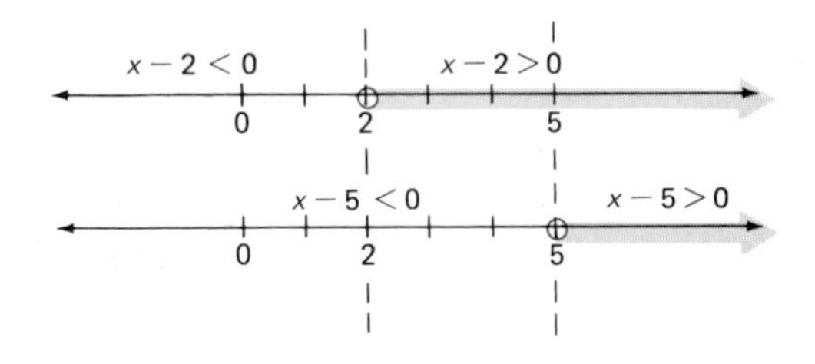

Since $(x - 2)(x - 5) < 0$ when $x - 2$ and $x - 5$ have *opposite* signs, the set of x for which this is the case are those between the vertical lines at 2 and 5. Thus the solution is $2 < x < 5$. Find the solution of $(x - 2)(x - 5) > 0$ from the preceding graph.

*40 Solve Exercise 30 by the method of Exercise 39.

▬ Throughout this text, this symbol will be used to identify exercises or groups of exercises that are of special value with respect to the later study of calculus. See the preface for further discussion.

1.6

Absolute value

What do the numbers -5 and $+5$ have in common? Obviously, they are different numbers and are the coordinates of two distinct points on the number line. However, they are both the same distance from 0, the **origin**, on the number line.

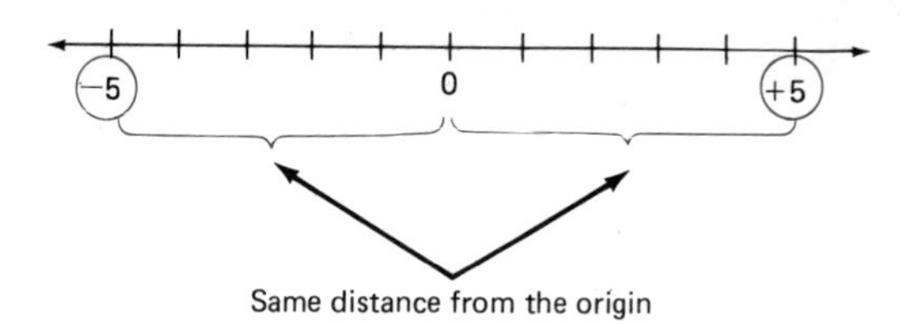

In other words, -5 is as far to the left of 0 as $+5$ is to the right of 0. We show this fact by using **absolute value notation** as follows:

$|-5| = 5$ read as "The *absolute value* of -5 is 5."
$|+5| = 5$ read as "The *absolute value* of $+5$ is 5."

Geometrically, for any real number x, $|x|$ is the distance (without regard to direction) that x is from the origin. Note that for a positive number, $|x| = x; |+6| = 6$. For a negative number, $|x| = -x; |-6| = -(-6) = 6$. Also, since 0 is the origin, it is natural to have $|0| = 0$.

We can summarize this with the following definition.

In words, if x is positive or zero, the absolute value of x is x. If x is negative, the absolute value is the opposite of x.

DEFINITION OF ABSOLUTE VALUE

For any real number x,

$$|x| = \begin{cases} x & \text{if } x \geq 0 \\ -x & \text{if } x < 0 \end{cases}$$

Some useful properties of absolute value follow, presented with illustrations but without formal proof. The variables represent any real numbers.

PROPERTY 1. If $|a| = k$, then $a = k$ or $-a = k$, that is, $a = k$ or $a = -k$; $|0| = 0$.

This property follows immediately from the definition of absolute value. For example, an equation such as $|x| = 2$ is just another way of saying that $x = 2$ or $x = -2$. The graph consists of the two points with coordinates 2 and -2; each of these points is 2 units from the origin.

EXAMPLE 1 Solve: $|5 - x| = 7$.

Solution

We use Property 1 for this solution, noting that $5 - x$ plays the role of a.

$$+(5 - x) = 7 \quad \text{or} \quad -(5 - x) = 7$$
$$5 - x = 7 \qquad -5 + x = 7$$
$$-x = 2 \qquad x = 12$$
$$x = -2$$

These two solutions, -2 and 12, can be checked by substitution into the original equation.
Check: $|5 - (-2)| = |7| = 7$; $|5 - 12| = |-7| = 7$.

Now consider the inequality $|x| < 2$. Here we are asked to find all the real numbers whose absolute value is less than 2. On the number line, we can think of the solution as consisting of all the points whose distance from the origin is less than 2 units, that is, all of the real numbers between -2 and 2. The solution of this inequality can be written in either of these two forms:

$$x < 2 \quad \text{and} \quad x > -2 \qquad\qquad -2 < x < 2$$

The graph of $|x| < 2$ is the interval $(-2, 2)$.

$|x| < 2$: −2 −1 0 1 2

Conversely, the graph also shows that if $-2 < x < 2$, then $|x| < 2$. If the endpoints are included, then we would have $-2 \leq x \leq 2$; that is, $|x| \leq 2$.

We can now generalize by means of the following property.

PROPERTY 2. $|a| < k$ if *and* only if $-k < a < k$.

Note that $-k < a < k$ is the same as saying that $a < k$ *and* $a > -k$.

EXAMPLE 2 Solve: $|x - 2| < 3$.

Solution Let $x - 2$ play the role of a in Property 2. Consequently, $|x - 2| < 3$ is equivalent to $-3 < x - 2 < 3$. Now add 2 to each member to isolate x in the middle.

$$\begin{array}{rcccl} -3 & < & x - 2 & < & 3 \\ +2 & & +\ 2 & & +2 \\ \hline -1 & < & x + 0 & < & 5 \end{array} \qquad \text{Thus } -1 < x < 5$$

In Example 2 we could also have solved the inequality by writing and solving two separate inequalities in this way:

$$|x - 2| < 3 \quad \text{means} \quad x - 2 < 3 \quad \text{and} \quad x - 2 > -3$$

Thus $x < 5$ and $x > -1$. The word "and" indicates that we are to consider the values for x that satisfy *both* conditions. Therefore, the solution consists

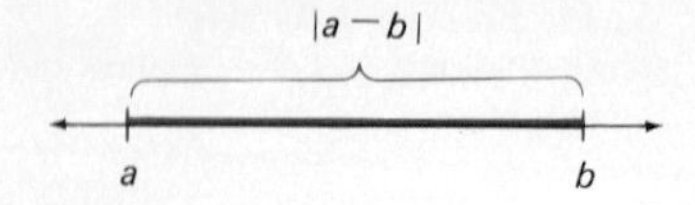

of those numbers that are both greater than -1 and less than 5; that is, $-1 < x < 5$.

The quantity $|a - b|$ also represents the distance between the points a and b on the number line. For example, the distance between 8 and 3 on the number line is $5 = |8 - 3| = |3 - 8|$. The idea of distance between points on a number line can be used to give an alternative way to solve Example 2. Think of the expression $|x - 2|$ as the distance between x and 2 on the number line, and then consider all the points x whose distance from 2 is less than 3 units.

Think: $|x - 2| < 3$ means that x is within 3 units of 2.

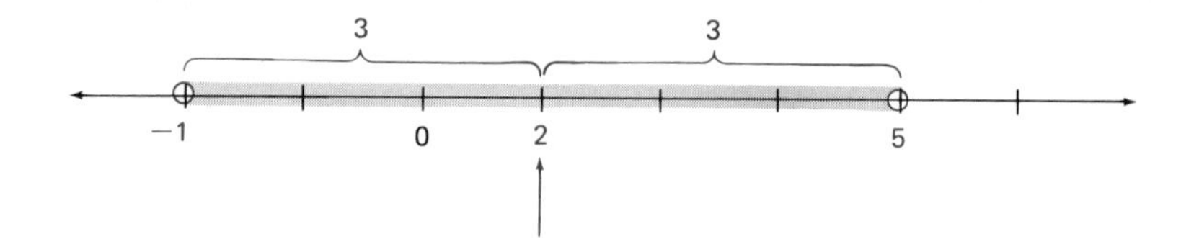

Observe that the *midpoint* or *center* of the interval in the preceding discussion has coordinate 2. This value can be found by taking the average of the numbers -1 and 5; $\frac{-1 + 5}{2} = 2$. In general, we have the following result.

MIDPOINT FORMULA FOR A LINE SEGMENT

If x_1 and x_2 are the endpoints of an interval, then the coordinate of the midpoint is

$$\frac{x_1 + x_2}{2}$$

EXAMPLE 3 Find the coordinate of the midpoint M of each line segment AB.

(a) −4 (A), 0, M, 6 (B)

(b) −10 (A), M, −2 (B), 0

Solution **(a)** $\frac{-4 + 6}{2} = 1$ **(b)** $\frac{-10 + (-2)}{2} = -6$

Throughout the text you will find sets of CAUTION items. These illustrate errors that are often made by students. Study these carefully so that you understand the errors shown and can avoid such mistakes.

CAUTION: LEARN TO AVOID MISTAKES LIKE THESE

WRONG	RIGHT
$\lvert \frac{3}{4} - 2 \rvert = \frac{3}{4} - 2 = -\frac{5}{4}$	$\lvert \frac{3}{4} - 2 \rvert = \lvert -\frac{5}{4} \rvert = \frac{5}{4}$
$\lvert 5 - 7 \rvert = \lvert 5 \rvert - \lvert 7 \rvert = -2$	$\lvert 5 - 7 \rvert = \lvert -2 \rvert = 2$

WRONG	RIGHT
If $|x| = -2$, then $x = -2$.	There is no solution; the absolute value of a number can never be negative.
If $|x - 1| < 3$, then $x < 4$.	If $|x - 1| < 3$, then $-3 < x - 1 < 3$; that is, $-2 < x < 4$

Next, consider the inequality $|x| > 3$. We know that $|x| < 3$ implies that $-3 < x < 3$. Therefore, $|x| > 3$ consists of those values of x for which $x < -3$ or $x > 3$. The graph of this set may be drawn as follows.

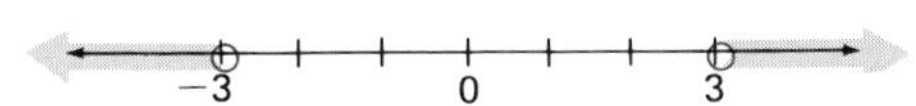

This graph shows the set of points whose distance from 0 is more than 3 units. The property involved is summarized in this way.

PROPERTY 3. $|a| > k$ if and only if $a < -k$ or $a > k$.

Also, $|a| \geq k$ if and only if $a \leq -k$ or $a \geq k$.

EXAMPLE 4 Graph: $|x + 1| > 2$.

Solution Think of $|x + 1|$ as $|x - (-1)|$ so as to have it in the form $|a - b|$. This represents the distance between x and -1, and we wish to find all the points that are more than 2 units away from the point -1.

CAUTION: A common error that students make is to write the solution for Example 4 as $1 < x < -3$. Explain why this is *not* correct.

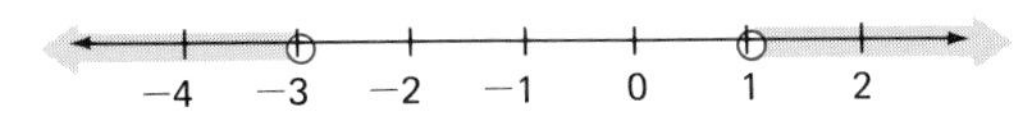

The solution is $x < -3$ or $x > 1$.

In Example 4 we could have solved the inequality by using Property 3.

$$|x + 1| > 2 \quad \text{means} \quad x + 1 < -2 \quad \text{or} \quad x + 1 > 2$$

Thus $x < -3$ or $x > 1$. The word "or" indicates that we are to consider the values for x that satisfy either one of these two conditions; that is, x may be less than -3 or x may be greater than 1.

TEST YOUR UNDERSTANDING

Solve for x.

1 $|x + 3| = 5$	**2** $|x - 2| = 3$	**3** $|4 - x| = 1$
4 $|2x - 1| = 3$	**5** $|3x + 2| = 5$	**6** $|1 - 2x| = 2$

Solve and graph.

7 $|x - 3| < 1$	**8** $|x + 3| > 1$	**9** $|2 - x| \leq 3$

The absolute value of a real number is the same as the absolute value of its negative.

PROPERTY 4. $|a| = |-a|$

For example, $|2| = |-2| = 2$. A useful application of this property is the result that for any real numbers x and y, $|x - y| = |y - x|$. Note that $x - y$ and $y - x$ are negatives of one another; for example, $|2 - \pi| = |\pi - 2| = \pi - 2$ because $\pi - 2 > 0$.

A real number is always between its absolute value and the negative of its absolute value.

PROPERTY 5. $-|a| \leq a \leq |a|$

For example, verify that each of the following is correct:

$$-|7| \leq 7 \leq |7| \qquad -|-7| \leq -7 \leq |-7|$$

The absolute value of a product is the product of the absolute values.

PROPERTY 6. $|ab| = |a| \cdot |b|$

For example:

$$|(-3)(5)| = |-15| = 15 \quad \text{and} \quad |-3| \cdot |5| = 3 \cdot 5 = 15$$

The absolute value of a quotient is the quotient of the absolute values.

PROPERTY 7. $\left|\frac{a}{b}\right| = \frac{|a|}{|b|}$; $b \neq 0$

For example:

$$\left|\frac{-9}{3}\right| = \left|-\frac{9}{3}\right| = |-3| = 3 \qquad \text{and} \qquad \frac{|-9|}{|3|} = \frac{9}{3} = 3$$

The absolute value of a sum is less than or equal to the sum of the absolute values.

PROPERTY 8. $|a + b| \leq |a| + |b|$ (the triangle inequality)

Verify that each of the following examples is correct.

$$|2 + 7| \leq |2| + |7| \qquad |(-2) + (-7)| \leq |-2| + |-7|$$
$$|(-2) + 7| \leq |-2| + |7| \qquad |2 + (-7)| \leq |2| + |-7|$$

EXERCISES 1.6

Classify each statement as true or false.

1 $-|-\frac{1}{3}| = \frac{1}{3}$

2 $|-1000| < 0$

3 $|-\frac{1}{2}| = 2$

4 $|\sqrt{2} - 5| = 5 - \sqrt{2}$

5 $\left|\frac{x}{y}\right| = |x| \cdot \frac{1}{|y|}$

6 $2 \cdot |0| = 0$

7 $||x|| = |x|$

8 $|-(-1)| = -1$

9 $|a| - |b| = a - b$

10 $|a - b| = b - a$

Solve for x.

11 $|x| = \frac{1}{2}$

12 $|3x| = 3$

13 $|x - 1| = 3$

14 $|3x - 4| = 0$

15 $|2x - 3| = 7$

16 $|6 - 2x| = 4$

17 $|4 - x| = 3$

18 $|3x - 2| = 1$

19 $|3x + 4| = 16$

20 $\left|\frac{1}{x - 1}\right| = 2$

21 $\frac{|x|}{x} = 1$

22 $\frac{|x|}{x} = -1$

Solve for x and graph.

23 $|x + 1| = 3$

24 $|x - 1| \leq 3$

25 $|x - 1| \geq 3$

26 $|x + 2| = 3$

27 $|x + 2| \leq 3$

28 $|x + 2| \geq 3$

29 $|-x| = 5$

30 $|x| \leq 5$

31 $|x| \geq 5$

32 $|x - 5| \neq 3$

33 $|x - 5| \leq 3$

34 $|x - 5| \geq 3$

35 $|x - 3| < 0.1$

36 $|2 - x| < 3$

37 $|2x - 1| < 7$

38 $|3x - 6| < 9$

39 $|4 - x| < 2$

40 $|1 + 5x| < 1$

41 $|x - 4| \not> 1$

42 $|x - 2| \ngeq 0$

43 $\frac{1}{|x - 3|} > 0$

Find the coordinate of the midpoint of a line segment with endpoints as given.

44 3 and 9

45 -8 and -2

46 -12 and 0

47 0 and 11

48 -5 and 8

49 -7 and 7

*50 Prove Property 5 of this section. (*Hint:* Consider the two cases $x \geq 0$ and $x < 0$.)

*51 Cite at least four different examples to confirm this inequality: $|x - y| \geq ||x| - |y||$.

*52 (a) Prove that $|xy| = |x| \cdot |y|$ for the case $x < 0$ and $y > 0$.
(b) Prove this product rule when $x < 0$ and $y < 0$.

*53 Prove $\left|\frac{x}{y}\right| = \frac{|x|}{|y|}$ for the case $x < 0$, $y > 0$.

*54 Prove that if x is a real number, then $x^2 = |x^2| = |x|^2$.

55 Write in absolute value notation.
(a) $-3 < x < 3$
(b) $-\frac{1}{2} < 2x < \frac{1}{2}$
(c) $-1 < x < 5$
(d) $-\frac{1}{2} < x < \frac{3}{2}$

56 Write an equivalent inequality without absolute values for $|x - a| < c$.

■ 57 Prove: $|x - 1| < \frac{1}{10}$ implies that $|y - 8| < \frac{3}{10}$, where $y = 3x + 5$, by completing the following steps.
(a) $|x - 1| < \frac{1}{10}$ implies that $-\frac{1}{10} < x - 1 < \frac{1}{10}$. Add 1 throughout to obtain:
_______ $< x <$ _______
(b) Multiply each member by 3:
_______ $< 3x <$ _______
(c) Add -3: _______ $< 3x - 3 <$ _______
(d) Now note that if $y = 3x + 5$, then $y - 8 = 3x - 3$. Therefore:
_______ $< y - 8 <$ _______
(e) Convert the last statement to absolute value form.

58 One endpoint of a line segment is located at -7. The midpoint is located at 3. What is the coordinate of the other endpoint of the line segment?

REVIEW EXERCISES FOR CHAPTER 1

The solutions to the following exercises can be found within the text of Chapter 1. Try to answer each question without referring to the text.

Section 1.1

1 Given two real numbers a and b, write in symbols the commutative property for addition and for multiplication.

2 Repeat Exercise 1 for the associative property.

3 Use the distributive property to find the product 7×61.

4 What basic property of the real numbers is illustrated by each of the following?
(a) $6 + (17 + 4) = (17 + 4) + 6$
(b) $\frac{3}{4} + (-\frac{3}{4}) = 0$
(c) $57 \times 1 = 57$
(d) $\frac{2}{3}(12 + 36) = \frac{2}{3}(12) + \frac{2}{3}(36)$

5 If $ab = 0$, what do you know about the numbers a and b?

6 State the definition of subtraction in terms of addition.

7 State the definition of division in terms of multiplication.

8 Explain what the phrase "if and only if" means.

9 Why is division by zero not possible?

Section 1.2

10 What is meant by the set of whole numbers?

11 Write the decimal form of $\frac{19}{22}$.

12 State the definition of a rational number.

13 Using straightedge and compass, locate the irrational number $\sqrt{2}$ on a number line.

14 State the relationship between the real numbers and their decimal representations.

15 Classify each number into as many different sets as possible:
(a) 5
(b) $\frac{2}{3}$
(c) $\sqrt{7}$
(d) -14

16 True or false: Every rational number is the coordinate of some point on the number line.

Section 1.3

17 State the addition property of equality.

18 State the multiplication property of equality.

19 Solve for x: $6x + 9 = 2x + 1$.

20 Solve for x: $2(x + 3) = x + 5$.

21 The formula relating degrees Fahrenheit and degrees Celsius is $\frac{5F - 160}{9} = C$. Solve for F in terms of C.

22 The length of a rectangle is 1 centimeter less than

twice the width. The perimeter is 28 centimeters. Find the dimensions of the rectangle.

23 A car leaves a certain town at noon, traveling due east at 40 miles per hour. At 1:00 P.M. a second car leaves the town traveling in the same direction at a rate of 50 miles per hour. In how many hours will the second car overtake the first car?

24 David has a total of $4.10 in nickels, dimes, and quarters. He has twice as many nickels as dimes, and two more quarters than dimes. How many dimes does he have?

Section 1.4

25 State the transitive property of order for real numbers a, b, and c.

26 True or false: If $a < b$ and $c < 0$, then $ac < bc$.

27 True or false: If $a < b$ and $ab > 0$, then $\frac{1}{a} > \frac{1}{b}$.

28 Prove: If $a < b$ and $b < c$, then $a < c$.

29 Translate each verbal statement into a statement of inequality.
(a) x is to the right of zero.
(b) x is between 2 and 3.
(c) x is greater than or equal to -1.
(d) x is at least 10 but less than 100.

30 Show each interval as a graph on a number line.
(a) $(-2, 3)$ (b) $[-2, 3]$
(c) $(-2, 3]$ (d) $[-2, 3)$

Section 1.5

31 Solve for x: $3x - 4 > 11$.

32 Solve for x and graph: $5(3 - 2x) \geq 10$.

33 Solve: $\frac{2}{x+4} < 0$.

34 Graph on a number line: $\frac{1-2x}{x-3} \leq 0$.

Section 1.6

35 State the definition of the absolute value of a number.

36 Solve: $|5 - x| = 7$.

37 Solve: $|x - 2| < 3$.

38 What does $|a - b|$ represent with respect to points a and b on a number line? Illustrate with two specific values.

39 Graph: $|x + 1| > 2$.

40 What is the coordinate of the midpoint of a line segment with endpoints x_1 and x_2?

SAMPLE TEST QUESTIONS FOR CHAPTER 1

Use these questions to test your knowledge of the basic skills and concepts of Chapter 1. Then check your answers with those given at the end of the book.

1 Check the boxes in the table to indicate the set to which each number belongs.

	Whole numbers	*Integers*	*Rational numbers*	*Irrational numbers*	*Negative integers*
$\frac{-6}{3}$					
0.231					
$\sqrt{12}$					
$\sqrt{\frac{9}{4}}$					
1983					

2 Explain how to locate $\sqrt{10}$ on a number line by using straightedge and compass.

3 Classify each statement as true or false.
(a) Negative irrational numbers are not real numbers.
(b) Every integer is a rational number.
(c) Some irrational numbers are integers.
(d) Zero is a rational number.
(e) If $x < y$, then $x - 5 > y - 5$.
(f) The absolute value of a sum equals the sum of the absolute values.
(g) If $-5x < -5y$, then $x > y$.
(h) $|x - 2| < 3$ means that x is within 2 units of 3 on the number line.

4 Name the property or definition illustrated by each statement.
(a) $3[(4 + 5) + 6] = 3(4 + 5) + 3(6)$
(b) $0 + 7 = 7$
(c) $-6 \cdot 1 = -6$
(d) $-5 - (-2) = -5 + (2)$

(e) $(8 \times 3) \times \sqrt{2} = 8 \times (3 \times \sqrt{2})$
(f) $-10 < -3$ since $-3 - (-10) > 0$

5 State, in symbols, each property:
(a) addition property of equality
(b) multiplication property of equality

6 (a) Find the coordinate of the midpoint of the line segment with endpoints at -4 and 6.
(b) One endpoint of a line segment is located at -7. The midpoint is located at -2. What is the coordinate of the other endpoint of the line segment?

Solve for x.

7 $\frac{3}{4}x - 1 = 11$ 8 $4x + 20 = 2x - 2$

9 $3(x + 2) = x - 3$

10 Find the dimensions of a rectangle whose perimeter is 52 inches if the length is 5 inches more than twice the width.

11 A car leaves from point B at noon traveling at the rate of 55 miles an hour. One hour later a second car leaves from the same point, traveling in the opposite direction at 45 miles per hour. At what time will they be 200 miles apart?

Show each interval on a number line.

12 $(-2, 2]$ 13 $[-3, 1)$

14 $(-2, 0)$ 15 $[-1, 1]$

Solve for x.

16 $|3 - 4x| = 2$ 17 $\frac{|x + 2|}{x + 2} = -1$

18 $\frac{|x - 2|}{x - 2} = 1$

Solve and graph each inequality.

19 $2(5x - 1) < x$ 20 $|2x - 1| \geq 3$

21 $\frac{3}{x - 2} > 0$ 22 $(2x - 1)(x + 5) < 0$

ANSWERS TO THE TEST YOUR UNDERSTANDING EXERCISES

Page 2

1 Associative property for addition.
2 Commutative property for addition.
3 Commutative property for addition.
4 Commutative property for multiplication.
5 Commutative property for multiplication.
6 Associative property for multiplication.
7 No; no.
8 $12 \div 3 \neq 3 \div 12$
9 No; no.
10 $(8 \div 4) \div 2 \neq 8 \div (4 \div 2)$
11 Yes; the closure properties for addition and multiplication of real numbers, respectively.

Page 8

1 $\{0, 1, 2, 3, 4\}$
2 $\{11, 12, 13, \ldots\}$
3 $\{2, 3, 4, 5, 6, 7, 8, 9\}$
4 $\{-2, -1, 0, 1, 2, 3\}$
5 $\{-4, -3, -2, -1\}$
6 $\{\ldots, -2, -1, 0, 1\}$
7 The sets in Exercises 1, 3, 4, 5.

Page 13

1 6	2 17	3 -4	4 6	5 13	6 10
7 1	8 2	9 5	10 -9	11 -3	12 -11

Page 20

1
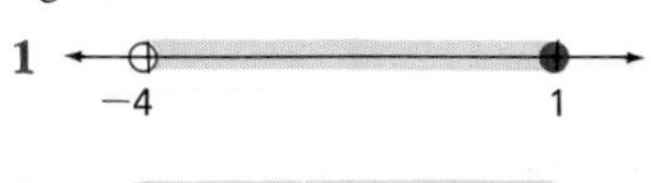

2
0 3

3

4
-2

5 $[-1, 2]$ 6 $(0, 3)$ 7 $(-\infty, 0]$ 8 $[-1, 4)$ 9 $[-3, \infty)$ 10 $(-5, -2)$

Page 23

1 $x < 3$

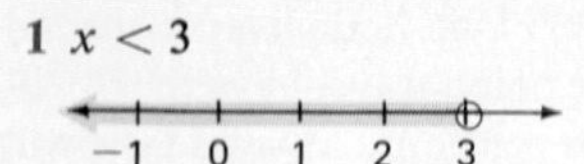

2 $x > -5$

3 $x < -4$

4 $x \geq \frac{5}{2}$

5 $x < 2$

6 $x \leq -\frac{5}{2}$

7 $x > -6$

8 $x \leq -6$

9 $x > 0$

10 $x > 3$

11 (**a**) Negative; (**b**) negative.

12 (**a**) Positive; (**b**) negative.

Page 29

1 $-8, 2$

2 $-1, 5$

3 $3, 5$

4 $-1, 2$

5 $-\frac{7}{3}, 1$

6 $-\frac{1}{2}, \frac{3}{2}$

7 $2 < x < 4$

8 $x < -4$ or $x > -2$

9 $-1 \leq x \leq 5$

FUNDAMENTALS OF ALGEBRA

2.1 Integral exponents

Much of mathematical notation can be viewed as efficient abbreviations of lengthier statements. For example:

$$4^9 = 4 \times 4 \times 4 \times 4 \times 4 \times 4 \times 4 \times 4 \times 4$$

This illustration is a specific example of the fundamental definition of a *positive integral exponent.*

DEFINITION OF POSITIVE INTEGRAL EXPONENT

If n is a positive integer and b is any real number, then

$$b^n = \underbrace{b \cdot b \cdot \cdots \cdot b}_{n \text{ factors}}$$

The number b is called the **base** and n is called the **exponent.**

Here are some illustrations of the definition:

The most common ways of referring to b^n are "b to the nth power," "b to the nth," or "the nth power of b."

$$b^1 = b$$

$$(a + b)^2 = (a + b)(a + b)$$

"Raising to a power" is not commutative or associative. Verify this by showing that $2^3 \neq 3^2$ and $(2^3)^4 \neq 2^{(3^4)}$.

$$(-2)^3 = (-2)(-2)(-2) = -8$$
$$(\tfrac{1}{3})^4 = \tfrac{1}{3} \cdot \tfrac{1}{3} \cdot \tfrac{1}{3} \cdot \tfrac{1}{3} = \tfrac{1}{81}$$
$$(-1)^5 = (-1)(-1)(-1)(-1)(-1) = -1$$
$$10^6 = 10 \cdot 10 \cdot 10 \cdot 10 \cdot 10 \cdot 10 = 1{,}000{,}000$$

A number of important rules concerning positive integral exponents are easily established on the basis of the preceding definition. Here is a list of these rules in which m and n are any positive integers, a and b are any real numbers, and with the usual understanding that denominators cannot be zero.

RULE 1. $b^m b^n = b^{m+n}$

Illustrations:

$$2^3 \cdot 2^4 = 2^{3+4} = 2^7$$
$$x^3 \cdot x^4 = x^7$$

RULE 2. $\dfrac{b^m}{b^n} = \begin{cases} b^{m-n} & \text{if } m > n \\ 1 & \text{if } m = n \\ \dfrac{1}{b^{n-m}} & \text{if } m < n \end{cases}$

Illustrations:

$(m > n)$ $\quad \dfrac{2^5}{2^2} = 2^{5-2} = 2^3 \qquad \dfrac{x^5}{x^2} = x^{5-2} = x^3$

$(m = n)$ $\quad \dfrac{5^2}{5^2} = 1 \qquad \dfrac{x^2}{x^2} = 1$

$(m < n)$ $\quad \dfrac{2^2}{2^5} = \dfrac{1}{2^{5-2}} = \dfrac{1}{2^3} \qquad \dfrac{x^2}{x^5} = \dfrac{1}{x^{5-2}} = \dfrac{1}{x^3}$

RULE 3. $(b^m)^n = b^{mn}$

Illustrations:

$$(2^3)^2 = 2^{3 \cdot 2} = 2^6$$
$$(x^3)^2 = x^{3 \cdot 2} = x^6$$

RULE 4. $(ab)^m = a^m b^m$

Illustrations:

$$(2 \cdot 3)^5 = 2^5 \cdot 3^5$$
$$(xy)^5 = x^5 y^5$$

Eventually, the rules for exponents will be extended to include all the real numbers (not just positive integers) as exponents. We will find that all the rules stated here will still apply.

RULE 5. $\left(\dfrac{a}{b}\right)^m = \dfrac{a^m}{b^m}$

Illustrations:

$$\left(\frac{3}{2}\right)^5 = \frac{3^5}{2^5}$$
$$\left(\frac{x}{y}\right)^5 = \frac{x^5}{y^5}$$

The proper use of these rules can simplify computations, as in the following example.

EXAMPLE 1 Evaluate: $12^3(\frac{1}{6})^3$.

Solution This expression can be evaluated without use of any of the rules of exponents.

$$\begin{aligned} 12^3(\tfrac{1}{6})^3 &= 1728(\tfrac{1}{216}) \\ &= \tfrac{1728}{216} \\ &= 8 \end{aligned}$$

Using Rule 4, the work proves to be much easier.

$$\begin{aligned} 12^3(\tfrac{1}{6})^3 &= (12 \cdot \tfrac{1}{6})^3 \\ &= 2^3 \\ &= 8 \end{aligned}$$

Here is a proof of Rule 4. You can try proving the other rules by using similar arguments.

$$\begin{aligned} (ab)^m &= (ab)(ab) \cdots (ab) && \text{(by definition)} \\ &= (a \cdot a \cdot \cdots \cdot a)(b \cdot b \cdot \cdots \cdot b) && \left\{\begin{array}{l}\text{(by repeated use of the} \\ \text{commutative and associative} \\ \text{laws for multiplication)}\end{array}\right. \\ &= a^m b^m && \text{(by definition)} \end{aligned}$$

EXAMPLE 2 Simplify: **(a)** $2x^3 \cdot x^4$ **(b)** $\dfrac{(x^3y)^2y^3}{x^4y^6}$

Solution

(a) $2x^3 \cdot x^4 = 2x^{3+4} = 2x^7$

(b) $\dfrac{(x^3y)^2y^3}{x^4y^6} = \dfrac{(x^3)^2y^2y^3}{x^4y^6} = \dfrac{x^6y^5}{x^4y^6} = \dfrac{x^2}{y}$

TEST YOUR UNDERSTANDING

Evaluate each of the following.

1 5^3 2 $(-\frac{1}{2})^5$ 3 $(-\frac{2}{3})^3 + \frac{8}{27}$

4 $(10^3)^2$ 5 $2^3(-2)^3$ 6 $(\frac{1}{2})^3 8^3$

7 $\dfrac{17^8}{17^9}$ 8 $\dfrac{(-2)^3 + 3^2}{3^3 - 2^2}$ 9 $\dfrac{(-12)^4}{4^4}$

10 $(ab^2)^3(a^2b)^4$ 11 $\dfrac{2^2 \cdot 16^3}{(-2)^8}$ 12 $\dfrac{(2x^3)^2(3x)^2}{6x^4}$

Our discussion of exponents has been restricted to the use of positive integers only. Now let us consider the meaning of 0 as an exponent. In particular, what is the meaning of 5^0? We know that 5^3 means that we are to

This discussion provides meaning for the use of 0 as an exponent. That is, an expression like 5^0 will now be defined.

use 5 three times as a factor. But certainly it makes no sense to use 5 zero times. The rules of exponents will help to resolve this dilemna.

We would like these laws for exponents to hold even if one of the exponents happens to be zero. That is, we would *like* Rule 2 to give

$$\frac{5^2}{5^2} = 5^{2-2} = 5^0$$

But it is already known that

$$\frac{5^2}{5^2} = \frac{25}{25} = 1$$

Thus 5^0 ought to be assigned the value 1. Consequently, in order to *preserve* the rules of exponents, we decide to let $5^0 = 1$. And, from now on, we agree to the following:

Notice that the definition calls for b to be a real number different from 0. That is, we do not define an expression such as 0^0; this is said to be undefined. (See Exercise 68.)

DEFINITION OF ZERO EXPONENT

If b is a real number different from 0, then

$$b^0 = 1$$

Let us explore the first part of Rule 2 for exponents in greater detail.

$$\frac{b^m}{b^n} = b^{m-n}, \qquad m > n, \qquad b \neq 0$$

Suppose that we try an example where $m < n$, and apply the same rule:

$$\frac{x^2}{x^5} = x^{2-5} = x^{-3}$$

What meaning should be assigned to x^{-3}? To answer this question we can return to the original meaning of an exponent, in this way:

This discussion provides the motivation for the definition of a negative exponent.

$$\frac{x^2}{x^5} = \frac{x \cdot x}{x \cdot x \cdot x \cdot x \cdot x} = \frac{x \cdot x}{x \cdot x} \cdot \frac{1}{x \cdot x \cdot x}$$

$$= \frac{x^2}{x^2} \cdot \frac{1}{x^3}$$

$$= 1 \cdot \frac{1}{x^3}$$

$$= \frac{1}{x^3}$$

It thus seems to make sense to *define* x^{-3} as $\frac{1}{x^3}$. Furthermore, this is consistent with Rule 1 for multiplication and the definition of x^0:

$$(x^3)(x^{-3}) = x^{3+(-3)} = x^0 = 1$$

Since $(x^3)(x^{-3}) = 1$, it also follows that $x^3 = \frac{1}{x^{-3}}$. In general, we make the following definition:

DEFINITION OF b^{-n}

If n is an integer and $b \neq 0$, then

$$b^{-n} = \frac{1}{b^n}$$

Note that in making definitions for b^0 and b^{-n}, the major guideline is to *preserve* our earlier rules for exponents. That is, we want those rules to hold regardless of the kind of integers m and n represent.

It follows from this definition that $\left(\frac{a}{b}\right)^{-1} = \frac{b}{a}$, since $\left(\frac{a}{b}\right)^{-1} = \frac{1}{\frac{a}{b}} = \frac{b}{a}$.

In other words, a fraction to the -1 power is the reciprocal of the fraction. Can you show that $\left(\frac{a}{b}\right)^{-2} = \left(\frac{b}{a}\right)^{2}$?

EXAMPLE 3 Evaluate:

(a) $(\frac{1}{7})^{-2}$ **(b)** $(\frac{1}{2})^3(2)^{-3}$ **(c)** $\frac{5^{-2}}{15^{-2}}$ **(d)** $\left(\frac{400 - 10^4}{80^2}\right)^0$

Solution

(a) $(\frac{1}{7})^{-2} = \frac{1}{(\frac{1}{7})^2} = \frac{1}{\frac{1}{49}} = 49$ or $(\frac{1}{7})^{-2} = (7^{-1})^{-2} = 7^2 = 49$

As in Example 3(a), more than one correct procedure is often possible. Finding the most efficient procedure depends largely on experience.

(b) $\left(\frac{1}{2}\right)^3(2)^{-3} = \left(\frac{1}{2}\right)^3\frac{1}{2^3} = \left(\frac{1}{2}\right)^3\left(\frac{1}{2}\right)^3 = \left(\frac{1}{2}\right)^6 = \frac{1}{64}$

(c) $\frac{5^{-2}}{15^{-2}} = \left(\frac{5}{15}\right)^{-2} = \left(\frac{1}{3}\right)^{-2} = (3^{-1})^{-2} = 3^2 = 9$

(d) $\left(\frac{400 - 10^4}{80^2}\right)^0 = 1$

It should be noted that the three cases of Rule 2 can now be condensed into this single form.

RULE 2. (revised): $\frac{b^m}{b^n} = b^{m-n}$

Illustrations:

$$\frac{3^4}{3^2} = 3^2 \qquad \frac{3^2}{3^4} = 3^{-2} \qquad \frac{3^2}{3^2} = 3^0 = 1$$

$$\frac{x^8}{x^2} = x^6 \qquad \frac{x^2}{x^8} = x^{-6} \qquad \frac{x^2}{x^2} = x^0 = 1$$

EXAMPLE 4 Simplify $\left(\frac{a^{-2}b^3}{a^3b^{-2}}\right)^5$ and express the answer using positive exponents only.

Solution There are several ways to proceed; here are two.

(a) $\left(\frac{a^{-2}b^3}{a^3b^{-2}}\right)^5 = (a^{-5}b^5)^5 = (a^{-5})^5(b^5)^5 = a^{-25}b^{25} = \frac{b^{25}}{a^{25}}$

(b) $\left(\frac{a^{-2}b^3}{a^3b^{-2}}\right)^5 = \left(\frac{b^5}{a^5}\right)^5 = \frac{(b^5)^5}{(a^5)^5} = \frac{b^{25}}{a^{25}}$

It can be shown that the rules presented in this section apply for all integral exponents. Here is a proof of Rule 4, for zero and negative integral exponents. The remaining rules can be proved in a similar manner.

If $m = 0$, then $(ab)^0 = 1$. Also, $a^0b^0 = 1 \cdot 1 = 1$. Thus $(ab)^0 = 1 = a^0b^0$. If $m < 0$, let $m = -k$, where k is the appropriate positive integer. Thus

$$\begin{aligned}
(ab)^m &= (ab)^{-k} && \text{(substitution)} \\
&= \frac{1}{(ab)^k} && \text{(definition)} \\
&= \frac{1}{a^kb^k} && \text{(Rule 4 since } k \text{ is a positive integer)} \\
&= \frac{1}{a^k} \cdot \frac{1}{b^k} && \text{(property of fractions)} \\
&= a^{-k}b^{-k} && \text{(definition)} \\
&= a^mb^m && \text{(substitution; } m = -k)
\end{aligned}$$

CAUTION: LEARN TO AVOID MISTAKES LIKE THESE

Many errors are made when working with exponents because of misuses of the basic rules and definitions. This list shows some of the common errors that you should try to avoid.

WRONG	RIGHT
$5^2 \cdot 5^4 = 5^8$ (Do not multiply exponents.) $5^2 \cdot 5^4 = 25^6$ (Do not multiply the base numbers.)	$5^2 \cdot 5^4 = 5^6$ (Rule 1)
$\frac{5^6}{5^2} = 5^3$ (Do not divide the exponents.) $\frac{5^6}{5^2} = 1^4$ (Do not divide the base numbers.)	$\frac{5^6}{5^2} = 5^4$ (Rule 2)
$(5^2)^6 = 5^8$ (Do not add the exponents.)	$(5^2)^6 = 5^{12}$ (Rule 3)
$(-2)^4 = -2^4$ (Misreading the parentheses)	$(-2)^4 = (-1)^4 2^4 = 2^4$ (Rule 4 or definition)
$(-5)^0 = -1$ (Misreading definition of b^0)	$(-5)^0 = 1$ (Definition of b^0)

WRONG	RIGHT
$2^{-3} = -\frac{1}{2^3}$ (Misreading definition of b^{-n})	$2^{-3} = \frac{1}{2^3}$ (Definition of b^{-n})
$\frac{2^3}{2^{-4}} = 2^{3-4} = 2^{-1}$ (Carelessness in subtracting exponents)	$\frac{2^3}{2^{-4}} = 2^{3-(-4)} = 2^7$ (Rule 2)
$5^3 + 5^3 = 5^6$ (Adding exponents does not apply because of plus sign.)	$5^3 + 5^3 = (1+1)5^3 = 2 \cdot 5^3$ (Distributive)
$(a+b)^{-1} = a^{-1} + b^{-1}$ (Wrong use of definition)	$(a+b)^{-1} = \frac{1}{a+b}$ (Definition)

EXERCISES 2.1

Classify each statement as true or false. If it is false, correct the right side of the equality to obtain a true statement.

1 $3^4 \cdot 3^2 = 3^8$

2 $(2^2)^3 = 2^8$

3 $2^5 \cdot 2^2 = 4^7$

4 $\frac{9^3}{9^3} = 1$

5 $\frac{10^4}{5^4} = 2^4$

6 $\left(\frac{2}{3}\right)^4 = \frac{2^4}{3}$

7 $(-27)^0 = 1$

8 $(2^0)^3 = 2^3$

9 $3^4 + 3^4 = 3^8$

10 $(a^2b)^3 = a^2b^3$

11 $(a+b)^0 = a+1$

12 $a^2 + a^2 = 2a^2$

13 $\frac{1}{2^{-3}} = -2^3$

14 $(2+\pi)^{-2} = \frac{1}{4} + \frac{1}{\pi^2}$

15 $\frac{2^{-5}}{2^3} = 2^{-2}$

Evaluate.

16 10^5

17 $2^0 + 2^1 + 2^2$

18 $(-3)^2(-2)^3$

19 $(\frac{2}{3})^0 + (\frac{2}{3})^1$

20 $[(\frac{1}{2})^3]^2$

21 $(\frac{1}{2})^4(-2)^4$

22 $\frac{3^2}{3^0}$

23 $\frac{(-2)^5}{(-2)^3}$

24 $\left(-\frac{3}{4}\right)^3$

25 $\frac{2^3 \cdot 3^4 \cdot 4^5}{2^2 \cdot 3^3 \cdot 4^4}$

26 $\frac{8^3}{16^2}$

27 $\frac{2^{10}}{2^5 \cdot 2^3}$

28 $\frac{3^{-3}}{4^{-3}}$

29 $(\frac{2}{3})^{-2} + (\frac{2}{3})^{-1}$

30 $[(-7)^2(-3)^2]^{-1}$

Express each answer using positive exponents only.

31 $(x^{-3})^2$

32 $(x^3)^{-2}$

33 $x^3 \cdot x^9$

34 $\frac{x^9}{x^3}$

35 $(2a)^3(3a)^2$

36 $(-2x^3y)^2(-3x^2y^2)^3$

37 $(-2a^2b^0)^4$

38 $(2x^3y^2)^0$

39 $\frac{(x^2y)^4}{(xy)^2}$

40 $\left(\frac{3a^2}{b^3}\right)^2\left(\frac{-2a}{3b}\right)^2$

41 $\left(\frac{x^3}{y^2}\right)^4\left(\frac{-y}{x^2}\right)^2$

42 $\frac{(x-2y)^6}{(x-2y)^2}$

43 $\frac{x^{-2}y^3}{x^3y^{-4}}$

44 $\frac{(x^{-2}y^2)^3}{(x^3y^{-2})^2}$

45 $\frac{5x^0y^{-2}}{x^{-1}y^{-2}}$

46 $\frac{(2x^3y^{-2})^2}{8x^{-3}y^2}$

47 $\frac{(-3a)^{-2}}{a^{-2}b^{-2}}$

48 $\frac{3a^{-3}b^2}{2^{-1}c^2d^{-4}}$

49 $\frac{(a+b)^{-2}}{(a+b)^{-8}}$

50 $\frac{8x^{-8}y^{-12}}{2x^{-2}y^{-6}}$

51 $\dfrac{-12x^{-9}y^{10}}{4x^{-12}y^7}$

52 $\dfrac{(2x^2y^{-1})^6}{(4x^{-6}y^{-5})^2}$

53 $\dfrac{(a+3b)^{-12}}{(a+3b)^{10}}$

54 $\dfrac{(-a^{-5}b^6)^3}{(a^8b^4)^2}$

55 $x^{-2}+y^{-2}$

56 $(a^{-2}b^3)^{-1}$

57 $\left(\dfrac{x^{-2}}{y^3}\right)^{-1}$

▬ 58 $-2(1+x^2)^{-3}(2x)$

▬ 59 $-2(4-5x)^{-3}(-5)$

▬ 60 $-7(x^2-3x)^{-8}(2x-3)$

Find a value of x to made each statement true.

61 $2^x \cdot 2^3 = 2^{12}$

62 $2^{-3} \cdot 2^x = 2^6$

63 $2^x \cdot 2^x = 2^{16}$

64 $2^x \cdot 2^{x-1} = 2^7$

▬ Recall that this symbol is used to identify exercises that are of special value with respect to the later study of calculus.

65 $\dfrac{2^x}{2^2} = 2^{-5}$

66 $\dfrac{2^{-3}}{2^x} = 2^4$

*67 Express as a single fraction using only positive exponents: $\dfrac{1}{a^{-1}+b^{-1}}$.

*68 We have said that 0^0 is undefined. The following shows why we have not defined this to be equal to 1. *Suppose* that $0^0 = 1$. Then

$$1 = \frac{1}{1} = \frac{1^0}{0^0}$$

$$= \left(\frac{1}{0}\right)^0$$

(a) What rule for exponents is being used in the last step? **(b)** What went wrong?

*69 Use the definition of a^{-n} to prove that $\dfrac{1}{a^{-n}} = a^n$.

2.2

Radicals and rational exponents

You may recall using radicals in the past as you worked with square roots. For example, $\sqrt{25}$ is called a *radical* and denotes the positive number whose square is 25. Now it is true that $5^2 = 25$ and $(-5)^2 = 25$, but the **radical sign** $\sqrt{}$ implies the positive or *principal square root of a number*. Thus we say that $\sqrt{25} = 5$. We denote the negative square root as $-\sqrt{25} = -5$.

In general, the *principal nth root* of a real number a is denoted by $\sqrt[n]{a}$, as in these illustrations:

In the preceding section b^n was defined only for integers n. The next stage is to extend this concept to include fractional exponents—that is, to give meaning to such expressions as $4^{1/2}$ and $(-8)^{2/3}$.

$$\sqrt[3]{64} = 4 \quad \text{because} \quad 4^3 = 64$$

$$\sqrt[5]{32} = 2 \quad \text{because} \quad 2^5 = 32$$

$$\sqrt[3]{-8} = -2 \quad \text{because} \quad (-2)^3 = -8$$

The expression $\sqrt[n]{a}$ does not always have meaning. For example, let us try to evaluate $\sqrt[4]{-16}$:

$$2^4 = 16 \qquad (-2)^4 = 16$$

It appears that there is no real number x such that $x^4 = -16$. In general, *there is no real number that is the even root of a negative number*. However, it is a fundamental property of real numbers that every positive real number a has exactly one positive nth root. Furthermore, every negative real number has a negative nth root provided that n is odd.

DEFINITION OF $\sqrt[n]{a}$; THE PRINCIPAL nTH ROOT OF a

Let a be a real number and n a positive integer, $n \geq 2$.

(i) If $a > 0$, then $\sqrt[n]{a}$ is the positive number x such that $x^n = a$.

(ii) $\sqrt[n]{0} = 0$

(iii) If $a < 0$ and n is odd, then $\sqrt[n]{a}$ is the negative number x such that $x^n = a$.

(iv) If $a < 0$ and n is even, then $\sqrt[n]{a}$ is not a real number.

The symbol $\sqrt[n]{a}$ is also said to be a *radical*; $\sqrt{}$ is the radical sign, n is the *index* or *root*, and a is called the *radicand*.

EXAMPLE 1 Evaluate: **(a)** $\sqrt[3]{\frac{1}{8}}$ **(b)** $\sqrt[5]{-32}$ **(c)** $\sqrt{0.09}$

Solution

(a) $\sqrt[3]{\frac{1}{8}} = \frac{1}{2}$ since $(\frac{1}{2})^3 = \frac{1}{8}$

(b) $\sqrt[5]{-32} = -2$ since $(-2)^5 = -32$

(c) $\sqrt{0.09} = 0.3$ since $(0.3)^2 = 0.09$

In order to multiply or divide radicals, the index must be the same. Here are illustrations that motivate Rules 2 and 3 stated below.

$$\left.\begin{aligned}\sqrt[3]{8}\sqrt[3]{-27} &= 2(-3) &= -6\\ \sqrt[3]{8(-27)} &= \sqrt[3]{-216} &= -6\end{aligned}\right\} \quad \text{so } \sqrt[3]{8} \cdot \sqrt[3]{-27} = \sqrt[3]{8(-27)}$$

$$\left.\begin{aligned}\frac{\sqrt{36}}{\sqrt{4}} &= \frac{6}{2} = 3\\ \sqrt{\frac{36}{4}} &= \sqrt{9} = 3\end{aligned}\right\} \quad \text{so } \frac{\sqrt{36}}{\sqrt{4}} = \sqrt{\frac{36}{4}}$$

In the following rules it is assumed that all radicals exist according to the definition of $\sqrt[n]{a}$ and, as usual, denominators are not zero.

RULES FOR RADICALS

1. $(\sqrt[n]{a})^n = \sqrt[n]{a^n} = a$
2. $\sqrt[n]{a} \cdot \sqrt[n]{b} = \sqrt[n]{ab}$
3. $\dfrac{\sqrt[n]{a}}{\sqrt[n]{b}} = \sqrt[n]{\dfrac{a}{b}}$
4. $\sqrt[m]{\sqrt[n]{a}} = \sqrt[mn]{a}$

The proofs of Rules 2 and 3 are called for in Exercises 70 and 71.

EXAMPLE 2 Simplify: **(a)** $(\sqrt[4]{7})^4$ **(b)** $\sqrt{9 \cdot 25}$ **(c)** $\sqrt[4]{\frac{16}{81}}$ **(d)** $\sqrt[2]{\sqrt[3]{64}}$ **(e)** $\sqrt{72}\sqrt{\frac{1}{2}}$

Solution

(a) $(\sqrt[4]{7})^4 = 7$ by Rule 1.

(b) $\sqrt{9 \cdot 25} = \sqrt{9} \cdot \sqrt{25} = 3 \cdot 5 = 15$ by Rule 2.

(c) $\sqrt[4]{\dfrac{16}{81}} = \dfrac{\sqrt[4]{16}}{\sqrt[4]{81}} = \dfrac{2}{3}$ by Rule 3. (*Note:* $2^4 = 16$ and $3^4 = 81$.)

(d) $\sqrt[2]{\sqrt[3]{64}} = \sqrt[6]{64} = 2$ by Rule 4. (*Note:* $2^6 = 64$.)
(e) $\sqrt{72} \cdot \sqrt{\frac{1}{2}} = \sqrt{(72)(\frac{1}{2})} = \sqrt{36} = 6$ by Rule 2.

CAUTION: Here is a common error you must avoid.

In applying the rules for radicals, care must be taken to avoid the type of error that results from the incorrect assumption of the existence of an nth root. For example, $\sqrt[2]{-4}$ is *not* a real number, but if this is not noticed, then *false* results such as the following occur:

$$4 = \sqrt{16} = \sqrt{(-4)(-4)} = \sqrt{-4} \cdot \sqrt{-4} = (\sqrt{-4})^2 = -4$$

TEST YOUR UNDERSTANDING

Evaluate each of the following.

1 $\sqrt{121}$	**2** $\sqrt[3]{-64}$	**3** $\sqrt[5]{32}$
4 $\sqrt{\sqrt{81}}$	**5** $\sqrt{(25)(49)}$	**6** $\sqrt[3]{\frac{27}{125}}$
7 $(\sqrt[4]{2})^4$	**8** $\sqrt[3]{(-1000)(343)}$	**9** $\sqrt[4]{\frac{256}{81}}$
10 $\sqrt{(9)(144)(225)}$	**11** $\sqrt[3]{\frac{(-8)(125)}{27}}$	**12** $\sqrt{\frac{144}{49}} \cdot \sqrt{\frac{196}{36}}$
13 $\sqrt[3]{-8} \cdot \sqrt[3]{-27}$	**14** $\sqrt[3]{\frac{81}{-3}}$	**15** $\sqrt[3]{\frac{1}{24}} \cdot \sqrt[3]{-81}$

We are now ready to make the promised extension of the exponential concept to include fractional exponents. Once again our guideline will be to *preserve* the earlier rules for integer exponents.

First let us consider exponents of the form $\frac{1}{n}$, where n is a positive integer. (Assume that $n \geq 2$ since the case $n = 1$ is trivial.) That is, we wish to give meaning to the expression $b^{1/n}$. If Rule 3 for exponents is to work, then

$$(b^{1/n})^n = b^{(1/n)(n)} = b$$

Thus $b^{1/n}$ is the nth root of b (provided that such a root exists). This leads us to the following definition for $b^{1/n}$.

Since $\sqrt{-1}$ is not a real number, $(-1)^{1/2}$ is not defined. In general, $b^{1/n}$ is not defined within the set of real numbers when $b < 0$ and n is even.

DEFINITION OF $b^{1/n}$

For a real number b and a positive integer n ($n \geq 2$),

$$b^{1/n} = \sqrt[n]{b}$$

provided that $\sqrt[n]{b}$ exists.

Illustrations:

$$\sqrt{15} = 15^{1/2} \qquad \sqrt[3]{-6} = (-6)^{1/3}$$

$$9^{-1/2} = \frac{1}{9^{1/2}} = \frac{1}{\sqrt{9}} = \frac{1}{3} \qquad (-27)^{1/3} = \sqrt[3]{-27} = -3$$

$(-16)^{1/4}$ is not defined

We now wish to give meaning to an expression such as $b^{m/n}$. To do so, we note that squaring $8^{1/3} = \sqrt[3]{8}$ produces $(8^{1/3})^2 = (\sqrt[3]{8})^2$. But to preserve the rules for exponents we write $(8^{1/3})^2 = 8^{2/3}$. Thus it is reasonable to let $8^{2/3} = (\sqrt[3]{8})^2$. Similarly, $8^{2/3} = (8^2)^{1/3} = \sqrt[3]{8^2}$. These observations lead to this definition:

DEFINTION OF $b^{m/n}$

If m/n is a rational number in lowest terms with $n > 0$, then

$$b^{m/n} = (\sqrt[n]{b})^m = \sqrt[n]{b^m}$$

provided that $\sqrt[n]{b}$ exists.

Note that a rational number can always be expressed with a positive denominator; for example, $\frac{2}{-3} = \frac{-2}{3}$.

Using only fractional exponents, we may also write

$$b^{m/n} = (b^{1/n})^m = (b^m)^{1/n}$$

EXAMPLE 3 Evaluate: $(-64)^{2/3}$.

Solution Using $b^{m/n} = (\sqrt[n]{b})^m$, we get

$$(-64)^{2/3} = (\sqrt[3]{-64})^2 = (-4)^2 = 16$$

Using $b^{m/n} = \sqrt[n]{b^m}$, we get

$$(-64)^{2/3} = \sqrt[3]{(-64)^2} = \sqrt[3]{4096} = 16$$

Obviously, the first approach is less work. For most such problems it is easier to first take the nth root and then raise to the mth power, rather than the reverse.

EXAMPLE 4 Evaluate: $8^{-2/3}$.

Solution First rewrite the example using positive exponents; then apply the definition.

Note: $8^{-2/3} = 8^{2/-3} = 8^{-(2/3)}$.

$$8^{-2/3} = \frac{1}{8^{2/3}} = \frac{1}{(\sqrt[3]{8})^2} = \frac{1}{4}$$

TEST YOUR UNDERSTANDING

Write each of the following as a radical.

1 $5^{1/2}$ 2 $(-9)^{1/3}$ 3 $10^{1/4}$ 4 $5^{2/3}$ 5 $2^{3/4}$

Write each of the following using fractional exponents.

6 $\sqrt{7}$ 7 $\sqrt[3]{-10}$ 8 $\sqrt[4]{7}$ 9 $\sqrt[3]{7^2}$ 10 $(\sqrt[4]{5})^3$

Evaluate.

11 $25^{1/2}$ 12 $64^{1/3}$ 13 $(\frac{1}{36})^{1/2}$ 14 $(\frac{1}{125})^{1/3}$ 15 $(\frac{9}{25})^{1/2}$

16 $49^{-1/2}$ 17 $4^{3/2}$ 18 $4^{-3/2}$ 19 $(-8)^{2/3}$ 20 $(-8)^{-2/3}$

EXAMPLE 5 Find what is wrong in the following "proof" that $2 = -2$.

The definition for $b^{m/n}$ calls for the fractional exponent to be in lowest terms so as to avoid situations as in Example 5.

$$-2 = \sqrt[3]{-8} = (-8)^{1/3} = (-8)^{2/6} = \sqrt[6]{(-8)^2} = \sqrt[6]{64} = 2$$

Solution Since $\frac{2}{6}$ is not in lowest terms, we are *not* permitted to write

$$(-8)^{2/6} = \sqrt[6]{(-8)^2}$$

If we restrict ourselves to the roots of only *positive* numbers, then the requirement that the exponent $\frac{m}{n}$ must be in lowest terms can be relaxed. For example:

$$2 = 8^{1/3} = 8^{2/6} = \sqrt[6]{8^2} = \sqrt[6]{64} = 2$$

Since the definition of fractional exponents was made on the basis of preserving the rules for integral exponents, it should come as no surprise that these rules apply. This is illustrated in the examples that follow.

EXAMPLE 6 Evaluate: $27^{2/3} - 16^{-1/4}$.

Solution

$$\begin{aligned} 27^{2/3} - 16^{-1/4} &= (\sqrt[3]{27})^2 - \frac{1}{16^{1/4}} \\ &= 3^2 - \frac{1}{\sqrt[4]{16}} \\ &= 9 - \tfrac{1}{2} \\ &= \frac{17}{2} \end{aligned}$$

EXAMPLE 7 Simplify, and express the result with positive exponents only.

$$\frac{x^{2/3}y^{-2}z^2}{x^{1/2}y^{1/2}z^{-1}}$$

In Example 7 you might find it easier to first rewrite the original problem using positive exponents only.

Solution

$$\begin{aligned} \frac{x^{2/3}y^{-2}z^2}{x^{1/2}y^{1/2}z^{-1}} &= x^{2/3-1/2}y^{-2-1/2}z^{2-(-1)} \\ &= x^{1/6}y^{-5/2}z^3 \\ &= \frac{x^{1/6}z^3}{y^{5/2}} \end{aligned}$$

CAUTION: LEARN TO AVOID MISTAKES LIKE THESE

WRONG	RIGHT
$\sqrt{25} = \pm 5$ (Misuse of the definition of $\sqrt{a}$)	$\sqrt{25} = 5$

WRONG	RIGHT
$16^{3/4} = (\sqrt[3]{16})^4$ (Misuse of the definition of $b^{m/n}$)	$16^{3/4} = (\sqrt[4]{16})^3$ or $\sqrt[4]{16^3}$
$(-2)^{-1/3} = 2^{1/3}$	$(-2)^{-1/3} = \frac{1}{(-2)^{1/3}} = \frac{1}{\sqrt[3]{-2}}$
$a^{-1/2} + b^{-1/2} = \frac{1}{\sqrt{a+b}}$	$a^{-1/2} + b^{-1/2} = \frac{1}{\sqrt{a}} + \frac{1}{\sqrt{b}}$

EXERCISES 2.2

Write with a fractional exponent.

1 $\sqrt{11}$
2 $\sqrt[3]{21}$
3 $\sqrt[4]{9}$
4 $\sqrt[3]{-10}$
5 $\sqrt[3]{6^2}$
6 $(\sqrt[3]{-7})^2$
7 $(\sqrt[5]{-\frac{1}{5}})^3$
8 $\frac{1}{\sqrt[3]{4^2}}$

Write in radical form.

9 $3^{1/2}$
10 $7^{1/3}$
11 $(-19)^{1/3}$
12 $6^{2/3}$
13 $2^{-1/2}$
14 $7^{-3/2}$
15 $(\frac{3}{4})^{-1/4}$
16 $(-\frac{1}{3})^{-2/3}$

Classify each statement as true or false. If it is false, correct the right side of the equality to obtain a true statement.

17 $\sqrt[3]{-27} = -3$
18 $4^{1/2} = -2$
19 $(-8)^{2/3} = 4$
20 $64^{3/4} = (\sqrt[3]{64})^4$
21 $(-\frac{1}{8})^{-1/3} = 2$
22 $(\sqrt{100})^{-1} = -10$
23 $\sqrt{1.44} = 0.12$
24 $(0.25)^{3/2} = \frac{1}{8}$
25 $\sqrt{\frac{49}{9}} = \frac{7}{3}$

Evaluate.

26 $121^{1/2}$
27 $125^{1/3}$
28 $\sqrt[3]{-64}$
29 $81^{-1/2}$
30 $(-64)^{1/3}$
31 $(64)^{-2/3}$
32 $\sqrt{5} \cdot \sqrt{20}$
33 $\sqrt[3]{9} \cdot \sqrt[3]{-3}$
34 $\frac{\sqrt{75}}{\sqrt{3}}$
35 $\frac{\sqrt[3]{-3}}{\sqrt[3]{-24}}$
36 $\frac{9^{1/2}}{\sqrt[3]{27}}$
37 $\frac{\sqrt{9}}{27^{-1/3}}$
38 $\sqrt[4]{\frac{16}{81}}$
39 $\sqrt[3]{(-125)(-1000)}$
40 $\sqrt{(4)(9)(49)(100)}$
41 $\left(\frac{1}{4}\right)^{3/2} \cdot \left(-\frac{1}{8}\right)^{2/3}$
42 $\sqrt{\frac{2}{3}} \cdot \sqrt{\frac{75}{98}}$
43 $\sqrt{\sqrt{625}}$
44 $\sqrt[5]{(-243)^2} \cdot (49)^{-1/2}$
45 $\sqrt{144 + 25}$
46 $\sqrt{144} + \sqrt{25}$
47 $\left(\frac{1}{8} + \frac{1}{27}\right)^{1/3}$
48 $\left(\frac{1}{8}\right)^{1/3} + \left(\frac{1}{27}\right)^{1/3}$
49 $\left(\frac{16}{81}\right)^{3/4} + \left(\frac{256}{625}\right)^{1/4}$
50 $\left(\frac{8}{27}\right)^{2/3} + \left(-\frac{32}{243}\right)^{2/5}$
51 $\left(-\frac{125}{8}\right)^{1/3} - \left(\frac{1}{64}\right)^{1/3}$
52 $\left(-\frac{125}{8} - \frac{1}{64}\right)^{1/3}$

Simplify, and express all answers with positive exponents. (Assume that all letters represent positive numbers.)

53 $(8a^3b^{-9})^{2/3}$
54 $(27a^{-3}b^9)^{-2/3}$
55 $(a^{-4}b^{-8})^{3/4}$
56 $(a^{2/3}b^{1/2})(a^{1/3}b^{-1/2})$
57 $(a^{-1/2}b^{1/3})(a^{1/2}b^{-1/3})$
58 $\frac{a^2b^{-1/2}c^{1/3}}{a^{-3}b^0c^{-1/3}}$
59 $\left(\frac{64a^6}{b^{-9}}\right)^{2/3}$
60 $\frac{(49a^{-4})^{-1/2}}{(81b^6)^{-1/2}}$
61 $\left(\frac{a^{-2}b^3}{a^4b^{-3}}\right)^{-1/2}\left(\frac{a^4b^{-5}}{ab}\right)^{-1/3}$
▬ 62 $\frac{2}{3}(3x-1)^{-1/3} \cdot 3$
▬ 63 $\frac{1}{2}(3x^2+2)^{-1/2} \cdot 6x$
▬ 64 $\frac{1}{3}(x^3+2)^{-2/3} \cdot 3x^2$
▬ 65 $\frac{1}{2}(x^2+4x)^{-1/2}(2x+4)$
▬ 66 $\frac{2}{3}(x^3-6x^2)^{-1/3}(3x^2-12x)$

Simplify, and express the answers without radicals, using only positive exponents. (Assume that n is a positive integer and that all other letters represent positive numbers.)

*67 $\sqrt{\dfrac{x^n}{x^{n-2}}}$

*68 $\left(\dfrac{x^n}{x^{n-2}}\right)^{-1/2}$

*69 $\sqrt[3]{\dfrac{x^{3n+1}y^n}{x^{3n+4}y^{4n}}}$

*70 Prove: $\sqrt[n]{a} \cdot \sqrt[n]{b} = \sqrt[n]{ab}$.
(*Hint:* Let $\sqrt[n]{a} = x$ and $\sqrt[n]{b} = y$. Then $x^n = a$ and $y^n = b$.)

*71 Prove: $\dfrac{\sqrt[n]{a}}{\sqrt[n]{b}} = \sqrt[n]{\dfrac{a}{b}}$.

2.3

Simplifying radicals

Computations with radicals cause no difficulty when the radicals represent rational numbers, as in the following.

$$\sqrt{36} + \sqrt[3]{-27} = 6 + (-3) = 3$$

When the radicals involve irrational numbers, however, special methods are needed. To *simplify* an expression such as $\sqrt{24}$ means that we are to rewrite the radical so that no perfect square appears as a factor under the radical sign. Thus

Note that the fundamental idea used here is that $\sqrt[n]{ab} = \sqrt[n]{a} \cdot \sqrt[n]{b}$.

$$\sqrt{24} = \sqrt{4 \cdot 6} = \sqrt{4} \cdot \sqrt{6} = 2\sqrt{6}$$

We say that $2\sqrt{6}$ is the *simplified form* of $\sqrt{24}$.

EXAMPLE 1 Simplify: **(a)** $\sqrt{50}$ **(b)** $\sqrt[3]{16}$ **(c)** $\sqrt[3]{-54}$

Solution

(a) $\sqrt{50} = \sqrt{25 \cdot 2} = \sqrt{25} \cdot \sqrt{2} = 5\sqrt{2}$

(b) Here we search for a perfect cube as a factor under the radical sign:

$$\sqrt[3]{16} = \sqrt[3]{8 \cdot 2} = \sqrt[3]{8} \cdot \sqrt[3]{2} = 2\sqrt[3]{2}$$

(c) $\sqrt[3]{-54} = \sqrt[3]{(-27)(2)} = \sqrt[3]{-27} \cdot \sqrt[3]{2} = -3\sqrt[3]{2}$

The rule for multiplication of radicals provides a way to find the product of any two radicals having the same index. For example,

$$\sqrt{4} \cdot \sqrt{9} = \sqrt{4 \cdot 9} = \sqrt{36} = 6$$

Does a similar pattern work for the addition of radicals? That is, is the sum of the square roots of two numbers equal to the square root of their sum? Does $\sqrt{4} + \sqrt{9} = \sqrt{4 + 9}$? This can easily be checked as follows:

$$\sqrt{4} + \sqrt{9} = 2 + 3 = 5$$

But $\sqrt{4 + 9} = \sqrt{13}$. Therefore, $\sqrt{4} + \sqrt{9} \neq \sqrt{4 + 9}$.

In general, $\sqrt{a} + \sqrt{b} \neq \sqrt{a + b}$.

This discussion tells you the conditions under which radicals can be combined.

Addition and subtraction of radicals is possible, under certain conditions, and is achieved through the use of the distributive property. Consider, for example, this sum:

$$3\sqrt{5} + 4\sqrt{5}$$

Although we will usually perform the computation mentally, we can complete this example by thinking of x to replace $\sqrt{5}$.

$$3x + 4x = (3 + 4)x = 7x$$

$$3\sqrt{5} + 4\sqrt{5} = (3 + 4)\sqrt{5} = 7\sqrt{5}$$

That is,

$$3\sqrt{5} + 4\sqrt{5} = 7\sqrt{5}$$

In order to be able to add or subtract radicals, they must have the same index and the same radicand.

At times a fraction can be simplified by a process known as **rationalizing the denominator**. This consists of eliminating a radical from the denominator of a fraction. For example, consider the fraction $4/\sqrt{2}$. To rationalize the denominator, multiply the numerator and denominator by $\sqrt{2}$.

$$\frac{4}{\sqrt{2}} = \frac{4}{\sqrt{2}} \cdot \frac{\sqrt{2}}{\sqrt{2}} \qquad \left(\frac{\sqrt{2}}{\sqrt{2}} = 1\right)$$

$$= \frac{4\sqrt{2}}{2}$$

$$= 2\sqrt{2}$$

One reason for rationalizing denominators is to make computations easier. For example, suppose that we wish to evaluate $\frac{4}{\sqrt{2}}$ ($\sqrt{2} = 1.414$) to three decimal places. It certainly is easier to multiply 1.414 by 2 than to divide 4 by 1.414. Another reason for rationalizing denominators is to achieve a standard form for radical expressions in which they are more easily combined or compared.

If a calculator is to be used, $\frac{4}{\sqrt{2}}$ is just as easy to evaluate as $2\sqrt{2}$ and would therefore be an acceptable form.

EXAMPLE 2 Combine: $\frac{6}{\sqrt{3}} + 2\sqrt{75} - \sqrt{3}$.

Solution Rationalize the denominator in the first term:

$$\frac{6}{\sqrt{3}} = \frac{6}{\sqrt{3}} \cdot \frac{\sqrt{3}}{\sqrt{3}} = \frac{6\sqrt{3}}{3} = 2\sqrt{3}$$

Simplify the second term:

$$2\sqrt{75} = 2\sqrt{25 \cdot 3} = 10\sqrt{3}$$

Now combine:

$$\frac{6}{\sqrt{3}} + 2\sqrt{75} - \sqrt{3} = 2\sqrt{3} + 10\sqrt{3} - 1\sqrt{3} = 11\sqrt{3}$$

TEST YOUR UNDERSTANDING

Simplify each expression, if possible.

1 $\sqrt{8} + \sqrt{32}$	2 $\sqrt{12} + \sqrt{48}$	3 $\sqrt{45} - \sqrt{20}$
4 $2\sqrt{9} - \sqrt{18}$	5 $\sqrt[3]{16} + \sqrt[3]{54}$	6 $\sqrt[3]{128} + \sqrt[3]{125}$
7 $\frac{8}{\sqrt{2}} + \sqrt{98}$	8 $\frac{9}{\sqrt{3}} + \sqrt{300}$	9 $2\sqrt{20} - \frac{5}{\sqrt{5}}$
10 $3\sqrt{63} - \frac{14}{\sqrt{7}}$	11 $\frac{2}{\sqrt{2}} - \sqrt{2}$	12 $\frac{1}{\sqrt{2}} + \frac{1}{\sqrt{8}}$

Do you think that $\sqrt{x^2} = x$? This is *not* always so. Study this explanation carefully; it is important for future work in mathematics.

Sometimes it is necessary to simplify a radical that contains a variable, such as $\sqrt{x^2}$. In this example, the usual reaction is to claim that $\sqrt{x^2} = x$. But suppose that x were negative, such as $x = -5$:

$$\text{If } \sqrt{x^2} = x, \quad \text{then } \sqrt{(-5)^2} = -5$$

$$\text{But } \sqrt{(-5)^2} = \sqrt{25} = 5$$

It was stated earlier that the radical sign, $\sqrt{\ }$, means the *positive* square root of a number. Therefore, it is necessary that $\sqrt{(-5)^2}$ be equal to 5, and not -5. That is, we must have each of the following:

$$\sqrt{5^2} = 5$$

$$\sqrt{(-5)^2} = 5$$

This leads to the following important result:

> For all real numbers a, $\sqrt{a^2} = |a|$

EXAMPLE 3 Simplify: $\sqrt{75x^2}$.

Solution

$$\begin{aligned}\sqrt{75x^2} &= \sqrt{25 \cdot 3}\sqrt{x^2} \\ &= 5\sqrt{3}\,|x|\end{aligned}$$

EXAMPLE 4 Simplify: $2\sqrt{8x^3} + 3x\sqrt{32x} - x\sqrt{18x}$.

Solution For this problem, $x \geq 0$. Thus we need not make use of absolute-value notation.

Note that the expressions under the radicals would be negative for $x < 0$. Since the index is even, we must assume that $x \geq 0$ in order for these to have meaning.

$$\begin{aligned}2\sqrt{8x^3} &= 2\sqrt{4 \cdot 2 \cdot x^2 \cdot x} = 4x\sqrt{2x} \\ 3x\sqrt{32x} &= 3x\sqrt{16 \cdot 2x} = 12x\sqrt{2x} \\ x\sqrt{18x} &= x\sqrt{9 \cdot 2x} = 3x\sqrt{2x}\end{aligned}$$

In each case the radicand and the index are the same, so that the distributive property can be used to simplify.

$$\begin{aligned}4x\sqrt{2x} + 12x\sqrt{2x} - 3x\sqrt{2x} &= (4x + 12x - 3x)\sqrt{2x} \\ &= 13x\sqrt{2x}\end{aligned}$$

CAUTION: LEARN TO AVOID MISTAKES LIKE THESE

WRONG	RIGHT
$\sqrt{9+16} = \sqrt{9} + \sqrt{16}$	$\sqrt{9+16} = \sqrt{25}$
$(a+b)^{1/3} = a^{1/3} + b^{1/3}$	$(a+b)^{1/3} = \sqrt[3]{a+b}$

WRONG	RIGHT
$\sqrt[3]{8}\cdot\sqrt[2]{8}=\sqrt[6]{64}$	$\sqrt[3]{8}\cdot\sqrt[2]{8}=2\cdot 2\sqrt{2}=4\sqrt{2}$
$2\sqrt{x+1}=\sqrt{2x+1}$	$2\sqrt{x+1}=\sqrt{4(x+1)}=\sqrt{4x+4}$
$2-\frac{1}{\sqrt{2}}=\frac{2-1}{\sqrt{2}}$	$2-\frac{1}{\sqrt{2}}=2-\frac{\sqrt{2}}{2}=\frac{4-\sqrt{2}}{2}$
$\sqrt{(x-1)^2}=x-1$	$\sqrt{(x-1)^2}=\lvert x-1\rvert$

EXERCISES 2.3

Simplify.

1 $\sqrt{2}+\sqrt{18}$

2 $\sqrt{48}-\sqrt{3}$

3 $\sqrt{25}+\sqrt{49}$

4 $\sqrt{64}-\sqrt{16}$

5 $\sqrt{12}-\sqrt{27}$

6 $\sqrt{32}+\sqrt{72}$

7 $\sqrt{6}\cdot\sqrt{12}$

8 $\sqrt[3]{4}\cdot\sqrt[3]{12}$

9 $2\sqrt{5}+3\sqrt{125}$

10 $-5\sqrt{24}-2\sqrt{54}$

11 $2\sqrt{200}-5\sqrt{8}$

12 $3\sqrt{45}-2\sqrt{20}$

13 $\sqrt[3]{128}+\sqrt[3]{16}$

14 $\sqrt[3]{24}+\sqrt[3]{81}$

15 $\sqrt{50}+\sqrt{32}-\sqrt{8}$

16 $\sqrt{12}-\sqrt{3}+\sqrt{108}$

17 $\frac{8}{\sqrt{2}}+2\sqrt{50}$

18 $\frac{12}{\sqrt{3}}-\sqrt{12}$

19 $\sqrt[3]{56x}+\sqrt[3]{7x}$

20 $\sqrt[3]{54x}+\sqrt[3]{250x}$

21 $3\sqrt{8x^2}-\sqrt{50x^2}$

22 $5\sqrt{75x^2}-2\sqrt{12x^2}$

23 $\sqrt[4]{32}+\sqrt[4]{162}$

24 $\sqrt[5]{32}+\sqrt[5]{64}$

25 $3\sqrt{10}+4\sqrt{90}-5\sqrt{40}$

26 $3\sqrt{24}-\sqrt{54}+2\sqrt{150}$

27 $\frac{1}{\sqrt{2}}+3\sqrt{72}-2\sqrt{2}$

28 $\frac{2}{\sqrt{3}}+10\sqrt{3}-2\sqrt{12}$

29 $\sqrt{24}+\sqrt{54}-\sqrt{18}$

30 $\sqrt{36}+\sqrt{28}+\sqrt{63}$

31 $\sqrt{18x}+\sqrt{50x}-\sqrt{2x}$

32 $10\sqrt{3x}-2\sqrt{75x}+3\sqrt{243x}$

33 $3\sqrt{9x^2}+2\sqrt{16x^2}-\sqrt{25x^2}$

34 $\sqrt{2x^2}+5\sqrt{32x^2}-2\sqrt{98x^2}$

35 $\sqrt{x^2y}+\sqrt{8x^2y}+\sqrt{200x^2y}$

36 $\sqrt{72xy}+2\sqrt{2xy}+\sqrt{128xy}$

37 $\sqrt{20a^3}+a\sqrt{5a}+\sqrt{80a^3}$

38 $\sqrt{12b^3}+\sqrt{27b^3}+2b\sqrt{3b}$

Rationalize the denominators and simplify.

39 $\frac{20}{\sqrt{5}}$

40 $\frac{24}{\sqrt{6}}$

41 $\frac{8x}{\sqrt{2}}$

42 $\frac{9y}{\sqrt{3}}$

43 $\frac{4}{\sqrt{2}}+\frac{6}{\sqrt{3}}$

44 $\frac{10}{\sqrt{5}}+\frac{8}{\sqrt{4}}$

45 $\frac{1}{\sqrt{18}}$

46 $\frac{1}{\sqrt{27}}$

47 $\frac{24}{\sqrt{3x^2}}$

48 $\frac{20x}{\sqrt{5x^3}}$

49 $\frac{8}{\sqrt[3]{2}}$

50 $\frac{12}{\sqrt[3]{-3}}$

*51 Use the result $\sqrt{a^2}=\lvert a\rvert$ and the rules for radicals to prove that $\lvert xy\rvert=\lvert x\rvert\cdot\lvert y\rvert$, where x and y are real numbers.

2.4

Fundamental operations with polynomials

The expression $5x^3 - 7x^2 + 4x - 12$ is called a **polynomial in the variable *x***. Its *degree* is 3, because 3 is the largest power of the variable x. The *terms* of this polynomial are $5x^3$, $-7x^2$, $4x$ and -12. The *coefficients* are 5, -7, 4, and -12.

All the exponents of the variable of a polynomial must be nonnegative integers. Therefore, $x^3 + x^{1/2}$ and $x^{-2} + 3x + 1$ are *not* polynomials because of the fractional and negative exponents.

A nonzero constant, like 7, is classified as a polynomial of degree zero, since $7 = 7x^0$. The number zero is also referred to as a constant polynomial, but it is not assigned any degree.

A polynomial is in *standard form* if its terms are arranged so that the powers of the variable are in descending or ascending order. Here are some illustrations.

Polynomial	*Degree*	*Standard form*
$x^2 - 3x + 12$	2	Yes
$\frac{2}{3}x^{10} - 4x^2 + \sqrt{2}x^4$	10	No
$32 - y^5 + 2y^3$	5	No
$6 + 2x - x^2 + x^3$	3	Yes

Some of the preceding polynomials have "missing" terms. For example, $x^3 - 3x + 12$ has no x^2 term, but it is still a third-degree polynomial. The highest power of x determines the degree, and any or all lesser powers may be missing.

In general, an nth-degree polynomial in the variable x may be written in either one of these standard forms:

You should become familiar with this notation, as you will encounter it frequently in future mathematics courses.

$$a_nx^n + a_{n-1}x^{n-1} + \cdots + a_2x^2 + a_1x + a_0$$

$$a_0 + a_1x + a_2x^2 + \cdots + a_{n-1}x^{n-1} + a_nx^n$$

The coefficients $a_n, a_{n-1}, \ldots, a_0$ are real numbers. The *leading coefficient* is $a_n \neq 0$, and a_0 is called the *constant term.*

In a polynomial like $3x^2 - x + 4$ the variable x represents a real number. Therefore, when a specific real value is substituted for x, the result will be a real number. For instance, using $x = -3$ in this polynomial gives

Note that polynomials represent real numbers regardless of the specific choice of the variable. Thus computations with polynomials are based on the fundamental properties for real numbers.

$$3(-3)^2 - (-3) + 4 = 34$$

Adding or subtracting polynomials involves the combining of **like terms** (those having the same exponent on the variable). This can be accomplished by first rearranging and regrouping the terms (associative and commutative properties) and then combining by using the distributive property.

EXAMPLE 1

(a) Add: $(4x^3 - 10x^2 + 5x + 8) + (12x^2 - 9x - 1)$.
(b) Subtract: $(4t^3 - 10t^2 + 5t + 8) - (12t^2 - 9t - 1)$.

Solution

(a) $(4x^3 - 10x^2 + 5x + 8) + (12x^2 - 9x - 1)$

$$= 4x^3 + (12x^2 - 10x^2) + (5x - 9x) + (8 - 1)$$
$$= 4x^3 + (12 - 10)x^2 + (5 - 9)x + 7$$
$$= 4x^3 + 2x^2 - 4x + 7$$

(b) $(4t^3 - 10t^2 + 5t + 8) - (12t^2 - 9t - 1)$

$$= 4t^3 - 10t^2 + 5t + 8 - 12t^2 + 9t + 1$$
$$= 4t^3 - 10t^2 - 12t^2 + 5t + 9t + 8 + 1$$
$$= 4t^3 - 22t^2 + 14t + 9$$

In Example 1(b), think of $a - b$ as $a - 1 \cdot b$ and use the distributive property to simplify.

An alternative method for Example 1 is to list the polynomials in column form, putting like terms in the same columns.

$$\text{Add:}\quad \begin{array}{r} 4x^3 - 10x^2 + 5x + 8 \\ 12x^2 - 9x - 1 \\ \hline 4x^3 + \ \ 2x^2 - 4x + 7 \end{array} \qquad \text{Subtract:}\quad \begin{array}{r} 4t^3 - 10t^2 + \ \ 5t + 8 \\ 12t^2 + \ \ 9t + 1 \\ \hline 4t^3 - 22t^2 + 14t + 9 \end{array}$$

The use of the distributive property is fundamental when multiplying polynomials. Perhaps the simplest situation calls for the product of a **monomial** (a polynomial having only one term) times a "lengthier" polynomial, as follows.

$$3x^2(4x^7 - 3x^4 - x^2 + 15) = 3x^2(4x^7) - 3x^2(3x^4) - 3x^2(x^2) + 3x^2(15)$$
$$= 12x^9 - 9x^6 - 3x^4 + 45x^2$$

In the first line we used an extended version of the distributive property, namely,

$$a(b - c - d + e) = ab - ac - ad + ae$$

Next, observe how the distributive property is used to multiply two **binomials** (polynomials having two terms).

$$(2x + 3)(4x + 5) = (2x + 3)4x + (2x + 3)5$$
$$= (2x)(4x) + (3)(4x) + (2x)(5) + (3)(5)$$
$$= 8x^2 + 12x + 10x + 15$$
$$= 8x^2 + 22x + 15$$

Here is a shortcut that can be used to multiply two binomials.

$8x^2$ 15

$(2x + 3)(4x + 5)$: $\quad (2x + 3)(4x + 5) = 8x^2 + 22x + 15$

$12x$

$10x$

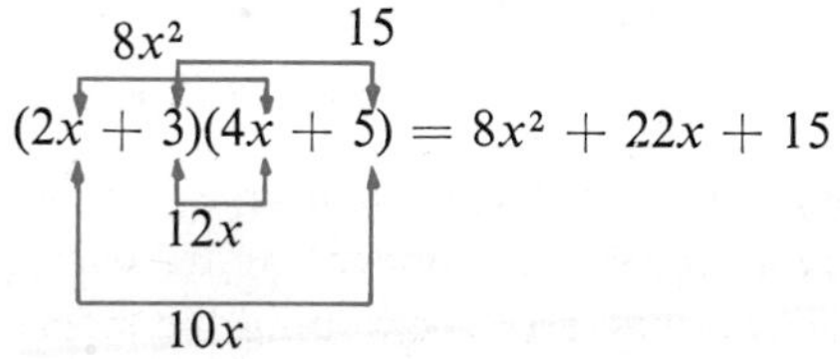

$8x^2$ is the product of the *first* terms in the binomials.

$10x$ and $12x$ are the products of the *outer* and *inner* terms;

$$10x + 12x = 22x.$$

15 is the product of the *last* terms in the binomials.

EXAMPLE 2 Find the product: $(5x - 9)(6x + 2)$.

Solution

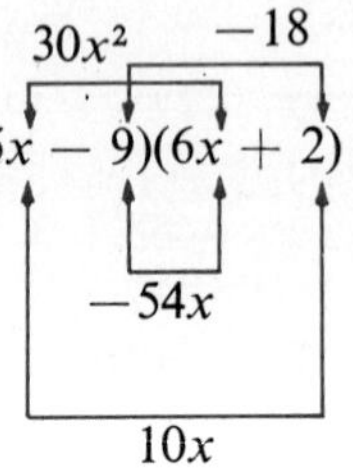

$$(5x - 9)(6x + 2) = 30x^2 + (10x - 54x) - 18$$
$$= 30x^2 - 44x - 18$$

Keep this diagram in mind as an aid to finding the product of two binomials mentally.

In general, we may write the product $(a + b)(c + d)$ in this way:

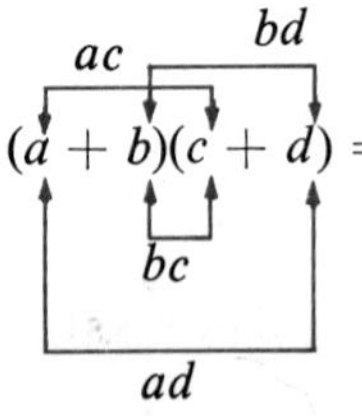

$$(a + b)(c + d) = ac + ad + bc + bd$$

The distributive property can be extended to multiply polynomials with three terms, **trinomials**, as in Example 3.

EXAMPLE 3 Multiply $3x^3 - 8x + 4$ by $2x^2 + 5x - 1$.

Solution

Note that the column method is a convenient way to organize your work. Be certain to keep like terms in the same column. Let $x = 2$ and check the solution.

$$\begin{array}{rrrrrrl}
 & & 3x^3 & & -8x & +4 & \\
 & & & 2x^2 & +5x & -1 & \\
\hline
 & & -3x^3 & & +8x & -4 & (-1 \text{ times } 3x^3 - 8x + 4) \\
\text{(add)} & 15x^4 & & -40x^2 & +20x & & (5x \text{ times } 3x^3 - 8x + 4) \\
6x^5 & & -16x^3 & +8x^2 & & & (2x^2 \text{ times } 3x^3 - 8x + 4) \\
\hline
6x^5 & +15x^4 & -19x^3 & -32x^2 & +28x & -4 &
\end{array}$$

TEST YOUR UNDERSTANDING

Combine.

1 $(x^2 + 2x - 6) + (-2x + 7)$
2 $(x^2 + 2x - 6) - (-2x + 7)$
3 $(5x^4 - 4x^3 + 3x^2 - 2x + 1) + (-5x^4 + 6x^3 + 10x)$
4 $(x^2 + x + 1) + (-3x - 4) + (6x - 5) - (x^3 + x^2)$

Find the products.

5 $(-3x)(x^3 + 2x^2 - 1)$
6 $-4x(\frac{1}{8} - \frac{1}{2}x - x^2)$

7 $(2x + 3)(3x + 2)$	8 $(6x - 1)(2x + 5)$
9 $(4x + 7)(4x - 7)$	10 $(2x - 3)(2x - 3)$
11 $(x^3 + 7x^2 - 4)(x + 2)$	12 $(x^2 - 3x + 5)(2x^3 + x^2 - 3x)$

More than two polynomials may be involved in a product. For example, here is a product of three polynomials:

$$\begin{aligned}(x + 2)(x + 3)(x + 4) &= [(x + 2)(x + 3)](x + 4)\\ &= (x^2 + 5x + 6)(x + 4)\\ &= x^3 + 9x^2 + 26x + 24\end{aligned}$$

Sometimes more than one operation is involved, as demonstrated in the next example.

Compare the areas of the two congruent squares in terms of the segments a and b.

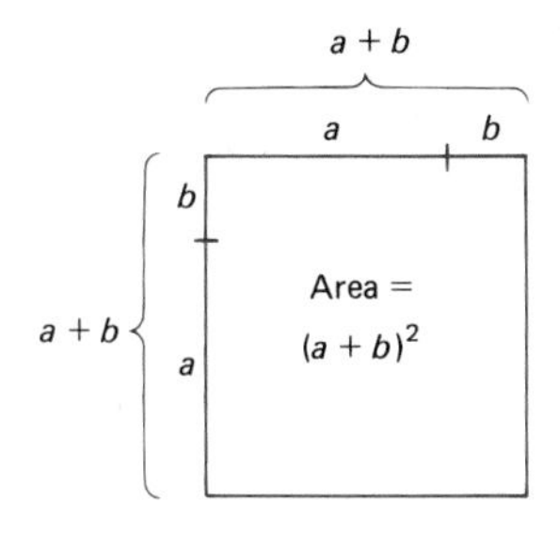

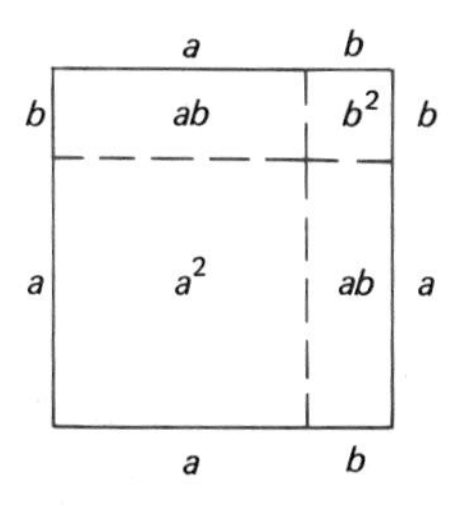

EXAMPLE 4 Simplify by performing the indicated operations:

$$(x^2 - 5x)(3x^2) + (x^3 - 1)(2x - 5)$$

Solution Multiply first and then combine like terms.

$$\begin{aligned}&(x^2 - 5x)(3x^2) + (x^3 - 1)(2x - 5)\\ &\qquad = (3x^4 - 15x^3) + (2x^4 - 5x^3 - 2x + 5)\\ &\qquad = 5x^4 - 20x^3 - 2x + 5\end{aligned}$$

The product of $a + b$ times itself is given by

$$(a + b)(a + b) = a^2 + 2ab + b^2$$

Using exponents, we have

$$(a + b)(a + b) = (a + b)^2$$

Thus

$$\boxed{(a + b)^2 = a^2 + 2ab + b^2}$$

We say that $a^2 + 2ab + b^2$ is the *expansion*, or expanded form, of $(a + b)^2$.

Caution: $(a + b)^2 \neq a^2 + b^2$.

Explain how these figures provide a geometric interpretation for the expansion of $(a + b)^2$.

EXAMPLE 5 Expand: $(a + b)^3$.

Solution First write

$$(a + b)^3 = (a + b)(a + b)^2$$

Then, use the expansion of $(a + b)^2$:

$$\begin{aligned}(a + b)^3 &= (a + b)(a^2 + 2ab + b^2)\\ &= a^3 + 2a^2b + ab^2 + a^2b + 2ab^2 + b^3\\ &= a^3 + 3a^2b + 3ab^2 + b^3\end{aligned}$$

The result of Example 5 provides a formula for the cube of a binomial, that is, for all expansions of the form $(a + b)^3$.

Can you find the product 38×42 mentally? Think of this as $(40 - 2)(40 + 2)$ and use this fact to multiply.

The next example makes use of $(a - b)(a + b) = a^2 - b^2$ to rationalize denominators. Each of the factors $a - b$ and $a + b$ is called the *conjugate* of the other.

EXAMPLE 6 Rationalize the denominators:

$$\text{(a) } \frac{5}{\sqrt{10} - 3} \qquad \text{(b) } \frac{x}{\sqrt{x} + \sqrt{y}}.$$

Solution

$\sqrt{10} + 3$ is the conjugate of $\sqrt{10} - 3$.

$$\text{(a) } \frac{5}{\sqrt{10} - 3} = \frac{5}{\sqrt{10} - 3} \cdot \frac{\sqrt{10} + 3}{\sqrt{10} + 3} = \frac{5(\sqrt{10} + 3)}{10 - 9} = 5(\sqrt{10} + 3)$$

$\sqrt{x} - \sqrt{y}$ is the conjugate of $\sqrt{x} + \sqrt{y}$.

$$\text{(b) } \frac{x}{\sqrt{x} + \sqrt{y}} = \frac{x}{\sqrt{x} + \sqrt{y}} \cdot \frac{\sqrt{x} - \sqrt{y}}{\sqrt{x} - \sqrt{y}} = \frac{x(\sqrt{x} - \sqrt{y})}{x - y}$$

EXERCISES 2.4

Add.

1 $\begin{array}{r} 3x^2 + 5x - 2 \\ 5x^2 - 7x + 9 \\ \hline \end{array}$

2 $\begin{array}{r} 5x^2 - 9x - 1 \\ 2x^2 + 2x + 7 \\ \hline \end{array}$

3 $\begin{array}{r} 3x^3 - 7x^2 - 8x + 12 \\ x^3 - 2x^2 + 8x - 9 \\ \hline \end{array}$

4 $\begin{array}{r} x^3 - 3x^2 + 2x - 5 \\ 5x^2 - x + 9 \\ \hline \end{array}$

5 $\begin{array}{r} 4x^2 + 9x - 17 \\ 2x^3 - 3x^2 + 2x - 11 \\ \hline \end{array}$

6 $\begin{array}{r} 2x^3 + x^2 - 7x + 1 \\ -2x^2 - x + 8 \\ \hline \end{array}$

Subtract.

7 $\begin{array}{r} 3x^3 - 2x^2 - 8x + 9 \\ 2x^3 + 5x^2 + 2x + 1 \\ \hline \end{array}$

8 $\begin{array}{r} x^3 - 2x^2 + 6x + 1 \\ -x^2 - 6x - 1 \\ \hline \end{array}$

9 $\begin{array}{r} 4x^3 + x^2 - 2x - 13 \\ 2x^2 + 3x + 9 \\ \hline \end{array}$

Simplify by performing the indicated operations.

10 $(3x + 5) + (3x - 2)$
11 $5x + (1 - 2x)$
12 $(7x + 5) - (2x + 3)$
13 $(y + 2) + (2y + 1) + (3y + 3)$
14 $h - (h + 2)$
15 $(x^3 + 3x^2 + 3x + 1) - (x^2 + 2x + 1)$
16 $7x - (3 + x) - 2x$
17 $5y - [y - (3y + 8)]$
18 $(x + 3) + (2x - 2) - (6x + 10)$
19 $(x^2 - 3x + 4) - (5x^2 + 2x + 1)$
20 $x(3x^2 - 2x + 5)$
21 $2x^2(2x + 1 - 10x^2)$
22 $-4t(t^4 - \frac{1}{4}t^3 + 4t^2 - \frac{1}{16}t + 1)$
23 $(x + 1)(x + 1)$
24 $(2x + 1)(2x - 1)$
25 $(4x - 2)(x + 7)$
26 $(5x + 3)(4x + 6)$
27 $(5x - 3)(4x - 6)$
28 $(5x + 3)(4x - 6)$
29 $(x + \frac{1}{2})(x - \frac{1}{4})$
30 $(12x - 8)(7x + 4)$

31 $(-2x + 3)(3x + 6)$

32 $(-2x - 3)(3x + 6)$

33 $(-2x - 3)(3x - 6)$

34 $(\frac{1}{2}x + 4)(\frac{1}{2}x - 4)$

35 $(\frac{2}{3}x + 6)(\frac{2}{3}x + 6)$

36 $(7 + 3x)(9 - 4x)$

37 $(7 - 3x)(4x - 9)$

38 $(15a + 30)(15a - 30)$

39 $(ax + b)(ax - b)$

40 $(ax - b)(ax - b)$

41 $(x - 0.1)(x + 0.1)$

42 $(x + \frac{3}{4})(x + \frac{3}{4})$

43 $(\frac{1}{5}x - \frac{1}{4})(\frac{1}{5}x - \frac{1}{4})$

44 $(x - \sqrt{3})(x + \sqrt{3})$

45 $(\sqrt{x} - 10)(\sqrt{x} + 10)$

46 $(\frac{1}{10}x - \frac{1}{100})(\frac{1}{10}x - \frac{1}{100})$

47 $(\sqrt{x} + \sqrt{2})(\sqrt{x} - \sqrt{2})$

48 $(x^2 - 3)(x^2 + 3)$

49 $(\sqrt{3} - \sqrt{2})(\sqrt{3} + \sqrt{2})$

50 $(5 - 3x)(x^4 + x^3 + x^2 + x + 1)$

51 $(x^2 + x + 9)(x^2 - 3x - 4)$

52 $(x^2 + 5x - 1)(36x^2 - 30x + 25)$

53 $(2x^2 - 3)(4x^2 + 6x + 9)$

54 $(x^{1/3} - 2)(x^{2/3} + 2x^{1/3} + 4)$

55 $(x - 2)(x^2 + 2x + 4)$

56 $(x - 2)(x^3 + 2x^2 + 4x + 8)$

57 $(x - 2)(x^4 + 2x^3 + 4x^2 + 8x + 16)$

58 $(x^n - 4)(x^n + 4)$

59 $(x^{2n} + 1)(x^{2n} - 2)$

60 $5(x + 5)(x - 5)$

61 $3x(1 - x)(1 - x)$

62 $(x + 3)(x + 1)(x - 4)$

63 $(2x + 1)(3x - 2)(3 - x)$

▬ 64 $(2x^2 + 3)(9x^2) + (3x^3 - 2)(4x)$

▬ 65 $(x^3 - 2x + 1)(2x) + (x^2 - 2)(3x^2 - 2)$

▬ 66 $(2x^3 - x^2)(6x - 5) + (3x^2 - 5x)(6x^2 - 2x)$

▬ 67 $(x^4 - 3x^2 + 5)(2x + 3) + (x^2 + 3x)(4x^3 - 6x)$

Expand each of the following and combine like terms.

68 $(a - b)^2$

69 $(x - 1)^3$

70 $(x + 1)^4$

71 $(a + b)^4$

72 $(a - b)^4$

73 $(2x + 3)^3$

74 $(\frac{1}{2}x - 4)^2$

75 $(\frac{1}{3}x + 3)^3$

76 $(\frac{1}{2}x - 1)^3$

Rationalize the denominator and simplify.

77 $\dfrac{12}{\sqrt{5} - \sqrt{3}}$

78 $\dfrac{20}{3 - \sqrt{2}}$

79 $\dfrac{14}{\sqrt{2} - 3}$

80 $\dfrac{\sqrt{x}}{\sqrt{x} + \sqrt{y}}$

*81 $\dfrac{\sqrt{x} + \sqrt{y}}{\sqrt{x} - \sqrt{y}}$

82 $\dfrac{1}{\sqrt{x} + 2}$

Rationalize the numerator in Exercises 83–85.

*83 $\dfrac{\sqrt{5} + 3}{\sqrt{5}}$

*84 $\dfrac{\sqrt{5} + \sqrt{3}}{\sqrt{2}}$

*85 $\dfrac{\sqrt{x} + \sqrt{y}}{\sqrt{x} - \sqrt{y}}$

▬ *In Exercises 86 and 87, show how to convert the first fraction given into the form of the second one.*

86 $\dfrac{\sqrt{x} - 2}{x - 4}$; $\dfrac{1}{\sqrt{x} + 2}$

87 $\dfrac{\sqrt{4 + h} - 2}{h}$; $\dfrac{1}{\sqrt{4 + h} + 2}$

*88 Find a geometric interpretation for the formula $(a - b)^2 = a^2 - 2ab + b^2$.

2.5 Factoring polynomials

Can you solve this equation: $x^2 - 3x^2 - 4x + 112 = 0$? This may not be an easy question to answer at first. The solution is not difficult, however, once you rewrite the equation in this form:

$$x^3 - 3x^2 - 4x + 12 = (x + 2)(x - 2)(x - 3) = 0$$

Therefore, $x^3 - 3x^2 - 4x + 12 = 0$ when $x + 2 = 0$ or when $x - 2 = 0$ or when $x - 3 = 0$. Consequently, $-2, 2, 3$ are the solutions to the original equation.

Recall that the product of two or more factors is zero whenever any one of the factors is zero.

The key to the preceding solution was the conversion of the polynomial $x^3 - 3x^2 - 4x + 12$ into the *factored form* $(x + 2)(x - 2)(x - 3)$. As you have just seen, it is precisely this conversion that made it possible to solve the original equation. And since solving equations is of vital importance throughout mathematics, it will be worthwhile to become familiar with some methods of factoring. In the preceding section, we multiplied polynomials like this:

$$\begin{aligned}(x + 2)(x - 2)(x - 3) &= [(x + 2)(x - 2)](x - 3)\\ &= (x^2 - 4)(x - 3)\\ &= x^3 - 3x^2 - 4x + 12\end{aligned}$$

In a sense, factoring is "unmultiplying."

Now we want to *begin* with $x^3 - 3x^2 - 4x + 12$ and factor (unmultiply) it into the form $(x + 2)(x - 2)(x - 3)$.

One of the basic methods of factoring is the reverse of multiplying by a monomial. Consider this multiplication problem:

$$6x^2(4x^7 - 3x^4 - x^2 + 15) = 24x^9 - 18x^6 - 6x^4 + 90x^2$$

As you read this equation from left to right it involves multiplication by $6x^2$. Reading from right to left, it involves "factoring out" the *common monomial factor* $6x^2$. These are the details of this factoring process:

$$\begin{aligned}24x^9 - 18x^6 - 6x^4 + 90x^2 &= 6x^2(4x^7) - 6x^2(3x^4) - 6x^2(x^2) + 6x^2(15)\\ &= 6x^2(4x^7 - 3x^4 - x^2 + 15)\end{aligned}$$

Note the use of the distributive property in this illustration.

Suppose that in the preceding illustration we were to use $3x$ as the common factor:

$$24x^9 - 18x^6 - 6x^4 + 90x^2 = 3x(8x^8 - 6x^5 - 2x^3 + 30x)$$

This is a correct factorization but it is not considered the *complete factored form* because $2x$ can still be factored out of the expression in the parentheses. In general, the instruction to factor calls for arriving at the complete factored form. We can extend factoring techniques to polynomials with more than one variable as in the example that follows.

EXAMPLE 1 Factor:

(a) $21x^4y - 14x^5y^2 - 63x^8y^3 + 91x^{11}y^4$
(b) $8x^2(x - 1) + 4x(x - 1) + 2(x - 1)$

Solution

(a) $21x^4y - 14x^5y^2 - 63x^8y^3 + 91x^{11}y^4$
$$= 7x^4y(3 - 2xy - 9x^4y^2 + 13x^7y^3)$$

(b) $8x^2(x - 1) + 4x(x - 1) + 2(x - 1) = 2(x - 1)(4x^2 + 2x + 1)$
Note: If you have trouble seeing this, then think of $x - 1$ as a single value and use $x - 1 = a$. Then

$$\begin{aligned}8x^2(x - 1) + 4x(x - 1) + 2(x - 1) &= 8x^2a + 4xa + 2a\\ &= 2a(4x^2 + 2x + 1)\\ &= 2(x - 1)(4x^2 + 2x + 1)\end{aligned}$$

We will now consider several basic procedures for factoring polynomials. You should check each of these by multiplication. These forms are very useful and you need to learn to recognize and apply each one.

THE DIFFERENCE OF TWO SQUARES

$$a^2 - b^2 = (a - b)(a + b)$$

Illustrations:

$$x^2 - 9 = x^2 - 3^2 = (x - 3)(x + 3)$$
$$49t^2 - 100 = (7t)^2 - 10^2 = (7t - 10)(7t + 10)$$
$$(x + 1)^2 - 4 = (x + 1)^2 - 2^2 = (x + 1 - 2)(x + 1 + 2)$$
$$= (x - 1)(x + 3)$$

In the second illustration, the second step may be omitted if you recognize that $a = 7t$ and $b = 10$.

Can $3x^2 - 75$ be factored by using the difference of two squares? It becomes possible if we first factor out the common factor 3.

$$3x^2 - 75 = 3(x^2 - 25) = 3(x - 5)(x + 5)$$

The trick is to look for a common factor first in each of the given terms. This step will not only save you work in some problems, but it may well be the difference between success and failure.

It is strongly recommended that whenever you attempt to factor, you *first look for common factors* in the given terms.

THE DIFFERENCE (SUM) OF TWO CUBES

$$a^3 - b^3 = (a - b)(a^2 + ab + b^2)$$
$$a^3 + b^3 = (a + b)(a^2 - ab + b^2)$$

Be sure to check each of these facts by multiplication.

Illustrations:

$$8x^3 - 27 = (2x)^3 - 3^3$$
$$= (2x - 3)(4x^2 + 6x + 9)$$
$$2x^3 + 128y^3 = 2(x^3 + 64y^3)$$
$$= 2[x^3 + (4y)^3]$$
$$= 2(x + 4y)(x^2 - 4xy + 16y^2)$$

TEST YOUR UNDERSTANDING

Factor out the common monomial.

1 $3x - 9$ 2 $-5x + 15$ 3 $5xy + 25y^2 + 10y^5$

Factor as the difference of squares.

4 $x^2 - 36$ 5 $4x^2 - 49$ 6 $(a + 2)^2 - 25b^2$

Factor as the difference (sum) of two cubes.

7 $x^3 - 27$ 8 $x^3 + 27$ 9 $8a^3 - 125$

Factor the following by first considering common monomial factors.

10 $3x^2 - 48$ 11 $ax^3 + ay^3$ 12 $2hx^2 - 8h^3$

It is not possible to factor $x^2 - 5$ as the difference of two squares by using integral coefficients. There are times, however, when it is desirable to allow coefficients other than integers. Consider, for example, these factorizations:

$$x^2 - 5 = x^2 - (\sqrt{5})^2 = (x + \sqrt{5})(x - \sqrt{5})$$

$$x^3 - 5 = x^3 - (\sqrt[3]{5})^3 = (x - \sqrt[3]{5})[x^2 + \sqrt[3]{5}\,x + (\sqrt[3]{5})^2]$$

When other than polynomial factors are allowed, it becomes possible to factor $x - 8$ as the difference of two squares as well as the difference of two cubes.

$$x - 8 = (\sqrt{x})^2 - (\sqrt{8})^2 = (\sqrt{x} + \sqrt{8})(\sqrt{x} - \sqrt{8})$$

$$x - 8 = (\sqrt[3]{x})^3 - 2^3 = (\sqrt[3]{x} - 2)[(\sqrt[3]{x})^2 + 2\sqrt[3]{x} + 4]$$

Using fractional exponents, this last line becomes

$$x - 8 = (x^{1/3})^3 - 2^3 = (x^{1/3} - 2)(x^{2/3} + 2x^{1/3} + 4)$$

In general, we follow this rule when factoring:

Unless otherwise indicated, use the same type of numerical coefficients and exponents in the factors as appear in the given unfactored form.

TEST YOUR UNDERSTANDING

Factor the following as the difference of two squares. Irrational numbers as well as radical expressions may be used.

1 $x^2 - 2$ 2 $7 - x^2$ 3 $x - 9$

Factor the following as the difference (sum) of two cubes. Irrational numbers and radical expressions may be used.

4 $x^3 - 4$ 5 $1 - h$ 6 $x + 27$

Just as we learned how to factor the difference of two squares or cubes, we can also learn how to factor the difference of two fourth powers, two fifth powers, and so on. All these situations can be collected into the single general form that gives the factorization of the difference of two nth powers, where n is an integer greater than 1.

The factorization of $a^n - b^n$ is one of the most useful in mathematics; it will be needed in calculus. Study it carefully here so that in later work you will be able to recall it with minimal effort.

THE DIFFERENCE OF TWO nTH POWERS

$$a^n - b^n = (a - b)(a^{n-1} + a^{n-2}b + a^{n-3}b^2 + \cdots + ab^{n-2} + b^{n-1})$$

To help describe the second factor, we may rewrite it like this:

$$a^{n-1}b^0 + a^{n-2}b^1 + a^{n-3}b^2 + \cdots + a^1b^{n-2} + a^0b^{n-1}$$

You can see that the exponents for a begin with $n - 1$ and decrease to 0, whereas for b they begin with 0 and increase to $n - 1$. Note also that the sum of the exponents for a and b, for each term, is $n - 1$.

EXAMPLE 2 Factor: $a^5 - b^5$.

Solution Use the general form for $a^n - b^n$ with $n = 5$.

$$a^5 - b^5 = (a - b)(a^4 + a^3b + a^2b^2 + ab^3 + b^4)$$

EXAMPLE 3 Use the result of Example 2 to factor $3y^5 - 96$.

Solution

$$\begin{aligned} 3y^5 - 96 &= 3(y^5 - 32) \\ &= 3(y^5 - 2^5) \\ &= 3(y - 2)(y^4 + 2y^3 + 4y^2 + 8y + 16) \end{aligned}$$

EXAMPLE 4 Factor $x^4 - 81$ as the difference of two fourth powers and also as the difference of two squares.

Solution

$$x^4 - 81 = x^4 - 3^4 = (x - 3)(x^3 + 3x^2 + 9x + 27)$$

$$\begin{aligned} x^4 - 81 = (x^2)^2 - (9)^2 &= (x^2 + 9)(x^2 - 9) \\ &= (x^2 + 9)(x + 3)(x - 3) \end{aligned}$$

Note: The second answer in Example 4 is the *complete factored form* of $x^4 - 81$ since none of its factors can be factored further (allowing only polynomials with integral coefficients). Unless otherwise directed, *always try to arrive at the complete factored form.*

The two answers in Example 4 imply that

$$(x - 3)(x^3 + 3x^2 + 9x + 27) = (x^2 + 9)(x + 3)(x - 3)$$

Divide each side by $x - 3$ to obtain

$$x^3 + 3x^2 + 9x + 27 = (x^2 + 9)(x + 3)$$

How can this factorization of $x^3 + 3x^2 + 9x + 27$ be found directly? The answer is to *group* the terms first and then factor, as follows.

$$\begin{aligned} x^3 + 3x^2 + 9x + 27 &= (x^3 + 3x^2) + (9x + 27) \\ &= x^2(x + 3) + 9(x + 3) \\ &= (x^2 + 9)(x + 3) \end{aligned}$$

CAUTION: $x^2(x + 3) + 9(x + 3)$ is *not* a factored form of the given expression.

Here are alternative groupings that lead to the same answer.

$$\begin{aligned} x^3 + 3x^2 + 9x + 27 &= (x^3 + 9x) + (3x^2 + 27) \\ &= x(x^2 + 9) + 3(x^2 + 9) \\ &= (x + 3)(x^2 + 9) \end{aligned}$$

$$x^3 + 3x^2 + 9x + 27 = (x^3 + 27) + (3x^2 + 9x)$$

$$= (x + 3)(x^2 - 3x + 9) + (x + 3)(3x)$$
$$= (x + 3)(x^2 - 3x + 9 + 3x)$$
$$= (x + 3)(x^2 + 9)$$

EXAMPLE 5 Factor $ax^2 - 15 - 5ax + 3x$ by grouping.

Solution

Not all groupings are productive. Thus in Example 5 the grouping $(ax^2 - 15) + (3x - 5ax)$ does not lead to a solution.

$$ax^2 - 15 - 5ax + 3x = (ax^2 - 5ax) + (3x - 15)$$
$$= ax(x - 5) + 3(x - 5)$$
$$= (ax + 3)(x - 5)$$

EXERCISES 2.5

Factor out the common monomial.

1 $5x - 5$
2 $3x^2 + 12x - 6$
3 $16x^4 + 8x^3 + 4x^2 + 2x$
4 $4a^2b - 6$
5 $4a^2b - 6ab$
6 $(a + b)x^2 + (a + b)y^2$
7 $2xy + 4x^2 + 8x^4$
8 $-12x^3y + 9x^2y^2 - 6xy^3$
9 $10x(a - b) + 5y(a - b)$

Factor as the difference of two squares.

10 $x^2 - 9$ 11 $81 - x^2$
12 $x^2 - 10{,}000$ 13 $4x^2 - 9$
14 $25x^2 - 144y^2$ 15 $a^2 - 121b^2$

Factor as the difference (sum) of two cubes.

16 $x^3 - 8$ 17 $x^3 + 64$
18 $8x^3 + 1$ 19 $125x^3 - 64$
20 $8 - 27a^3$ 21 $8x^3 + 343y^3$

Factor as the difference of two squares, allowing irrational numbers as well as radical expressions. (All letters represent positive numbers.)

22 $a^2 - 15$ 23 $3 - 4x^2$
24 $x - 1$ 25 $x - 36$
26 $2x - 9$ 27 $8 - 3x$

Factor as the difference (sum) of two cubes, allowing irrational numbers as well as radical expressions.

28 $x^3 - 2$ 29 $7 + a^3$
30 $1 - h$ 31 $27x + 1$
32 $27x - 64$ 33 $3x - 4$

Factor by grouping.

34 $a^2 - 2b + 2a - ab$ 35 $x^2 - y - x + xy$
36 $x + 1 + y + xy$ 37 $-y - x + 1 + xy$
38 $ax + by + ay + bx$ 39 $2 - y^2 + 2x - xy^2$

Factor completely.

40 $8a^2 - 2b^2$ 41 $7x^3 + 7h^3$
42 $81x^3 - 3y^3$ 43 $x^3y - xy^3$
44 $5 - 80x^4$ 45 $x^4 - y^4$
46 $81x^4 - 256y^4$ 47 $a^8 - b^8$
48 $40ab^3 - 5a^4$ 49 $a^5 - 32$
50 $3x^5 - 96y^5$ 51 $1 - h^7$
52 $(2a + b)a^2 - (2a + b)b^2$
53 $3(a + 1)x^3 + 24(a + 1)$
54 $x^6 + x^2y^4 - x^4y^2 - y^6$
55 $a^3x - b^3y + b^3x - a^3y$
56 $7a^2 - 35b + 35a - 7ab$
57 $x^5 - 16xy^4 - 2x^4y + 32y^5$

▪ *Simplify by factoring.*

58 $(1 + x)^2(-1) + (1 - x)(2)(1 + x)$
59 $(x + 2)^3(2) + (2x + 1)(3)(x + 2)^2$
60 $(x^2 + 2)^2(3) + (3x - 1)(2)(x^2 + 2)(2x)$
61 $(x^3 + 1)^3(2x) + (x^2 - 1)(3)(x^3 + 1)^2(3x^2)$
62 Find these products.
(a) $(a + b)(a^2 - ab + b^2)$
(b) $(a + b)(a^4 - a^3b + a^2b^2 - ab^3 + b^4)$

(c) $(a+b)(a^6 - a^5b + a^4b^2 - a^3b^3 + a^2b^4 - ab^5 + b^6)$

63 Use the results of Exercise 62 to factor the following:

(a) $x^5 + 32$ (b) $128x^8 + xy^7$

*64 Write the factored form of $a^n + b^n$, where n is an odd positive integer greater than 1.

2.6 Factoring trinomials

By multiplication we can establish the following products:

$$(a+b)^2 = (a+b)(a+b) = a^2 + 2ab + b^2$$
$$(a-b)^2 = (a-b)(a-b) = a^2 - 2ab + b^2$$

Each product is called a *perfect trinomial square.* In each case the first and last terms are squares of a and b, respectively, and the middle term is twice their product. Reversing the procedure gives us two more general factoring forms.

PERFECT TRINOMIAL SQUARES

$$a^2 + 2ab + b^2 = (a+b)^2$$
$$a^2 - 2ab + b^2 = (a-b)^2$$

Observe that in $a^2 \pm 2ab + b^2$ the middle term (ignoring signs) is twice the product of the square roots of the end terms. Hence the factored form is the square of the sum (or difference) of these square roots.

Illustrations:

$$x^2 + 10x + 25 = x^2 + 2(x \cdot 5) + 5^2$$
$$= (x+5)^2$$
$$25t^2 - 30t + 9 = (5t)^2 - 2(5t \cdot 3) + 3^2$$
$$= (5t-3)^2$$

EXAMPLE 1 Factor: $1 + 18b + 81b^2$.

Solution

$$1 + 18b + 81b^2 = (1 + 9b)^2$$

EXAMPLE 2 Factor: $9x^3 - 42x^2 + 49x$.

Solution

$$9x^3 - 42x^2 + 49x = x(9x^2 - 42x + 49)$$
$$= x[(3x)^2 - 2(3x)(7) + 7^2]$$
$$= x(3x-7)^2$$

Always remember, as in Example 2, to search first for a common monomial factor.

Another factoring technique that we will consider deals with trinomials that are not necessarily perfect squares. Let us factor $x^2 + 7x + 12$.

From our experience with multiplying binomials we can anticipate that the factors will be of this form:

$$(x + \underline{?})(x + \underline{?})$$

We need to fill in the blanks with two integers whose product is 12. Furthermore, the middle term of the product must be $+7x$. The possible choices for the two integers are

12 and 1 6 and 2 4 and 3

To find the correct pair is now a matter of trial and error. These are the possible factorizations:

With a little luck, and much more experience, you can often avoid exhausting *all* the possibilities before finding the correct factors. You will then find that such work can often be shortened significantly.

$$(x + 12)(x + 1)$$
$$(x + 6)(x + 2)$$
$$(x + 4)(x + 3)$$

Only the last form gives the correct middle term of $7x$. Therefore, we conclude that $x^2 + 7x + 12 = (x + 4)(x + 3)$.

EXAMPLE 3 Factor: $x^2 - 10x + 24$.

Solution The final term, $+24$, must be the product of two positive numbers or two negative numbers. (Why?) Since the middle term, $-10x$, has a minus sign, the factorization must be of this form:

$$(x - \underline{?})(x - \underline{?})$$

Now try all the pairs of integers whose product is 24: 24 and 1, 12 and 2, 8 and 3, 6 and 4. Only the last of these gives $-10x$ as the middle term. Thus

$$x^2 - 10x + 24 = (x - 6)(x - 4)$$

Let us now consider a more complicated factoring problem. If we wish to factor the trinomial $15x^2 + 43x + 8$, we need to consider possible factors both of 15 and of 8. Because of the "+" signs in the trinomial, the factorization will be of this form:

$$(\underline{?}\,x + \underline{?})(\underline{?}\,x + \underline{?})$$

Here are the different possibilities for factoring 15 and 8:

$$15 = 15 \cdot 1 \qquad 15 = 5 \cdot 3$$
$$8 = 8 \cdot 1 \qquad 8 = 4 \cdot 2$$

Using $15 \cdot 1$, write the form

$$(15x + \underline{?})(x + \underline{?})$$

Try 8 and 1 in the blanks, *both ways*, namely:

$$(15x + 8)(x + 1)$$
$$(15x + 1)(x + 8)$$

Neither gives a middle term of $43x$; so now try 4 and 2 in the blanks both ways. Again you see that neither case works. Next consider the form

$$(5x + \underline{?})(3x + \underline{?})$$

Once again try 4 with 2 both ways, and 8 with 1 both ways. Here is the correct answer:

$$15x^2 + 43x + 8 = (5x + 1)(3x + 8)$$

Factoring $15x^2 - 43x + 8$ is very similar to factoring $15x^2 + 43x + 8$. The only difference is that the binomial factors have minus signs instead of plus signs; that is,

$$15x^2 - 43x + 8 = (5x - 1)(3x - 8)$$

EXAMPLE 4 Factor: $12x^2 - 9x + 2$.

Example 4 shows us that not every trinomial can be factored.

Solution Consider the forms

$$(12x - \underline{?})(x - \underline{?})$$
$$(6x - \underline{?})(2x - \underline{?})$$
$$(4x - \underline{?})(3x - \underline{?})$$

In each form try 2 with 1 both ways. None of these produces a middle term of $-9x$; hence we say that $12x^2 - 9x + 2$ is *not factorable* with integral coefficients.

TEST YOUR UNDERSTANDING

Factor the following trinomial squares.

1 $a^2 + 6a + 9$	2 $x^2 - 10x + 25$	3 $4x^2 + 12xy + 9y^2$

Factor each trinomial if possible.

4 $x^2 + 8x + 15$	5 $a^2 - 12a + 20$	6 $x^2 + 3x + 4$
7 $10x^2 - 39x + 14$	8 $6x^2 - 11x - 10$	9 $6x^2 + 6x - 5$

In some problems you may find it easier to leave out the signs in the two binomial forms as you try various cases. Thus, to factor $6x^2 - 7x - 20$ consider the forms

$$(6x \quad \underline{?})(x \quad \underline{?}) \qquad (3x \quad \underline{?})(2x \quad \underline{?})$$

Then as you try a pair, like 5 with 4, keep in mind that the signs must be opposite. As soon as you see that the difference of the two partial products is $\pm 7x$, then insert the signs to produce the desired term $-7x$. In this problem the answer is $(3x + 4)(2x - 5)$.

EXAMPLE 5 Factor: $15x^2 + 7x - 8$.

Solution Try these forms:

$$(5x \quad \underline{?})(3x \quad \underline{?})$$
$$(15x \quad \underline{?})(x \quad \underline{?})$$

In each form try 4 with 2 and 8 with 1 both ways. The factored form is $(15x - 8)(x + 1)$.

All the trinomials we have considered were of degree 2. The same methods can be modified to factor certain trinomials of higher degree, as in Example 6.

EXAMPLE 6 Factor each trinomial:
(a) $x^4 + x^2 - 12$ **(b)** $a^6 - 3a^3b - 18b^2$

Solution

(a) Note that $x^4 = (x^2)^2$ and let $u = x^2$. Then

The substitution step in Example 6(a) is a very useful one to learn for future use.

$$
\begin{aligned}
x^4 + x^2 - 12 &= u^2 + u - 12 \\
&= (u + 4)(u - 3) \\
&= (x^2 + 4)(x^2 - 3)
\end{aligned}
$$

(b) $a^6 - 3a^3b - 18b^2 = (a^3 - 6b)(a^3 + 3b)$

CAUTION: LEARN TO AVOID MISTAKES LIKE THESE

WRONG	RIGHT
$(x + 2)3 + (x + 2)y = (x + 2)3y$	$(x+2)3+(x+2)y=(x+2)(3+y)$
$3x + 1 = 3(x + 1)$	$3x + 1$ is not factorable by using integers.
$x^3 - y^3 = (x - y)(x^2 + y^2)$	$x^3 - y^3 = (x - y)(x^2 + xy + y^2)$
$x^3 + 8$ is not factorable.	$x^3 + 8 = (x + 2)(x^2 - 2x + 4)$
$x^2 + y^2 = (x + y)(x + y)$	$x^2 + y^2$ is not factorable by using real numbers.
$4x^2 - 6xy + 9y^2 = (2x - 3y)^2$	$4x^2 - 6xy + 9y^2$ is not a perfect trinomial square and cannot be factored using real numbers.

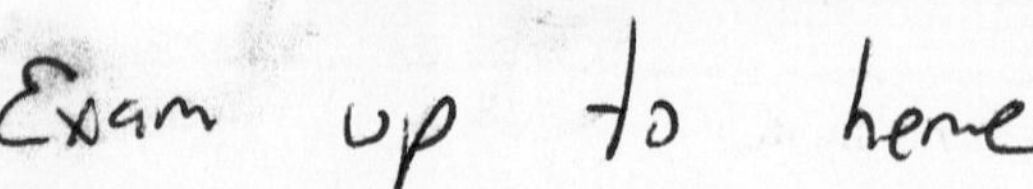

EXERCISES 2.6

Factor each of the following perfect trinomial squares.

1 $x^2 + 4x + 4$
2 $x^2 - 8x + 16$
3 $a^2 - 14a + 49$
4 $r^2 - 2r + 1$
5 $1 + 2b + b^2$
6 $100 - 20x + x^2$
7 $4a^2 + 8a + 4$
8 $4a^2 - 8a + 4$
9 $9x^2 - 18xy + 9y^2$
10 $64a^2 + 64a + 16$

11 $9x^2 + 12xy + 4y^2$

12 $4x^2 - 12xy + 9y^2$

Factor each trinomial.

13 $x^2 + 5x + 6$

14 $x^2 - 7x + 10$

15 $x^2 + 20x + 51$

16 $12a^2 - 13a + 1$

17 $20a^2 - 9a + 1$

18 $4 - 5b + b^2$

19 $x^2 + 20x + 36$

20 $a^2 - 24a + 63$

21 $9x^2 + 6x + 1$

22 $x^2 - 20x + 64$

23 $25a^2 - 10a + 1$

24 $8x^2 + 14x + 3$

25 $3x^2 + 20x + 12$

26 $5x^2 + 31x + 6$

27 $14x^2 + 37x + 5$

28 $9x^2 - 18x + 5$

29 $8x^2 - 9x + 1$

30 $30a^2 - 17a + 1$

31 $4a^2 + 20a + 25$

32 $a^2 - 9a + 18$

33 $b^2 + 18b + 45$

34 $6x^2 + 12x + 6$

35 $8x^2 - 16x + 6$

36 $12x^2 + 92x + 15$

37 $12a^2 - 25a + 12$

38 $18t^2 - 67t + 14$

39 $15x^2 - 7x - 2$

40 $15x^2 + 7x - 2$

41 $6a^2 + 5a - 21$

42 $6a^2 - 5a - 21$

43 $4x^2 + 4x - 3$

44 $15x^2 + 19x - 56$

45 $24a^2 + 25ab + 6b^2$

Factor each trinomial when possible. When appropriate, first factor out the common monomial.

46 $3a^2 + 6a + 3$

47 $5x^2 + 25x + 20$

48 $18x^2 - 24x + 8$

49 $4ax^2 + 4ax + a$

50 $x^2 + x + 1$

51 $a^2 - 2a + 2$

52 $49r^2s - 42rs + 9s$

53 $6x^2 + 2x - 20$

54 $50a^2 - 440a - 90$

55 $6a^2 + 4a - 9$

56 $36x^2 - 96x + 64$

57 $2x^2 - 2x - 112$

58 $15 + 5y - 10y^2$

59 $2b^2 + 12b + 16$

60 $4a^2x^2 - 4abx^2 + b^2x^2$

61 $a^3b - 2a^2b^2 + ab^3$

62 $12x^2y + 22xy^2 - 60y^3$

63 $16x^2 - 24x + 8$

64 $16x^2 - 24x - 8$

65 $25a^2 + 50ab + 25b^2$

66 $x^4 - 2x^2 + 1$

67 $a^6 - 2a^3 + 1$

68 $2x^4 + 8x^2 - 42$

69 $6x^5y - 3x^3y^2 - 30xy^3$

70 $3x^4 + 6x^2 + 3$

71 $a^6 - 2a^3b^3 + b^6$

▪ 72 Factor completely: $3(x^2 - 2x + 1)^2(2x - 2)$

2.7 Fundamental operations with rational expressions

A **rational expression** is a ratio of polynomials. Rational expressions are the "algebraic extensions" of rational numbers, so that the fundamental rules for operating with rational numbers extend to rational expressions.

The important rules for operating with rational expressions, also referred to as *algebraic fractions*, will now be considered. In each case we shall give an example of the rule under discussion in terms of work with arithmetic fractions so that you can compare the procedures being used. Also, in each case, we exclude values of the variable for which the denominator is equal to zero.

Negative of a fraction.

RULE 1. $-\frac{a}{b} = \frac{-a}{b} = \frac{a}{-b} \quad \left[-\frac{2}{3} = \frac{-2}{3} = \frac{2}{-3}\right]$

The negative of a fraction is the same as the fraction obtained by taking the negative of the numerator or of the denominator.

Illustration:

$$-\frac{2-x}{x^2-5} = \frac{-(2-x)}{x^2-5} = \frac{2-x}{-(x^2-5)}$$

Note that since $-(2 - x) = x - 2$, each of these fractions is also equal to $\frac{x-2}{x^2-5}$.

Reducing fractions. RULE 2. $\frac{ac}{bc} = \frac{a}{b}$ $\left[\frac{2 \cdot 3}{5 \cdot 3} = \frac{2}{5}\right]$

Reading this formula from left to right, it shows how to reduce a fraction to *lowest terms* so that the numerator and denominator of the resulting fraction have no common polynomial factors.

Illustration:

Any nonzero number divided by itself is equal to 1. $\frac{x+6}{x+6} = 1$

$$\frac{x^2+5x-6}{x^2+6x} = \frac{(x-1)(x+6)}{x(x+6)}$$

$$= \frac{x-1}{x} \quad \text{(Rule 2)}$$

Multiplication of fractions. RULE 3. $\frac{a}{b} \cdot \frac{c}{d} = \frac{ac}{bd}$ $\left[\frac{2}{3} \cdot \frac{4}{5} = \frac{2 \cdot 4}{3 \cdot 5} = \frac{8}{15}\right]$

Illustration:

Whenever we are working with rational expressions it is taken for granted that final answers have been reduced to lowest terms.

$$\frac{x-1}{x+1} \cdot \frac{x^2-x-2}{5x} = \frac{(x-1)(x^2-x-2)}{(x+1)5x} \quad \text{(Rule 3)}$$

$$= \frac{(x-1)(x+1)(x-2)}{(x+1)5x} \quad \text{(factoring)}$$

$$= \frac{(x-1)(x-2)}{5x} \quad \text{(Rule 2)}$$

This work can be shortened.

$$\frac{x-1}{x+1} \cdot \frac{x^2-x-2}{5x} = \frac{x-1}{x+1} \cdot \frac{(x+1)(x-2)}{5x} = \frac{(x-1)(x-2)}{5x}$$

Note that $\frac{x^2-3x+2}{5x}$ is an alternative form of the answer.

Division of fractions. RULE 4. $\frac{a}{b} \div \frac{c}{d} = \frac{a}{b} \cdot \frac{d}{c} = \frac{ad}{bc}$ $\left[\frac{3}{5} \div \frac{2}{3} = \frac{3}{5} \cdot \frac{3}{2} = \frac{3 \cdot 3}{5 \cdot 2} = \frac{9}{10}\right]$

Illustration:

There are a number of alternative forms for Rule 4. For example:

$$\frac{\frac{a}{b}}{\frac{c}{d}} = \frac{a}{b} \div \frac{c}{d} = \frac{ad}{bc}$$

$$\frac{\frac{a}{b}}{c} = \frac{a}{b} \div \frac{c}{1} = \frac{a}{bc}$$

$$\frac{a}{\frac{b}{c}} = \frac{a}{1} \div \frac{b}{c} = \frac{ac}{b}$$

$$\frac{(x+1)^2}{x^2-6x+9} \div \frac{3x+3}{x-3} = \frac{(x+1)^2}{x^2-6x+9} \cdot \frac{x-3}{3x+3} \quad \text{(Rule 4)}$$

$$= \frac{(x+1)^2}{(x-3)^2} \cdot \frac{x-3}{3(x+1)} \quad \text{(factoring)}$$

$$= \frac{(x+1)^2(x-3)}{(x-3)^2 3(x+1)} \quad \text{(Rule 3)}$$

$$= \frac{(x+1)(x+1)(x-3)}{3(x-3)(x+1)(x-3)} \quad \text{(rewriting)}$$

$$= \frac{x+1}{3(x-3)} \quad \text{(Rule 2)}$$

The preceding example can be shortened as follows:

$$\frac{(x+1)^2}{x^2-6x+9} \div \frac{3x+3}{x-3} = \frac{(x+1)^2}{(x-3)^2} \cdot \frac{x-3}{3(x+1)} = \frac{x+1}{3(x-3)}$$

TEST YOUR UNDERSTANDING

Simplify each expression by reducing to lowest terms.

1 $\frac{x^2 + 2x}{x}$

2 $\frac{x^2}{x^2 + 2x}$

3 $\frac{x^2 - 9}{x^2 - 5x + 6}$

4 $\frac{x^2 + 6x + 5}{x^2 - x - 2}$

5 $\frac{x^2 - 4}{x^4 - 16}$

6 $\frac{3x^2 + x - 10}{5x - 3x^2}$

Perform the indicated operation and simplify.

7 $\frac{2x - 4}{2} \cdot \frac{x + 2}{x^2 - 4}$

8 $\frac{x - 2}{3(x + 1)} \div \frac{x^2 + 2x}{x + 2}$

9 $\frac{a}{b} \div \frac{a^2 - ab}{ab + b^2}$

10 $\frac{x + y}{x - y} \cdot \frac{x^2 - 2xy + y^2}{x^2 - y^2}$

Addition and subtraction of fractions—same denominators.

RULE 5. $\frac{a}{d} + \frac{c}{d} = \frac{a + c}{d}$ $\left[\frac{3}{7} + \frac{2}{7} = \frac{3 + 2}{7} = \frac{5}{7}\right]$

RULE 6. $\frac{a}{d} - \frac{c}{d} = \frac{a - c}{d}$ $\left[\frac{7}{9} - \frac{2}{9} = \frac{7 - 2}{9} = \frac{5}{9}\right]$

Illustration:

$$\frac{6x^2}{2x^2 - x - 10} + \frac{-15x}{2x^2 - x - 10} = \frac{6x^2 + (-15x)}{2x^2 - x - 10} \quad \text{(Rule 5)}$$

$$= \frac{6x^2 - 15x}{2x^2 - x - 10}$$

$$= \frac{3x(2x - 5)}{(x + 2)(2x - 5)} \quad \text{(factoring)}$$

$$= \frac{3x}{x + 2} \quad \text{(Rule 2)}$$

Addition and subtraction of fractions—different denominators.

RULE 7. $\frac{a}{b} + \frac{c}{d} = \frac{ad + bc}{bd}$ $\left[\frac{2}{3} + \frac{3}{4} = \frac{2 \cdot 4 + 3 \cdot 3}{3 \cdot 4} = \frac{8 + 9}{12} = \frac{17}{12}\right]$

RULE 8. $\frac{a}{b} - \frac{c}{d} = \frac{ad - bc}{bd}$ $\left[\frac{4}{5} - \frac{2}{3} = \frac{4 \cdot 3 - 5 \cdot 2}{5 \cdot 3} = \frac{12 - 10}{15} = \frac{2}{15}\right]$

Illustration:

$$\frac{3}{x^2 + x} + \frac{2}{x^2 - 1} = \frac{3(x^2 - 1) + 2(x^2 + x)}{(x^2 + x)(x^2 - 1)} \quad \text{(Rule 7)}$$

$$= \frac{5x^2 + 2x - 3}{(x^2 + x)(x^2 - 1)} \quad \text{(combining terms)}$$

$$= \frac{(5x - 3)(x + 1)}{x(x + 1)(x^2 - 1)} \quad \text{(factoring)}$$

$$= \frac{5x - 3}{x(x^2 - 1)} \quad \text{(Rule 2)}$$

Here is an alternative method for the preceding example. It makes use of the **least common denominator (LCD)** of the two fractions.

The LCD of the two fractions is $x(x+1)(x-1)$. We express each fraction using this common denominator and then add the numerators.

$$\frac{3}{x^2+x}+\frac{2}{x^2-1}=\frac{3}{x(x+1)}+\frac{2}{(x+1)(x-1)}$$

$$=\frac{3(x-1)}{x(x+1)(x-1)}+\frac{2x}{(x+1)(x-1)x} \qquad \text{(Rule 2)}$$

$$=\frac{3(x-1)+2x}{x(x+1)(x-1)} \qquad \text{(Rule 5)}$$

$$=\frac{5x-3}{x(x^2-1)}$$

EXAMPLE 1 Find the LCD of the fractions whose denominators are the following:

(a) $x^2+x,\quad x^2-1$ **(b)** $x^4-x,\quad x^2-2x+1,\quad x^4+2x^3$

To find the LCD, take *each* factor that appears in the factored forms of the denominators the *maximum* number of times it appears in any *one* of the factored forms.

Solution

(a) $x^2+x=x(x+1)$

$x^2-1=(x+1)(x-1)$

$\text{LCD}=x(x+1)(x-1)$

(b) $x^4-x=x(x^3-1)=x(x-1)(x^2+x+1)$

$x^2-2x+1=(x-1)^2$

$x^4+2x^3=x^3(x+2)$

$\text{LCD}=x^3(x-1)^2(x+2)(x^2+x+1)$

EXAMPLE 2 Combine and simplify: $\frac{3}{2}-\frac{4}{3x(x+1)}-\frac{x-5}{3x^2}$.

Solution The least common denominator of the fractions is $6x^2(x+1)$.

$$\frac{3}{2}-\frac{4}{3x(x+1)}-\frac{x-5}{3x^2}=\frac{3\cdot 3x^2(x+1)}{2\cdot 3x^2(x+1)}-\frac{4\cdot 2x}{3x(x+1)\cdot 2x}-\frac{(x-5)\cdot 2(x+1)}{3x^2\cdot 2(x+1)}$$

$$=\frac{(9x^3+9x^2)-8x-(2x^2-8x-10)}{6x^2(x+1)}$$

$$=\frac{9x^3+7x^2+10}{6x^2(x+1)}$$

TEST YOUR UNDERSTANDING

Combine and simplify.

1 $\frac{x}{5}+\frac{2x}{3}$

2 $\frac{4x}{3}-\frac{x}{2}$

3 $\frac{2}{3x^2}-\frac{1}{2x}$

4 $\frac{3}{2x}+\frac{5}{3x}+\frac{1}{x}$

5 $\frac{7}{x-2}+\frac{3}{x+2}$

6 $\frac{9}{x-3}+\frac{7}{x^2-9}$

7 $\frac{5}{(x-1)(x+2)}-\frac{8}{x^2-4}$

8 $\frac{5x}{(2x+5)^2}+\frac{3x-2}{4x^2-25}$

The fundamental properties of fractions can be used to simplify rational expressions whose numerators and denominators may themselves also contain fractions.

EXAMPLE 3 Simplify: $\dfrac{\dfrac{1}{5+h}-\dfrac{1}{5}}{h}$.

Solution Combine the fractions in the numerator and then divide.

The expression in Example 3 is a type that you will encounter in calculs. Be certain that you understand each step in the solution.

$$\frac{\dfrac{1}{5+h}-\dfrac{1}{5}}{h}=\frac{\dfrac{5-(5+h)}{5(5+h)}}{h}=\frac{\dfrac{-h}{5(5+h)}}{h}$$

$$=\frac{-h}{5(5+h)}\cdot\frac{1}{h}$$

$$=\frac{-h}{5h(5+h)}=-\frac{1}{5(5+h)}$$

EXAMPLE 4 Simplify: $\dfrac{x^{-2}-y^{-2}}{\dfrac{1}{x}-\dfrac{1}{y}}$.

Solution

$$\frac{x^{-2}-y^{-2}}{\dfrac{1}{x}-\dfrac{1}{y}}=\frac{\dfrac{1}{x^2}-\dfrac{1}{y^2}}{\dfrac{1}{x}-\dfrac{1}{y}}$$

$$=\frac{\left(\dfrac{1}{x^2}-\dfrac{1}{y^2}\right)(x^2y^2)}{\left(\dfrac{1}{x}-\dfrac{1}{y}\right)(x^2y^2)} \qquad \text{(Rule 2)}$$

$$=\frac{y^2-x^2}{xy^2-x^2y}$$

$$=\frac{(y-x)(y+x)}{xy(y-x)}$$

$$=\frac{y+x}{xy}$$

Note that we multiplied both the numerator and denominator by x^2y^2 to simplify. Can you simplify this fraction in an alternative way without this step?

CAUTION: LEARN TO AVOID MISTAKES LIKE THESE

WRONG	RIGHT
$\dfrac{2}{3}+\dfrac{x}{5}=\dfrac{2+x}{3+5}$	$\dfrac{2}{3}+\dfrac{x}{5}=\dfrac{2\cdot 5+3\cdot x}{3\cdot 5}=\dfrac{10+3x}{15}$

Working with fractions often creates difficulties for many students. Study this list; it may help you avoid some common pitfalls.

WRONG	RIGHT
$\frac{1}{a}+\frac{1}{b}=\frac{1}{a+b}$	$\frac{1}{a}+\frac{1}{b}=\frac{b+a}{ab}$
$\frac{2x+5}{4}=\frac{x+5}{2}$	$\frac{2x+5}{4}=\frac{2x}{4}+\frac{5}{4}=\frac{x}{2}+\frac{5}{4}$
$2+\frac{x}{y}=\frac{2+x}{y}$	$2+\frac{x}{y}=\frac{2y+x}{y}$
$3\left(\frac{x+1}{x-1}\right)=\frac{3(x+1)}{3(x+1)}$	$3\left(\frac{x+1}{x-1}\right)=\frac{3(x+1)}{x-1}$
$a\div\frac{b}{c}=\frac{1}{a}\cdot\frac{b}{c}$	$a\div\frac{b}{c}=a\cdot\frac{c}{b}=\frac{ac}{b}$
$\frac{1}{a^{-1}+b^{-1}}=a+b$	$\frac{1}{a^{-1}+b^{-1}}=\frac{1}{\frac{1}{a}+\frac{1}{b}}=\frac{ab}{b+a}$

EXERCISES 2.7

Classify each statement as true or false. If it is false, correct the right side to get a correct equality.

1 $\frac{5}{7}-\frac{2}{3}=\frac{3}{4}$

2 $\frac{2x+y}{y-2x}=-2\left(\frac{x+y}{x-y}\right)$

3 $\frac{3ax-5b}{6}=\frac{ax-5b}{2}$

4 $\frac{x+x^{-1}}{xy}=\frac{x+1}{x^2y}$

5 $x^{-1}+y^{-1}=\frac{y+x}{xy}$

6 $\frac{2}{\frac{3}{4}}=\frac{8}{3}$

Simplify, if possible.

7 $\frac{8xy}{12yz}$

8 $\frac{24abc^2}{36bc^2d}$

9 $\frac{x^2-5x}{5-x}$

10 $\frac{n-1}{n^2-1}$

11 $\frac{n+1}{n^2+1}$

12 $\frac{(x+1)^2}{x^2-1}$

13 $\frac{3x^2+3x-6}{2x^2+6x+4}$

14 $\frac{x^3-x}{x^3-2x^2+x}$

15 $\frac{4x^2+12x+9}{4x^2-9}$

16 $\frac{x^2+2x+xy+2y}{x^2+4x+4}$

17 $\frac{a^2-16b^2}{a^3+64b^3}$

18 $\frac{a^2-b^2}{a^2-6b-ab+6a}$

Perform the indicated operations and simplify.

19 $\frac{2x^2}{y}\cdot\frac{y^2}{x^3}$

20 $\frac{3x^2}{2y^2}\div\frac{3x^3}{y}$

21 $\frac{2a}{3}\cdot\frac{3}{a^2}\cdot\frac{1}{a}$

22 $\left(\frac{a^2}{b^2}\cdot\frac{b}{c^2}\right)\div a$

23 $\frac{3x}{2y}-\frac{x}{2y}$

24 $\frac{a+2b}{a}+\frac{3a+b}{a}$

25 $\frac{a-2b}{2}-\frac{3a+b}{3}$

26 $\frac{7}{5x}-\frac{2}{x}+\frac{1}{2x}$

27 $\frac{x-1}{3}\cdot\frac{x^2+1}{x^2-1}$

28 $\frac{x^2-x-6}{x^2-3x}\cdot\frac{x^3+x^2}{x+2}$

29 $\frac{x-1}{x+2}\div\frac{x^2-x}{x^2+2x}$

30 $\frac{x^2+3x}{x^2+4x+3}\div\frac{x^2-2x}{x+1}$

31 $\frac{2}{x}-y$

32 $\dfrac{x^2}{x-1} - \dfrac{1}{x-1}$

33 $\dfrac{3y}{y+1} + \dfrac{2y}{y-1}$

34 $\dfrac{2a}{a^2-1} - \dfrac{a}{a+1}$

35 $\dfrac{2x^2}{x^2+x} + \dfrac{x}{x+1}$

36 $\dfrac{3x+3}{2x^2-x-1} + \dfrac{1}{2x+1}$

37 $\dfrac{5}{x^2-4} - \dfrac{3-x}{4-x^2}$

38 $\dfrac{1}{a^2-4} + \dfrac{3}{a-2} - \dfrac{2}{a+2}$

39 $\dfrac{2x}{x^2-9} + \dfrac{x}{x^2+6x+9} - \dfrac{3}{x+3}$

40 $\dfrac{a^2+2ab+b^2}{a^2-b^2} \div \dfrac{a^2+3ab+2b^2}{a^2-3ab+2b^2}$

41 $\dfrac{x^3+x^2-12x}{x^2-3x} \cdot \dfrac{3x^2-10x+3}{3x^2+11x-4}$

42 $\dfrac{n^2+n}{2n^2+7n-4} \cdot \dfrac{4n^2-4n+1}{2n^2-n-3} \cdot \dfrac{2n^2+5n-12}{2n^3-n^2}$

43 $\dfrac{n^3-8}{n+2} \cdot \dfrac{2n^2+8}{n^3-4n} \cdot \dfrac{n^3+2n^2}{n^3+2n^2+4n}$

44 $\dfrac{a^3-27}{a^2-9} \div \left(\dfrac{a^2+2ab+b^2}{a^3+b^3} \cdot \dfrac{a^3-a^2b+ab^2}{a^2+ab}\right)$

Simplify.

45 $\dfrac{\dfrac{5}{x^2-4}}{\dfrac{10}{x-2}}$

46 $\dfrac{x+\dfrac{3}{2}}{x-\dfrac{1}{2}}$

47 $\dfrac{\dfrac{1}{x}-\dfrac{1}{4}}{x-4}$

48 $\dfrac{\dfrac{1}{x}+\dfrac{1}{5}}{x+5}$

49 $\dfrac{\dfrac{1}{4+h}-\dfrac{1}{4}}{h}$

50 $\dfrac{\dfrac{1}{x^2}-\dfrac{1}{9}}{x-3}$

51 $\dfrac{\dfrac{1}{x+3}-\dfrac{1}{3}}{x}$

52 $\dfrac{\dfrac{1}{4}-\dfrac{1}{x^2}}{x-2}$

53 $\dfrac{\dfrac{1}{x^2}-\dfrac{1}{16}}{x+4}$

54 $\dfrac{\dfrac{1}{x^2}-\dfrac{1}{4}}{\dfrac{1}{x}-\dfrac{1}{2}}$

55 $\dfrac{\dfrac{1}{x}-\dfrac{1}{y}}{\dfrac{1}{x^2}-\dfrac{1}{y^2}}$

56 $\dfrac{\dfrac{4}{x^2}-\dfrac{1}{y^2}}{\dfrac{2}{x}-\dfrac{1}{y}}$

▬57 $\dfrac{(1+x^2)(-2x)-(1-x^2)(2x)}{(1+x^2)^2}$

▬58 $\dfrac{(x^2-9)(2x)-x^2(2x)}{(x^2-9)^2}$

▬59 $\dfrac{x^2(4-2x)-(4x-x^2)(2x)}{x^4}$

▬60 $\dfrac{(x+1)^2(2x)-(x^2-1)(2)(x+1)}{(x+1)^4}$

Simplify, and express as a single fraction without negative exponents.

*61 $\dfrac{a^{-1}-b^{-1}}{a-b}$

*62 $\dfrac{(a+b)^{-1}}{a^{-1}+b^{-1}}$

*63 $\dfrac{x^{-2}-y^{-2}}{xy}$

*64 There are three tests and a final examination given in a mathematics course. Let a, b, and c be the numerical grades of the tests, and let d represent the examination grade.

(a) If the final grade is computed by allowing the average of the three tests and the exam to count the same, show that the final average is given by the expression $\dfrac{a+b+c+3d}{6}$.

(b) Assume that the average of the three tests accounts for 60% of the final grade, and that the examination accounts for 40%. Show that the final average is given by the expression $\dfrac{a+b+c+2d}{5}$.

*65 Some calculators require that certain calculations be performed in a different manner to accommodate the machine. Show that in each case the expression on the left can be computed by using the equivalent expression on the right.

(a) $\dfrac{A}{B}+\dfrac{C}{D} = \dfrac{\dfrac{A \cdot D}{B}+C}{D}$

(b) $A \cdot B + C \cdot D + E \cdot F$

$$= \left[\frac{\left(\frac{A \cdot B}{D}+C\right) \cdot D}{F} + E\right] \cdot F$$

▬66 If $x^2+y^2=4$, show that $\dfrac{-y-x\left(-\dfrac{x}{y}\right)}{y^2} = -\dfrac{4}{y^3}$.

▬67 If $y^3-x^3=8$,

show that $\dfrac{2xy^2-2x^2y\left(\dfrac{x^2}{y^2}\right)}{y^4} = \dfrac{16x}{y^5}$.

REVIEW EXERCISES FOR CHAPTER 2

The solutions to the following exercises can be found within the text of Chapter 2. Try to answer each question without referring to the text.

Section 2.1

Simplify.

1 $12^3(\frac{1}{6})^3$

2 $2x^3 \cdot x^4$

3 $\frac{(x^3y)^2y^3}{x^4y^6}$

4 $(\frac{1}{2})^3(2)^{-3}$

5 Simplify and write without negative exponents: $\left(\frac{a^{-2}b^3}{a^3b^{-2}}\right)^5$.

6 Prove that $(ab)^m = a^mb^m$, where m is a negative integer. (*Hint:* Let $m = -k$ and apply the rules for positive integral exponents.)

7 What is the motivation behind the definition $b^0 = 1$?

Section 2.2

8 What is the definition of $\sqrt[n]{a}$, where $a > 0$ and n is a positive integer ≥ 2?

Evaluate.

9 $\sqrt[3]{\frac{1}{8}}$

10 $\sqrt[5]{-32}$

11 $\sqrt{0.09}$

12 $(\sqrt[4]{7})^4$

13 $\sqrt{9 \cdot 25}$

14 $\sqrt[4]{\frac{16}{81}}$

15 $\sqrt[2]{\sqrt[3]{64}}$

16 $\sqrt{72} \cdot \sqrt{\frac{1}{2}}$

17 $(-27)^{1/3}$

18 $(-64)^{2/3}$

19 $8^{-2/3}$

20 $27^{2/3} - 16^{-1/4}$

21 Simplify, and express the result with positive exponents only: $\frac{x^{2/3}y^{-2}z^2}{x^{1/2}y^{1/2}z^{-1}}$.

Section 2.3

Simplify.

22 $\sqrt{50}$

23 $\sqrt[3]{16}$

24 $\sqrt[3]{-54}$

25 $3\sqrt{5} + 4\sqrt{5}$

26 $\frac{6}{\sqrt{3}} + 2\sqrt{75} - \sqrt{3}$

27 $\sqrt{75x^2}$

28 $2\sqrt{8x^3} + 3x\sqrt{32x} - x\sqrt{18x}$

Section 2.4

29 Add: $(4x^3 - 10x^2 + 5x + 8) + (12x^2 - 9x - 1)$.

30 Subtract: $(4t^3 - 10t^2 + 5t + 8) - (12t^2 - 9t - 1)$.

Multiply.

31 $3x^2(4x^7 - 3x^4 - x^2 + 15)$

32 $(5x - 9)(6x + 2)$

33 $(a + b)(c + d)$

34 Multiply: $3x^3 - 8x + 4$ by $2x^2 + 5x - 1$.

35 Simplify by performing the indicated operations: $(x^2 - 5x)(3x^2) + (x^3 - 1)(2x - 5)$.

36 Expand: **(a)** $(a + b)^2$ **(b)** $(a + b)^3$

37 Rationalize the denominators:

(a) $\frac{5}{\sqrt{10} - 3}$ **(b)** $\frac{x}{\sqrt{x} + \sqrt{y}}$

Section 2.5

Factor.

38 $24x^9 - 18x^6 - 6x^4 + 90x^2$

39 $8x^2(x - 1) + 4x(x - 1) + 2(x - 1)$

40 $3x^2 - 75$

41 $8x^3 - 27$

42 $(x + 1)^2 - 4$

43 $2x^3 + 128y^3$

44 $a^5 - b^5$

45 $3y^5 - 96$

46 $x^4 - 81$

47 $ax^2 - 15 - 5ax + 3x$

48 Factor as the difference of squares using irrational numbers: $x^2 - 5$.

49 Factor as the difference of cubes using irrational numbers: $x - 8$.

Section 2.6

Factor.

50 $a^2 + 2ab + b^2$

51 $a^2 - 2ab + b^2$

52 $1 + 18b + 81b^2$

53 $9x^3 - 42x^2 + 49x$

54 $x^2 - 10x + 24$

55 $15x^2 + 43x + 8$

56 $15x^2 - 43x + 8$

57 $15x^2 + 7x - 8$

58 $x^4 + x^2 - 12$

59 $a^6 - 3a^3b - 18b^2$

Section 2.7

60 Reduce to lowest terms: $\frac{x^2 + 5x - 6}{x^2 + 6x}$.

61 Multiply: $\frac{x - 1}{x + 1} \cdot \frac{x^2 - x - 2}{5x}$.

62 Divide: $\dfrac{(x+1)^2}{x^2-6x+9} \div \dfrac{3x+3}{x-3}$.

63 Add: $\dfrac{3}{x^2+x} + \dfrac{2}{x^2-1}$.

64 Combine: $\dfrac{3}{2} - \dfrac{4}{3x(x+1)} - \dfrac{x-5}{3x^2}$.

65 Simplify: $\dfrac{\dfrac{1}{5+h} - \dfrac{1}{5}}{h}$.

66 Simplify: $\dfrac{x^{-2} - y^{-2}}{\dfrac{1}{x} - \dfrac{1}{y}}$.

SAMPLE TEST QUESTIONS FOR CHAPTER 2

Use these questions to test your knowledge of the basic skills and concepts of Chapter 2. Then check your answers with those given at the end of the book.

Classify each statement as true or false.

1 $\dfrac{x^3(-x)^2}{x^5} = x$

2 $\left(\dfrac{3}{2+a}\right)^{-1} = \dfrac{2}{3} + \dfrac{a}{3}$

3 $(-27)^{-1/3} = 3$

4 $(x+y)^{3/5} = (\sqrt[3]{x+y})^5$

5 $\sqrt{9x^2} = 3|x|$

6 $(8+a^3)^{1/3} = 2 + a$

In Exercises 7–9, find the correct choice. There is only one correct answer in each case.

7 $\dfrac{4^2(-3)^3 2^0}{6^3(12^2)} =$

(a) $-\frac{1}{216}$ (b) $-\frac{1}{72}$ (c) $-\frac{1}{36}$ (d) $(-\frac{1}{6})^5$ (e) $\dfrac{1}{6^3}$

8 $\left(\dfrac{27a^6b^{-3}}{c^{-2}}\right)^{-2/3} =$

(a) $\dfrac{b^2}{9a^4c^{4/3}}$ (b) $\dfrac{b^2}{3a^4\sqrt[3]{c}}$ (c) $\dfrac{c^2}{18ab}$ (d) $-\dfrac{9c^3b^2}{a^4}$ (e) $\dfrac{-3b^2}{a^4c^{4/3}}$

9 $\dfrac{30}{\sqrt{20}} - 2\sqrt{45} =$

(a) $\dfrac{15-3\sqrt{5}}{\sqrt{5}}$ (b) $-3\sqrt{5}$ (c) $-15\sqrt{45}$ (d) $\dfrac{3\sqrt{2}}{2} - 6\sqrt{5}$ (e) $-\frac{5}{2}$

Simplify.

10 $(ab^2 + 5cd) - (3ab^2 - 2cd + a^2) + (2ab^2 - 5a^2)$

11 $(5x+3)(2x-4) - (3-x)(3+x)$

12 $(x^2+3x)(3x^2) + (x^3-1)(2x+3)$

Factor completely.

13 $64 - 27b^3$

14 $3x^5 - 48x$

15 $6x^2 - 7x - 3$

16 $x^5 - 32$

17 $2x^2 - 6xy - 3y^3 + xy^2$

18 $ax^2 - 2x + 3ax - 6$

Perform the indicated operations and simplify.

19 $\dfrac{x^2-9}{x^3+4x^2+4x} \cdot \dfrac{2x^2+4x}{x^2+2x-15}$

20 $\dfrac{x^3+8}{x^2-4x-12} \div \dfrac{x^3-2x^2+4x}{x^3-6x^2}$

21 $\dfrac{\dfrac{1}{x^2} - \dfrac{1}{49}}{x-7}$

22 $\dfrac{4x-1}{2x^2-x-3} + \dfrac{2}{3-2x}$

23 Combine and simplify:

$\dfrac{1}{x+3} - \dfrac{2}{x^2-9} + \dfrac{x}{2x^2+x-15}$.

ANSWERS TO THE TEST YOUR UNDERSTANDING EXERCISES

Page 37

1 125	2 $-\frac{1}{32}$	3 0	4 1,000,000	5 −64	6 64
7 $\frac{1}{17}$	8 $\frac{1}{23}$	9 81	10 $a^{11}b^{10}$	11 64	12 $6x^4$

Page 44

1 11 2 -4 3 2 4 3 5 35 6 $\frac{3}{5}$
7 2 8 -70 9 $\frac{4}{3}$ 10 540 11 $-\frac{10}{3}$ 12 4
13 6 14 -3 15 $-\frac{3}{2}$

Page 45

1 $\sqrt{5}$ 2 $\sqrt[3]{-9}$ 3 $\sqrt[4]{10}$ 4 $\sqrt[3]{25}$ 5 $\sqrt[4]{8}$ 6 $7^{1/2}$
7 $(-10)^{1/3}$ 8 $7^{1/4}$ 9 $7^{2/3}$ 10 $5^{3/4}$ 11 5 12 4
13 $\frac{1}{6}$ 14 $\frac{1}{5}$ 15 $\frac{3}{5}$ 16 $\frac{1}{7}$ 17 8 18 $\frac{1}{8}$
19 4 20 $\frac{1}{4}$

Page 49

1 $6\sqrt{2}$ 2 $6\sqrt{3}$ 3 $\sqrt{5}$ 4 $6 - 3\sqrt{2}$ 5 $5\sqrt[3]{2}$ 6 $4\sqrt[3]{2} + 5$
7 $11\sqrt{2}$ 8 $13\sqrt{3}$ 9 $3\sqrt{5}$ 10 $7\sqrt{7}$ 11 0 12 $\frac{3\sqrt{2}}{4}$

Page 54

1 $x^2 + 1$ 2 $x^2 + 4x - 13$
3 $2x^3 + 3x^2 + 8x + 1$ 4 $-x^3 + 4x - 8$
5 $-3x^4 - 6x^3 + 3x$ 6 $-\frac{1}{2}x + 2x^2 + 4x^3$
7 $6x^2 + 13x + 6$ 8 $12x^2 + 28x - 5$
9 $16x^2 - 49$ 10 $4x^2 - 12x + 9$
11 $x^4 + 9x^3 + 14x^2 - 4x - 8$ 12 $2x^5 - 5x^4 + 4x^3 + 14x^2 - 15x$

Page 59

1 $3(x - 3)$ 2 $-5(x - 3)$ 3 $5y(x + 5y + 2y^4)$
4 $(x + 6)(x - 6)$ 5 $(2x + 7)(2x - 7)$ 6 $(a + 2 + 5b)(a + 2 - 5b)$
7 $(x - 3)(x^2 + 3x + 9)$ 8 $(x + 3)(x^2 - 3x + 9)$ 9 $(2a - 5)(4a^2 + 10a + 25)$
10 $3(x + 4)(x - 4)$ 11 $a(x + y)(x^2 - xy + y^2)$ 12 $2h(x + 2h)(x - 2h)$

Page 60

1 $(x + \sqrt{2})(x - \sqrt{2})$ 2 $(\sqrt{7} + x)(\sqrt{7} - x)$
3 $(\sqrt{x} + 3)(\sqrt{x} - 3)$ 4 $(x - \sqrt[3]{4})(x^2 + \sqrt[3]{4}\,x + (\sqrt[3]{4})^2)$
5 $(1 - \sqrt[3]{h})(1 + \sqrt[3]{h} + (\sqrt[3]{h})^2)$ 6 $(\sqrt[3]{x} + 3)[(\sqrt[3]{x})^2 - 3\sqrt[3]{x} + 9]$

Page 65

1 $(a + 3)^2$ 2 $(x - 5)^2$ 3 $(2x + 3y)^2$
4 $(x + 3)(x + 5)$ 5 $(a - 10)(a - 2)$ 6 Not factorable.
7 $(5x - 2)(2x - 7)$ 8 $(3x + 2)(2x - 5)$ 9 Not factorable.

Page 69

1 $x + 2$ 2 $\frac{x}{x + 2}$ 3 $\frac{x + 3}{x - 2}$ 4 $\frac{x + 5}{x - 2}$ 5 $\frac{1}{x^2 + 4}$
6 $-\frac{x + 2}{x}$ 7 1 8 $\frac{x - 2}{3x(x + 1)}$ 9 $\frac{a + b}{a - b}$ 10 1

Page 70

1 $\frac{13x}{15}$ 2 $\frac{5x}{6}$ 3 $\frac{4 - 3x}{6x^2}$ 4 $\frac{25}{6x}$
5 $\frac{2(5x + 4)}{(x - 2)(x + 2)}$ 6 $\frac{9x + 34}{x^2 - 9}$ 7 $-\frac{3x + 2}{(x - 1)(x^2 - 4)}$ 8 $\frac{2(8x^2 - 7x - 5)}{(2x + 5)^2(2x - 5)}$

LINEAR FUNCTIONS AND EQUATIONS

3

3.1 Introduction to the function concept

Suppose that you are riding in a car that is averaging 40 miles per hour. Then the distance traveled is determined by the time traveled.

$$\text{distance} = \text{rate} \times \text{time}$$

Symbolically, this relationship can be expressed by the equation

$$s = 40t$$

where s is the distance traveled in time t (measured in hours). For $t = 2$ hours, the distance traveled is

$$s = 40(2) = 80 \text{ miles}$$

Similarly, for each specific value of $t \geq 0$ the equation produces *exactly one* value for s. This correspondence between the distance s and the time t is an example of a *functional relationship*. More specifically, we say that the equation $s = 40t$ defines s as a *function* of t.

We say that $s = 40t$ defines s as a function of t because for *each* choice of t there corresponds *exactly one* value for s. We first choose a value of t. Then there is a corresponding value of s that depends on t; s is the *dependent variable* and t is the *independent variable* of the function defined by $s = 40t$.

In contrast, the equation $y^2 = 12x$ does *not* define y as a function of x; for a given value of x there is *more than one* corresponding value for y. If $x = 3$, for example, then $y^2 = 36$ and $y = 6$ or $y = -6$.

Because the variable t represents time in the equation $s = 40t$, it is reasonable to say that $t \geq 0$. This set of allowable values for the independent variable is called the **domain** of the function. The set of corresponding values for the dependent variable is called the **range** of the function.

DEFINITION OF FUNCTION

This is an important definition. Much of the work in calculus deals with the study of functions.

A **function** is a correspondence between two sets, the domain and the range, such that for each value in the domain there corresponds exactly one value in the range.

The specific letters used for the independent and dependent variables are of no consequence. Usually, we will use x for the independent variable and y for the dependent variable. Thus the equation $y = 40x$ can be used to define the same function as $s = 40t$. However, letters that are suggestive, such as t for time, can prove to be helpful.

Most of the expressions we encountered earlier in this text can be used to define functions. Here are some illustrations. Note that in each case *the domain of the function is taken as the largest set of real numbers for which the defining expression in x leads to a real value.* Unless otherwise indicated, this will be our policy throughout this text.

Explain why x cannot equal 2 or -2.

Explain why x cannot be negative.

Function given by	*Domain*
$y = 6x^4 - 3x^2 + 7x + 1$	All real numbers
$y = \dfrac{2x}{x^2 - 4}$	All reals except 2 and -2
$y = \lvert x \rvert$	All real numbers
$y = \sqrt{x}$	All real $x \geq 0$

EXAMPLE 1 Explain why the following equation defines y as a function of x and find the domain: $y = \dfrac{1}{\sqrt{x - 1}}$.

Solution For each allowable x the expression $\dfrac{1}{\sqrt{x-1}}$ produces just one y-value. Therefore, the given equation defines a function. To find the domain observe that $x - 1$ cannot be negative because the even root of a negative value is not a real number. Thus $x - 1 \geq 0$, or $x \geq 1$. But $x = 1$ produces division by zero, so we must exclude it as well. Hence the domain consists of all $x > 1$.

EXAMPLE 2 Find the domain of the function given by $y = \sqrt{1 - x^2}$.

See Section 1.5 for methods to solve such inequalities.

Solution To avoid square roots of negative numbers we must have $1 - x^2 \geq 0$, or $x^2 - 1 \leq 0$. Now factor $x^2 - 1$ and solve the inequality

$(x+1)(x-1) \leq 0$. The domain consists of all values of x from -1 through 1; that is, $-1 \leq x \leq 1$.

TEST YOUR UNDERSTANDING

Decide whether the given equation defines y to be a function of x. For each function, find the domain.

1 $y = (x+2)^2$

2 $y = \dfrac{1}{(x+2)^2}$

3 $y = \dfrac{1}{x^2+2}$

4 $y = \pm 3x$

5 $y = \sqrt{x^2+x-6}$

6 $y^2 = x^2$

7 $y = \dfrac{x}{|x|}$

8 $y = \dfrac{1}{\sqrt{x^2+2x+1}}$

Thus far only equations have been used to defined functions. One could gain the impression that equations are the only way to state functions. This impression is *not* correct. Since a function defined by an equation is the *correspondence* between the variables, such correspondences can be stated in many ways. Here, for example, is a *table of values*. The table defines y to be a function of x because for each domain value there corresponds *exactly one* value for y.

Sometimes we say that an equation, such as $y = 40x$, is a function. Such informal language is commonly used and should not cause difficulty.

x	1	2	5	-7	23	$\sqrt{2}$
y	6	-6	6	-4	0	5

Instead of using a single equation to define a function, there will be times when a function is defined in terms of more than one equation. For instance, the following three equations define a function whose domain is the set of all real numbers.

$$y = \begin{cases} 1 & \text{for } x \leq -6 \\ x^2 & \text{for } -6 < x < 0 \\ 2x+1 & \text{for } x \geq 0 \end{cases}$$

Note that if $x = 5$ the corresponding y-value comes from $y = 2x + 1$, namely 11; if $x = -5$, $y = (-5)^2 = 25$; and if $x = -7$, $y = 1$.

EXAMPLE 3 Decide whether these two equations define y to be a function of x.

$$y = \begin{cases} 3x-1 & \text{if } x \leq 1 \\ 2x+1 & \text{if } x \geq 1 \end{cases}$$

Solution If $x = 1$, the first equation gives $y = 3(1) - 1 = 2$. For $x = 1$ the second equation gives $y = 2(1) + 1 = 3$. Since we have two different y-values for the same x-value, the equations do *not* define a function.

EXAMPLE 4 Use the definition of absolute value (Section 1.6) to explain why the equation $y = |x|$ defines a function whose domain consists of all real numbers x.

Solution The definition of absolute value says that $|x| = x$ for $x \geq 0$ and $|x| = -x$ for $x < 0$. Thus $y = |x|$ defines a function because each real value x produces just one corresponding y-value.

A useful way to refer to a function is to name it by using a specific letter, such as f, g, F, and the like. For example, the function given by $y = \frac{1}{x - 3}$ may be referred to as f. The domain of f is the set of all real numbers not equal to 3; that is, $x \neq 3$. We write

$$f(x) = \frac{1}{x - 3}$$

to mean the value of the function f at x is $\frac{1}{x - 3}$. For example:

$$f(4) = 1 \text{ means that when } x = 4,\ y = 1$$

$$f(4) = 1 \text{ is read as "} f \text{ of 4 is 1" or "} f \text{ at 4 is 1"}$$

CAUTION: Note that $f(x)$ does *not* mean that we are to multiply f by x; f does not stand for a number.

We use $f(x)$ to represent the range value for the specific value of x given in the parentheses. In this illustration f stands for the function that is given by $y = \frac{1}{x - 3}$. For $x \neq 3$, $f(x) = \frac{1}{x - 3}$. Then

Note that $f(3)$ is undefined. Can you explain why?

$$f(0) = \frac{1}{0 - 3} = -\frac{1}{3} \text{ and}$$

$$f(9) = \frac{1}{9 - 3} = \frac{1}{6} \leftarrow \begin{cases} 9 \text{ is the } \textit{input} \\ \frac{1}{6} \text{ is the } \textit{output} \end{cases}$$

Let us explore the function notation with another example. Let g be the function defined by $y = g(x) = x^2$. Then we have

$$g(1) = 1^2 = 1$$

$$g(2) = 2^2 = 4$$

$$g(3) = 3^2 = 9$$

Note that $g(1) + g(2) \neq g(3)$. To write $g(1) + g(2) = g(1 + 2)$ would be to assume, *incorrectly*, that the distributive property holds for the functional notation. This is not true in general, which comes as no great surprise since g is not a number.

Keep in mind that the variable x in $g(x) = x^2$ is only a *placeholder*. Any letter could serve the same purpose. For example, $g(t) = t^2$ and $g(z) = z^2$ both define the same function with domain all real numbers.

EXAMPLE 5 For the function g defined by $g(x) = \frac{1}{x}$, find: **(a)** $3g(x)$ **(b)** $g(3x)$ **(c)** $3 + g(x)$ **(d)** $g(3) + g(x)$ **(e)** $g(3 + x)$ **(f)** $g\left(\frac{1}{x}\right)$

Solution **(a)** $3g(x) = 3 \cdot \frac{1}{x} = \frac{3}{x}$ **(b)** $g(3x) = \frac{1}{3x}$ **(c)** $3 + g(x) = 3 + \frac{1}{x}$ **(d)** $g(3) + g(x) = \frac{1}{3} + \frac{1}{x}$ **(e)** $g(3 + x) = \frac{1}{3 + x}$
(f) $g\left(\frac{1}{x}\right) = \frac{1}{\frac{1}{x}} = x$

EXAMPLE 6 Let $g(x) = x^2$. Evaluate and simplify the *difference quotient:* $\frac{g(x) - g(4)}{x - 4}$; $x \neq 4$.

Difference quotients will be used in the study of calculus.

Solution $g(x) = x^2$ and $g(4) = 16$

$$\begin{aligned}\frac{g(x) - g(4)}{x - 4} &= \frac{x^2 - 16}{x - 4} \\ &= \frac{(x - 4)(x + 4)}{x - 4} \\ &= x + 4\end{aligned}$$

EXAMPLE 7 Let $g(x) = \frac{1}{x}$. Evaluate and simplify the *difference quotient:* $\frac{g(4 + h) - g(4)}{h}$; $h \neq 0$.

Solution $g(4 + h) = \frac{1}{4 + h}$ and $g(4) = \frac{1}{4}$

$$\begin{aligned}\frac{g(4 + h) - g(4)}{h} &= \frac{\frac{1}{4 + h} - \frac{1}{4}}{h} \\ &= \frac{\frac{4 - (4 + h)}{4(4 + h)}}{h} \\ &= \frac{-h}{4h(4 + h)} \\ &= -\frac{1}{4(4 + h)}\end{aligned}$$

CAUTION: LEARN TO AVOID MISTAKES LIKE THESE

In each of the following, the function f is defined by $f(x) = 3x^2 - 4$.

WRONG	RIGHT
$f(0) = 0$	$f(0) = 3(0)^2 - 4 = -4$
$f(-2) = -f(2)$	$f(-2) = 3(-2)^2 - 4 = 8$ $-f(2) = -[3(2)^2 - 4] = -8$

In each case $f(x) = 3x^2 - 4$.

WRONG	RIGHT
$f\left(\frac{1}{2}\right) = \frac{1}{f(2)}$	$f\left(\frac{1}{2}\right) = 3\left(\frac{1}{2}\right)^2 - 4 = -\frac{13}{4}$ $\frac{1}{f(2)} = \frac{1}{8}$
$[f(2)]^2 = f(4)$	$[f(2)]^2 = 8^2 = 64$ $f(4) = 3(4)^2 - 4 = 44$
$2 \cdot f(5) = f(10)$	$2 \cdot f(5) = 2[3(5)^2 - 4] = 142$ $f(10) = 3(10)^2 - 4 = 296$
$f(5) + f(2) = f(7)$	$f(5) + f(2) = 3(5)^2 - 4 + 3(2)^2 - 4 = 79$ $f(7) = 3(7)^2 - 4 = 143$

EXERCISES 3.1

Decide whether the given equation defines y to be a function of x. For each function, find the domain.

1 $y = x^3$

2 $y = \sqrt[3]{x}$

3 $y = \dfrac{1}{\sqrt{x}}$

4 $y = |2x|$

5 $y^2 = 2x$

6 $y = x \pm 3$

7 $y = \dfrac{1}{x+1}$

8 $y = \dfrac{x-2}{x^2+1}$

9 $y = \dfrac{1}{1 \pm x}$

10 $y = \sqrt{x^2 - 4}$

11 $y = \dfrac{1}{\sqrt{x^2 - 4}}$

12 $y = \dfrac{1}{\sqrt[3]{x^2 - 4}}$

Decide whether the given table defines a function with domain x and range y. If it does not define a function, give a reason.

13

x	y
3	4
2	3
1	2
3	1

14

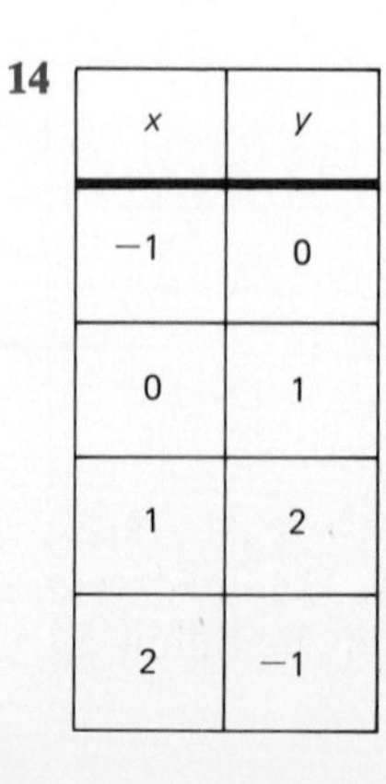

x	y
−1	0
0	1
1	2
2	−1

Classify each statement as true or false. If it is false, correct the right side to get a correct equation. For each of these statements, use $f(x) = -x^2 + 3$.

15 $f(3) = -6$

16 $f(2)f(3) = -33$

17 $3f(2) = -33$

18 $f(3) + f(-2) = 2$

19 $f(3) - f(2) = -5$

20 $f(2) - f(3) = 11$

21 $f(x) - f(4) = -(x-4)^2 + 3$

22 $f(x) - f(4) = x^2 + 19$

23 $f(4+h) = -h^2 - 8h - 13$

24 $f(4+h) = -h^2 - 10$

In Exercises 25–36, find (a) $f(-1)$; (b) $f(0)$; and (c) $f(\frac{1}{2})$, if they exist.

25 $f(x) = 2x - 1$

26 $f(x) = -5x + 6$

27 $f(x) = x^2$

28 $f(x) = x^2 - 5x + 6$

29 $f(x) = x^3 - 1$

30 $f(x) = (x-1)^2$

31 $f(x) = x^4 + x^2$

32 $f(x) = -3x^3 + \frac{1}{2}x^2 - 4x$

33 $f(x) = \dfrac{1}{x-1}$

34 $f(x) = \sqrt{x}$

35 $f(x) = \dfrac{1}{\sqrt[3]{x}}$

36 $f(x) = \dfrac{1}{3|x|}$

37 For $g(x) = x^2 - 2x + 1$, find:
(a) $g(10)$ (b) $5g(2)$ (c) $g(\frac{1}{2}) + g(\frac{1}{3})$ (d) $g(\frac{1}{2} + \frac{1}{3})$.

38 Let h be given by $h(x) = x^2 + 2x$. Find $h(3)$ and $h(1) + h(2)$ and compare.

39 Let h be given by $h(x) = x^2 + 2x$. Find $3h(2)$ and $h(6)$ and compare.

40 Let h be given by $h(x) = x^2 + 2x$. Find **(a)** $h(2x)$ **(b)** $h(2 + x)$ **(c)** $h\left(\frac{1}{x}\right)$ **(d)** $h(x^2)$.

▬ *In Exercises 41–46, find the difference quotient* $\frac{f(x) - f(3)}{x - 3}$ *and simplify for the given function f.*

41 $f(x) = x^2$

42 $f(x) = x^2 - 1$

43 $f(x) = \frac{1}{x}$

44 $f(x) = \sqrt{x}$

45 $f(x) = 2x + 1$

46 $f(x) = -x^3 + 1$

▬ *In Exercises 47–52, find the difference quotient* $\frac{f(2 + h) - f(2)}{h}$ *and simplify.*

47 $f(x) = x$

48 $f(x) = -x + 3$

49 $f(x) = -x^2$

*50 $f(x) = \sqrt{x + 2}$

*51 $f(x) = \frac{1}{x^2}$

*52 $f(x) = \frac{1}{x - 1}$

3.2 The rectangular coordinate system

A great deal of information can be learned about a functional relationship by studying its graph. A fundamental objective of this course is to acquaint you with the graphs of some important functions, as well as to develop basic graphing procedures. First we need to review the structure of a **rectangular coordinate system.**

In a plane take any two lines that intersect at right angles and call their point of intersection the **origin**. Let each of these lines be a number line with the origin corresponding to zero for each line. Unless otherwise specified, the unit length is the same on both lines. On the horizontal line the positive direction is taken to be to the right of the origin, and on the vertical line it is taken to be above the origin. Each of these two lines will be referred to as an **axis** of the system (plural: **axes**).

The union of algebra and geometry, credited to French mathematician René Descartes (1596–1650), led to the development of analytic geometry. In his honor, we often refer to this as the *Cartesian coordinate system,* or simply the *Cartesian plane.*

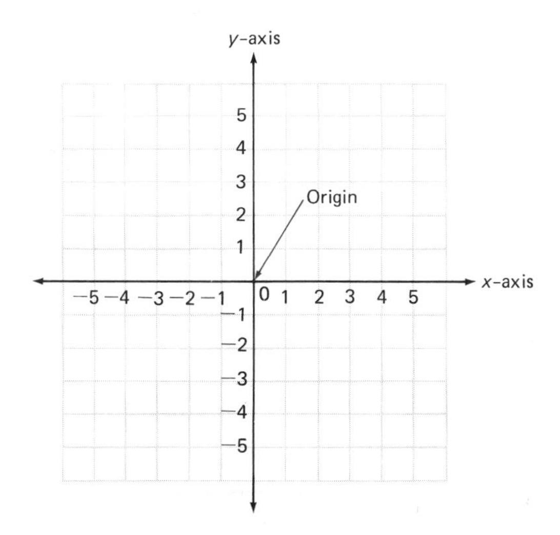

The horizontal line is usually called the ***x*-axis**, and the vertical line the ***y*-axis**. The axes divide the plane into four regions called **quadrants**. The quadrants are numbered in a counterclockwise direction as shown in the following figure.

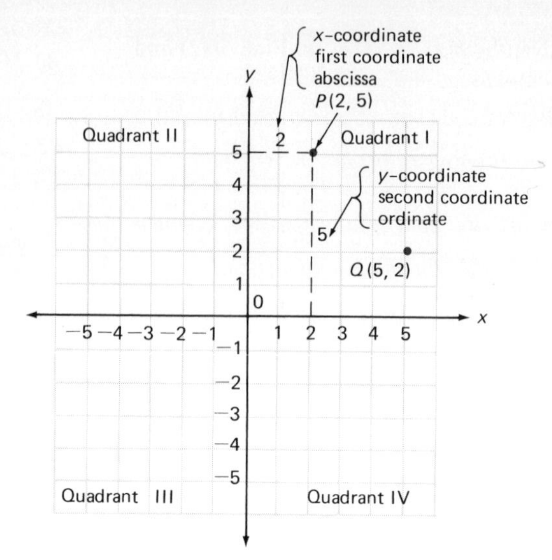

Note that the ordered pair (2, 5) is not the same as the pair (5, 2). Each gives the coordinates of a different point on the plane.

The points in the plane (denoted by the capital letters) are matched with pairs of numbers, referred to as the **coordinates** of these points. For example, starting at the origin, P can be reached by moving 2 units to the right, parallel to the x-axis; then 5 units up, parallel to the y-axis. Thus the first coordinate, 2, of P is called the ***x*-coordinate** (another name is "abscissa") and the second coordinate, 5, is the ***y*-coordinate** (also called "ordinate"). We say that the *ordered pair of numbers* (2, 5) is the coordinates of P.

All points in the first quadrant are to the right and above the origin, and therefore have positive coordinates. Any point in quadrant II is to the left and above the origin and therefore has a negative x-coordinate and a positive y-coordinate. In quadrant III both coordinates are negative, and in the fourth quadrant they are positive and negative, respectively.

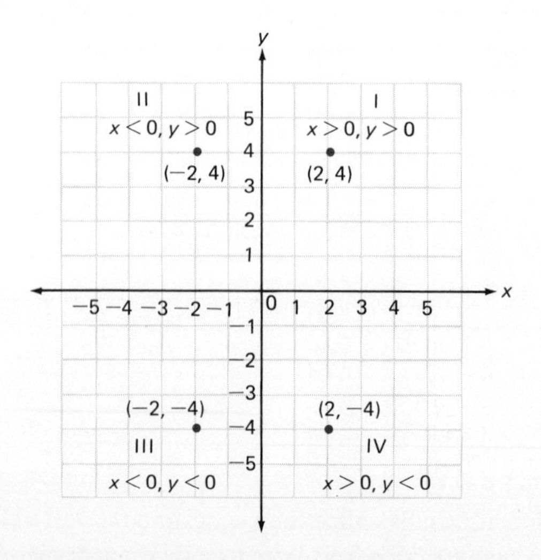

EXAMPLE 1 Find the coordinates of the given points. Also state the quadrant or axis in which each point is located.

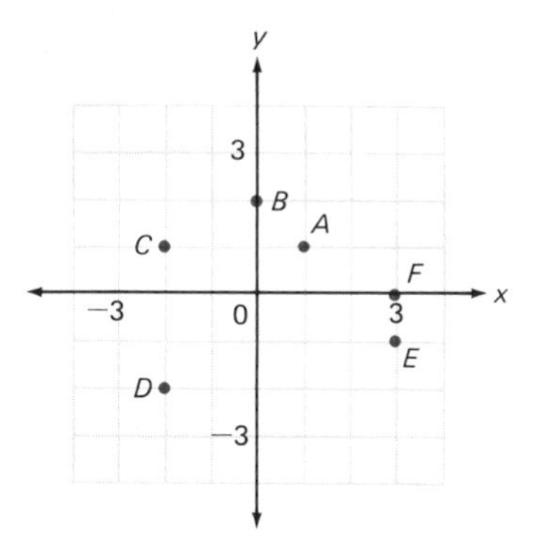

Usually, the points on the axes are labeled with just a single number. In that case it is understood that the missing coordinate is 0. Thus in Example 1, B has coordinates (0, 2) and the coordinates of point F are (3, 0).

Solution Point A is located at (1, 1), in the first quadrant. Point B is at (0, 2), on the y-axis. Point C is at (−2, 1), in the second quadrant. Point D is at (−2, −2), in the third quadrant. Point E is at (3, −1), in the fourth quadrant. Point F is at (3, 0), on the x-axis.

The equality $y = x + 2$ is an equation in two variables. When a specific value for x is substituted into this equation, we get a corresponding y-value. For example, substituting 3 for x gives $y = 3 + 2 = 5$. We therefore say that the ordered pair (3, 5) *satisfies* the equation $y = x + 2$.

In the following table of values there are six more ordered pairs that satisfy this equation. Note that the ordered pairs in the table have been written without the usual parentheses. These ordered pairs have also been *plotted* in a rectangular system.

x	$y = x + 2$
−3	−1
−2	0
−1	1
0	2
1	3
2	4

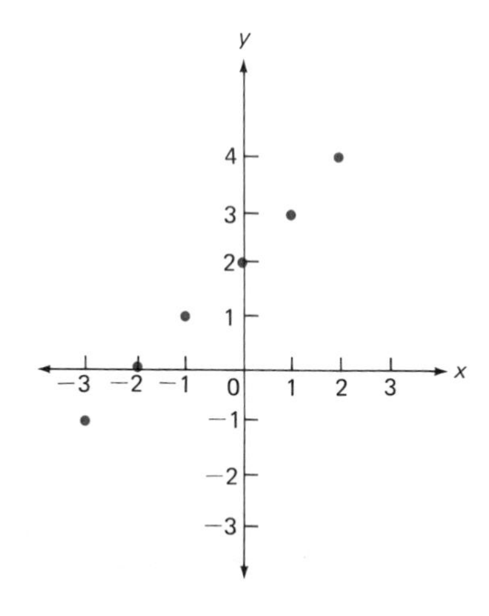

To graph an equation in the variables x and y means to locate all the points in a rectangular system whose coordinates satisfy the given equation. There are an infinite number of ordered pairs that satisfy $y = x + 2$, and all are located on the same straight line. Since a line is endless, we draw a partial graph by joining the specific points previously located.

The arrowheads in the figure suggest that the line continues endlessly in both directions.

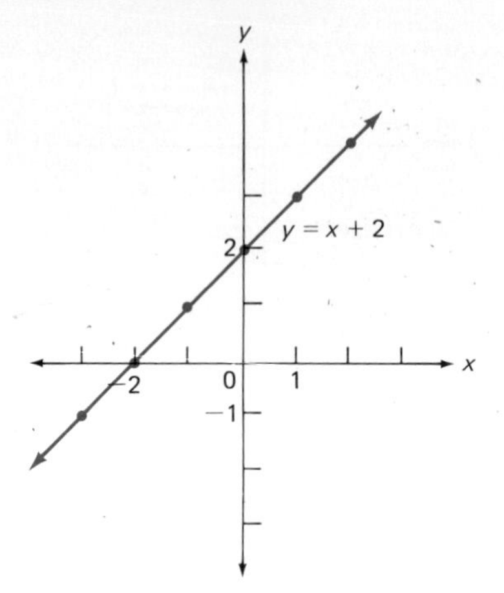

The straight line contains exactly those points whose ordered pairs (x, y) satisfy the equation $y = x + 2$. Any point not on the line has an ordered pair (x, y) where $y \neq x + 2$. Thus (1, 5) is not on the line, since $5 \neq 1 + 2$.

The graph of $y = x + 2$ is a straight line and the equation is called a *linear equation*. Since two points determine a line, a convenient way to graph a line is to locate its **intercepts**. The ***x*-intercept** for $y = x + 2$ is -2, the abscissa of the point where the line crosses the x-axis. The ***y*-intercept** is 2, the ordinate of the point where the line crosses the y-axis.

EXAMPLE 2 Graph the linear equation $y = 2x - 1$ using the intercepts.

Solution To find the x-intercept, let $y = 0$.

When the line crosses the x-axis, the y-value is 0.

When the line crosses the y-axis, the x-value is 0.

$$2x - 1 = 0$$

$$2x = 1$$

$$x = \tfrac{1}{2} \longleftarrow \text{the } x\text{-intercept}$$

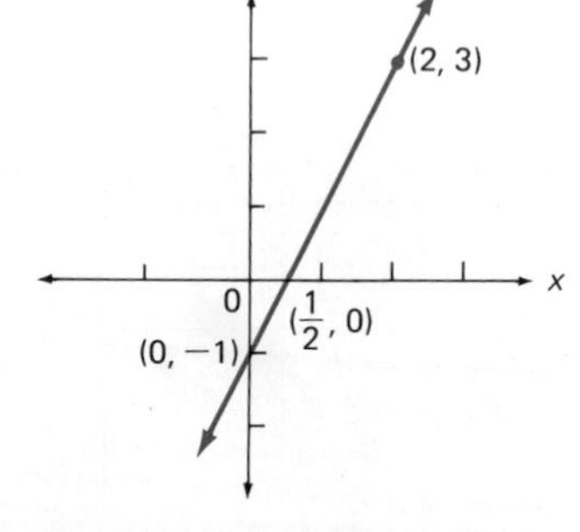

To find the y-intercept, let $x = 0$.

It is generally wise to locate a third point to verify your work. Thus for $x = 2$, $y = 3$, and the line passes through the point (2, 3).

$$y = 2(0) - 1$$

$$y = -1 \longleftarrow \text{the } y\text{-intercept}$$

Plot the points $(\frac{1}{2}, 0)$ and $(0, -1)$ and draw the line through them to determine the graph.

Note: If the equation $y = 2x - 1$ were given in the equivalent form $y - 2x = -1$, the intercepts can be found as before. Thus let $y = 0$ to obtain $0 - 2x = -1$ or $x = \frac{1}{2}$, the x-intercept. Likewise, if $x = 0$, then $y - 2(0) = -1$ or $y = -1$, the y-intercept.

An equation such as $y = 2x - 1$ defines y as a *linear function* of x. Usually the domain of such a function is the set of all real numbers. However, at times we may wish to limit the domain of a function. For example, the graph of $y = 2x - 1$ for $-2 \leq x \leq 1$ is a line segment with endpoints at $(-2, -5)$ and $(1, 1)$ as in the following figure.

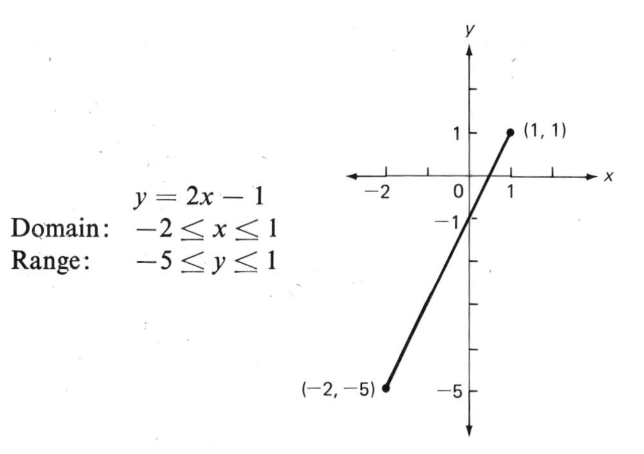

The equation $y = 2x - 1$ can also be used to identify two-dimensional regions in the plane. First note that the graph of $y = 2x - 1$ in Example 2 divides the plane into two *half-planes.* These two half-planes represent the graphs for the two statements of inequality, $y < 2x - 1$ (below the line) and $y > 2x - 1$ (above the line). To show these graphs we would use a dashed line for $y = 2x - 1$ and shade the appropriate half-plane, as in the following figures.

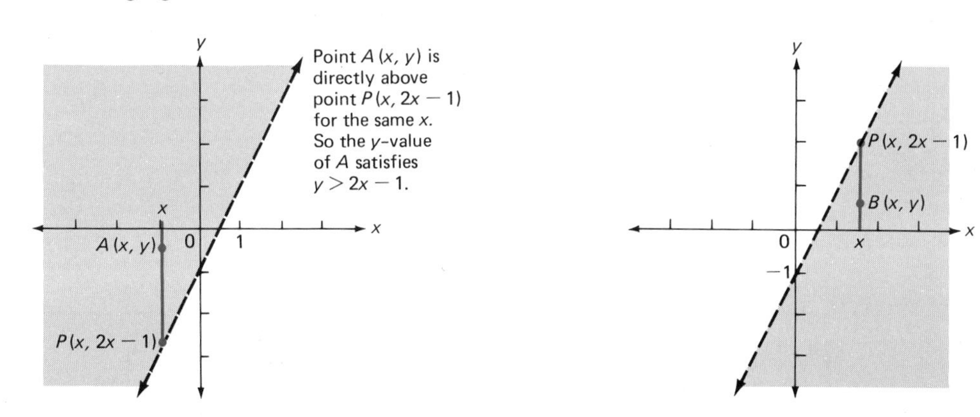

Point $B\,(x, y)$ is directly below point $P\,(x, 2x - 1)$ for the same x. So the y-value of B satisfies $y < 2x - 1$.

To show a graph for an inequality such as $y \leq 2x - 1$, we would draw both the line and the half-plane.

EXERCISES 3.2

Copy and complete each table of values. Then graph the line given by the equation.

1 $y = x - 2$

x	−3	−2	−1	0	1	2
y						

2 $y = -x + 1$

x	−3	−2	−1	0	1	2	3
y							

3 $y = 2x - 4$

x	-2	-1	0	1	2
y					

4 $y = -2x + 3$

x	-2	-1	0	1	2
y					

Find the x- and y-intercepts and use these to graph each of the following lines.

5 $x + 2y = 4$ **6** $2x + y = 4$
7 $x - 2y = 4$ **8** $2x - y = 4$
9 $2x - 3y = 6$ **10** $3x + y = 6$
11 $y = -3x - 9$ **12** $y + 2x = -5$
13 $y = 2x - 1$

14 Sketch the following on the same set of axes.
(a) $y = x$
(b) By adding 1 to each y-value (ordinate) in part (a), graph $y = x + 1$. In other words, *shift* each point of $y = x$ one unit up.
(c) By subtracting 1 from each y-value in part (a), graph $y = x - 1$. That is, shift each point of $y = x$ one unit down.

15 Repeat Exercise 14 for (a) $y = -x$; (b) $y = -x + 1$; (c) $y = -x - 1$.

16 Sketch the following on the same set of axes.
(a) $y = x$
(b) By multiplying each y-value in part (a) by 2, graph $y = 2x$. In other words, *stretch* each y-value of $y = x$ to twice its size.
(c) Graph $y = 2x + 3$ by shifting $y = 2x$ three units upward.

Graph the points that satisfy each equation for the given values of x.

17 $y = \frac{1}{2}x; \quad -6 \leq x \leq 6$
18 $y = -2x + 1; \quad -2 \leq x \leq 2$
19 $y = 3x - 5; \quad 1 \leq x \leq 4$
20 $y = x^2; \quad x = -3, -2, -1, 0, 1, 2, 3$
21 (a) $y = x^2; \quad x = -2, -1, -\frac{1}{2}, 0, \frac{1}{2}, 1, 2$
(b) $y = x^2; \quad -2 \leq x \leq 2$
22 (a) $y = |x|; \quad x = -3, -1, 0, 2, 4$
(b) $y = |x|; \quad -3 \leq x \leq 4$
23 (a) $y = \frac{1}{x}; \quad x = \frac{1}{2}, 1, \frac{3}{2}, 2, \frac{5}{2}, 3$
(b) $y = \frac{1}{x}; \quad \frac{1}{2} \leq x \leq 3$
24 (a) $y = \sqrt{x}; \quad x = 0, \frac{1}{4}, 1, \frac{25}{16}, \frac{9}{4}, 4$
(b) $y = \sqrt{x}; \quad 0 \leq x \leq 4$

Graph each inequality.

25 $y > x + 2$ **26** $y \leq x + 2$
27 $y \geq x - 1$ **28** $y < x - 1$
29 $2x + y - 4 > 0$ **30** $x - 2y + 2 \leq 0$

3.3

Slope

The adjoining figure shows the graph of the linear equation $y = 2x - 4$, including the coordinates of four specific points. From the diagram you will see that the y-value increases 2 units each time that the x-value increases by 1 unit. The ratio of this change in y compared to the corresponding change in x is $\frac{2}{1} = 2$. Using the coordinates of the points (3, 2) and (2, 0), we have the following:

For convenience, the change in y may be referred to as "Δy" (read "delta y"); the change in x is denoted as "Δx" (read "delta x").

$$\frac{\Delta y}{\Delta x} = \frac{2-0}{3-2} = \frac{2}{1} = 2$$

$$\frac{\text{change in } y\text{-values}}{\text{change in } x\text{-values}} = \frac{2-0}{3-2} = \frac{2}{1} = 2$$

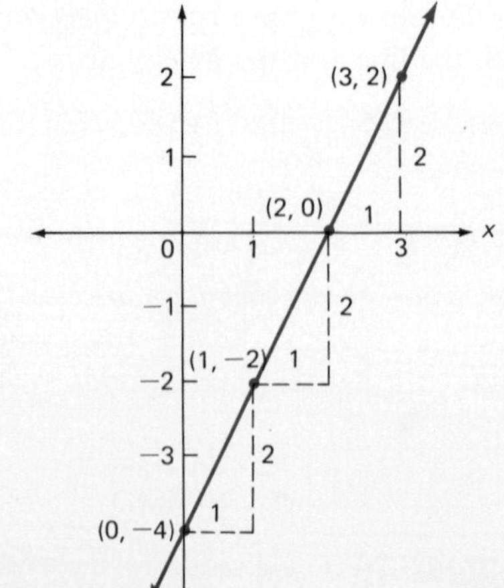

We call this ratio the *slope* of the line, defined as follows.

DEFINITION OF SLOPE

If two points (x_1, y_1) and (x_2, y_2) are on a line ℓ, then the slope m of line ℓ is defined by

$$m = \frac{y_2 - y_1}{x_2 - x_1}, \qquad x_2 \neq x_1$$

Notice that in the definition $x_2 - x_1$ cannot be zero; that is, $x_2 \neq x_1$. The only time that $x_2 = x_1$ is when the line is vertical.

In the following figure the coordinates of two points A and B have been labeled (x_1, y_1) and (x_2, y_2). The change in the y direction from A to B is given by the difference $y_2 - y_1$; the change in the x direction is $x_2 - x_1$. If a different pair of points is chosen, such as P and Q, then the ratio of these differences is still the same because the resulting triangles (ABC and PQR) are similar. That is, since corresponding sides of similar triangles are proportional, we have

$$m = \frac{y_2 - y_1}{x_2 - x_1} = \frac{AC}{CB} = \frac{PR}{RQ}$$

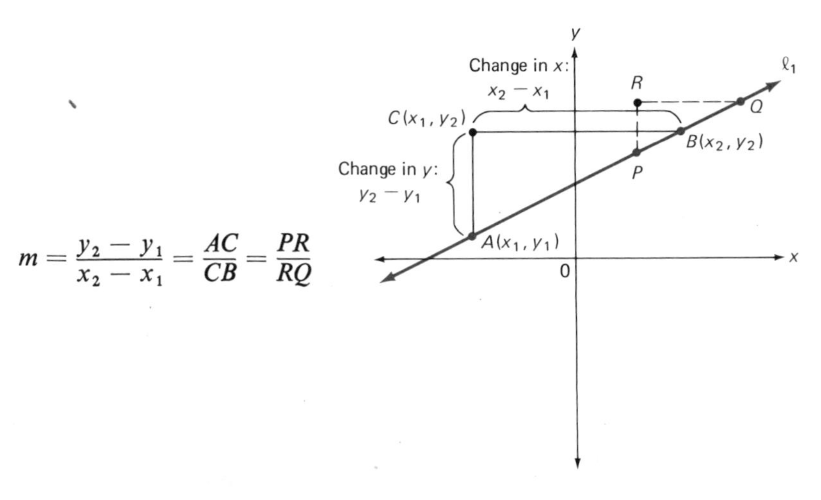

This discussion shows that there can only be one slope for a given line.

It may be helpful to think of the slope of a line in any of these ways:

$$m = \frac{y_2 - y_1}{x_2 - x_1} = \frac{\text{change in } y}{\text{change in } x} = \frac{\text{vertical change}}{\text{horizontal change}}$$

Another descriptive language for slope is $m = \frac{\text{rise}}{\text{run}}$, where *rise* is the vertical change and *run* is the horizontal change.

EXAMPLE 1 Find the slope of line ℓ determined by the points $(-3, 4)$ and $(1, -6)$.

Solution Use $(x_1, y_1) = (-3, 4)$; $(x_2, y_2) = (1, -6)$. Then

$$m = \frac{-6 - 4}{1 - (-3)} = -\frac{10}{4} = -\frac{5}{2}$$

Note: It makes no difference which of the two points is called (x_1, y_1) or (x_2, y_2) since the ratio will still be the same. If $(x_1, y_1) = (1, -6)$ and $(x_2, y_2) = (-3, 4)$, for example, then $\frac{y_2 - y_1}{x_2 - x_1} = \frac{4 - (-6)}{-3 - 1} = -\frac{5}{2} = m$.

CAUTION: Do not mix up the coordinates like this:

$$\frac{y_2 - y_1}{x_1 - x_2} = \frac{4 - (-6)}{1 - (-3)} = \frac{5}{2}$$

This is the *negative* of the slope.

Reading from left to right, a rising line has a positive slope and a falling line has a negative slope.

EXAMPLE 2 Graph the line with slope 3 that passes through the point $(-2, -2)$.

Solution Write $3 = \frac{3}{1} = \frac{\text{change in } y}{\text{change in } x}$. Now start at $(-2, -2)$ and move 3 units up and 1 unit to the right. This locates the point $(-1, 1)$. Draw the straight line through these two points.

Alternatively, start at $(-2, -2)$ and move 3 units down and 1 unit to the left to locate $(-3, -5)$.

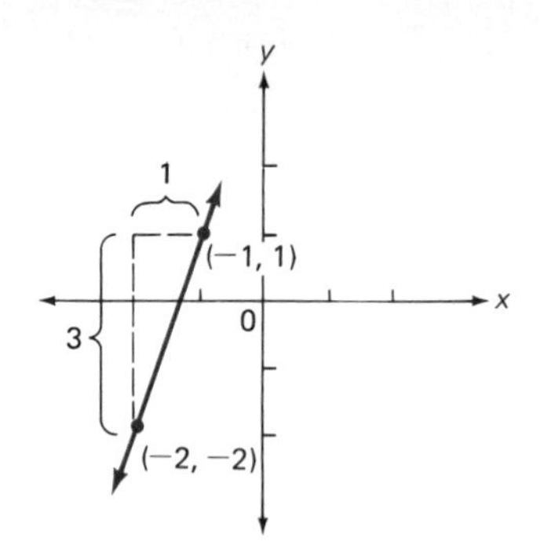

TEST YOUR UNDERSTANDING

Find the slopes of the lines determined by the given pairs of points.

1 $(1, 5); (4, 6)$

2 $(3, -5); (-3, 3)$

3 $(-2, -3); (-1, 1)$

4 $(-1, 0); (0, 1)$

Draw the line through the given point and with the given slope.

5 $(0, 0); m = 2$

6 $(-3, 4); m = -\frac{3}{2}$

In the adjoining figure ℓ is a horizontal line. Since ℓ is parallel to the x-axis, $y_1 = y_2$ or $y_1 - y_2 = 0$, and the slope is therefore 0:

$$m = \frac{y_2 - y_1}{x_2 - x_1} = \frac{0}{x_2 - x_1} = 0$$

CAUTION: Do not confuse a slope of 0 for horizontal lines with no slope for vertical lines.

Now consider a vertical line ℓ. Since ℓ is parallel to the y-axis, $x_1 = x_2$ or $x_1 - x_2 = 0$. Then, since division by 0 is undefined, the formula $m = \frac{y_2 - y_1}{x_2 - x_1}$ does not apply. Consequently, it is agreed that *vertical lines do not have slope*, as in the following figure.

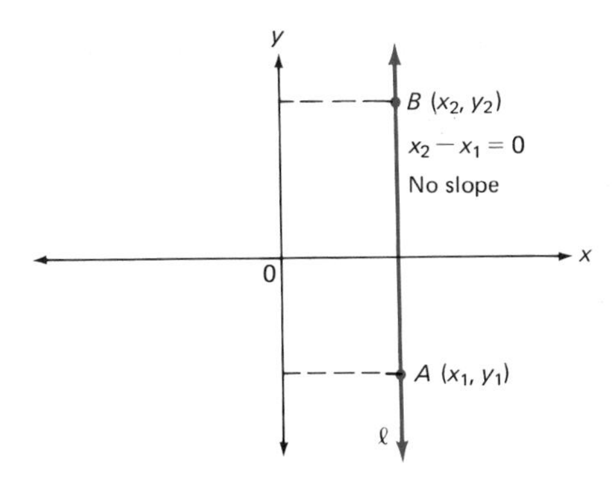

Two nonvertical lines are parallel if and only if they have the same slope. The slope property for perpendicular lines is not as obvious. The adjoining figure suggests the following (see Exercise 35):

Two lines not parallel to the coordinate axes are perpendicular if and only if their slopes are negative reciprocals of one another.

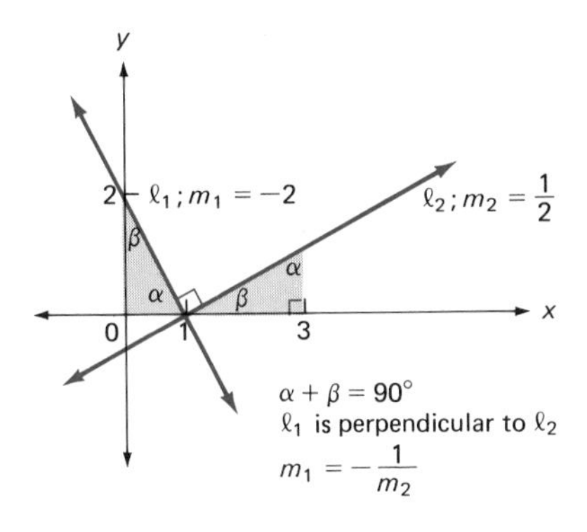

EXAMPLE 3 In the figure, line ℓ_1 has slope $\frac{2}{3}$ and is perpendicular to ℓ_2. If the lines intersect at $P(-1, 4)$, use the slope of ℓ_2 to find the coordinates of another point on ℓ_2.

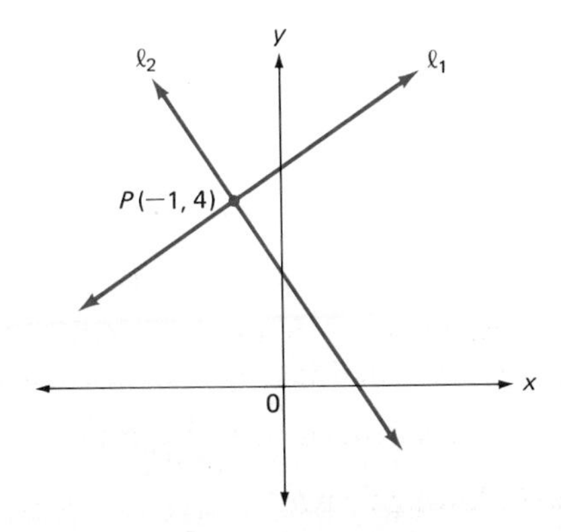

Solution Since the lines are perpendicular, the slope of ℓ_2 is $-\frac{3}{2}$. Now start at P, count 3 units downward and 2 units to the right to reach point (1, 1) on ℓ_2. Other solutions are possible.

EXERCISES 3.3

1 Use the coordinates of each of the following pairs of points to find the slope of ℓ.

(a) A, C **(b)** B, D
(c) C, D **(d)** A, E
(e) B, E **(f)** C, E

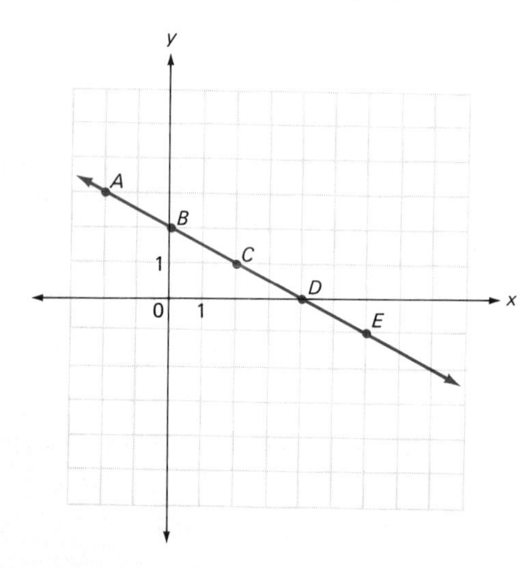

Compute the slope, if it exists, for the line determined by the given pair of points.

2 (3, 4); (2, −5) **3** (4, 3); (−5, 2)
4 (−7, 6); (−7, 106) **5** (6, −7); (106, −7)
6 $(5\sqrt{2}, 3\sqrt{8})$; $(\sqrt{2}, \sqrt{8})$
7 $(2, -\frac{3}{4})$; $(-\frac{1}{3}, \frac{2}{3})$ **8** $(\frac{1}{2}, 0.1)$; $(-9, 0.1)$
9 $(\sqrt{3}, 8)$; $(-1, 6)$ **10** $(-9, \frac{1}{2})$; $(2, \frac{1}{2})$

In Exercises 11–20, draw the line through the given point having slope m.

11 (0, 0); $m = 2$ **12** (0, 0); $m = -\frac{1}{2}$
13 (0, 2); $m = \frac{3}{4}$ **14** $(-\frac{1}{2}, 0)$; $m = -1$
15 (−3, 4); $m = -\frac{1}{4}$ **16** (3, −4); $m = 4$
17 $(-2, \frac{3}{2})$; $m = 0$ **18** (1, 1); $m = 2$
19 (5, −3); no slope **20** $(-\frac{3}{4}, -\frac{1}{2})$; $m = 1$

21 In the same coordinate system draw five lines, each with slope −2, through points (−2, 0), (−1, 0), (0, 0), $(\frac{1}{2}, 0)$, and (1, 1), respectively.

22 Follow the instructions of Exercise 21 using $m = \frac{1}{3}$ and (2, 2), (−2, 2), (−2, −2), (2, −2), (0, 0).

23 In the same coordinate system, draw the line:
(a) through (1, 0) with $m = -1$
(b) through (0, 1) with $m = 1$
(c) through (−1, 0) with $m = -1$
(d) through (0, −1) with $m = 1$

24 Use the instructions of Exercise 23 for the following:
(a) (1, 1); slope 0 **(b)** $(0, \frac{1}{2})$; slope 0
(c) (1, 1); slope $\frac{1}{2}$ **(d)** $(2, \frac{1}{2})$; slope $\frac{1}{2}$

25 Graph each of the lines with the following slopes through the point (5, −3):
$m = -2$ $m = -1$ $m = 0$ $m = 1$ $m = 2$

26 Line ℓ passes through (−4, 5) and (8, −2).
(a) Draw the line through (0, 0) perpendicular to ℓ.
(b) What is the slope of any line perpendicular to line ℓ?
(c) Draw the four lines, each perpendicular to ℓ, through the points (−4, 5), $(4, \frac{1}{3})$, $(0, \frac{8}{3})$, and (8, −2).

27 Why is the line determined by the points (6, −5) and (8, −8) parallel to the line through (−3, 12) and (1, 6)?

28 Verify that the points $A(1, 2)$, $B(4, -1)$, $C(2, -2)$, and $D(-1, 1)$ are the vertices of a parallelogram. Sketch.

29 Consider the four points $P(5, 11)$, $Q(-7, 16)$, $R(-12, 4)$, and $S(0, -1)$. Show that the four angles of the quadrilateral $PQRS$ are right angles. Also show that the diagonals are perpendicular.

30 Lines ℓ_1 and ℓ_2 are perpendicular and intersect at point (−2, −6). ℓ_1 has slope $-\frac{2}{3}$. Use the slope of ℓ_2 to find the y-intercept of ℓ_2.

31 Any horizontal line is perpendicular to any vertical line. Why were such lines excluded from the result, which states that lines are perpendicular if and only if their slopes are negative reciprocals?

32 Find t if the line through $(-1, 1)$ and $(3, 2)$ is parallel to the line through $(0, 6)$ and $(-8, t)$.

33 Find t if the line through $(-1, 1)$ and $(1, \frac{1}{2})$ is perpendicular to the line through $(1, \frac{1}{2})$ and $(7, t)$. Use the fact that two perpendicular lines have slopes that are negative reciprocals of one another.

*34 (a) Prove that nonvertical parallel lines have equal slopes by considering two parallel lines ℓ_1, ℓ_2 as in the figure. On ℓ_1 select points A and B, and choose A' and B' on ℓ_2. Now form the appropriate right triangles ABC and $A'B'C'$ using points C and C' on the x-axis. Prove they are similar and write a proportion to show the slopes of ℓ_1 and ℓ_2 are equal.

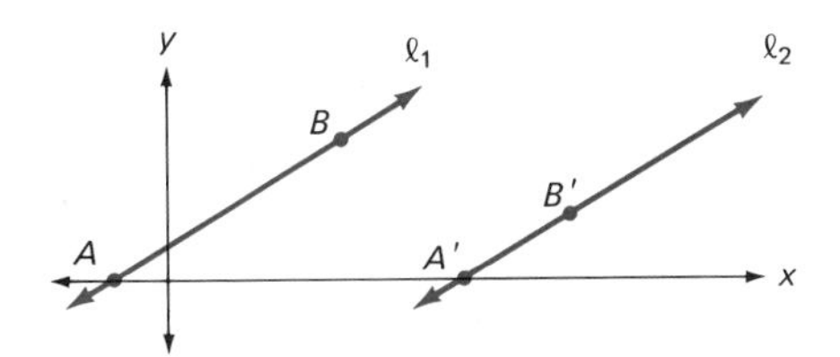

(b) Why is the converse of the fact given in part (a) also true?

*35 This exercise gives a proof that if two lines are perpendicular, then they have slopes that are negative reciprocals of one another.

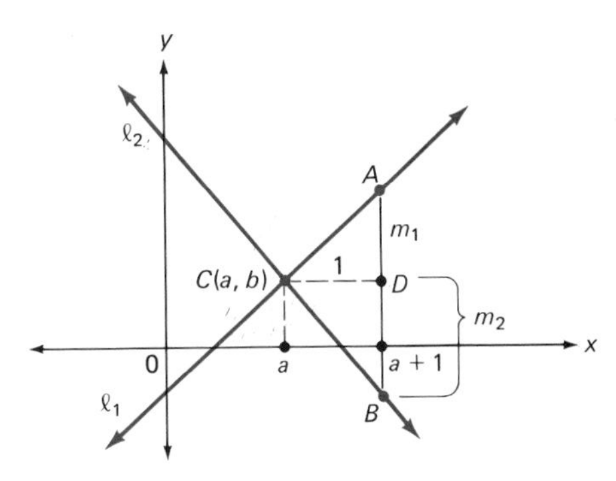

Let line ℓ_1 be perpendicular to line ℓ_2 at the point $C(a, b)$. Use m_1 for the slope of ℓ_1, and m_2 for ℓ_2. We want to show that

$$m_1 m_2 = -1 \qquad \text{or} \qquad m_1 = -\frac{1}{m_2}$$

Add 1 to the x-coordinate a of point C and draw the vertical line through $a + 1$ on the x-axis. This vertical line will meet ℓ_1 at some point A and ℓ_2 at some point B, forming right triangle ABC with right angle at C. Draw the perpendicular from C to AB meeting AB at D. Then CD has length 1.

(a) Using the right triangle CDA, show that $m_1 = DA$.

(b) Show that $m_2 = DB$. Is m_2 positive or negative?

(c) For right triangle ABC, CD is the mean proportional between segments BD and DA on the hypotenuse. Use this fact to conclude that $\frac{m_1}{1} = \frac{1}{-m_2}$, or $m_1 m_2 = -1$.

3.4 Linear functions

Pictured below is a line with slope m and y-intercept k. To find the equation of ℓ we begin by considering any point $P(x, y)$ on ℓ other than $(0, k)$.

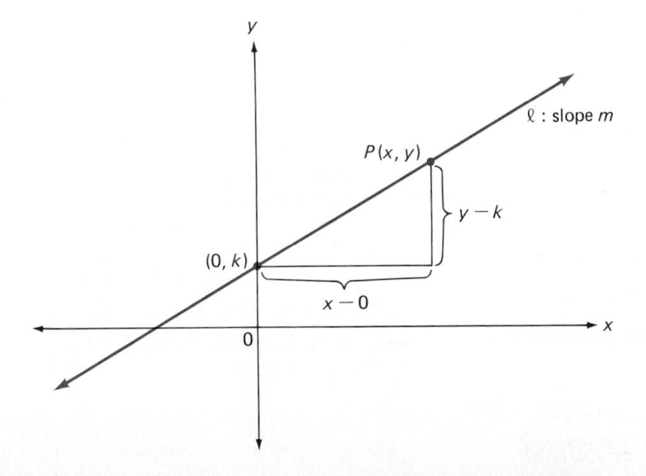

Since the slope of ℓ is given by any two of its points, we may use $(0, k)$ and (x, y) to write

$$m = \frac{y - k}{x - 0} = \frac{y - k}{x}$$

If both sides are multiplied by x, then

$$y - k = mx$$

or

$$y = mx + k \longleftarrow$$ Note that the point $(0, k)$ also satisfies this final form.

This leads to the following y-form of the equation of a line.

SLOPE-INTERCEPT FORM OF A LINE

$$y = mx + k$$

where m is the slope and k is the y-intercept.

The equation $y = mx + k$ also defines a function; thus we may think of $y = f(x) = mx + k$ as the *general linear function* with domain consisting of all the real numbers.

EXAMPLE 1 Graph the linear function f defined by $y = f(x) = 2x - 1$ by using the slope and y-intercept. Also indicate the domain and range of f, and display $f(2) = 3$ geometrically; that is, show the point corresponding to $f(2) = 3$.

Solution The y-intercept is -1. Locate $(0, -1)$ and use $m = 2 = \frac{2}{1}$ to reach $(1, 1)$, another point on the line.

Both the domain and range of f consist of all real numbers.

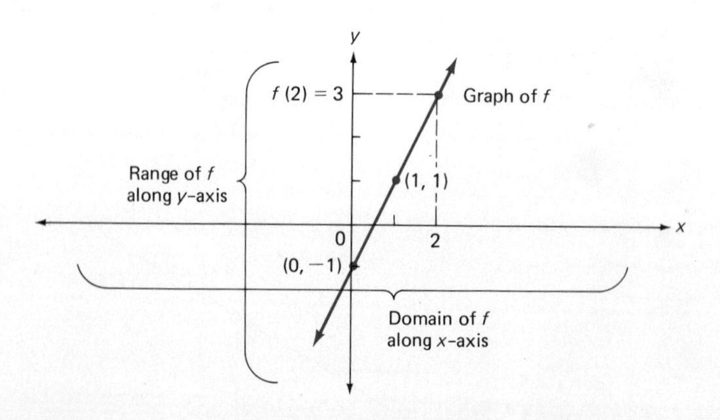

EXAMPLE 2 Find the equation of the line with slope $\frac{2}{3}$ passing through the point $(0, -5)$.

Solution Since $m = \frac{2}{3}$ and $k = -5$, the slope-intercept form gives

$$y = \tfrac{2}{3}x + (-5)$$
$$= \tfrac{2}{3}x - 5$$

A special case of $y = f(x) = mx + k$ is obtained when $m = 0$. Then

$$y = f(x) = k$$

This says that for *each* input x, the output $f(x)$ is always the same value k.

Domain: all reals
Range: the single value k

Since $f(x) = k$ is constant for all x, this linear function is also referred to as a *constant function*. Its graph is a horizontal line.

Can a vertical line be the graph of a function where y depends on x? Explain.

Now let ℓ be a line with slope m that passes through (x_1, y_1). We wish to determine the conditions on the coordinates of any point $P(x, y)$ that is on ℓ.

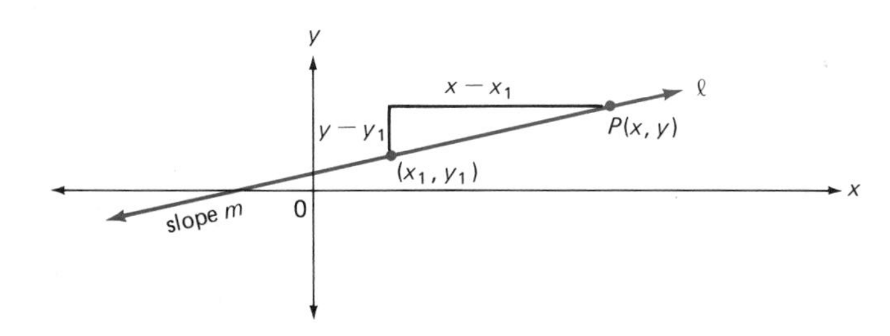

From the figure you can see that $P(x, y)$ will be on ℓ if and only if the ratio $\dfrac{y - y_1}{x - x_1}$ is the same as m. This is, P is on ℓ if and only if

$$m = \frac{y - y_1}{x - x_1}$$

Multiply both sides of this equation by $x - x_1$.

$$m(x - x_1) = y - y_1$$

This step leads to the following form for the equation of a straight line.

This is the form of a line that is most frequently used in calculus.

POINT-SLOPE FORM OF A LINE

$$y - y_1 = m(x - x_1)$$

where m is the slope and (x_1, y_1) is a point on the line.

EXAMPLE 3 Write the point-slope form of the line ℓ with slope $m = 3$ that passes through the point $(-1, 1)$. Verify that $(-2, -2)$ is on the line.

CAUTION: Pay attention to the minus signs on the coordinates when used in the point-slope form. Note the substitution of $x_1 = -1$ in this example.

Solution Since $m = 3$, any (x, y) on ℓ satisfies this equation:

$$y - 1 = 3[x - (-1)]$$
$$y - 1 = 3(x + 1)$$

Let $x = -2$:

$$y - 1 = 3(-2 + 1) = -3$$
$$y = -2$$

Thus $(-2, -2)$ is on the line.

TEST YOUR UNDERSTANDING

Write the slope-intercept form of the line with the given slope and y-intercept.

1 $m = 2; k = -2$ **2** $m = -\frac{1}{2}; k = 0$ **3** $m = \sqrt{2}; k = 1$

Write the point-slope form of the line through the given point with slope m.

4 $(2, 6); m = -3$ **5** $(-1, 4); m = \frac{1}{2}$ **6** $(5, -\frac{2}{3}); m = 1$

7 $(0, 0); m = -\frac{1}{4}$ **8** $(-3, -5); m = 0$ **9** $(1, -1); m = -1$

10 Which, if any, of the preceding produce a constant linear function? State its range.

EXAMPLE 4 Write the slope-intercept form of the line through the two points $(6, -4)$ and $(-3, 8)$.

See Exercise 64 for the *two-point form* for the equation of a line. Use that form to complete this example in another way.

Solution First compute the slope.

$$m = \frac{-4 - 8}{6 - (-3)} = \frac{-12}{9} = -\frac{4}{3}$$

Use either point to write the point-slope form, and then convert to the slope-intercept form. Thus, using the point $(6, -4)$,

$$y - (-4) = -\tfrac{4}{3}(x - 6)$$
$$y + 4 = -\tfrac{4}{3}x + 8$$
$$y = -\tfrac{4}{3}x + 4$$

Show that the same final form is obtained using the point $(-3, 8)$.

EXAMPLE 5 Write the equation of the line that is perpendicular to the line $5x - 2y = 2$ and that passes through the point $(-2, -6)$.

Solution First find the slope of the given line by writing it in slope-intercept form.

$$5x - 2y = 2$$
$$-2y = -5x + 2$$
$$y = \tfrac{5}{2}x - 1 \qquad \text{The slope is } \tfrac{5}{2}.$$

The perpendicular line has slope $-\dfrac{1}{\frac{5}{2}} = -\dfrac{2}{5}$. Since this line also goes through $P(-2, -6)$, the point-slope form gives

Recall that two perpendicular lines have slopes that are negative reciprocals of one another.

$$y + 6 = -\tfrac{2}{5}(x + 2)$$
$$y = -\tfrac{2}{5}x - \tfrac{34}{5}$$

A linear equation such as $y = -\frac{2}{3}x + 4$ can be converted to other equivalent forms. In particular, when this equation is multiplied by 3, and the variable terms are put on the same side of the equation, we then have $2x + 3y = 12$. This is an illustration of the *general linear equation.*

GENERAL LINEAR EQUATION

$$ax + by = c$$

where a, b, c are constants and a, b are not both 0.

The general linear equation $ax + by = c$ is said to define y as a function of x *implicitly* if $b \neq 0$. In other words, we have these equivalent forms:

$$ax + by = c \quad \longleftarrow \text{ implicit form of the linear function}$$

$$by = -ax + c$$

$$y = -\frac{a}{b}x + \frac{c}{b} \quad \longleftarrow \text{ explicit form of the linear function; slope-intercept form}$$

Note here that the slope is $-\frac{a}{b}$ and the y-intercept is $\frac{c}{b}$.

EXAMPLE 6 Find the slope and y-intercept of the line $-6x + 2y = 5$.

Solution Convert from the implicit to the explicit form of the linear function.

$$-6x + 2y = 5$$
$$2y = 6x + 5$$
$$y = 3x + \tfrac{5}{2} \longleftarrow \text{slope-intercept form}$$

Thus $m = 3$ and $k = \frac{5}{2}$.

CAUTION: LEARN TO AVOID MISTAKES LIKE THESE

WRONG	RIGHT
The slope of the line through (2, 3) and (5, 7) is $m = \frac{7-3}{2-5} = -\frac{4}{3}$	The slope of the line through (2, 3) and (5, 7) is $m = \frac{7-3}{5-2} = \frac{4}{3}$
The slope of the line $2x - 3y = 7$ is $-\frac{3}{2}$.	The slope is $\frac{2}{3}$.
The line through $(-4, -3)$ with slope 2 is $y - 3 = 2(x - 4)$.	The equation is $y - (-3) = 2[x - (-4)]$ or $y + 3 = 2(x + 4)$.

EXERCISES 3.4

Write the equation of the line with the given slope m and y-intercept k.

1 $m = 2, k = 3$
2 $m = -2, k = 1$
3 $m = 1, k = 1$
4 $m = -1, k = 2$
5 $m = 0, k = 5$
6 $m = 0, k = -5$
7 $m = \frac{1}{2}, k = 3$
8 $m = -\frac{1}{2}, k = 2$
9 $m = \frac{1}{4}, k = -2$

Write the point-slope form of the line through the given point with the indicated slope.

10 $(3, 4)$; $m = 2$
11 $(2, 3)$; $m = 1$
12 $(1, -2)$; $m = 0$
13 $(-2, 3)$; $m = 4$
14 $(-3, 5)$; $m = -2$
15 $(-3, 5)$; $m = 0$
16 $(8, 0)$; $m = -\frac{2}{3}$
17 $(2, 1)$; $m = \frac{1}{2}$
18 $(-6, -3)$; $m = \frac{4}{3}$
19 $(0, 0)$; $m = 5$
20 $(-\frac{3}{4}, \frac{2}{5})$; $m = 1$
21 $(\sqrt{2}, -\sqrt{2})$; $m = 10$

22 (a) Find the slope of the line determined by the points $A(-3, 5)$ and $B(1, 7)$, and write its equation in point-slope form, using the coordinates of A.
(b) Do the same as in part (a) using the coordinates of B.
(c) Verify that the equations obtained in parts (a) and (b) give the same slope-intercept form.

Write each equation in y-form; give the slope and y-intercept.

23 $3x + y = 4$
24 $2x - y = 5$
25 $6x - 3y = 1$
26 $4x + 2y = 10$
27 $3y - 5 = 0$
28 $x = \frac{3}{2}y + 3$
29 $4x - 3y - 7 = 0$
30 $5x - 2y + 10 = 0$
31 $\frac{1}{4}x - \frac{1}{2}y = 1$

Write the equation of the line through the two given points.

32 $(-1, 2), (2, -1)$
33 $(2, 3), (3, 2)$
34 $(1, 1), (-1, -1)$
35 $(3, 0), (0, -3)$
36 $(3, -4), (0, 0)$
37 $(-1, -13), (-8, 1)$
38 $(\frac{1}{2}, 7), (-4, -\frac{3}{2})$
39 $(10, 27), (12, 27)$
40 $(\sqrt{2}, 4\sqrt{2}), (-3\sqrt{2}, -10\sqrt{2})$

41 Two lines, parallel to the coordinate axes, intersect at the point $(5, -7)$. What are their equations?

42 Write the equation of the line that is parallel to $y = -3x - 6$ and with y-intercept 6.

43 Write the equation of the line parallel to $2x + 3y = 6$ that passes through the point $(1, -1)$.

In Exercises 44–47, write the equation of the line that is perpendicular to the given line and passes through the indicated point.

44 $y = -10x$; $(0, 0)$
45 $y = 3x - 1$; $(4, 7)$

46 $3x + 2y = 6$; $(6, 7)$ **47** $y - 2x = 5$; $(-5, 1)$

Graph the linear function f by using the slope and y-intercept. Display the point corresponding to $y = f(-2)$ on the graph.

48 $y = f(x) = -2x + 1$

49 $y = f(x) = x + 3$

50 $y = f(x) = 3x - \frac{1}{2}$

51 $y = f(x) = \frac{1}{2}x - 3$

In Exercises 52–55, the domain is given for the function defined by the equation $y = 3x - 7$. Graph each function.

52 All $x \leq 4$. **53** All $x \geq 0$.

54 All x where $-1 \leq x \leq 3$.

55 $x = -1, 0, 1, 2, 3$

Write the equation for each graph. State the domain and range in case it is the graph of a function.

56 **57**

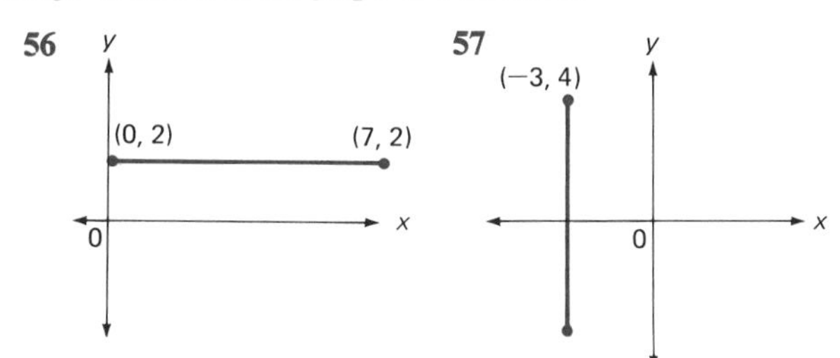

58 **59**

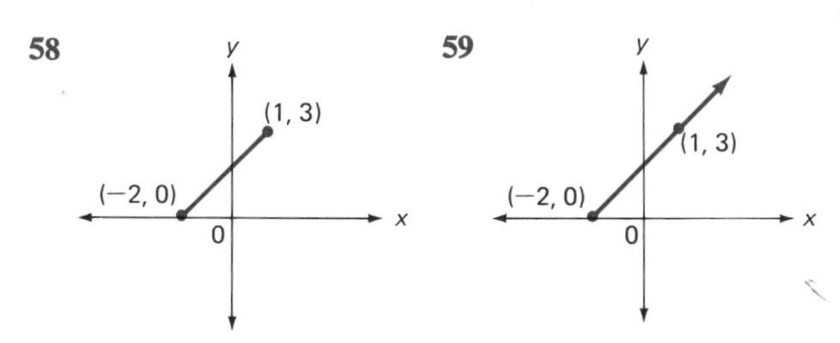

60

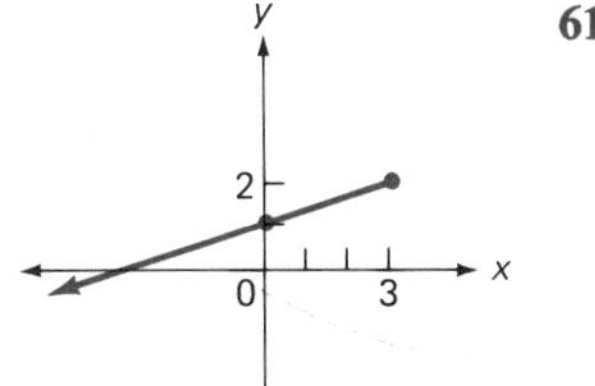

61

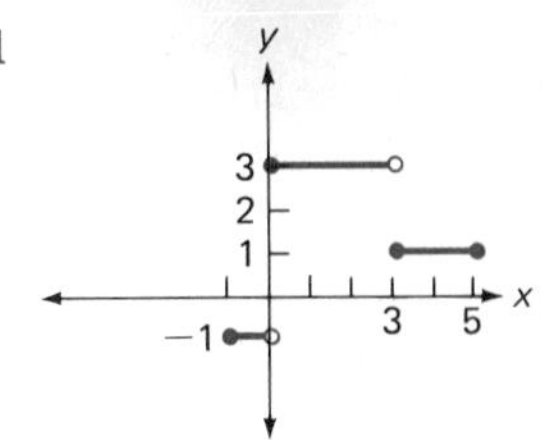

62 The sides of the parallelogram with vertices $(-1, 1)$, $(0, 3)$, $(2, 1)$, and $(3, 3)$ are the graphs of four different functions. In each case find the equation that defines the function and state the domain.

***63** Any line having a nonzero slope that does not pass through the origin always has both an x- and a y-intercept. Let ℓ be such a line having equation $ax + by = c$.

(a) Why is $c \neq 0$?

(b) What are the x- and y-intercepts?

(c) Derive the equation

$$\frac{x}{q} + \frac{y}{p} = 1$$

where q and p are the x- and y-intercepts, respectively. This is known as the *intercept form* of a line.

(d) Use the intercept form to write the equation of the line passing through $(\frac{3}{2}, 0)$ and $(0, -5)$.

(e) Use the two points in part (d) of find the slope, write the slope-intercept form, and compare with the result in part (d).

***64** Replace m in the point-slope form of a line through points (x_1, y_1) and (x_2, y_2) by $\frac{y_2 - y_1}{x_2 - x_1}$. Show that this gives the *two-point form* for the equation of a line:

$$\frac{y - y_1}{y_2 - y_1} = \frac{x - x_1}{x_2 - x_1}$$

65 Use the result of Exercise 64 to find the equation of the line through $(-2, 3)$ and $(5, -2)$. Write the equation in point-slope form.

3.5 Some special functions

We have seen that the graph of a linear function is a straight line. Linear functions can also be used to define other functions which, in themselves, are not linear but may be described as being "partly" or "piecewise" linear. An important example is the **absolute-value function**.

See Section 1.6 to review the meaning of absolute value.

$$f(x) = |x| = \begin{cases} x & \text{if } x \geq 0 \\ -x & \text{if } x < 0 \end{cases}$$

To graph this function, first draw the line $y = -x$ and eliminate all those points on it for which x is positive. Then draw the line $y = x$ and eliminate the part for which x is negative. Now join these two parts to get the graph of $y = |x|$.

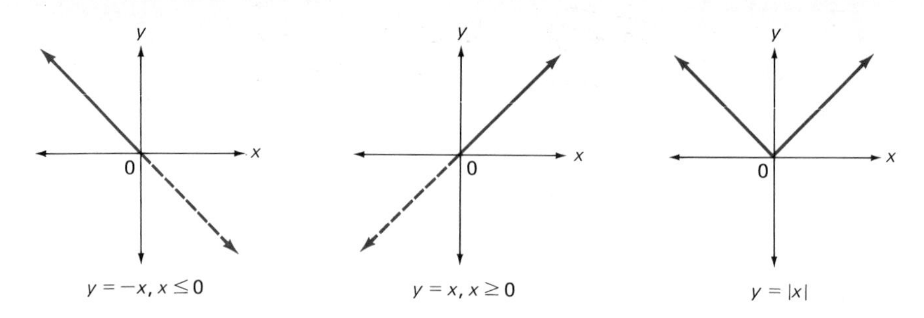

The graph of $y = |x|$ consists of two perpendicular rays intersecting at the origin. Now $y = |x|$ is not a linear function, but it is linear in parts; the two halves $y = x$, $y = -x$ are linear.

Note that the graph is symmetric about the y-axis. (If the paper were folded along the y-axis, the two parts would coincide.) This symmetry can be observed by noting that the y-values for x and $-x$ are the same. That is,

$$|x| = |-x| \qquad \text{for all } x$$

EXAMPLE 1 Graph: $y = |x - 2|$.

See Exercise 29 for an alternative method to draw this graph.

Solution For $x \geq 2$, we find that $x - 2 \geq 0$, which implies that $y = |x - 2| = x - 2$. This is the ray through (and to the right of) the point (2, 0), with slope 1. For $x < 2$, we get $x - 2 < 0$, which implies that $y = |x - 2| = -(x - 2) = -x + 2$. This gives the left half of the graph shown.

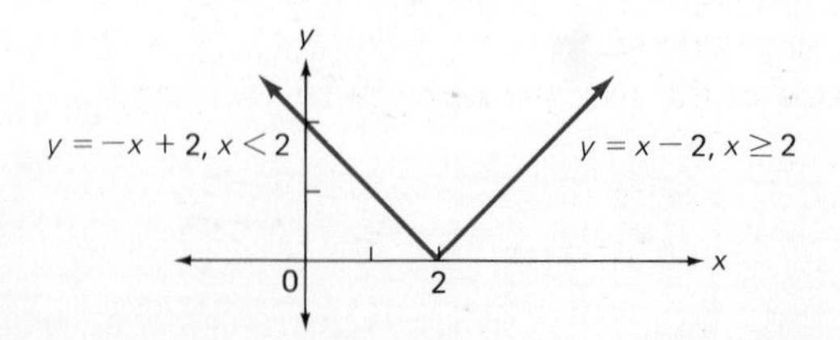

EXAMPLE 2 Graph the function f defined by the following two equations.

$$y = f(x) = \begin{cases} 2x & \text{if } 0 \leq x \leq 1 \\ -x + 2 & \text{if } 1 < x \end{cases}$$

What are the domain and range of f?

Solution The domain of f is all $x \geq 0$. From the graph we see that the range consists of all $y \leq 2$.

Note: The open dot at (1, 1) means that this point is not part of the graph; but the point (1, 2) is on the graph since $f(1) = 2 \cdot 1 = 2$.

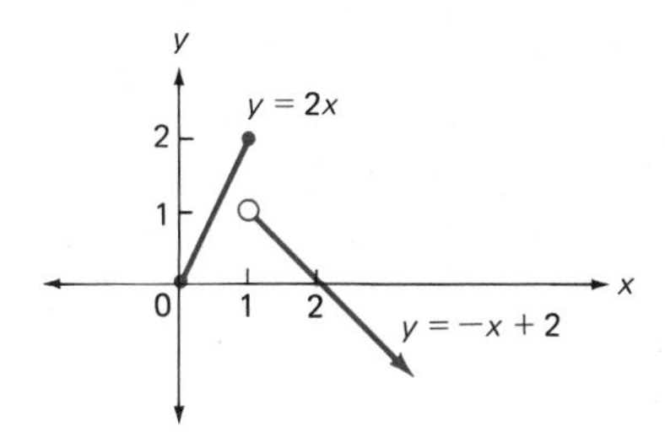

EXAMPLE 3 Graph the function f given by $y = f(x) = \dfrac{|x|}{x}$. What is the domain of f?

Solution The domain consists of all $x \neq 0$. When $x > 0$, $|x| = x$ and

$$f(x) = \frac{|x|}{x} = \frac{x}{x} = 1$$

Thus $f(x)$ is the constant 1, for all positive x. Similarly, for $x < 0$, $|x| = -x$ and

$$f(x) = \frac{|x|}{x} = \frac{-x}{x} = -1$$

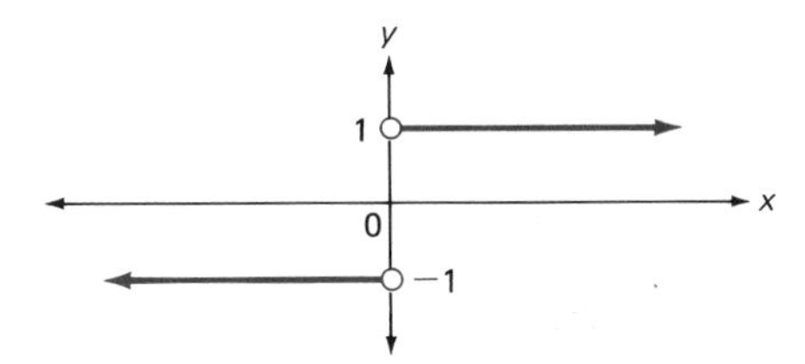

Example 3 is an illustration of a **step function**. Such a function may be described as a function whose graph consists of parts of horizontal lines. Here is another step function defined for $-2 \leq x < 4$. Note that each step is the graph of one of the six equations used to define f.

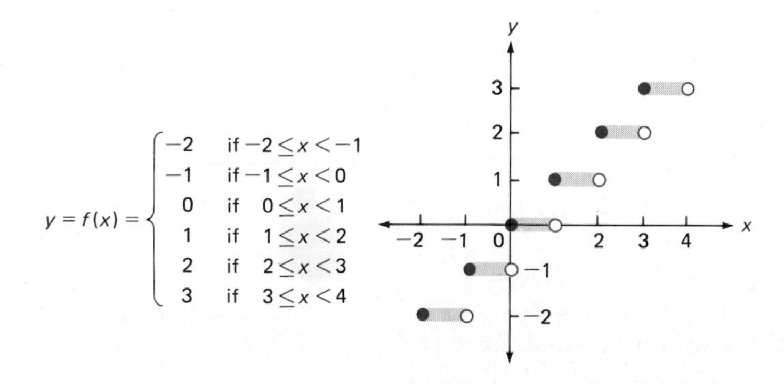

Observe that in each case, say $2 \leq x < 3$, the integer 2 at the left of the inequality is also the corresponding y-value for each x within this inequality. Putting it another way, we say that the y-value 2 is *the greatest integer less than or equal to x.*

For each number x there is an integer n such that $n \leq x < n + 1$. Therefore, the greatest integer less than or equal to x equals n. In other words, the preceding step function may be extended to a step function with domain of *all* real numbers; its graph would consist of an infinite number of steps.

We use the symbol $[x]$ to mean "greatest integer less than or equal to x." Thus

$[x]$ is the greatest integer not exceeding x itself.

$$y = [x] = \text{greatest integer} \leq x$$

Here are a few illustrations:

$[2\frac{1}{2}] = 2$ because 2 is the greatest integer $\leq 2\frac{1}{2}$

$[0.64] = 0$ because zero is the greatest integer ≤ 0.64

$[-2.8] = -3$ because -3 is the greatest integer ≤ -2.8

$[-5] = -5$ because -5 is the greatest integer ≤ -5

TEST YOUR UNDERSTANDING

Evaluate.

1 $[12.3]$	**2** $[12.5]$	**3** $[12.9]$	**4** $[13]$
5 $[-3\frac{3}{4}]$	**6** $[-3\frac{1}{4}]$	**7** $[-3]$	**8** $[0]$
9 $[-0.25]$	**10** $[0.25]$	**11** $[-0.75]$	**12** $[0.75]$

EXERCISES 3.5

Evaluate.

1 $[99.1]$

2 $[100.1]$

3 $[-99.1]$

4 $[-100.1]$

5 $[-\frac{7}{2}]$

6 $[10^3]$

7 $[\sqrt{2}]$

8 $[\pi]$

Graph each function and state the domain and range.

9 $y = |x - 1|$

10 $y = |x + 3|$

11 $y = |2x|$

12 $y = |2x - 1|$

13 $y = |3 - 2x|$

14 $y = |\frac{1}{2}x + 4|$

15 $y = \begin{cases} 3x & \text{if } -1 \leq x \leq 1 \\ -x & \text{if } 1 < x \end{cases}$

16 $y = \begin{cases} -2x + 3 & \text{if } x < 2 \\ x + 1 & \text{if } x > 2 \end{cases}$

17 $y = \begin{cases} x & \text{if } -2 < x \leq 0 \\ 2x & \text{if } 0 < x \leq 2 \\ -x + 3 & \text{if } 2 < x \leq 3 \end{cases}$

18 $y = \begin{cases} x & \text{if } 0 \leq x < 1 \\ x - 1 & \text{if } 1 \leq x < 2 \\ x - 2 & \text{if } 2 \leq x < 3 \\ x - 3 & \text{if } 3 \leq x \leq 4 \end{cases}$

Graph each step function for its given domain.

19 $y = \dfrac{x}{|x|}$; all $x \neq 0$

20 $y = \dfrac{|x - 2|}{x - 2}$; all $x \neq 2$

21 $y = \dfrac{x + 3}{|x + 3|}$; all $x \neq -3$

22 $y = \begin{cases} -1 & \text{if } -1 \leq x \leq 0 \\ 0 & \text{if } 0 < x \leq 1 \\ 1 & \text{if } 1 < x \leq 2 \\ 2 & \text{if } 2 < x \leq 3 \end{cases}$

23 $y = [x]$; $-3 \leq x \leq 3$

*24 $y = [2x]$; $0 \leq x \leq 2$

*25 $y = 2[x]$; $0 \leq x \leq 2$

*26 $y = [3x]$; $-1 \leq x \leq 1$

*27 $y = [x - 1]$; $-2 \leq x \leq 3$

*28 (a) The postage for mailing packages depends on the weight and destination. Let the rates for a certain destination be as follows:

x = *weight* (*pounds*)	y = *postage* (*cost*)
under 1	\$0.80
1 or more but under 2	\$0.90
2 or more but under 3	\$1.00
3 or more but under 4	\$1.10
4 or more but under 5	\$1.20
5 or more but under 6	\$1.30
6 or more but under 7	\$1.40
7 or more but under 8	\$1.50
8 or more but under 9	\$1.60
9 or more but under 10	\$1.70

This table defines y to be a function of x. If we use P (for postage) we may write $y = P(x)$, and the table gives $P(x) = 1.20$ for $4 \leq x < 5$. Graph this function on its domain $0 < x < 10$. To achieve clarity, you may want to use a larger unit along the vertical axis then on the x-axis.

(b) A formula for $P(x)$ can be given in terms of the greatest integer function. Find such a formula.

(c) Draw a graph that shows the cost of sending up to 6 ounces of first-class mail at the current postage rate. Can you think of other examples of step functions in daily life situations?

29 The graph of $y = |x - 2|$ was found in Example 1. Here is an alternative procedure. First graph $y = x - 2$ using a dashed line for the part below the x-axis. Now reflect the negative part (the dashed part) through the x-axis to get the final graph of $y = |x - 2|$.

30 Follow the procedure in Exercise 29 to graph $y = |2x + 5|$.

3.6 Systems of linear equations

Any two nonparallel lines intersect in exactly one point. Our objective is to learn how to find the coordinates of this point from the equations of the lines.

Let us consider a linear system of two equations as follows:

$$(1)\ 2x + 3y = 12$$

$$(2)\ 3x + 2y = 12$$

We may graph this system by graphing each line on the same axes, using the x- and y-intercepts of each line. For instance, letting $x = 0$ in Equation (1) gives the y-intercept $y = 4$. Similarly, letting $y = 0$ produces the x-intercept $x = 6$. The line through (0, 4) and (6, 0) is the graph of Equation (1). Equation (2) may be graphed similarly. Can you find the point of intersection of the two lines?

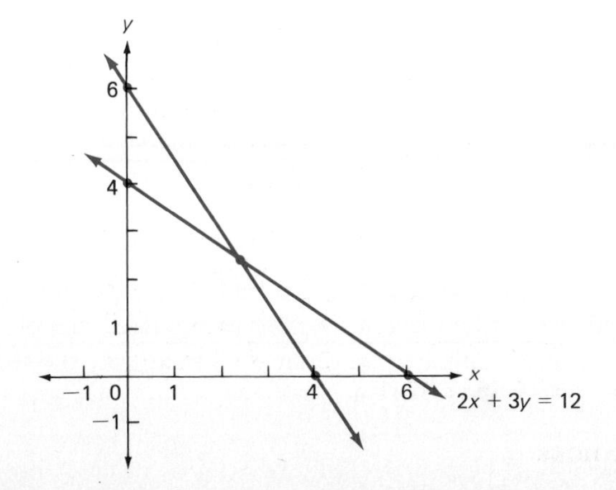

The coordinates of the point of intersection must satisfy *both* equations.

Finding the coordinates of the point of intersection of the two lines is also described as *solving the system*

$$2x + 3y = 12$$
$$3x + 2y = 12$$

Two procedures will now be described to solve such systems. The underlying idea for each of them is that the coordinates of the point of intersection must satisfy each equation. The **substitution method** will be taken up first.

We begin by letting (x, y) be the coordinates of the point of intersection of the preceding system. Then these x- and y-values fit both equations. Hence either equation may be solved for x or y and then substituted into the other equation. For example, solve the second equation for y:

$$y = -\tfrac{3}{2}x + 6$$

Substitute this expression into the first equation and solve for x.

$$2x + 3y = 12$$
$$2x + 3(-\tfrac{3}{2}x + 6) = 12$$
$$2x - \tfrac{9}{2}x + 18 = 12$$
$$-\tfrac{5}{2}x = -6$$
$$x = (-\tfrac{2}{5})(-6) = \tfrac{12}{5}$$

To find the y-value, substitute $x = \frac{12}{5}$ into either of the given equations. It is easiest to use the equation that was written in y-form as shown:

$$y = -\tfrac{3}{2}x + 6$$
$$y = -\tfrac{3}{2}(\tfrac{12}{5}) + 6 = \tfrac{12}{5}$$

Check the solution by substituting these values into each of the given equations for this system.

The solution is $(\frac{12}{5}, \frac{12}{5})$.

TEST YOUR UNDERSTANDING

Use the substitution method to solve each linear system.

1 $y = 3x - 1$
$y = -5x + 7$

2 $y = 4x + 16$
$y = -\frac{2}{5}x + \frac{14}{5}$

3 $4x - 3y = 11$
$y = 6x - 13$

4 $2x + 2y = \frac{4}{5}$
$-7x + 2y = -1$

5 $x + 7y = 3$
$5x + 12y = -8$

6 $4x - 2y = 40$
$-3x + 3y = 45$

Now we will solve the same system as before, this time using the **multiplication–addition** (or **multiplication–subtraction**) **method**.

$$2x + 3y = 12$$
$$3x + 2y = 12$$

The idea here is to alter the equations so that the coefficients of one of the variables are either negatives of one another or equal to each other. We may, for example, multiply the first equation by 3 and multiply the second by -2.

The resulting system looks like this:

$$6x + 9y = 36$$
$$-6x - 4y = -24$$

Keep in mind that we are looking for the pair (x, y) that fits both equations. Thus for these x- and y-values we may add equals to equals to eliminate the variable x.

$$5y = 12$$
$$y = \frac{12}{5}$$

As before, x is now found by substituting $y = \frac{12}{5}$ into one of the given equations.

If in the preceding solution the second equation is multiplied by 2 instead of -2, the system becomes

$$6x + 9y = 36$$
$$6x + 4y = 24$$

Now x can be eliminated by subtracting the equations.

As illustrated in the next example, this method can be condensed into a compact procedure.

EXAMPLE 1 Solve the system by the multiplication–addition method.

$$\tfrac{1}{3}x - \tfrac{2}{5}y = 4$$
$$7x + 3y = 27$$

Solution

$$15(\tfrac{1}{3}x - \tfrac{2}{5}y = 4) \Longrightarrow 5x - 6y = 60$$
$$2(7x + 3y = 27) \Longrightarrow 14x + 6y = 54$$
$$\text{Add:} \quad 19x = 114$$
$$x = 6$$

Multiply both sides of the first equation by 15, and both sides of the second equation by 2.

Substitute to solve for y.

$$7x + 3y = 27 \Longrightarrow 7(6) + 3y = 27$$
$$3y = -15$$
$$y = -5$$

The solution is $(6, -5)$.

Check: $\frac{1}{3}(6) - \frac{2}{5}(-5) = 4$; $7(6) + 3(-5) = 27$

TEST YOUR UNDERSTANDING

Use the multiplication–addition (or subtraction) method to solve each linear system.

1 $3x + 4y = 5$
$5x + 6y = 7$

2 $-8x + 5y = -19$
$4x + 2y = -4$

3 $\frac{1}{7}x + \frac{5}{2}y = 2$
$\frac{1}{2}x - 7y = -\frac{17}{4}$

4 $10x + 9y = 0$
$\frac{2}{3}x = -6y$

There is an important feature that all these procedures have in common. They all begin by eliminating one of the two variables. Thus you soon reach one equation in one unknown. This basic strategy of reducing the number of unknowns can be applied to "larger" linear systems (as well as to systems that are not linear). For instance, here is a system of three linear equations in three variables:

$$\begin{aligned} &(1) & 2x - 5y + z &= -10 \\ &(2) & x + 2y + 3z &= 26 \\ &(3) & -3x - 4y + 2z &= 5 \end{aligned}$$

To solve this system we may begin by eliminating the variable x from the first two equations.

$$\left.\begin{aligned} 2x - 5y + z &= -10 \\ -2(x + 2y + 3z &= 26) \end{aligned}\right\} \Longrightarrow \begin{aligned} 2x - 5y + z &= -10 \\ -2x - 4y - 6z &= -52 \end{aligned}$$

$$\text{Add:} \qquad -9y - 5z = -62$$

Another equation in y and z can be obtained from Equations (2) and (3) by eliminating x:

$$\left.\begin{aligned} 3(x + 2y + 3z &= 26) \\ -3x - 4y + 2z &= 5 \end{aligned}\right\} \Longrightarrow \begin{aligned} 3x + 6y + 9z &= 78 \\ -3x - 4y + 2z &= 5 \end{aligned}$$

$$\text{Add:} \qquad 2y + 11z = 83$$

Now we have this system in two variables:

$$\begin{aligned} 9y + 5z &= 62 \\ 2y + 11z &= 83 \end{aligned}$$

You can solve this system as before to find $y = 3$ and $z = 7$. Now substitute these values into an earlier equation, say (2).

$$\begin{aligned} x + 2y + 3z &= 26 \\ x + 2(3) + 3(7) &= 26 \\ x &= -1 \end{aligned}$$

The remaining equations can be used for checking:

$$\begin{aligned} &(1) & 2(-1) - 5(3) + 7 &= -10 \\ &(3) & -3(-1) - 4(3) + 2(7) &= 5 \end{aligned}$$

The solution is $x = -1$, $y = 3$, $z = 7$.

There are verbal problems that can be solved by using systems of linear equations. In some cases only one equation in one variable can also be used. However, it is worthwhile to learn how to use more than one variable in order to simplify the process of translating from the verbal to the mathematical form. Unfortunately, because of the variety of problems available, as well as the numerous ways in which a problem can be stated in ordinary language, there are no fixed methods of translation that apply to all situations.

EXAMPLE 2 A field goal in basketball is worth 2 points and a free throw is worth 1 point. In a recent game the school basketball team scored 85 points. If there were twice as many field goals as free throws, how many of each were there?

The guidelines given in Section 1.3 to solve verbal problems leading to one equation in one variable can be adjusted and used here. The solution to Example 2 is based on these guidelines.

Solution

(a) Read the problem once or twice.
(b) The quantities involved are numbers of field goals and of free throws.
(c) Let x be the number of field goals and y the number of free throws. Then the points due to field goals is $2x$, and y is the number of points due to free throws.
(d) There were twice as many field goals as free throws, so

$$x = 2y$$

The total points in the game was 85; therefore,

$$2x + y = 85$$

(e) The solution for the system

$$x = 2y$$
$$2x + y = 85$$

is (34, 17).
(f) *Check:* 34 field goals (at 2 points each) gives 68 points; and 17 free throws (1 point each) produces a total of $68 + 17 = 85$ points. Also, there were twice as many field goals as free throws.

Recall that the check for a verbal problem should be done by direct substitution of the answer into the original statement of the problem.

EXAMPLE 3 For her participation in a recent "walk-for-poverty," Kathi collected $2 per mile for a total of $52. She recorded that for a certain time she walked at the rate of 3 miles per hour and the rest at 4 miles per hour. Afterward she mentioned that it was too bad that she did not have the energy to reverse the rates. If she could have walked 4 miles per hour for the same time that she actually walked 3 miles per hour, and vice versa, she would have collected a total of $60. How long did her walk take?

Solution The problem asks for the total time of the walk. This time is broken into two parts: the time she walked at 3 miles per hour and the time at 4 miles per hour.

Let $x =$ time at 3 miles per hour
and $y =$ time at 4 miles per hour

We want to find $x + y$.

Since *distance* = *rate* × *time*, $3x$ is the distance at the 3-mile-per-hour rate and $4y$ is the distance at 4 miles per hour; the total distance is the sum $3x + 4y$. This total must equal 26 because she earned $52 at $2 per mile. Thus

$$3x + 4y = 26$$

The \$60 she could have earned would have required walking 30 miles, and this 30 miles, she said, would have been possible by reversing the rates for the actual times that she did walk. This says that $4x + 3y = 30$. We now have the system

$$3x + 4y = 26$$
$$4x + 3y = 30$$

The common solution is (6, 2), and therefore the total time for the walk is $6 + 2$, or 8 hours.

Check: She walked $3 \times 6 = 18$ miles at 3 miles per hour, and $4 \times 2 = 8$ miles at miles per hour, for a total of 26 miles. At \$2 per mile, she collected \$52. Reversing the rates gives $(4 \times 6) + (3 \times 2) = 30$ miles, for which she would have earned \$60.

EXERCISES 3.6

Solve each system by the substitution method.

1 $2x + y = -10$
$6x - 3y = 6$

2 $-3x + 6y = 0$
$4x + y = 9$

3 $v - w = 14$
$3v + w = 2$

4 $4x - y = 6$
$2x + 3y = 10$

5 $x + 5y = -9$
$4x - 3y = -13$

6 $s + 2t = 5$
$-3s + 10t = -7$

Solve each system by the multiplication–addition method.

7 $-3x + y = 16$
$2x - y = 10$

8 $2x + 4y = 24$
$-3x + 5y = -25$

9 $3y - 9x = 30$
$8x - 4y = 24$

10 $2u - 6v = -16$
$5u - 3v = 8$

11 $4x - 5y = 3$
$16x + 2y = 3$

12 $\frac{1}{2}x + 3y = 6$
$-x - 8y = 18$

Solve the systems given in Exercises 13–28 by using any method.

13 $x - 2y = 3$
$y - 3x = -14$

14 $2x + y = 6$
$3x - 4y = 12$

15 $-3x + 8y = 16$
$16x - 5y = 103$

16 $3x - 8y = -16$
$7x + 19y = -188$

17 $16x - 5y = 103$
$7x + 19y = -188$

18 $4x = 7y - 6$
$9y = -12x + 12$

19 $3s + t - 3 = 0$
$2s - 3t - 2 = 0$

20 $\frac{1}{2}x - \frac{1}{3}y = 2$
$\frac{3}{4}x + \frac{2}{5}y = -1$

21 $\frac{1}{4}x + \frac{1}{3}y = \frac{5}{12}$
$\frac{1}{2}x + y = 1$

22 $0.1x + 0.2y = 0.7$
$0.01x - 0.01y = 0.04$

23 $\frac{x}{2} + \frac{y}{6} = \frac{1}{2}$
$0.2x - 0.3y = 0.2$

24 $2(x + y) = 4 - 3y$
$\frac{1}{2}x + y = \frac{1}{2}$

25 $2(x - y - 1) = 1 - 2x$
$6(x - y) = 4 - 3(3y - x)$

26 $\frac{x-2}{5} + \frac{y+1}{10} = 1$
$\frac{x+2}{3} - \frac{y+3}{2} = 4$

27 $2x = 7$
$y = 4$

28 $2x - 10c = -y$
$7x - 2y = 2c$
(c is constant)

Solve each linear system.

29 $x + y + z = 2$
$x - y + 3z = 12$
$2x + 5y + 2z = -2$

30 $x + 2y + 3z = 5$
$-4x + z = 6$
$3x - y = -3$

31 $-3x + 3y + z = -10$
$4x + y + 5z = 2$
$x - 8y - 2z = 12$

32 $x + 2y + 3z = -4$
$4x + 5y + 6z = -4$
$7x - 15y - 9z = 4$

33 Solve the system

$$ax + by = c$$
$$dx + ey = f$$

for x and y, where a, b, c, d, e, and f are constants with $ae - bd \neq 0$. The solution to this exercise provides a formula for finding the solution of a system of equations where each is written in the standard form shown.

34 Why do we not allow both a and b to be zero in the linear equation $ax + by = 0$?

Some of the following exercises can be solved by using just one equation with one variable. However, in most cases it is easier to use two equations and two variables.

35 Points $(-8, -16)$, $(0, 10)$, and $(12, 14)$ are three vertices of a parallelogram. Find the coordinates of the fourth vertex if it is located in the third quadrant.

36 Find the point of intersection for the diagonals of the parallelogram in Exercise 35.

37 The total points that a basketball team scored was 96. If there were $2\frac{1}{2}$ times as many field goals as free throws, how many of each were there? (Field goals count 2 points; free throws count 1 point.)

38 During a round of golf a player scored only fours and fives per hole. If he played 18 holes and his total score was 80, how many holes did he play in four strokes and how many in five?

39 The tuition fee at a college plus the room and board comes to $5300 per year. The room and board is $100 less than half the tuition. How much is the tuition, and what does the room and board cost?

40 A college student had a work–study scholarship that paid $3.50 per hour. He also made $1 per hour babysitting. His income one week was $61 for a total of $23\frac{1}{2}$ hours of employment. How many hours did he spend on each of the two jobs?

*41 An airplane, flying with a tail wind, takes 2 hours and 40 minutes to travel 1120 miles. The plane makes the return trip against the wind in 2 hours and 48 minutes. What is the wind velocity and what is the speed of the plane in still air? (Assume that both velocities are constant; add the velocities for the downwind trip, and subtract them for the return trip.)

42 The perimeter of a rectangle is 72 inches. The length is $3\frac{1}{2}$ times as large as the width. Find the dimensions.

*43 A wholesaler has two grades of oil that ordinarily sell for 94¢ per quart and 74¢ per quart. He wants a blend of the two oils to sell at 85¢ per quart. If he antitipates selling 400 quarts of the new blend, how much of each grade should he use? (One of the equations makes use of the fact that the total income will be $400(0.85) = \$340$.)

44 A store paid $226.19 for a recent mailing. Some of the letters cost 20¢ postage and the rest needed 37¢ postage. How many letters at each rate were mailed if the total number sent out was 887?

45 The annual return on two investments totals $464. One investment gives 8% interest and the other $7\frac{1}{2}$%. How much money is invested at each rate if the total investment is $6000?

*46 A salesman said that it did not matter to him whether he sold one pair of shoes for $21 or two pairs for $35, because he made the same profit on each sale. How much does one pair of shoes cost the salesman, and what is the profit?

*47 To go to work a commuter first averages 36 miles per hour driving his car to the train station and then rides the train, which averages 60 miles per hour. The entire trip takes 1 hour and 22 minutes. It costs the commuter 12¢ per mile to drive the car and 6¢ per mile to ride the train. If the total cost is $5.04, find the distances traveled by car and by train.

*48 A line with equation $y = mx + k$ passes through the points $(-\frac{1}{3}, -6)$ and $(2, 1)$. Find m and k by substituting the coordinates into the equation and solving the resulting system.

*49 A line with equation $ax + by = 3$ passes through $(6, 3)$ and $(-1, -1)$. Find a and b without finding the slope.

*50 A student in a chemistry laboratory wants to form a 32-milliliter mixture of two solutions to contain 30% acid. Solution A contains 42% acid and solution B contains 18% acid. How many milliliters of each solution must be used? (*Hint:* Use the fact that the final mixture will have $0.30(32) = 9.6$ milliliters of acid.)

3.7 Classification of linear systems

Can you solve the following system?

$$\begin{aligned} 39x - 91y &= -28 \\ 6x - 14y &= 7 \end{aligned}$$

No matter which technique you use, something unexpected happens. Let us try the substitution method. As usual, assume that there is an (x, y) that satisfies both equations. Solve the second equation for y to get

$$y = \tfrac{3}{7}x - \tfrac{1}{2}$$

Now substitute into the first equation.

Arriving at a false result, as shown here, is not a wasted effort. As long as there are no computational errors, such a false conclusion tells us that the given system has no solution. That is, *the system has been solved by learning that it has no solution.*

$$39x - 91(\tfrac{3}{7}x - \tfrac{1}{2}) = -28$$
$$39x - 39x + \tfrac{91}{2} = -28$$
$$\tfrac{91}{2} = -28$$

Something went wrong; yet each step appears to be correct. What is the trouble?

Graphing the two lines reveals the difficulty. The lines are parallel and obviously cannot have a point in common. There is no pair (x, y) that satisfies the given system. But we started by assuming that there was a pair (x, y) that fits both equations. It is now clear that this initial assumption is false; therefore, it should not be surprising to get a false conclusion.

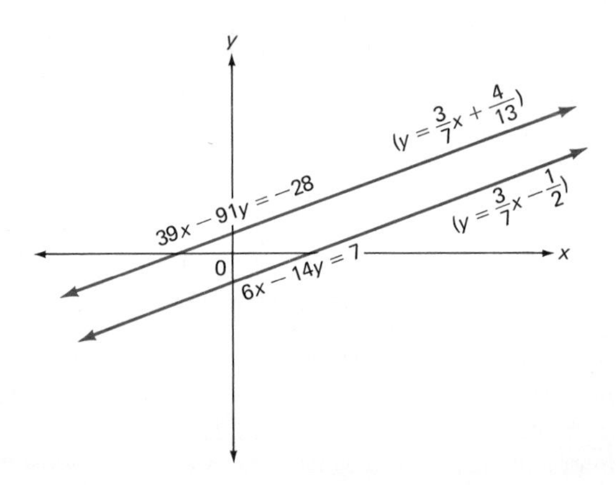

Two lines in the plane either intersect or they are parallel. In the first case the corresponding system of equations has a unique solution, and we say that such a system is **consistent**. When the lines are parallel the corresponding equations do not have a common solution, and such a system is said to be **inconsistent.**

EXAMPLE 1 Decide if each of the systems is consistent or inconsistent.

(a) $-6x + 3y = 9$
$\quad 10x + 5y = -1$

(b) $2x - y = -3$
$\quad -\tfrac{1}{6}x + \tfrac{1}{12}y = \tfrac{1}{2}$

Solution **(a)** Solve the first equation for y ($y = 2x + 3$) and substitute into the second:

$$10x + 5(2x + 3) = -1$$
$$20x + 15 = -1$$

$$20x = -16$$
$$x = -\tfrac{4}{5}$$

Now solve for y, using $y = 2x + 3$:

$$y = 2(-\tfrac{4}{5}) + 3 = \tfrac{7}{5}$$

The solution is $(-\frac{4}{5}, \frac{7}{5})$, and thus system (a) is consistent.

(b) Multiply the second equation by 12. Then the system becomes

$$\begin{aligned} 2x - y &= -3 \\ -2x + y &= 6 \end{aligned}$$

By addition we obtain this result:

$$0 = 3$$

This false result tells us that the equations have no common solution; system (b) is inconsistent.

Consistency and inconsistency of a system can also be determined by placing both equations in slope-intercept form. If the slopes are unequal, the lines intersect; that is, there is a common solution, and the system is consistent. In case the slopes are equal and the y-intercepts are different, the lines are parallel; therefore, there is no solution, and the system is inconsistent. This approach is shown in Example 2.

EXAMPLE 2 Decide if the systems in Example 1 are consistent or inconsistent by using the slope-intercept form.

Solution System (a) is consistent because in slope-intercept form the equations are

$$\begin{aligned} y &= 2x + 3 \\ y &= -2x - \tfrac{1}{5} \end{aligned}$$

and the slopes are unequal. For system (b),

$$\begin{aligned} y &= 2x + 3 \\ y &= 2x + 6 \end{aligned}$$

The slopes are equal but not the y-intercepts; the system is inconsistent.

It might happen that a pair of equations really represents the same straight line. The following system gives two different ways of naming the same straight line. You can see this by changing either equation into the form of the other.

$$\begin{aligned} y &= \tfrac{2}{3}x - 5 \\ 2x - 3y &= 15 \end{aligned}$$

Such systems are said to be **dependent**. The graph for a dependent system, therefore, is always a single line.

In a dependent system each pair (x, y) that satisfies one equation must also satisfy the other. Any attempt at "solving" the system will result in some statement that is always true (an *identity*). For example:

$$3(y = \tfrac{2}{3}x - 5) \Longrightarrow 3y = 2x - 15 \Longrightarrow -2x + 3y = -15$$
$$2x - 3y = 15 \Longrightarrow 2x - 3y = 15 \Longrightarrow \quad 2x - 3y = \quad 15$$
$$\text{Add:} \qquad 0 = \quad 0$$

The given system has an infinite number of solutions and it is therefore a dependent system.

TEST YOUR UNDERSTANDING

Decide whether the given systems are consistent, inconsistent, or dependent.

1 $x + y = 2$ $x + y = 3$	**2** $3x - y = 7$ $-9x + 3y = -21$	**3** $2x - 3y = 8$ $-8x + 12y = 33$
4 $\frac{1}{2}x + 5y = -4$ $7x - 3y = 17$	**5** $-6x + 3y = 9$ $10x + 5y = -1$	**6** $20x + 36y = -27$ $\frac{5}{27}x + \frac{1}{3}y = -\frac{1}{4}$

See page 87 for graphing a single linear inequality in two variables.

The lines of a linear system, whether the system is consistent, inconsistent, or dependent, separate the plane into regions. Such a region will be the graph of a *system of linear inequalities*. To graph a system of linear inequalities means to locate all points (x, y) that satisfy each inequality in the system. Consider, for example, this system:

$$2x - y + 4 \leq 0$$
$$x + y - 2 \leq 0$$

It is convenient to write each in *y-form* first; that is, express y in terms of x. Verify that the following is correct.

$$y \geq 2x + 4$$
$$y \leq -x + 2$$

Can you write a system of inequalities whose graph is the unshaded region of the plane, including the boundaries?

Now graph each statement of inequality on the same coordinate system. In the graph that follows, the solution for $y \geq 2x + 4$ is shown with horizontal shading; the solution for $y \leq -x + 2$ is shown with vertical shading. The solution for the given system is the region shaded with *both* horizontal and vertical lines, including the boundaries.

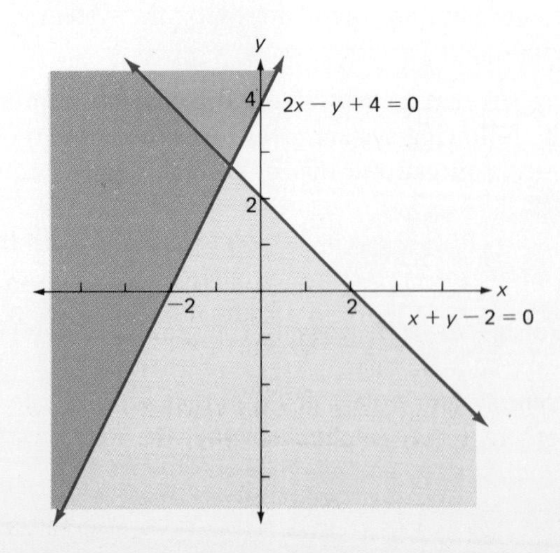

Classify each system by converting the equations into the slope-intercept form.

1 $5x - 3y = 3$
$3y + 5x = 6$

2 $4x + 6y = 30$
$\frac{2}{3}x = 5 - y$

3 $4x + 5y = 6$
$\frac{5}{2}y + 3 = -2x$

4 $\frac{1}{2}x + 3 = \frac{2}{3}y$
$2x + 3y = 15$

5 $\frac{1}{2}x - \frac{2}{3}y = 4$
$3x - 4y = 24$

6 $3x - \frac{1}{2}y = 1$
$2y - 6 = 12x$

In Exercises 7–16, decide if the given system is consistent, inconsistent, or dependent by using any one of the algebraic methods of solving a system of two linear equations. You should arrive at either a common solution (consistent), a false statement (inconsistent), or an identity (dependent).

7 $x + 2y = 3$
$6x + 5y = 4$

8 $x + 2y = 3$
$10x + 20y = 30$

9 $x + 2y = 3$
$-x - 2y = 3$

10 $8x + 4y = 12$
$y = -2x + 3$

11 $4x - 12y = 3$
$x + \frac{1}{3}y = 3$

12 $-7x + y = 2$
$28x - 4y = -2$

13 $2x + 5y = -20$
$x = -\frac{5}{2}y - 10$

14 $x - y = 3$
$-\frac{1}{3}x + \frac{1}{3}y = 1$

15 $x - 5y = 15$
$0.01x - 0.05y = 0.5$

16 $4y = 3x + 2$
$2x = 3y - 3$

Decide whether each of the systems gives parallel or perpendicular lines.

17 $2x - 3y = 4$
$6y = 4x - 30$

18 $2x - 3y = 4$
$3x + 2y = 1$

Graph each system of inequalities.

19 $y > x + 1$
$y < 1 - x$

20 $y > 2x + 2$
$y > -x$

21 $x + 2y - 2 \geq 0$
$x - 2y + 2 \geq 0$

22 $2x - y + 1 \geq 0$
$x - 2y + 2 \leq 0$

23 $y \geq -4x + 1$
$y \leq 4x + 1$

24 $y \leq -4x + 1$
$y \geq 4x + 1$

25 $x - y \leq -1$
$2x + y \geq 7$
$4x - y \leq 11$

26 $x - y \geq -1$
$2x + y \geq 7$
$4x - y \leq 11$

27 $x - y \geq -1$
$2x + y \leq 7$
$4x - y \leq 11$
$x \geq 0$
$y \geq 0$

28 Suppose that someone asked you to find the two numbers in the following puzzle:

The larger of two numbers is 16 more than twice the smaller. The difference between $\frac{1}{4}$ of the larger and $\frac{1}{2}$ the smaller is 2.

Why can you say that there is no answer possible for this puzzle?

29 How many answers are there for the following puzzle?

The difference between two numbers is 3. The larger number decreased by 1 is the same as $\frac{1}{3}$ the sum of the smaller plus twice the larger.

30 How many answers are there to the puzzle in Exercise 29 if the difference between the two numbers is 2?

***31** A bag containing a mixture of 6 oranges and 12 tangerines sold for \$2.34. A smaller bag containing 2 oranges and 4 tangerines sold for 77¢. An alert shopper asked the salesclerk if it was a better buy to purchase the larger bag. The clerk was not sure, but said that it really made no difference because the price of each package was based on the same unit price for each kind of fruit. Why was the clerk wrong?

▪ 3.8 Extracting functions from geometric figures

You have recently learned how linear functions correspond to straight lines. This connection and many other interrelationships between algebra and geometry are vital to the study of mathematics. The purpose of this section is to learn how to "extract" algebraic functions from a variety of geometric

▪ The use of this symbol indicates that the material in this section is particularly useful as background for the study of calculus.

A summary of some useful geometric formulas is given on the back inside cover of the text.

situations, many of which make use of straight lines and linear functions. This process will prove to be of great value to you when you study applications of calculus.

EXAMPLE 1 In the figure, right triangle ABE is similar to triangle ACD; $CD = 8$ and $BC = 10$; h and x are the measures of the altitude and base of triangle ABE. Express h as a function of x.

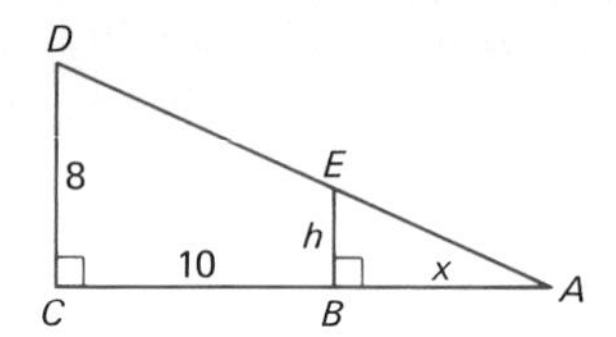

Solution Since corresponding sides of similar triangles are proportional, $\frac{BE}{AB} = \frac{CD}{AC}$. Substitute as follows:

$$\frac{h}{x} = \frac{8}{x + 10} \qquad AC = AB + BC = x + 10$$

$$h = \frac{8x}{x + 10} \qquad \text{multiply by } x$$

To emphasize that h is a function of x, use the functional notation to write the answer in this form:

$$h(x) = \frac{8x}{x + 10}$$

EXAMPLE 2 A water tank is in the shape of a right circular cone with altitude 30 feet and radius 8 feet. The tank is filled to a depth of h feet. Let x be the radius of the circle at the top of the water level. Solve for h in terms of x and use this to express the volume of water as a function of x.

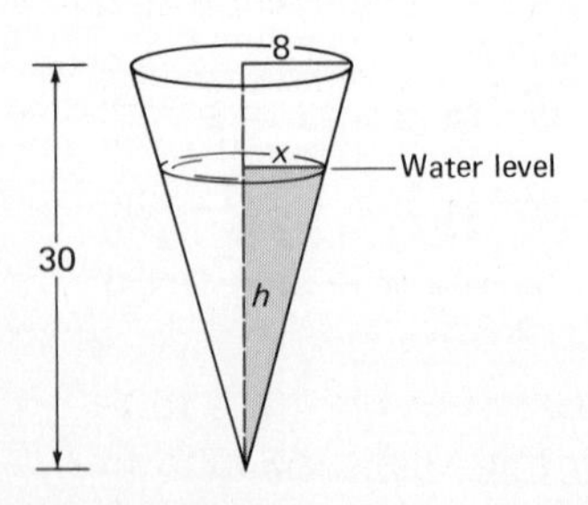

Solution The shaded right triangle is similar to the larger right triangle having base 8 and altitude 30. Therefore,

$$\frac{h}{x} = \frac{30}{8}$$

$$h = \frac{15}{4}x \qquad \text{solve for } h$$

Now substitute for h in the formula for the volume of a right circular cone.

$$V = \tfrac{1}{3}\pi r^2 h \qquad \text{volume of a right circular cone}$$
$$V(x) = \tfrac{1}{3}\pi x^2(\tfrac{15}{4}x) \qquad \text{substitute for } h \text{ and } r$$
$$= \tfrac{5}{4}\pi x^3$$

EXAMPLE 3 A window is in the shape of a rectangle with a semicircular top as shown. The perimeter of the window is 15 feet. Use r as the radius of the semicircle and express the area of the window as a function of r.

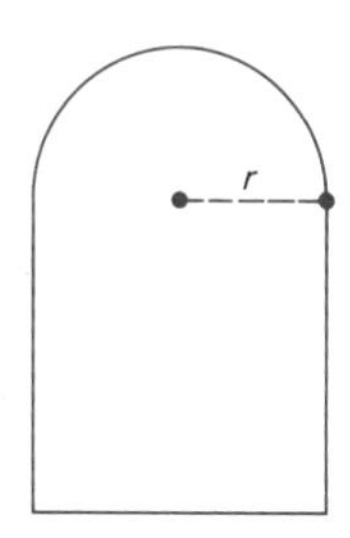

Solution The length of the semicircle is $\frac{1}{2}(2\pi r) = \pi r$. Now subtract this from 15 to get

$$15 - \pi r \longleftarrow \text{total length of the three straight sides}$$

The base of the rectangle has length $2r$. Then

$$\frac{(15 - \pi r) - 2r}{2} \qquad \text{length of each vertical side}$$

Using A for the area of the window, we have

$$A(r) = \text{area of semicircle} + \text{area of rectangle}$$
$$= \tfrac{1}{2}\pi r^2 + 2r\left(\frac{15 - \pi r - 2r}{2}\right)$$
$$= \tfrac{1}{2}\pi r^2 + 15r - \pi r^2 - 2r^2$$
$$= 15r - \tfrac{1}{2}\pi r^2 - 2r^2$$

EXAMPLE 4 A 50-inch piece of wire is to be cut into two parts AP and PB as shown. If part AP is to be used to form a square and PB is used to form a circle, express the total area enclosed by these figures as a function of x.

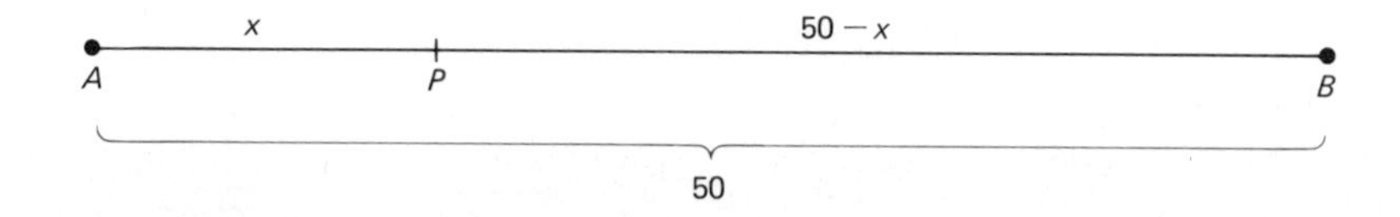

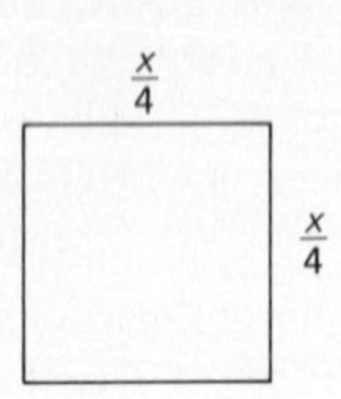

Solution Since the perimeter of the square is x, one of its sides is $\frac{x}{4}$. Then

$$(\tfrac{1}{4}x)(\tfrac{1}{4}x) = \frac{x^2}{16} \qquad \text{area of square}$$

The circumference of the circle is $50 - x$, which can be used to find the radius of the circle.

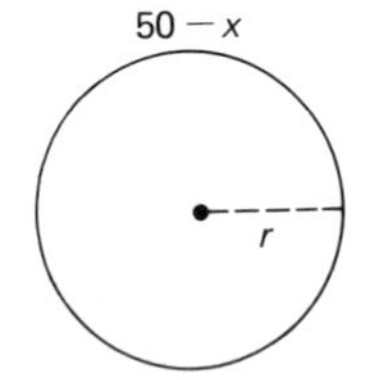

$$2\pi r = 50 - x$$

$$r = \frac{50 - x}{2\pi}$$

Then

$$\pi r^2 = \pi\left(\frac{50 - x}{2\pi}\right)^2 \qquad \text{area of circle}$$

$$= \frac{(50 - x)^2}{4\pi}$$

Using F to represent the sum of the areas, we have

$$F(x) = \frac{x^2}{16} + \frac{(50 - x)^2}{4\pi}$$

EXERCISES 3.8

1 In the figure, the shaded right triangle, with altitude x, is similar to the larger triangle that has altitude h. Express h as a function of x.

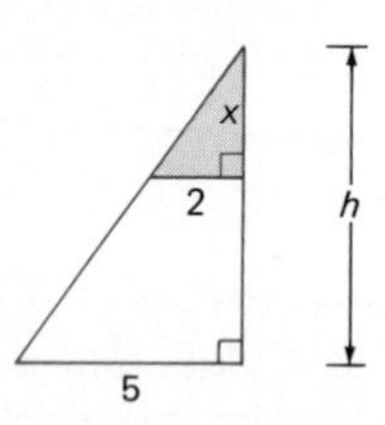

2 Use the result in Exercise 1 to express the area A of the larger triangle as a function of x.

3 In the figure, the shaded triangle is similar to triangle ABC. If $BC = 20$ and the altitude of triangle $ABC = 9$, express w as a function of the altitude h of the shaded triangle.

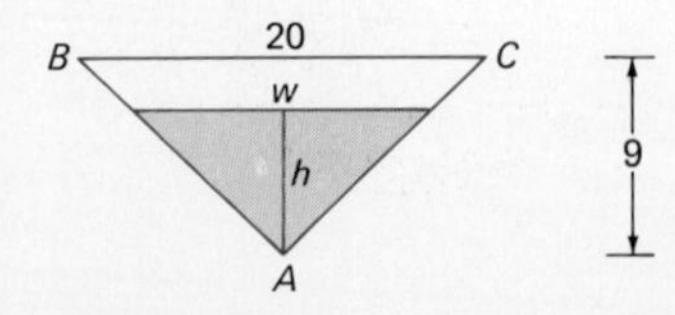

4 In Exercise 3, express the area of the shaded triangle as a function of w.

5 In the figure, s is the length of the shadow cast by a 6-foot person standing x feet from a light source that is 24 feet above the level ground. Express s as a function of x.

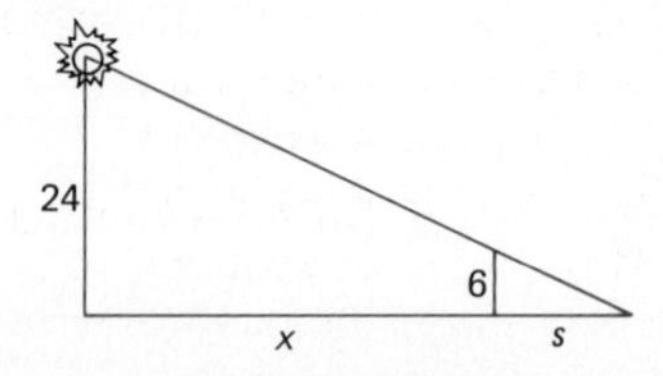

6 Triangle ABC is an isosceles right triangle with right angle at C. h is the measure of the perpendicular from C to side AB. Express the area of triangle ABC as a function of h.

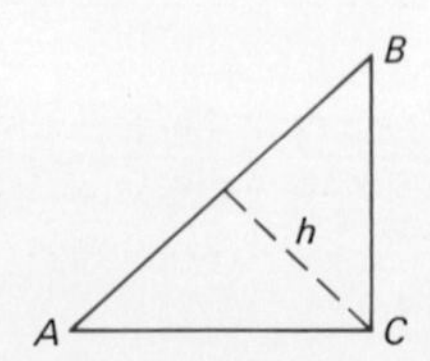

7 Express the area of rectangle $PQRS$ as a function of $x = OP$.

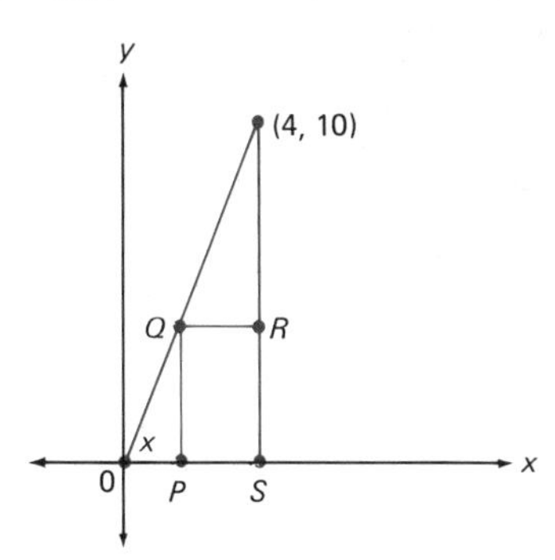

8 Express the area of triangle OPQ in Exercise 7 as a function of x.

9 A square piece of tin is 50 centimeters on a side. Congruent squares are cut from the four corners of the square so that when the sides are folded up a rectangular box (without a top) if formed. If the four congruent squares are x centimeters on a side, what is the volume of the box?

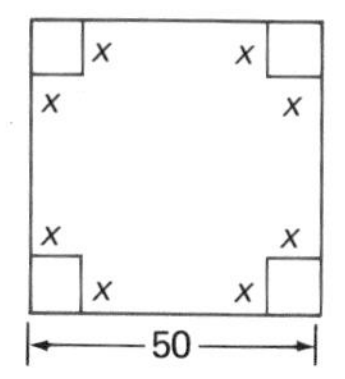

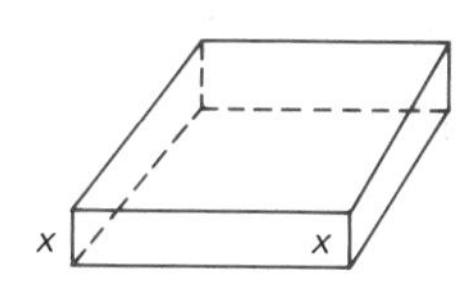

10 Replace the square piece of tin in Exercise 9 by a rectangular piece of tin with dimensions 30 centimeters by 60 centimeters and find the volume of the resulting box in terms of x.

11 An athletic field is semicircular at each end as shown. If the radius of each semicircle is r, and if the total perimeter of the field is 400 meters, express the area of the field in terms of r.

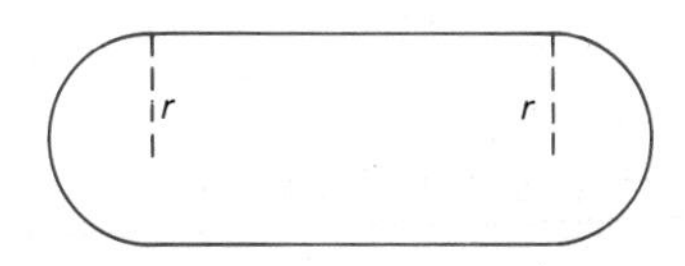

12 A closed tin can with height h and radius r has volume 5 cubic centimeters. Solve for h in terms of r, and express the surface area of the tin can as a function of r.

13 A closed rectangular shaped box is x units wide and it is twice as long. Let h be the altitude of the box. If the total surface area of the box is 120 square units, express the volume of the box as a function of x. (*Hint:* First solve for h in terms of x.)

***14** The vertices of the right triangle are $(0, y)$, $(0, 0)$, and $(x, 0)$. The hypotenuse passes through $(3, 1)$. Express the area of the triangle as a function of x.

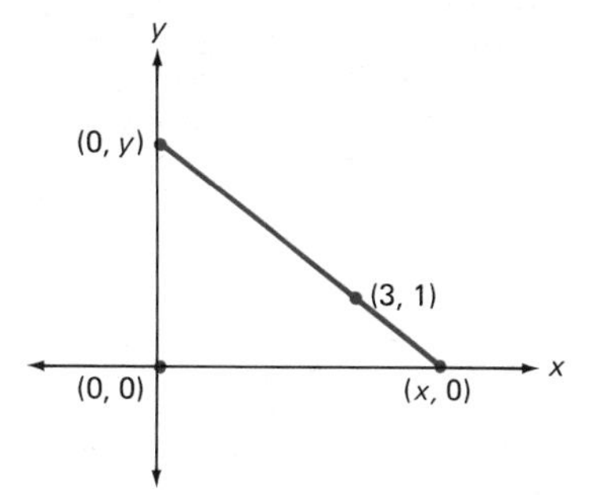

***15** A shaded rectangle is inscribed in an isosceles triangle as shown. Express the area of the rectangle as a function of x.

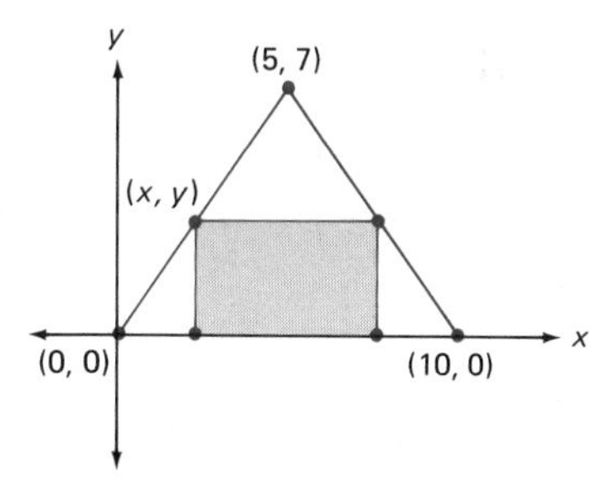

***16** The figure represents a solid with a circular base having radius 2. A plane cutting the solid perpendicular to the xy-plane and to the x-axis, between -2 and 2, forms a cross section in the shape of a square. Express the area of the cross-section in terms of the variable x. (*Hint:* The length of a side of the square is $2y$.)

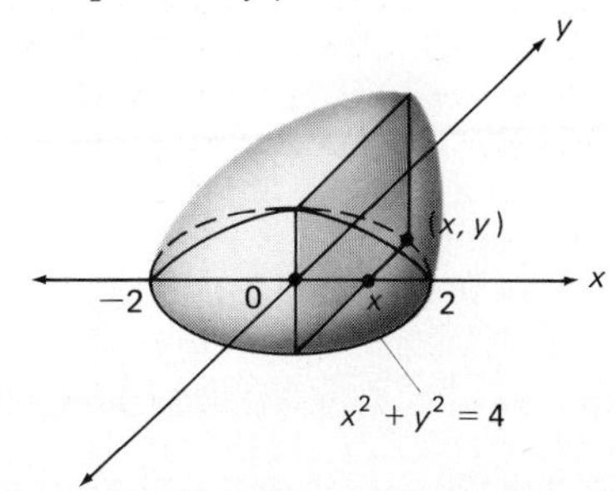

REVIEW EXERCISES FOR CHAPTER 3

The solutions to the following exercises can be found within the text of Chapter 3. Try to answer each question without referring to the text.

Section 3.1

1 State the definition of a function.

2 Explain why the equation $y = \frac{1}{\sqrt{x-1}}$ defines y to be a function of x.

3 Decide whether these two equations define y to be a function of x.

$$y = \begin{cases} 3x - 1 & \text{if } x \leq 1 \\ 2x + 1 & \text{if } x \geq 1 \end{cases}$$

Find the domain.

4 $y = \frac{2x}{x^2 - 4}$

5 $y = |x|$

6 $y = \sqrt{x}$

7 $y = 6x^4 - 3x^2 + 7x + 1$

8 $y = \sqrt{1 - x^2}$

9 $y = \frac{1}{\sqrt{x-1}}$

10 For $g(x) = \frac{1}{x}$, find: **(a)** $3g(x)$; **(b)** $g(3x)$; **(c)** $g(3) + g(x)$; **(d)** $g(3 + x)$.

11 For $g(x) = x^2$, evaluate and simplify the difference quotient $\frac{g(x) - g(4)}{x - 4}$.

12 For $g(x) = \frac{1}{x}$, evaluate and simplify the difference quotient $\frac{g(4 + h) - g(4)}{h}$.

Section 3.2

13 Graph the equation $y = x + 2$.

14 Find the x- and y-intercepts for $y = x + 2$.

15 Graph the linear equation $y = 2x - 1$ using the intercepts.

16 Graph $y = 2x - 1$ for $-2 \leq x \leq 1$.

17 Graph: **(a)** $y < 2x - 1$; **(b)** $y > 2x - 1$.

Section 3.3

18 State the definition of the slope of a line.

19 What is the slope of the line through the points (x_1, y_1) and (x_2, y_2)?

20 Find the slope of the line determined by the points $(-3, 4)$ and $(1, -6)$.

21 Graph the line with slope 3 that passes through the point $(-2, -2)$.

22 What is the slope of a horizontal line? Of a vertical line?

23 What is the relationship between the slopes of two perpendicular lines?

Section 3.4

24 Write the slope-intercept form of a line where m is the slope and k is the y-intercept.

25 Write the point-slope form of a line where m is the slope and (x_1, y_1) is a point on the line.

26 Graph the linear function f defined by $y = f(x) = 2x - 1$ by using the slope and y-intercept.

27 Find the equation of the line with slope $\frac{2}{3}$ passing through the point $(0, -5)$.

28 Describe the graph of a constant function.

29 Write the point-slope form of a line with $m = 3$ that passes through the point $(-1, 1)$.

30 Write the slope-intercept form of the line through the points $(6, -4)$ and $(-3, 8)$.

31 Write the equation of the line that is perpendicular to the line $5x - 2y = 2$ and that passes through the point $(-2, -6)$.

32 Find the slope and y-intercept of the line $-6x + 2y = 5$.

Section 3.5

33 Graph $y = |x - 2|$.

34 Graph the function f defined by the following two equations.

$$y = f(x) = \begin{cases} 2x & \text{if } 0 \leq x \leq 1 \\ -x + 2 & \text{if } 1 < x \end{cases}$$

35 Graph the function f given by $y = f(x) = \frac{|x|}{x}$. What is the domain of f?

Section 3.6

36 Solve and graph this system:

$$2x + 3y = 12$$
$$3x + 2y = 12$$

Solve each system.

37 $\frac{1}{3}x - \frac{2}{5}y = 4$
$7x + 3y = 27$

38
$$\begin{aligned} 2x - 5y + z &= -10 \\ x + 2y + 3z &= 26 \\ -3x - 4y + 2z &= 5 \end{aligned}$$

39 A field goal in basketball is worth 2 points and a free throw is worth 1 point. In a recent game the school basketball team scored 85 points. If there were twice as many field goals as free throws, how many of each were there?

Section 3.7

Classify each system as consistent, inconsistent, or dependent.

40 $\begin{aligned} 39x - 91y &= -28 \\ 6x - 14y &= 7 \end{aligned}$

41 $\begin{aligned} -6x + 3y &= 9 \\ 10x + 5y &= -1 \end{aligned}$

42 $\begin{aligned} y &= \tfrac{2}{3}x - 5 \\ 2x - 3y &= 15 \end{aligned}$

43 Decide if the system in Exercise 41 is consistent or inconsistent by using the slope-intercept form.

44 Graph this system of inequalities:

$$2x - y + 4 \leq 0$$
$$x + y - 2 \leq 0$$

Section 3.8

45 A water tank is in the shape of a right circular cone with altitude 30 feet and radius 8 feet. The tank is filled to a depth of h feet. Let x be the radius of the circle at the top of the water level. Solve for h in terms of x and use this to express the volume of water as a function of x.

46 A 50-inch piece of wire is to be cut into two parts. If one part is used to form a square and the other part to form a circle, express the total areas of these figures as a function of one of the parts x.

SAMPLE TEST QUESTIONS FOR CHAPTER 3

Use these questions to test your knowledge of the basic skills and concepts of Chapter 3. Then check your answers with those given at the end of the book.

1 Find the domain of the function given by $y = \sqrt{x^2 - 1}$.

2 Let $g(x) = \dfrac{3}{x}$. Find **(a)** $g(2 + x)$; **(b)** $g(2) + g(x)$.

3 Let $g(x) = \dfrac{2}{x}$. Evaluate and simplify:

$$\frac{g(3 + h) - g(3)}{h}.$$

4 Find the x- and y-intercepts and use these to graph $3x - 2y = 6$.

5 Graph $y = 2 - x$ for $-1 \leq x \leq 2$.

6 Find the slope of the line determined by the points $(2, -3)$ and $(-1, 4)$.

7 Find the equation of the line with slope $-\frac{3}{4}$ passing through the point $(-2, 3)$.

8 Write the slope-intercept form of the line through the points $(3, -5)$ and $(-2, 4)$.

9 Find the slope and y-intercept of $-3x + 4y = 2$.

10 Write an equation of the line through the point $(2, 8)$ and perpendicular to the line $y = -\frac{2}{5}x + 3$.

11 Graph; state the domain and range for $y = 2x + 1$.

12 Graph the step function $y = \dfrac{|x + 1|}{x + 1}$.

Solve each system.

13 $\begin{aligned} 2x - y &= 7 \\ -3x + 2y &= -11 \end{aligned}$

14 $\begin{aligned} 2x + 3y - z &= -1 \\ 3x - y + 2z &= 5 \\ -3x + 4y - 2z &= 1 \end{aligned}$

15 A firm mailed 40 letters at a total cost of \$10.55. Some of the letters cost 20¢ postage and the rest needed 37¢ postage. How many letters were mailed at each of these rates?

In Exercises 16–18, decide whether the system is consistent, inconsistent, or dependent. If it is consistent, find the solution.

16 $\begin{aligned} 4x - 5y &= 12 \\ -2x + \tfrac{5}{2}y &= -6 \end{aligned}$

17 $\begin{aligned} 7x + 4y &= 1 \\ -3x + 2y &= -19 \end{aligned}$

18 $\begin{aligned} -6x + 4y &= -24 \\ 9x - 6y &= 14 \end{aligned}$

19 Graph this system of inequalities:

$$x + y - 2 \leq 0$$
$$2x - y + 2 \geq 0$$

20 A square piece of tin is 40 centimeters on each side. Congruent squares of length x centimeters are cut from the four corners, and the sides are then folded up to form a rectangular box without a top. What is the volume of the box in terms of x?

ANSWERS TO THE TEST YOUR UNDERSTANDING EXERCISES

Page 79

1 Function; all reals.

2 Function; all real $x \neq -2$.

3 Function; all reals.

4 Not a function.

5 Function; $x \geq 2$ or $x \leq -3$.

6 Not a function.

7 Function; all $x \neq 0$.

8 Function; all $x \neq -1$.

Page 90

1 $\frac{6-5}{4-1} = \frac{1}{3}$

2 $\frac{3-(-5)}{-3-3} = -\frac{8}{6} = -\frac{4}{3}$

3 $\frac{1-(-3)}{-1-(-2)} = \frac{4}{1} = 4$

4 $\frac{1-0}{0-(-1)} = \frac{1}{1} = 1$

5

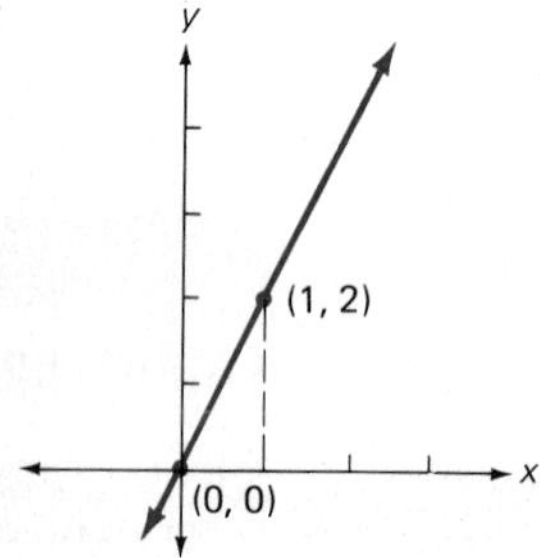

6

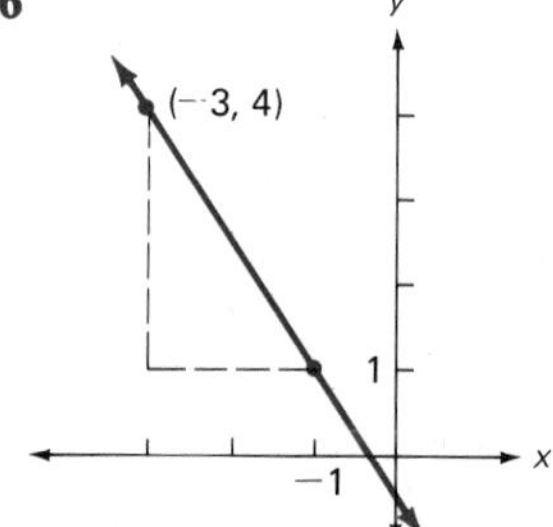

Page 96

1 $y = 2x - 2$

2 $y = -\frac{1}{2}x$

3 $y = \sqrt{2}x + 1$

4 $y - 6 = -3(x - 2)$

5 $y - 4 = \frac{1}{2}(x + 1)$

6 $y + \frac{2}{3} = 1(x - 5)$

7 $y = -\frac{1}{4}x$

8 $y + 5 = 0(x + 3)$

9 $y + 1 = -(x - 1)$

10 Exercise 8; $y = -5$ and the range is -5.

Page 102

1 12	**2** 12	**3** 12	**4** 13	**5** −4	**6** −4
7 −3	**8** 0	**9** −1	**10** 0	**11** −1	**12** 0

Page 104

1 (1, 2) **2** (−3, 4) **3** (2, −1) **4** $(\frac{1}{5}, \frac{1}{5})$ **5** (−4, 1) **6** (35, 50)

Page 105

1 (−1, 2) **2** $(\frac{1}{2}, -3)$ **3** $(\frac{3}{2}, \frac{5}{7})$ **4** (0, 0)

Page 112

1 Inconsistent.

2 Dependent.

3 Inconsistent.

4 Consistent.

5 Consistent.

6 Dependent.

QUADRATIC FUNCTIONS AND EQUATIONS

4

4.1 Graphing quadratic functions

A function defined by a polynomial expression of degree 2 is referred to as a **quadratic function** in x. Thus the following are all examples of quadratic functions in x:

$$f(x) = -3x^2 + 4x + 1$$
$$g(x) = 7x^2 - 4$$
$$h(x) = x^2$$

The most general form of such a quadratic function is

$$f(x) = ax^2 + bx + c$$

where a, b, and c represent constants, with $a \neq 0$. If $a = 0$, then the resulting polynomial no longer represents a quadratic function; $f(x) = bx + c$ is a linear function.

The simplest quadratic function is given by $f(x) = x^2$. The graph of this quadratic function will serve as the basis for drawing the graph of any quadratic function $f(x) = ax^2 + bx + c$. We can save some labor by noting the *symmetry* that exists. For example, note the following:

$$f(-3) = f(3) = 9$$
$$f(-1) = f(1) = 1$$

$$f(-\tfrac{1}{2}) = f(\tfrac{1}{2}) = \tfrac{1}{4}$$

In general, for this function,

$$f(-x) = (-x)^2 = x^2 = f(x)$$

Note: When $f(-x) = f(x)$, the graph is said to be *symmetric* with respect to the y-axis.

Greater accuracy can be obtained by using more points. But since we can never locate an infinite number of points, we must admit that there is a certain amount of faith involved in connecting the points as we did.

The accompanying table of values gives several ordered pairs of numbers that are coordinates of points on the graph of $y = x^2$. When these points are located on a rectangular system and connected by a smooth curve, the graph of $y = f(x) = x^2$ is obtained.

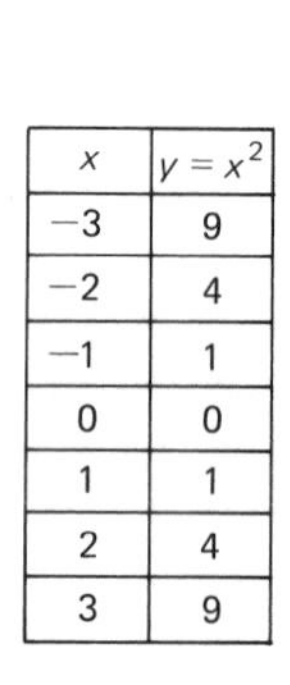

x	$y = x^2$
-3	9
-2	4
-1	1
0	0
1	1
2	4
3	9

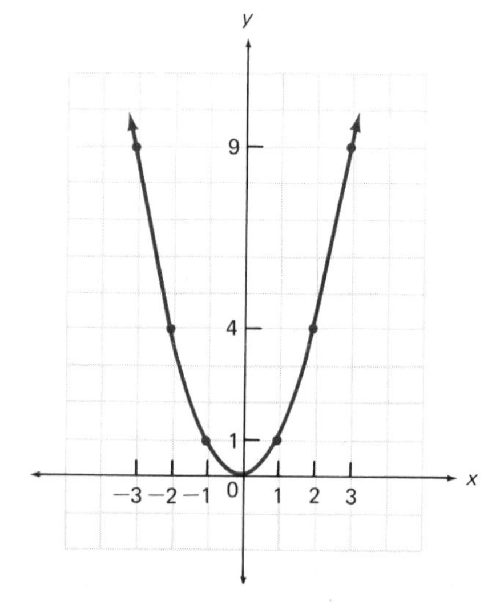

The curve is called a **parabola**, and every quadratic function $y = ax^2 + bx + c$ has such a parabola as its graph. The domain of the function is the set of all real numbers.

An important feature of such a parabola is that it is symmetric about a vertical line called its **axis of symmetry**. The graph of $y = x^2$ is symmetric with respect to the y-axis. This symmetry is due to the fact that $(-x)^2 = x^2$.

The parabola has a *turning point*, called the **vertex**, which is located at the intersection of the parabola with its axis of symmetry. For the preceding graph the coordinates of the vertex are (0, 0).

From the graph you can see that, reading from left to right, the curve is "falling" down to the origin and then is "rising." These features are technically described as f **decreasing** and f **increasing**.

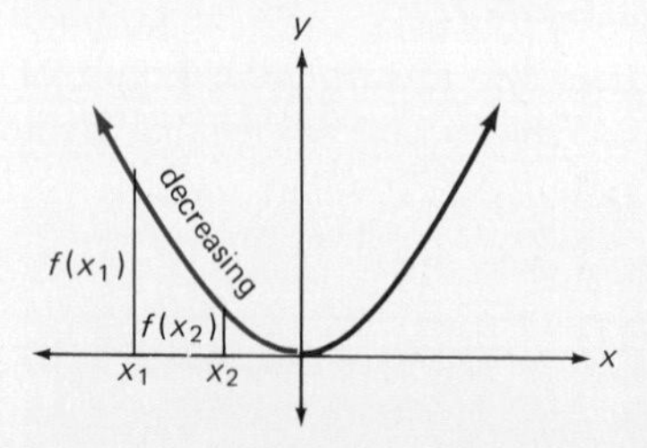

$f(x) = x^2$ is decreasing on $(-\infty, 0]$, because for *each* pair x_1, x_2 in this interval, if $x_1 < x_2$, then $f(x_1) > f(x_2)$.

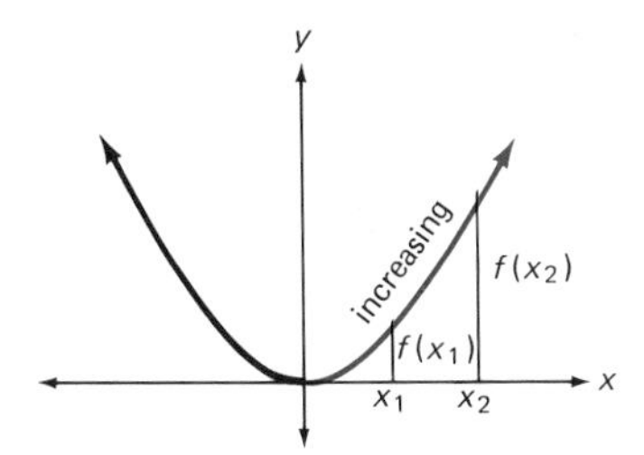

$f(x) = x^2$ is increasing in $[0, \infty)$, because for *each* pair x_1, x_2 in this interval, if $x_1 < x_2$, then $f(x_1) < f(x_2)$.

The graph of $y = -x^2$ may be obtained by multiplying each of the ordinates of $y = x^2$ by -1. This step has the effect of "flipping" the parabola $y = x^2$ downward, a *reflection* in the x-axis. Since the graph of $y = x^2$ bends "upward," we say that the curve is **concave up**. Also, since $y = -x^2$ bends "downward," we say that the curve is **concave down**.

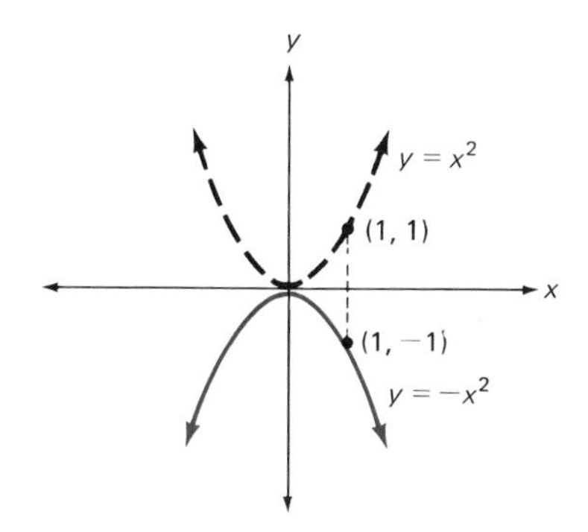

Next consider $y = 2x^2$. It is clear from this equation that the y-values can be obtained by multiplying x^2 by 2. So we may take the graph of $y = x^2$ and double or multiply (stretch) its ordinates by 2 to locate the points on the parabola $y = 2x^2$. Similarly, to obtain the graph of $y = \frac{1}{2}x^2$ we divide (shrink) the y-values of $y = x^2$ by 2.

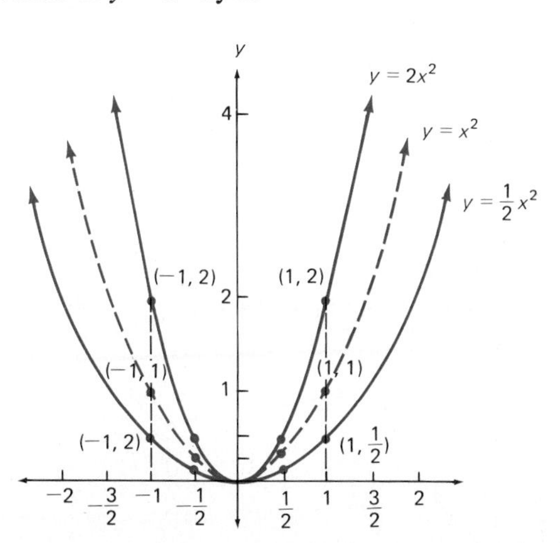

This figure compares the graphs of $y = 2x^2$ and $y = \frac{1}{2}x^2$ to the graph of $y = x^2$.

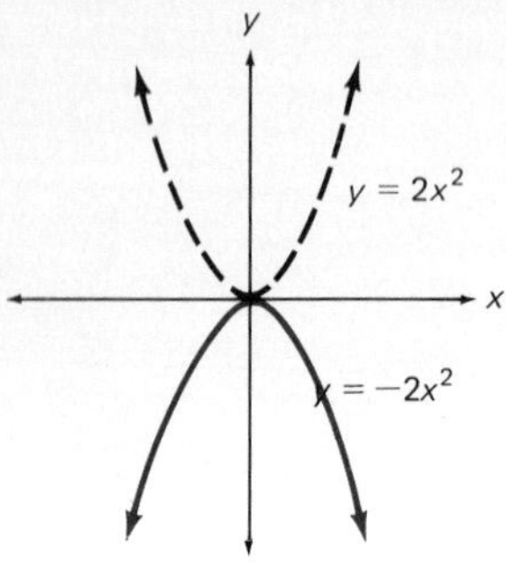

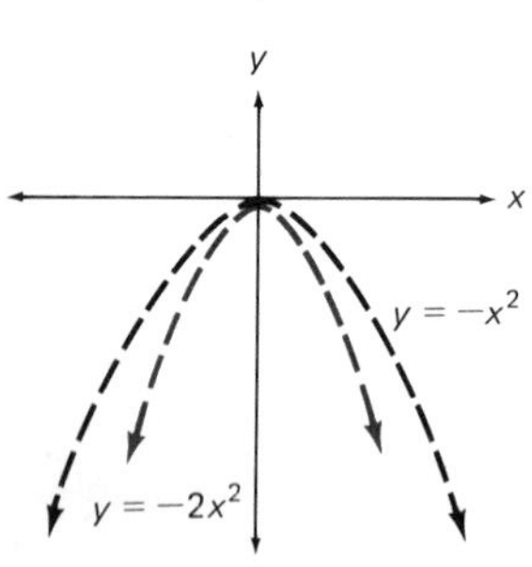

To graph $y = -2x^2$ you may first graph $y = 2x^2$ as before and then draw the reflection in the x-axis, or you may first graph $y = -x^2$ and then multiply by 2, as shown in the figures in the margin.

Now consider the quadratic function $y = g(x) = (x - 1)^2$. If we write x^2 as $(x - 0)^2$, then a useful comparison between these two functions can be made:

$$y = f(x) = (x - 0)^2$$

$$y = g(x) = (x - 1)^2$$

Just as $x = 0$ is the axis of symmetry for the graph of f, so $x = 1$ is the axis of symmetry for g. Similarly, the parabola $y = (x + 1)^2 = [x - (-1)]^2$ has $x = -1$ as an axis of symmetry.

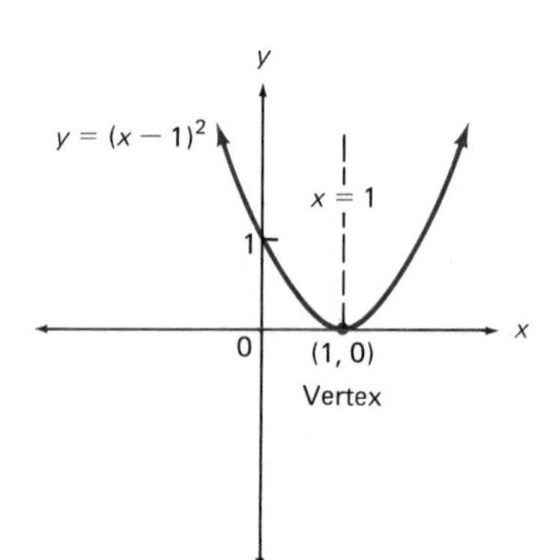

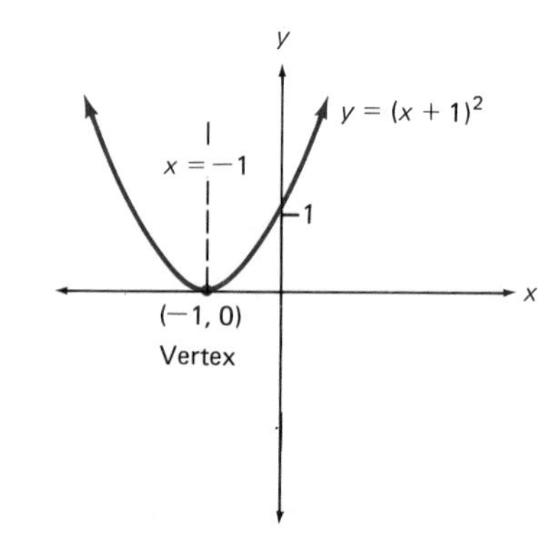

Both of these parabolas are congruent to the basic parabola $y = x^2$. Each may be graphed by *translating* (shifting) the parabola $y = x^2$ by 1 unit, to the right for $y = (x - 1)^2$ and to the left for $y = (x + 1)^2$.

TEST YOUR UNDERSTANDING

Match each graph with one of the given quadratic equations.

1

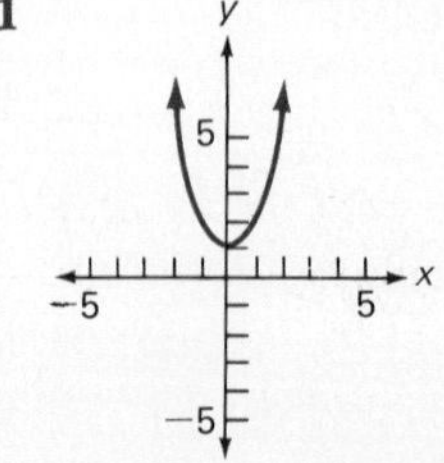

2

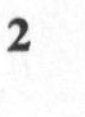

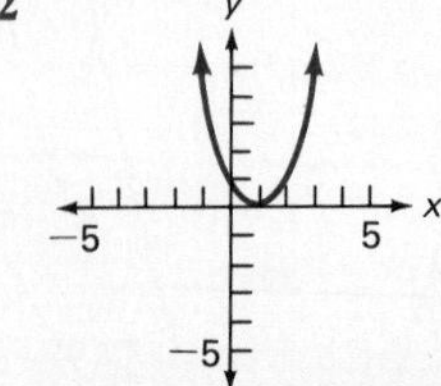

(a) $y = x^2 - 1$
(b) $y = x^2 + 1$
(c) $y = (x - 1)^2$
(d) $y = (x + 1)^2$
(e) $y = -x^2 + 1$
(f) $y = -(x + 1)^2$

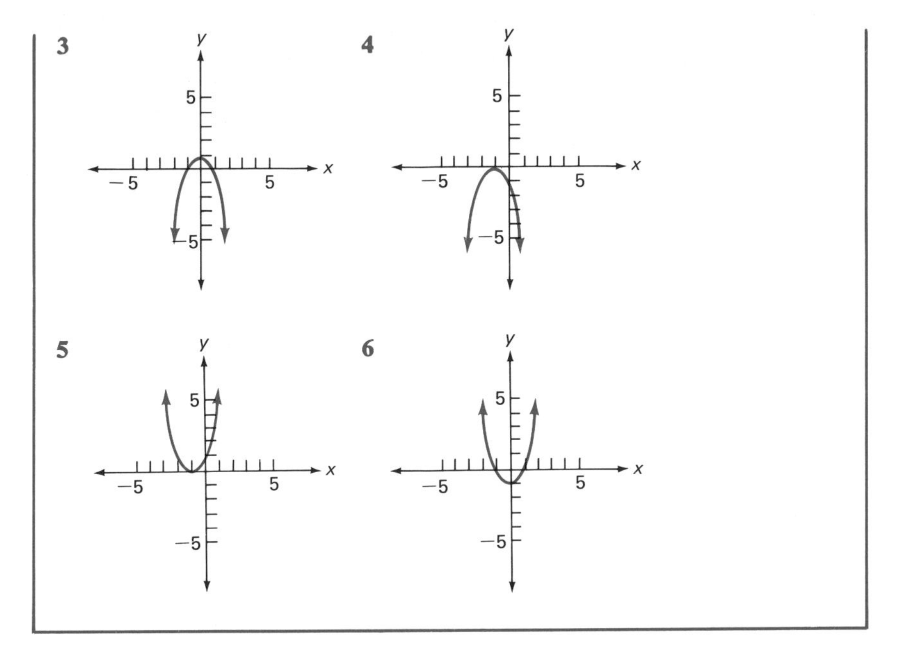

Let us now put several ideas together and draw the graph of this function:

$$y = f(x) = (x + 2)^2 - 2$$

An effective way to do this is to begin with the graph of $y = x^2$, shift the graph 2 units to the left for $y = (x + 2)^2$, and then 2 units down for the graph of $f(x) = (x + 2)^2 - 2$.

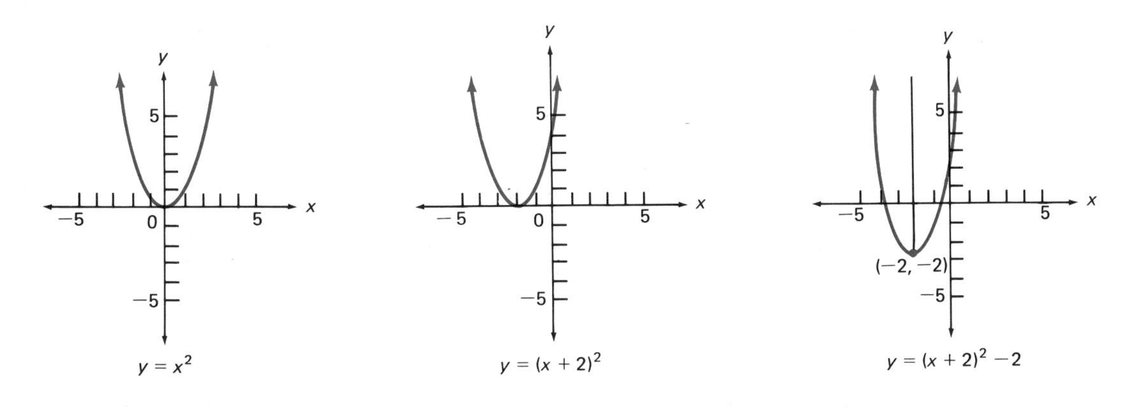

Note that the graph of $y = (x + 2)^2 - 2$ is congruent to the graph of $y = x^2$. The vertex of the curve is at $(-2, -2)$, and the axis of symmetry is the line $x = -2$. The minimum value of the function, -2, occurs at the vertex. Also observe that the domain consists of all numbers x, and the range consists of all numbers $y \geq -2$.

We may generalize our results thus far as follows:

The graph of $y = a(x - h)^2 + k$ is congruent to the graph of $y = ax^2$, but is shifted h units horizontally, and k units vertically.
The horizontal shift is to the right if $h > 0$, and to the left if $h < 0$.
The vertical shift is upward if $k > 0$, and downward if $k < 0$.
The vertex is at (h, k) and the axis of symmetry is the line $x = h$.
If $a < 0$, the parabola opens downward, and k is the maximum value.
If $a > 0$, the parabola opens upward, and k is the minimum value.

EXAMPLE 1 Graph the parabola $y = -2(x - 3)^2 + 4$.

Solution The graph will be a parabola congruent to $y = -2x^2$, with vertex at (3, 4) and with $x = 3$ as axis of symmetry. A brief table of values, together with the graph, is shown.

The function is increasing on $(-\infty, 3]$, decreasing on $[3, \infty)$, and the curve is concave down.

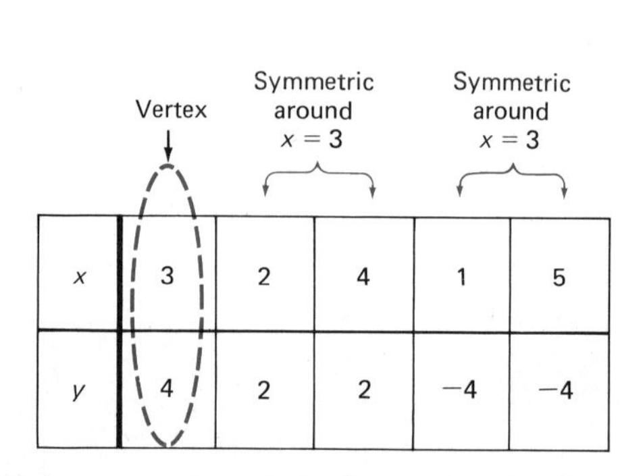

x	3	2	4	1	5
y	4	2	2	−4	−4

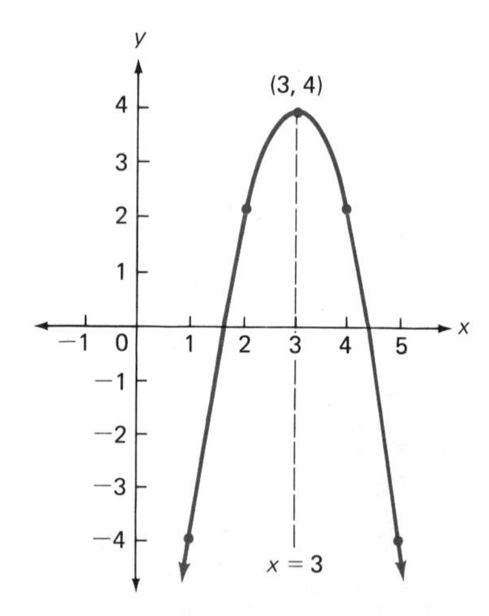

We can extend these ideas to statements of inequality as well. The graph of $y \geq x^2$ is shown below. It consists of all the points where $y = x^2$, as well as the points above the curve where $y > x^2$. (The unshaded region of the plane represents the graph of $y < x^2$.)

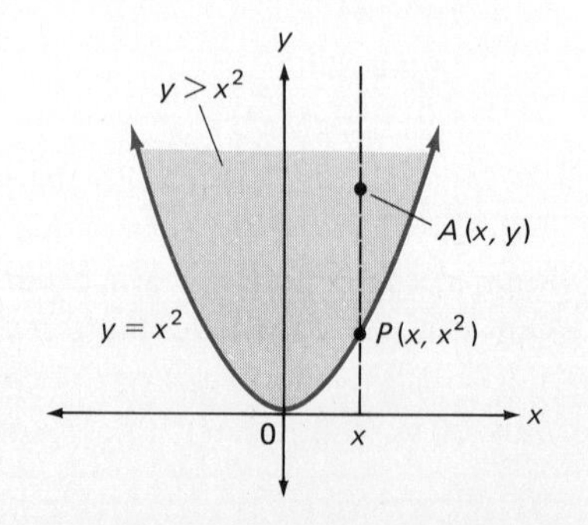

Point $A(x, y)$ is directly above point $P(x, x^2)$ for the same x. So the y-value of A satisfies $y > x^2$.

EXERCISES 4.1

Draw each set of graphs on the same axes.

1 (a) $y = x^2$ (b) $y = (x - 1)^2$
(c) $y = (x - 1)^2 + 3$

2 (a) $y = x^2$ (b) $y = (x + 1)^2$
(c) $y = (x + 1)^2 - 3$

3 (a) $y = -x^2$ (b) $y = -(x - 1)^2$
(c) $y = -(x - 1)^2 + 3$

4 (a) $y = -x^2$ (b) $y = -(x + 1)^2$
(c) $y = -(x + 1)^2 - 3$

5 (a) $y = x^2$ (b) $y = 2x^2$
(c) $y = 3x^2$

6 (a) $y = -x^2$ (b) $y = -\frac{1}{2}x^2$
(c) $y = -\frac{1}{2}x^2 + 1$

Draw the graph of each function.

7 $y = x^2 + 3$
8 $y = (x + 3)^2$
9 $y = -x^2 + 3$
10 $y = -(x + 3)^2$
11 $y = 3x^2$
12 $y = 3x^2 + 1$
13 $y = \frac{1}{4}x^2$
14 $y = \frac{1}{4}x^2 - 1$
15 $y = \frac{1}{4}x^2 + 1$
16 $y = -2x^2$
17 $y = -2x^2 + 2$
18 $y = -2x^2 - 2$

Graph each of the following functions. Where is the function increasing and decreasing? Discuss concavity.

19 $f(x) = (x - 1)^2 + 2$
20 $f(x) = (x + 1)^2 - 2$
21 $f(x) = -(x + 1)^2 + 2$
22 $f(x) = -(x + 1)^2 - 2$
23 $y = f(x) = 2(x - 3)^2 - 1$
24 $y = f(x) = 2(x + \frac{5}{4})^2 + \frac{5}{4}$

For Exercises 25–30, state (a) the coordinates of the vertex, (b) the equation of the axis of symmetry, (c) the domain, and (d) the range.

25 $y = (x - 3)^2 + 5$
26 $y = (x + 3)^2 - 5$
27 $y = -(x - 3)^2 + 5$
28 $y = -(x + 3)^2 - 5$
29 $y = 2(x + 1)^2 - 3$
30 $y = \frac{1}{2}(x - 4)^2 + 1$

31 Graph the function f where

$$f(x) = \begin{cases} x^2 & \text{if } -2 \leq x \leq 1 \\ x & \text{if } \ 1 < x \leq 3 \end{cases}$$

*32 Graph f where

$$f(x) = \begin{cases} x^2 - 9 & \text{if } -2 \leq x < 4 \\ -3x + 15 & \text{if } \ 4 \leq x < 6 \\ 3 & \text{if } \ x = 6 \end{cases}$$

33 Graph $x = y^2$. Why is y not a function of x?

34 Compare the graphs of $y = x^2$ and $y = |x|$. In what ways are they alike?

*35 What is the relationship between the graph of $y = x^2 - 4$ on a plane and of $x^2 - 4 > 0$ on a line?

*36 Repeat Exercise 35 for $y = x^2 - 9$ and for $x^2 - 9 < 0$.

*37 The graph of $y = ax^2$ passes through the point $(1, -2)$. Find a.

*38 The graph of $y = ax^2 + c$ has its vertex at $(0, 4)$ and passes through the point $(3, -5)$. Find the values for a and c.

*39 Find the value for k so that the graph of $y = (x - 2)^2 + k$ will pass through the point $(5, 12)$.

*40 Find the value for h so that the graph of $y = (x - h)^2 + 5$ will pass through the point $(3, 6)$.

Graph each of the following inequalities.

41 $y \geq (x + 2)^2$
42 $y \leq (x + 2)^2$
43 $y \leq -x^2$
44 $y \leq -(x + 2)^2$
45 $y \geq x^2 - 4$
46 $y \geq -x^2 + 4$

4.2 Completing the square

When a quadratic function is given in the form $f(x) = ax^2 + bx + c$, the properties of the graph are not evident. However, if this function is converted to the standard form $f(x) = a(x - h)^2 + k$, then the methods of Section 4.1 can be used to sketch the parabola. Our objective here is to learn how to make this algebraic conversion.

Let us begin with the quadratic function given by $y = x^2 + 4x + 3$. First rewrite the equation in this way:

$$y = (x^2 + 4x + \underline{?}) + 3$$

Note that if the question mark is replaced by 4, then we will have a perfect square within the parentheses. However, since this changes the given equation, we must also subtract 4. The completed work looks like this:

$$\begin{aligned} y &= x^2 + 4x + 3 \\ &= (x^2 + 4x + 4) + 3 - 4 \\ &= (x + 2)^2 - 1 \end{aligned}$$

From this last form you should recognize the graph to be a parabola with vertex at $(-2, -1)$, and $x = -2$ as axis of symmetry.

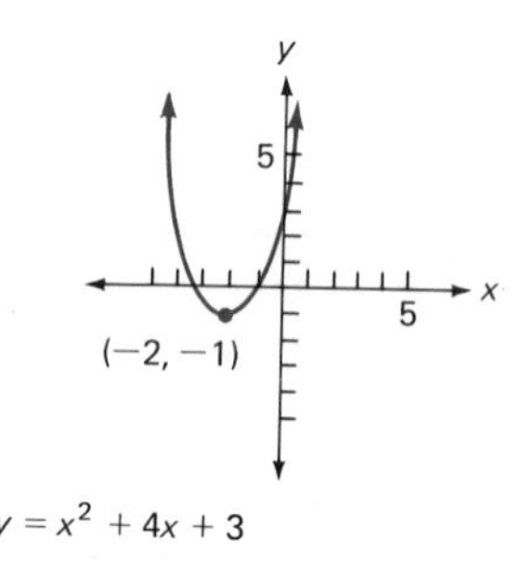

$y = x^2 + 4x + 3$

The technique that we have just used is called **completing the square.** Study these illustrations of perfect squares that have been completed. Note that in each case the coefficient of the x^2-term is 1.

$$x^2 + 8x + \underline{?} \longrightarrow x^2 + 8x + \underline{16} = (x + 4)^2$$

$$\frac{1}{2} \cdot 8 = 4$$

$$4^2 = 16$$

$$x^2 - 3x + \underline{?} \longrightarrow x^2 - 3x + \underline{\frac{9}{4}} = \left(x - \frac{3}{2}\right)^2$$

$$\frac{1}{2}(-3) = -\frac{3}{2}$$

$$\left(-\frac{3}{2}\right)^2 = \frac{9}{4}$$

$$x^2 + \frac{b}{a}x + \underline{?} \longrightarrow x^2 + \frac{b}{a}x + \underline{\frac{b^2}{4a^2}} = \left(x + \frac{b}{2a}\right)^2$$

$$\frac{1}{2}\left(\frac{b}{a}\right) = \frac{b}{2a}$$

$$\left(\frac{b}{2a}\right)^2 = \frac{b^2}{4a^2}$$

The process of completing the square makes use of one of these two identities:

$$(x + h)^2 = x^2 + 2hx + h^2$$

$$(x - h)^2 = x^2 - 2hx + h^2$$

In the trinomials, h^2 is the square of one-half the coefficient of the x-term (without regard to sign). That is, the third term $= [\frac{1}{2}(2h)]^2 = h^2$.

The process of completing the square can be extended to the case where the coefficient of x^2 is a number different than 1. Consider the equation $y = 2x^2 - 12x + 11$. The first step is to factor the coefficient of x^2 from the first two terms only.

$$y = 2x^2 - 12x + 11$$
$$= 2(x^2 - 6x + \underline{\ ?\ }) + 11$$

Next, add 9 within the parentheses to form the square $x^2 - 6x + 9 = (x - 3)^2$. However, because of the coefficient in front of the parentheses, we are really adding $2 \times 9 = 18$; thus 18 must also be subtracted.

$$y = 2(x^2 - 6x + 9) + 11 - 18$$
$$= 2(x - 3)^2 - 7$$

This is a parabola with vertex at $(3, -7)$.

EXAMPLE 1 Convert $-\frac{1}{3}x^2 - 2x + 1$ into the form $a(x - h)^2 + k$.

Example 1 illustrates the procedure for completing the square when the coefficient of x^2 is a negative number or a fraction.

Solution First factor $-\frac{1}{3}$ from the first two terms only.

$$-\tfrac{1}{3}x^2 - 2x + 1 = -\tfrac{1}{3}(x^2 + 6x) + 1$$

Next add 9 inside the parentheses to form the perfect square $x^2 + 6x + 9 = (x + 3)^2$. Because of the coefficient in front of the parentheses, however, we will really be adding $-\frac{1}{3}(9) = -3$. Thus 3 must also be added outside the parentheses.

$$-\tfrac{1}{3}x^2 - 2x + 1 = -\tfrac{1}{3}(x^2 + 6x + 9) + 1 + 3$$
$$= -\tfrac{1}{3}(x + 3)^2 + 4$$

The graph of the function $y = -\frac{1}{3}x^2 - 2x + 1$ is a parabola with vertex at $(-3, 4)$ and with $x = -3$ as axis of symmetry.

To match the general form $a(x - h)^2 + k$, the answer may be written as

$$-\tfrac{1}{3}[x - (-3)]^2 + 4$$

TEST YOUR UNDERSTANDING

Complete so as to express y as a perfect square trinomial.

1 $y = x^2 + 8x + \_$ **2** $y = x^2 + 10x + \_$

3 $y = x^2 - 6x + \_$ **4** $y = x^2 - 12x + \_$

5 $y = x^2 + 3x + \_$ **6** $y = x^2 - 5x + \_$

Write in standard form: $y = a(x - h)^2 + k$.

7 $y = x^2 + 4x - 3$ **8** $y = x^2 - 6x + 7$

9 $y = x^2 - 2x + 9$ **10** $y = 2x^2 + 8x - 1$

11 $y = -x^2 + x - 2$ **12** $y = \frac{1}{2}x^2 - 3x + 2$

We have seen that any quadratic expression $ax^2 + bx + c$ may be written in the form $a(x - h)^2 + k$. From this form we can identify the vertex, the axis of symmetry, and other information to help us graph the parabola, as illustrated in the next example.

EXAMPLE 2 Write $y = 3x^2 - 4x - 2$ in the form $y = a(x - h)^2 + k$. Find the vertex, axis of symmetry, and graph. On which interval is the function increasing or decreasing? What is the concavity?

Solution First complete the square.

Note: Since the original equation is in the form $y = ax^2 + bx + c$, it is very easy to find the y-intercept by letting $x = 0$. This gives the point $(0, -2)$. Since this point is $\frac{2}{3}$ unit to the left of the axis of symmetry $x = \frac{2}{3}$, we quickly find the *symmetric point* $(\frac{4}{3}, -2)$. Can you locate the x-intercepts?

$$
\begin{aligned}
y &= 3x^2 - 4x - 2 \\
&= 3(x^2 - \tfrac{4}{3}x) - 2 \\
&= 3(x^2 - \tfrac{4}{3}x + \tfrac{4}{9}) - 2 - \tfrac{4}{3} \\
&= 3(x - \tfrac{2}{3})^2 - \tfrac{10}{3}
\end{aligned}
$$

Vertex: $(\frac{2}{3}, -\frac{10}{3})$
Axis of symmetry: $x = \frac{2}{3}$

The function is decreasing on $(-\infty, \frac{2}{3}]$ and increasing on $[\frac{2}{3}, \infty)$, and the curve is concave up.

EXAMPLE 3 Graph the function $y = f(x) = |x^2 - 4|$. Find the range.

Solution First graph the parabola $y = x^2 - 4$. Then take the part of this curve that is below the x-axis (these are the points for which $x^2 - 4$ is negative) and reflect it through the x-axis.

The curve is decreasing on $(-\infty, -2]$ and on $[0, 2]$. It is increasing on $[-2, 0]$ and on $[2, \infty)$. The curve is concave up on $(-\infty, -2]$ and on $[2, \infty)$, and concave down on $[-2, 2]$.

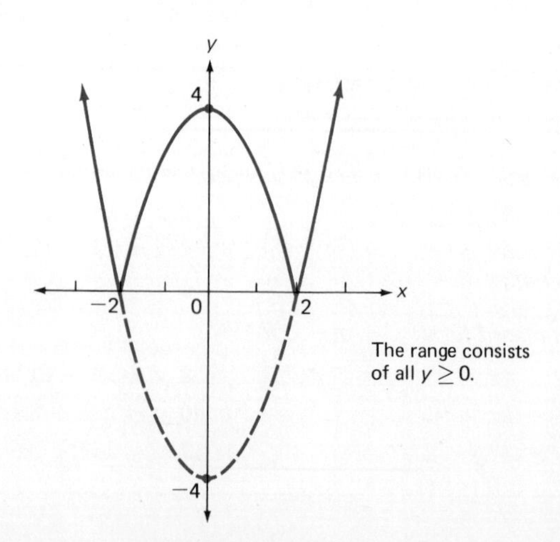

EXAMPLE 4 State the conditions on the values a and k so that the parabola $y = a(x - h)^2 + k$ opens downward and intersects the x-axis in two points. What are the domain and range of this function?

Solution In order for the parabola to open downward we must have $a < 0$. If $k > 0$, the parabola will intersect the x-axis in two distinct points. The domain is the set of all real numbers and the range consists of $y \leq k$.

EXERCISES 4.2

Write in standard form: $a(x - h)^2 + k$.

1 $x^2 + 2x - 5$

2 $x^2 - 2x + 5$

3 $-x^2 - 6x + 2$

4 $x^2 - 3x + 4$

5 $-x^2 + 3x - 4$

6 $x^2 - 5x - 2$

7 $x^2 + 5x - 2$

8 $2x^2 - 4x + 3$

9 $2x^2 + 4x - 3$

10 $5 - 6x + 3x^2$

11 $-5 + 6x + 3x^2$

12 $-3x^2 - 6x + 5$

13 $x^2 - \frac{1}{2}x + 1$

14 $-x^2 - \frac{1}{2}x + 1$

15 $\frac{3}{4}x^2 - x - \frac{1}{3}$

16 $-\frac{3}{4}x^2 + x - \frac{1}{3}$

17 $-5x^2 - 2x + \frac{4}{5}$

***18** $ax^2 + bx + c$

***19** $(x + 1)^2 - 3(x + 1) - \frac{3}{4}$

***20** $ax^2 - 2ahx + ah^2 + k$

***21** Compare Exercises 18 and 20 and express h and k in terms of a, b, and c.

Write each of the following in standard form. Identify the coordinates of the vertex, the equation of the axis of symmetry, the y-intercept, and check by graphing.

22 $y = x^2 + 2x - 1$

23 $y = x^2 - 4x + 7$

24 $y = -x^2 + 4x - 1$

25 $y = x^2 - 6x + 5$

26 $y = 3x^2 + 6x - 3$

27 $y = 2x^2 - 4x - 4$

28 By Exercise 21, $h = -\dfrac{b}{2a}$ is the first coordinate of the vertex of the parabola $y = ax^2 + bx + c$. Show that $(2h, c)$ is on the parabola and explain why this point is the reflection of $(0, c)$ through the axis of symmetry $x = h$.

In Exercises 29–34, graph the indicated parabola using the three points shown. (Note: $y = ax^2 + bx + c = a(x - h)^2 + k$.)

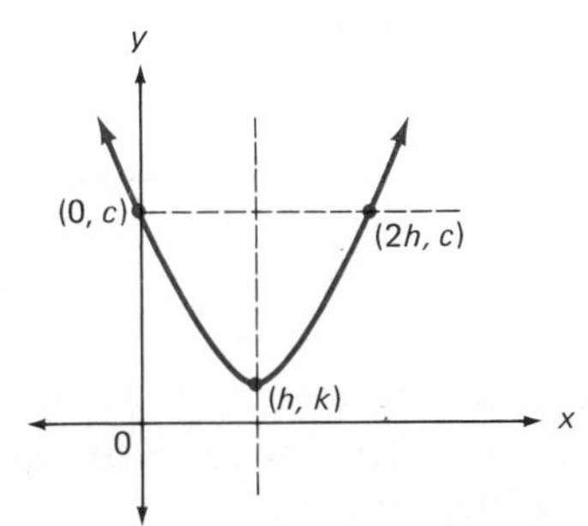

$(0, c)$ ← coordinates of y-intercept

(h, k) ← vertex

$(2h, c)$ ← reflection of $(0, c)$ through axis of symmetry $x = h$

29 $y = x^2 - 3x + \frac{9}{4}$

30 $y = -x^2 + 2x$

31 $y = \frac{1}{2}x^2 + \frac{5}{2}x + 5$

32 $y = 3x^2 - 12x + \frac{29}{2}$

33 $y = -2x^2 - 6x - \frac{9}{2}$

34 $y = 3x^2 + 3x + \frac{3}{4}$

35 Graph the function f where $f(x) = |9 - x^2|$.

36 Graph the function f where $f(x) = |x^2 - 1|$.

37 Graph the function f where $f(x) = |x^2 - x - 6|$.

38 Graph the function f where

$$f(x) = \begin{cases} -x^2 + 4 & \text{if } -2 \leq x < 2 \\ x^2 - 10x + 21 & \text{if } \ 2 \leq x \leq 7 \end{cases}$$

39 Graph the function f where

$$f(x) = \begin{cases} 1 & \text{if } -3 \leq x < 0 \\ x^2 - 4x + 1 & \text{if } \ 0 \leq x < 5 \\ -2x + 16 & \text{if } \ 5 \leq x < 9 \end{cases}$$

In Exercises 40–43, state the conditions on the values a and k so that the parabola $y = a(x - h)^2 + k$ has the following properties.

40 Concave down and has range $y \leq 0$.

41 Concave up and has vertex at $(h, 0)$.

42 Concave up and does not intersect the x-axis.

43 Concave up and has range $y \geq 2$.

***44** This exercise supports (but does not prove) the claim that only three points are needed to determine a parabola. Show that points $(-1, 3)$, $(2, 1)$, and $(4, 8)$ determine a unique parabola with equation $y = ax^2 + bx + c$ by proving that the system

$$\begin{aligned} a(-1)^2 + b(-1) + c &= 3 \\ a(2)^2 + b(2) + c &= 1 \\ a(4)^2 + b(4) + c &= 8 \end{aligned}$$

produces a unique solution for a, b, and c.

***45** Follow the procedure in Exercise 44 to find the equation of the parabola that is determined by the points $(-2, -7)$, $(1, 8)$, and $(3, -2)$.

4.3

The quadratic formula

The graph of a quadratic function may or may not intersect the x-axis. Here are some typical cases:

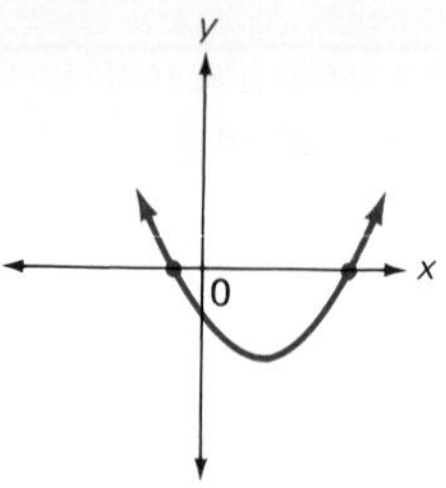

Two x-intercepts; two solutions for $y = ax^2 + bx + c = 0$

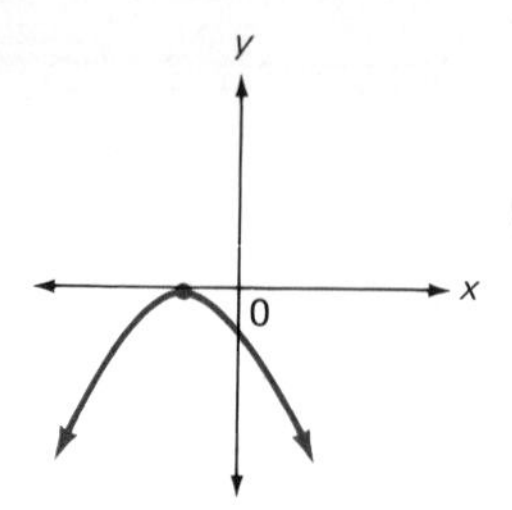

One x-intercept; one solution for $y = ax^2 + bx + c = 0$

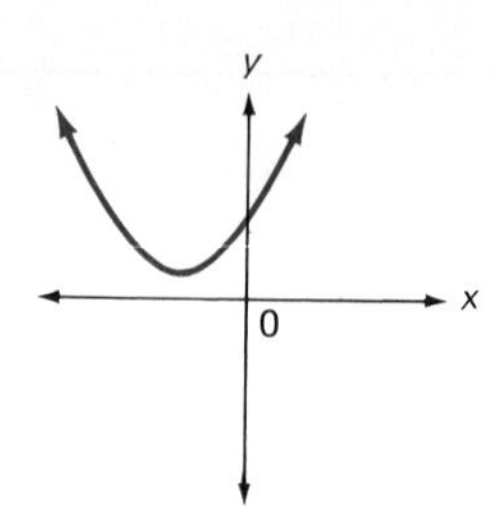

No x-intercepts; no solutions for $y = ax^2 + bx + c = 0$

It is clear from these figures that if there are x-intercepts, then these values of x are the solutions to the equation $y = ax^2 + bx + c = 0$. If there are no x-intercepts, then the equation will not have any solutions. In this section we learn procedures to handle all cases.

As a first example let us find the x-intercepts of the parabola $y = f(x) = x^2 - x - 6$. This calls for those values x for which $y = 0$. That is, we need to solve the equation $y = f(x) = 0$ for x. This can be done by factoring:

$$x^2 - x - 6 = 0$$
$$(x + 2)(x - 3) = 0$$

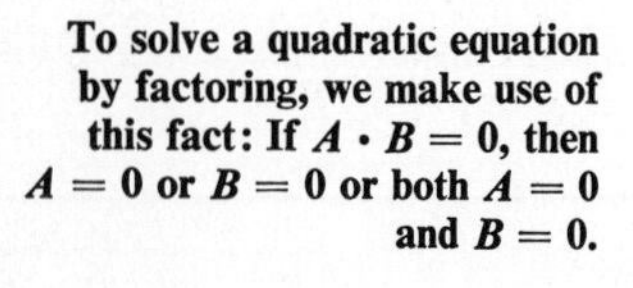
To solve a quadratic equation by factoring, we make use of this fact: If $A \cdot B = 0$, then $A = 0$ or $B = 0$ or both $A = 0$ and $B = 0$.

Since the product of two factors is zero only when one or both of them is zero, it follows that

$$x + 2 = 0 \quad \text{or} \quad x - 3 = 0$$
$$x = -2 \quad \text{or} \quad x = 3$$

The x-intercepts are -2 and 3. The x-intercepts of the parabola are also called the *roots* of the equation $f(x) = 0$.

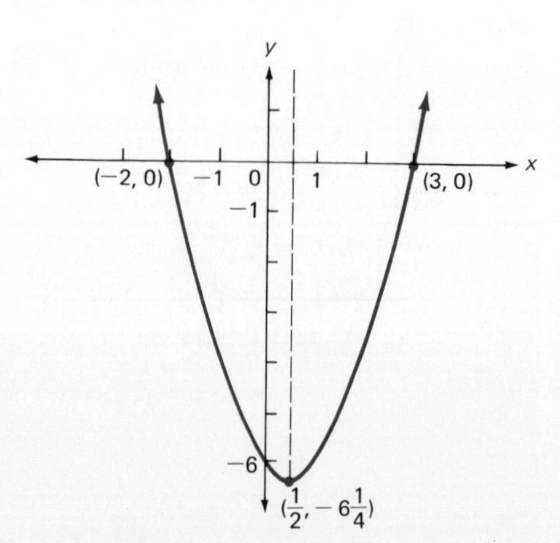

EXAMPLE 1 Find the x-intercepts of $f(x) = 25x^2 + 30x + 9$.

Solution Let $f(x) = 0$:

$$25x^2 + 30x + 9 = 0$$
$$(5x + 3)(5x + 3) = 0$$
$$5x + 3 = 0$$
$$x = -\tfrac{3}{5}$$

Check: $25(\frac{9}{25}) + 30(-\frac{3}{5}) + 9 = 9 - 18 + 9 = 0$.

Note: Since there is only *one* answer to this quadratic equation, it follows that the parabola $y = 25x^2 + 30x + 9$ has only one x-intercept; the x-axis is *tangent* to the parabola at its vertex $(-\frac{3}{5}, 0)$.

The number $-\frac{3}{5}$ is referred to as a *double root* of $25x^2 + 30x + 9 = 0$.

When a quadratic is difficult to factor, the x-intercepts can be found, if there are any, by completing the square. The following illustration shows the procedure for solving $x^2 - 2x - 4 = 0$.

$x^2 - 2x - 4 = 0$

$x^2 - 2x = 4$

$x^2 - 2x + 1 = 4 + 1$ (Complete the square by adding 1 to each side.)

$(x - 1)^2 = 5$

$x - 1 = \pm\sqrt{5}$ (Take the square root of each side; if $x^2 = a$, then $x = \pm a$.)

$x = 1 \pm \sqrt{5}$

We use $x = 1 \pm \sqrt{5}$ as an abbreviation for $x = 1 + \sqrt{5}$ or $x = 1 - \sqrt{5}$.

EXAMPLE 2 Find the roots of $2x^2 - 9x - 18 = 0$.

Solution

Try to explain each step in this solution.

$$2x^2 - 9x - 18 = 0$$
$$2x^2 - 9x = 18$$
$$x^2 - \tfrac{9}{2}x = 9$$
$$x^2 - \tfrac{9}{2}x + \tfrac{81}{16} = 9 + \tfrac{81}{16}$$
$$(x - \tfrac{9}{4})^2 = \tfrac{225}{16}$$
$$x - \tfrac{9}{4} = \pm\tfrac{15}{4}$$
$$x = \tfrac{9}{4} \pm \tfrac{15}{4}$$
$$x = 6 \quad \text{or} \quad x = -\tfrac{3}{2}$$

Check: $2(6)^2 - 9(6) - 18 = 72 - 54 - 18 = 0$

$2(-\frac{3}{2})^2 - 9(-\frac{3}{2}) - 18 = \frac{9}{2} + \frac{27}{2} - 18 = 0$

As another check, solve this quadratic equation by factoring.

TEST YOUR UNDERSTANDING

Find the x-intercepts (if any). Use the factoring method or complete the square.

1 $y = f(x) = 2x^2 + 13x - 24$

2 $y = f(x) = 5x^2 - 3x$

3 $y = f(x) = 4x^2 - 1$

4 $y = f(x) = -4x^2 + 4x - 1$

5 $y = f(x) = x^2 - 8$

6 $f(x) = x^2 + 4$

7 $f(x) = x^2 + 2x - 2$

8 $f(x) = 1 + 4x - 3x^2$

9 $f(x) = x^2 + x + 1$

10 $f(x) = 9 - 12x + 4x^2$

Another way to solve quadratic equations is by use of a formula. When the general quadratic equation $ax^2 + bx + c = 0$ is solved for x in terms of a, b, and c, we will have a formula that applies to any specific quadratic equation written in this form.

$$ax^2 + bx + c = 0$$

Add $-c$:

$$ax^2 + bx = -c$$

Divide by a ($a \neq 0$):

$$x^2 + \frac{b}{a}x = -\frac{c}{a}$$

Add $\left[\frac{1}{2}\left(\frac{b}{a}\right)\right]^2 = \frac{b^2}{4a^2}$:

$$x^2 + \frac{b}{a}x + \frac{b^2}{4a^2} = \frac{b^2}{4a^2} - \frac{c}{a}$$

Factor on the left and combine on the right:

$$\left(x + \frac{b}{2a}\right)^2 = \frac{b^2 - 4ac}{4a^2}$$

If $b^2 - 4ac$ is not negative, take square roots and solve for x.

$$x + \frac{b}{2a} = \pm\sqrt{\frac{b^2 - 4ac}{4a^2}}$$

$$x + \frac{b}{2a} = \pm\frac{\sqrt{b^2 - 4ac}}{2a}$$

$$x = -\frac{b}{2a} \pm \frac{\sqrt{b^2 - 4ac}}{2a}$$

Note: $\sqrt{\frac{b^2 - 4ac}{4a^2}} = \frac{\sqrt{b^2 - 4ac}}{\sqrt{4a^2}} = \frac{\sqrt{b^2 - 4ac}}{\pm 2a}$ $\begin{cases} +, \text{ if } a > 0 \\ -, \text{ if } a < 0 \end{cases}$

Combine terms to obtain the **quadratic formula.**

QUADRATIC FORMULA

If $ax^2 + bx + c = 0$, $a \neq 0$,

then $x = \dfrac{-b \pm \sqrt{b^2 - 4ac}}{2a}$

The values $x = \frac{-b + \sqrt{b^2 - 4ac}}{2a}$ and $x = \frac{-b - \sqrt{b^2 - 4ac}}{2a}$ are the **roots** of the quadratic equation. These are also the x-intercepts of the parabola $y = ax^2 + bx + c$.

This formula now allows you to solve any quadratic equation in terms of the constants used. Let us apply it to the equation $2x^2 - 5x + 1 = 0$:

$$2x^2 - 5x + 1 = 0 \qquad x = \frac{-b \pm \sqrt{b^2 - 4ac}}{2a}$$

$$a = 2$$

$$b = -5 \qquad = \frac{-(-5) \pm \sqrt{(-5)^2 - 4(2)(1)}}{2(2)}$$

$$c = 1$$

$$= \frac{5 \pm \sqrt{17}}{4}$$

Thus

$$x = \frac{5 + \sqrt{17}}{4} \quad \text{or} \quad x = \frac{5 - \sqrt{17}}{4}$$

You can use $\sqrt{17} = 4.1$ from the square root table (Appendix Table I) or from a calculator to obtain rational approximations for the solutions.

EXAMPLE 3 Solve for x: $2x^2 = x - 1$.

Solution First rewrite the equation in the general form $ax^2 + bx + c = 0$.

$$2x^2 - x + 1 = 0$$

Use the quadratic formula with $a = 2$, $b = -1$, and $c = 1$.

$$x = \frac{-(-1) \pm \sqrt{(-1)^2 - 4(2)(1)}}{2(2)}$$

$$= \frac{1 \pm \sqrt{-7}}{4}$$

In Section 4.8 complex numbers are introduced and used to solve equations that have no real solutions.

In all cases, unless otherwise stated, assume that all solutions are to be in the set of real numbers. Since $\sqrt{-7}$ is not a real number, there are no real solutions to the given equation.

TEST YOUR UNDERSTANDING

Use the quadratic formula to solve for x.

1 $x^2 + 3x - 10 = 0$

2 $x^2 - x - 12 = 0$

3 $x^2 - 9 = 0$

4 $x^2 - 2x - 2 = 0$

5 $x^2 + 6x + 6 = 0$

6 $x^2 + 6x + 12 = 0$

7 $2x^2 - 2x + 5 = 0$

8 $2x^2 + x - 4 = 0$

As in Example 3, if $b^2 - 4ac < 0$, then no real square roots are possible. Geometrically, this means that the parabola $y = ax^2 + bx + c$ does not meet the x-axis; there are no real solutions for $ax^2 + bx + c = 0$.

When $b^2 - 4ac = 0$, only the solution $x = -\frac{b}{2a}$ is possible; the x-axis is tangent to the parabola. Finally, when $b^2 - 4ac > 0$, we have two solutions that are the x-intercepts of the parabola.

Since $b^2 - 4ac$ tells us how many (if any) solutions $ax^2 + bx + c = 0$ has, it is called the **discriminant**. The use of the discriminant is illustrated in the following table.

Quadratic function	*Value of* $b^2 - 4ac$	*Real solutions of* $y = 0$	*Number of* x*-intercepts*
$y = x^2 - x - 6$	25	two	two
$y = x^2 - 4x + 4$	0	one	one
$y = x^2 - 4x + 5$	-4	none	none

Note also that when the discriminant $b^2 - 4ac > 0$, then the two solutions of $ax^2 + bx + c = 0$ will be rational numbers if $b^2 - 4ac$ is a perfect square. In case $b^2 - 4ac$ is not a perfect square, then the roots are irrational. We summarize as follows:

USING THE DISCRIMINANT

1. If $b^2 - 4ac > 0$, then $ax^2 + bx + c = 0$ has two real solutions and the graph of $y = ax^2 + bx + c$ crosses the x-axis at two points. If the discriminant is a perfect square, these roots will be rational numbers; if not, they will be irrational.
2. If $b^2 - 4ac = 0$, the solution for $ax^2 + bx + c = 0$ is only one number (a double root), and the graph of $y = ax^2 + bx + c$ touches the x-axis at one point. (The x-axis is said to be *tangent* to the graph.)
3. If $b^2 - 4ac < 0$, $ax^2 + bx + c = 0$ has no real solutions, and the graph of $y = ax^2 + bx + c$ does not cross the x-axis.

The x-intercepts of a parabola can also be used to solve a **quadratic inequality**. Consider, for example, the inequality $x^2 - x - 6 < 0$. To solve this inequality, first examine the graph of $y = x^2 - x - 6$. This is a parabola with x-intercepts at -2 and 3, as can be seen by writing the equation in the factored form $y = (x + 2)(x - 3)$. Note that $y = x^2 - x - 6$ is below the x-axis between the x-intercepts. That is, $x^2 - x - 6 < 0$ for $-2 < x < 3$. Also, $x^2 - x - 6 > 0$ for $x < -2$ or $x > 3$.

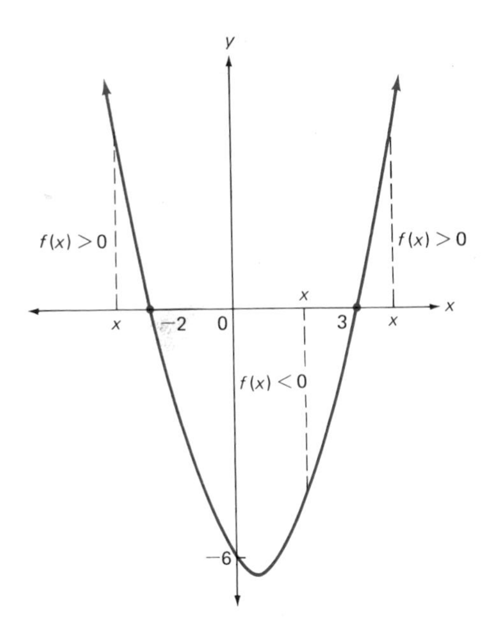

In practice this method can be used without constructing the graph. All you really need to know are the x-intercepts and whether the parabola opens up or down.

EXERCISES 4.3

Use the quadratic formula to solve for x.

1 $x^2 - 3x - 10 = 0$ **2** $2x^2 + 3x - 2 = 0$

3 $x^2 - 6x + 9 = 0$ **4** $9 - 4x^2 = 0$

5 $3x^2 + 7x + 2 = 0$ **6** $6 - 6x + x^2 = 0$

7 $x^2 = 2x + 4$ **8** $6x - 14 = x^2$

9 $-2x^2 + 3x = -1$

Find the x-intercepts.

10 $y = x^2 - 3x - 4$ **11** $y = 2x^2 - 5x - 3$

12 $y = x^2 - 10x + 25$ **13** $y = x^2 - x + 3$

14 $y = 9x^2 - 4$ **15** $y = 2x^2 - 7x + 6$

16 $y = x^2 + 4x + 1$ **17** $y = -x^2 + 4x - 7$

18 $y = 3x^2 + x - 1$

Find the value of $b^2 - 4ac$. Then state if there are (a) no solutions, (b) one solution, (c) two rational solutions, or (d) two irrational solutions.

19 $x^2 - 8x + 16 = 0$ **20** $x^2 + 3x + 5 = 0$

21 $x^2 + 2x - 8 = 0$ **22** $4x^2 - 4x + 1 = 0$

23 $x^2 + 3x - 1 = 0$ **24** $2x + 15 = x^2$

25 $2x^2 + x = 5$ **26** $6x^2 + 7x = 3$

27 $1 = x - 2x^2$

Use the discriminant to predict how many times, if any, the parabola will cross the x-axis. Then find (a) the vertex, (b) the y-intercept, and (c) the x-intercepts.

28 $y = x^2 - 6x + 13$ **29** $y = 9x^2 - 6x + 1$

30 $y = -x^2 - 4x + 3$

Solve each quadratic inequality.

31 $x^2 - 2x - 3 < 0$ **32** $8 + 2x - x^2 < 0$

33 $x^2 + 3x - 10 \geq 0$ **34** $2x^2 + 3x - 2 \leq 0$

35 $x^2 + 6x + 9 < 0$ **36** $3 - x^2 > 0$

Graph. Show all intercepts.

37 $f(x) = (x - 3)(x + 1)$

38 $f(x) = (x - \frac{1}{2})(5 - x)$

39 $f(x) = 4 - 4x - x^2$

40 $f(x) = x^2 + 1$

41 $f(x) = 3x^2 - 4x + 1$

42 $f(x) = -x^2 + 1$

43 $f(x) = -x^2 + 2x - 4$

44 $f(x) = x^2 + 2x$

In Exercises 45–48 find the values of b so that the x-axis will be tangent to the parabola.

45 $y = x^2 + bx + 9$

46 $y = 4x^2 + bx + 25$

47 $y = x^2 - bx + 7$

48 $y = 9x^2 + bx + 14$

Find the values of k so that equation will have two real roots. (Hint: Let $b^2 - 4ac > 0$.)

49 $-x^2 + 4x + k = 0$

50 $2x^2 - 3x + k = 0$

51 $kx^2 - x - 1 = 0$

52 $kx^2 + 3x - 2 = 0$

Find the values of a so that the parabola will not intersect the x-axis.

53 $y = ax^2 - x - 1$

54 $y = ax^2 + 3x + 7$

*55 Consider the two roots of $ax^2 + bx + c = 0$, where $b^2 - 4ac \geq 0$. Find **(a)** the sum and **(b)** the product of these roots.

In Exercises 56–61, use the results of Exercise 55 to find the sum and product of the roots of the given equation. Then verify your answers by solving for the roots and forming their sum and product.

56 $x^2 - 3x - 10 = 0$

57 $6x^2 + 5x - 4 = 0$

58 $x^2 = 25$

59 $3x^2 + 35 = 26x$

60 $x^2 + 2x - 5 = 0$

61 $2x^2 + 6x - 9 = 0$

*62 Solve for x: $x^4 - 5x^2 + 4 = 0$. (*Hint:* Use $u = x^2$ and solve $u^2 - 5u + 4 = 0$.)

*63 Solve for x: $x^3 + 3x^2 - 4x - 12 = 0$. (*Hint:* Factor by grouping.)

*64 Solve for x: $a^3x^2 - 2ax - 1 = 0$ $(a > 0)$.

4.4 Applications of quadratic functions

The parabola with equation

$$y = ax^2 + bx + c = a(x - h)^2 + k$$

opens upward or downward depending on the sign of a. When $a > 0$, the vertex is the lowest point on the parabola; when $a < 0$, it is the highest point. These special points will be useful in solving certain applied problems.

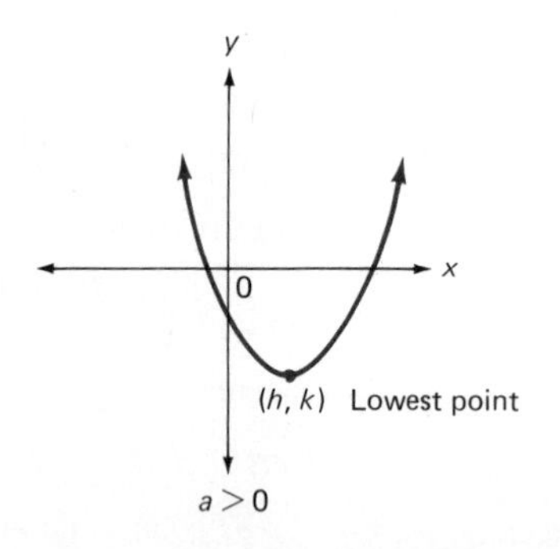

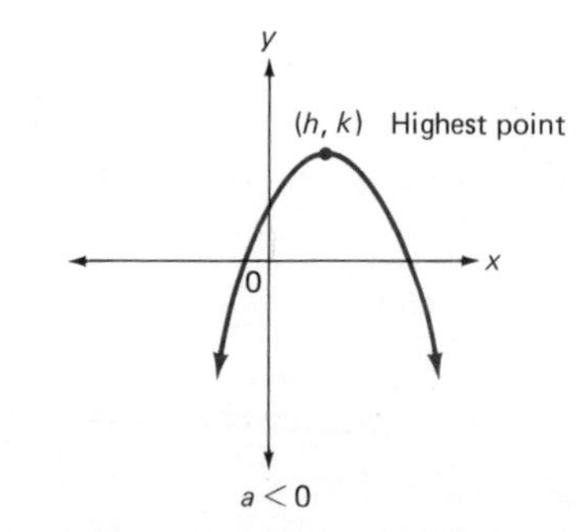

The conversion to the form $y = a(x - h)^2 + k$ instantly identifies (h, k) as this extreme point. We say that the y-value k is the **minimum value** of f when $a > 0$; it is the **maximum value** of f when $a < 0$.

EXAMPLE 1 Find the maximum value or minimum value of the function $f(x) = 2(x + 3)^2 + 5$.

Solution Since $2(x + 3)^2 + 5 = 2[x - (-3)]^2 + 5$, we note that $(-3, 5)$ is the turning point. Also since $a = 2 > 0$, the parabola opens upward and $f(-3) = 5$ is the minimum value.

EXAMPLE 2 Find the maximum value of the quadratic function $f(x) = -\frac{1}{3}x^2 + x + 2$. At which value x does f achieve this maximum?

Solution Convert to the form $a(x-h)^2 + k$:

The purpose of converting to the form $a(x-h)^2 + k$ is to find the turning point (h, k).

$$\begin{aligned} y = f(x) &= -\tfrac{1}{3}(x^2 - 3x) + 2 \\ &= -\tfrac{1}{3}(x^2 - 3x + \tfrac{9}{4}) + 2 + \tfrac{3}{4} \\ &= -\tfrac{1}{3}(x - \tfrac{3}{2})^2 + \tfrac{11}{4} \end{aligned}$$

From this form we have $a = -\frac{1}{3}$. Since $a < 0$, $(\frac{3}{2}, \frac{11}{4})$ is the highest point of the parabola. Thus f has a maximum value of $\frac{11}{4}$ when $x = \frac{3}{2}$; $f(\frac{3}{2}) = \frac{11}{4}$ is the maximum.

TEST YOUR UNDERSTANDING

Find the maximum or minimum value of each quadratic function and state the x-value at which this occurs.

1 $f(x) = x^2 - 10x + 21$
2 $f(x) = x^2 + \frac{4}{3}x - \frac{7}{18}$
3 $f(x) = 10x^2 - 20x + \frac{21}{2}$
4 $f(x) = -8x^2 - 64x + 3$
5 $f(x) = -2x^2 - 1$
6 $f(x) = x^2 - 6x + 9$
7 $f(x) = 25x^2 + 70x + 49$
8 $f(x) = (x-3)(x+4)$

EXAMPLE 3 Suppose that 60 meters of fencing is available to enclose a rectangular garden, one side of which will be against the side of a house. What dimensions of the garden will guarantee a maximum area?

Examples 3, 4, and 5 illustrate how the concepts developed in this section can be used to solve applied problems.

Solution From the sketch you can see that the 60 meters need only be used for three sides, two of which are of the same length x.

For each x between 0 and 30, such a rectangle is possible. Here are a few.

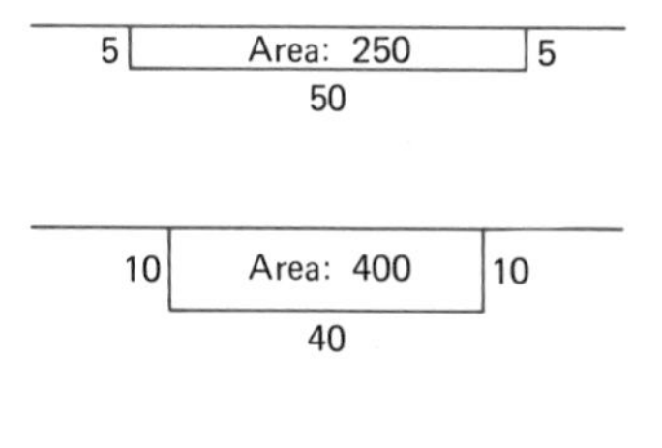

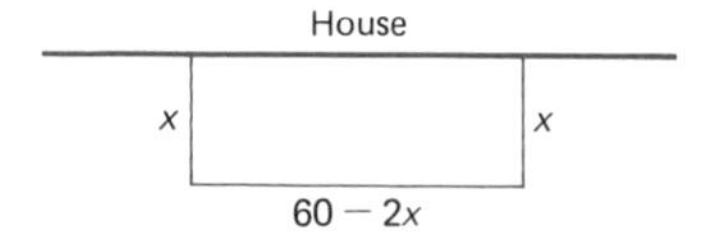

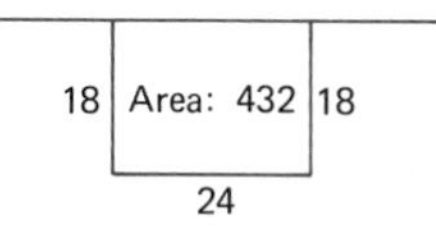

The remaining side has length $60 - 2x$, and the area A is given by

$$\begin{aligned} A(x) &= x(60 - 2x) \\ &= 60x - 2x^2 \end{aligned}$$

To "maximize" A, convert to the form $a(x-h)^2 + k$. Thus

$$\begin{aligned} A(x) &= -2(x^2 - 30x) \\ &= -2(x^2 - 30x + 225) + 450 \\ &= -2(x - 15)^2 + 450 \end{aligned}$$

Therefore, the maximum area of 450 square meters is obtained when the dimensions are $x = 15$ meters by $60 - 2x = 30$ meters.

Example 3 shows how to select the rectangle of maximum area from such a vast collection of possibilities. Can you explain why the domain of $A(x)$ is $0 < x < 30$?

EXAMPLE 4 The sum of two numbers is 24. Find the two numbers if their product is to be a maximum.

Solution Let x represent one of the numbers. Since the sum is 24, the other number is $24 - x$. Now let p represent the product of these numbers.

$$\begin{aligned} p &= x(24 - x) \\ &= -x^2 + 24x \\ &= -(x^2 - 24x) \\ &= -(x^2 - 24x + 144) + 144 \\ &= -(x - 12)^2 + 144 \end{aligned}$$

Try to solve Example 4 if the product is to be a minimum.

Since $a = -1$, the product has a maximum value of 144 when $x = 12$. Hence the numbers are 12 and 12.

EXAMPLE 5 A ball is thrown straight upward from ground level with an initial velocity of 32 feet per second. The formula $s = 32t - 16t^2$ gives its height in feet, s, after t seconds. **(a)** What is the maximum height reached by the ball? **(b)** When does the ball return to the ground?

Solution **(a)** First complete the square in t.

$$\begin{aligned} s &= 32t - 16t^2 \\ &= -16t^2 + 32t \\ &= -16(t^2 - 2t) \\ &= -16(t^2 - 2t + 1) + 16 \\ &= -16(t - 1)^2 + 16 \end{aligned}$$

You should now recognize this as describing a parabola with vertex at (1, 16). Because the coefficient of t^2 is negative, the curve opens downward. The maximum height, 16 feet, is reached after 1 second.

The motion of the ball is straight up and down.

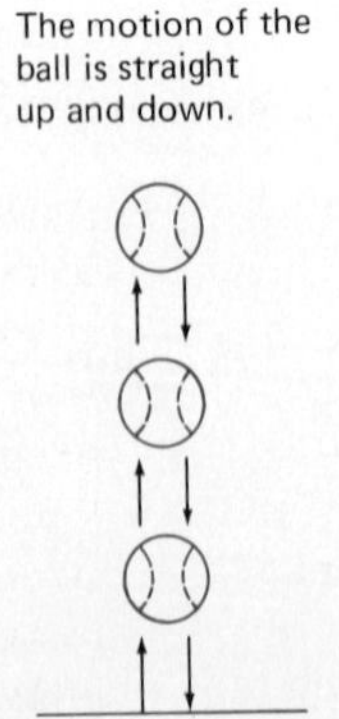

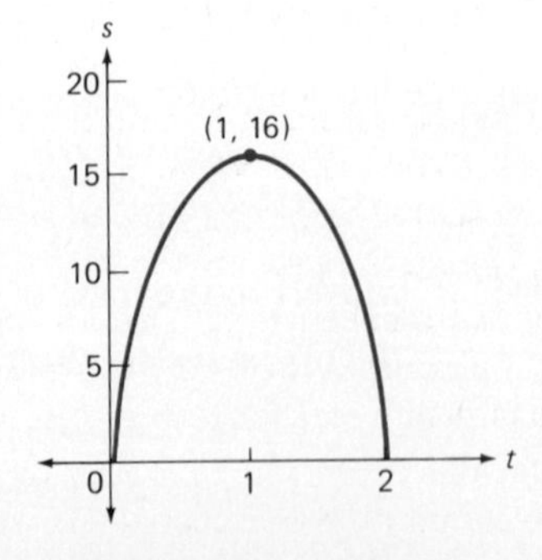

This parabolic arc is the graph of the relation between time t and distance s. It is *not* the path of the ball.

(b) The ball hits the ground when the distance $s = 0$ feet. Thus

$$s = 32t - 16t^2 = 0$$
$$16t(2 - t) = 0$$
$$t = 0 \quad \text{or} \quad t = 2$$

Since time $t = 0$ is the starting time, the ball returns to the ground 2 seconds later, when $t = 2$.

There is no easy way to become a problem solver. You must continue to solve many problems until they no longer become an object of fear! Do not let the list of problems that follows intimidate you. Try them; you may find that you really enjoy them!

EXERCISES 4.4

Find the coordinates of the highest or lowest point of the given quadratic and sketch the graph.

1 $y = x^2 - 4x + 7$

2 $y = 1 - 6x - x^2$

3 $y = 1 - 4x + 4x^2$

4 $y = -2x^2 + 10x - 5$

Find the maximum or minimum value of the quadratic function and state the x-value at which this occurs.

5 $f(x) = -x^2 + 10x - 18$

6 $f(x) = x^2 + 18x + 49$

7 $f(x) = 16x^2 - 64x + 100$

8 $f(x) = -\frac{1}{2}x^2 + 3x - 6$

9 $f(x) = 49 - 28x + 4x^2$

10 $f(x) = x(x - 10)$

11 $f(x) = -x(\frac{2}{3} + x)$

12 $f(x) = (x - 4)(2x - 7)$

13 A manufacturer is in the business of producing small statues called Heros. He finds that the daily cost in dollars, C, of manufacturing n Heros is given by the formula $C = n^2 - 120n + 4200$. How many Heros should be produced per day so that the cost will be minimum? What is the minimal daily cost?

14 A company's daily profit, P, in dollars, is given by $P = -2x^2 + 120x - 800$, where x is the number of articles produced per day. Find x so that the daily profit is a maximum.

15 The sum of two numbers is 12. Find the two numbers if their product is to be a maximum. (*Hint:* Find the maximum value for $y = x(12 - x)$.)

16 The sum of two numbers is n. Find the two numbers such that their product is a maximum.

17 In Exercise 16, are there two numbers that will give a minimum product? Explain.

18 The difference of two numbers is 22. Find the numbers if their product is to be a minimum and also find this product.

19 A homeowner has 40 feet of wire and wishes to use it to enclose a rectangular garden. What should be the dimensions of the garden so as to enclose the largest possible area?

20 Repeat Exercise 19, but this time assume that the side of the house is to be used as one boundary for the garden. Thus the wire is only needed for the other three sides.

21 The formula $h = 128t - 16t^2$ gives the distance in feet above the ground, h, reached by an object in t seconds. What is the maximum height reached by the object? How long does it take to reach this height?

22 In Exercise 21, how long does it take for the object to return to the ground?

23 In Exercise 21, after how many seconds will the object be at a height of 192 feet? (There are two possible answers.)

24 Suppose it is known that if 65 apple trees are planted in a certain orchard, the average yield per tree will be 1500 apples per year. For each additional tree planted in the same orchard, the annual yield per tree drops by 20 apples. How many trees should be planted in order to produce the maximum crop of apples per year? (*Hint:* If n trees are added to the 65 trees, then the yield per tree is $1500 - 20n$.)

25 It is estimated that 14,000 people will attend a basketball game when the admission price is $7.00. For each 25¢ added to the price, the attendance will decrease by 280. What admission price will produce the largest gate receipts? (*Hint:* If x quarters are added, the attendance will be $14{,}000 - 280x$.)

26 The sum of the lengths of the two perpendicular sides of a right triangle is 30 centimeters. What are their lengths if the square of the hypotenuse is a minimum?

27 Each point P on the line segment between endpoints (0, 4) and (2, 0) determines a rectangle with dimensions x by y as shown in the figure. Find the coordinates of P that give the rectangle of maximum area.

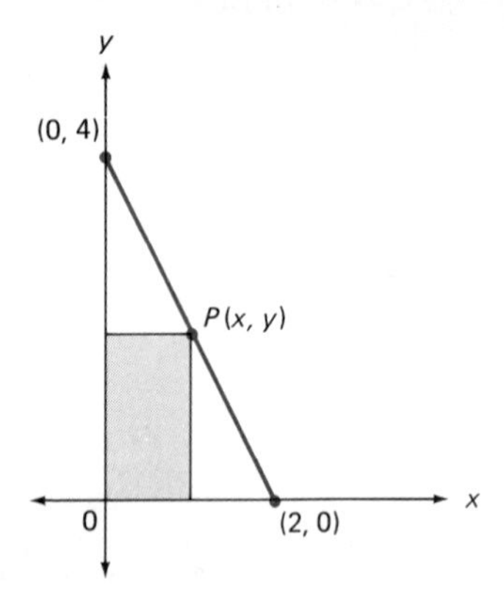

28 **(a)** Let s be the square of the distance from the origin to point $P(x, y)$ on the line through points (0, 4) and (2, 0). Find the coordinates of P such that s is a minimum value.

(b) The answer to part (a) is also the intersection of the given line with the perpendicular through the origin to this line. Using the result that perpendicular lines have slopes that are negative reciprocals, find this point by solving the appropriate linear system.

29 Find the maximum or minimum value of $y = ax^2 + bx + c$, $a \neq 0$, and state at which x-value this occurs.

From Exercise 29, we have that the vertex of the parabola with equation $y = ax^2 + bx + c$ *has coordinates* $\left(-\frac{b}{2a}, \frac{4ac - b^2}{4a}\right)$. *Use this general result to find the coordinates of the vertex of each parabola, and decide whether* $\frac{4ac - b^2}{4a}$ *is a maximum or minimum value.*

30 $y = 2x^2 - 6x + 9$

31 $y = -3x^2 + 24x - 41$

32 $y = -\frac{1}{2}x^2 - \frac{1}{3}x + 1$

33 $y - x^2 + 5x = 0$

34 $y + \frac{2}{3}x^2 = 9$

35 $y = 10x^2 + 100x + 1000$

Exercises 36–43 call for the solution of a quadratic equation in one variable.

36 The sum of two consecutive positive integers is subtracted from their product to obtain a difference of 71. What are the integers?

37 The measures of the legs of a right triangle are consecutive odd integers. The hypotenuse is $\sqrt{130}$. Find the lengths of the legs. (*Hint:* Use the Pythagorean theorem.)

38 One positive integer is 3 greater than another. The difference of their reciprocals is $\frac{1}{6}$. Find the integers.

39 The sum of a number and twice the square of that number is 3. Find all such numbers.

40 If the length of one pair of opposite sides of a square is increased by 3 centimeters, and the length of the other pair is decreased by 1 centimeter, the area of the new figure will be 7 square centimeters greater than that of the original square. What is the length of the side of the square?

41 If the length of the sides of a square are increased by 2 centimeters, the newly formed square will have an area that is 36 square centimeters greater than the original one. Find the length of a side of the original square.

42 Wendy is 5 years older than Sharon. In 5 years the product of their ages will be $1\frac{1}{2}$ times as great as the product of their present ages. How old is Sharon now? (*Hint:* Let Sharon's age be represented by x and Wendy's age by $x + 5$. Then in 5 years their ages will be $x + 5$ and $x + 10$, respectively.)

43 A square piece of tin is to be used to form a box without a top by cutting off a 2-inch square from each corner, and then folding up the sides. The volume of the box will be 128 cubic inches. Find the length of a side of the original square.

4.5

Circles and their equations

Let $A(x_1, y_1)$ and $B(x_2, y_2)$ be two points in a rectangular system. We will use the symbol AB to represent the distance between the points A and B. Then AB is given by the **distance formula**:

$$AB = \sqrt{(x_1 - x_2)^2 + (y_1 - y_2)^2}$$

You can verify this result by studying this figure:

If A and B are on the same horizontal line, then $y_1 = y_2$ and $AB = \sqrt{(x_1 - x_2)^2} = |x_1 - x_2|$ (see page 28).

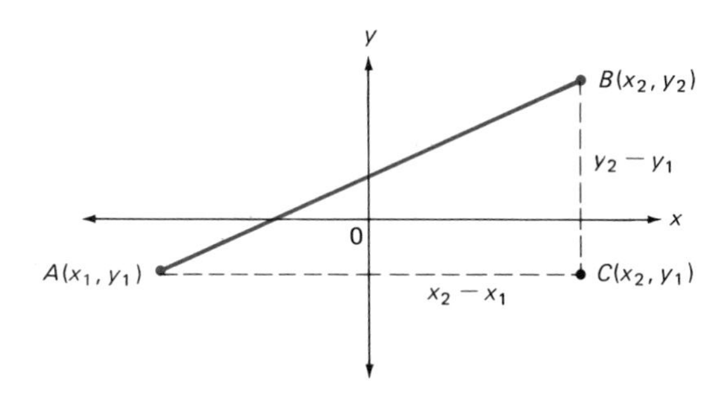

Since AB is the hypotenuse of the right triangle ABC, the Pythagorean theorem gives

$$AB^2 = AC^2 + CB^2$$

But $AC = x_2 - x_1$ and $CB = y_2 - y_1$. Thus

$$AB^2 = (x_2 - x_1)^2 + (y_2 - y_1)^2$$

Taking the positive square root gives the stated result.

Note that the diagram was set up so that $x_2 - x_1 > 0$ and $y_2 - y_1 > 0$. Other situations may have negative values, but it makes no difference because $(x_2 - x_1)^2 = (x_1 - x_2)^2$ and $(y_2 - y_1)^2 = (y_1 - y_2)^2$.

EXAMPLE 1 Find the length of the line segment determined by points $A(-2, 2)$ and $B(6, -4)$.

Solution

$$\begin{aligned} AB &= \sqrt{(-2 - 6)^2 + [2 - (-4)]^2} \\ &= \sqrt{(-8)^2 + 6^2} \\ &= \sqrt{64 + 36} \\ &= \sqrt{100} \\ &= 10 \end{aligned}$$

The distance formula can be used to obtain the equation of a circle. *A circle consists of all points in the plane, each of which is a fixed distance r from a given point called the* **center** *of the circle; r is the* **radius** *of the circle.* ($r > 0$).

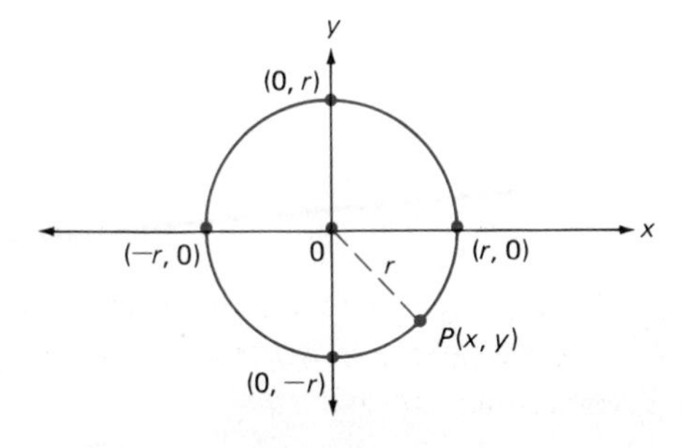

The words "if and only if" here mean that if P is on the circle, then $OP = r$ and if $OP = r$ then P is on the circle.

The preceding figure is a circle with center at the origin and radius r. A point will be on this circle if and only if its distance from the origin is equal to r. That is, $P(x, y)$ is on this circle if and only if $OP = r$. Since the origin has coordinates (0, 0), the distance formula gives

$$\begin{aligned} r &= \sqrt{(x - 0)^2 + (y - 0)^2} \\ &= \sqrt{x^2 + y^2} \end{aligned}$$

Squaring produces this result:

$$r^2 = x^2 + y^2$$

We conclude that $P(x, y)$ is on the circle with center O and radius r if and only if the coordinates of P satisfy the preceding equation.

A circle with center at the origin and radius r has the equation

$$x^2 + y^2 = r^2$$

EXAMPLE 2 What is the equation of the circle with center O and radius 3?

Solution Using the equation $x^2 + y^2 = r^2$, we get

$$x^2 + y^2 = 3^2 = 9$$

Now consider any circle of radius r, not necessarily one with the origin as center. Let the center C have coordinates (h, k). Then, using the distance formula, a point $P(x, y)$ is on this circle if and only if

$$CP = r = \sqrt{(x - h)^2 + (y - k)^2}$$

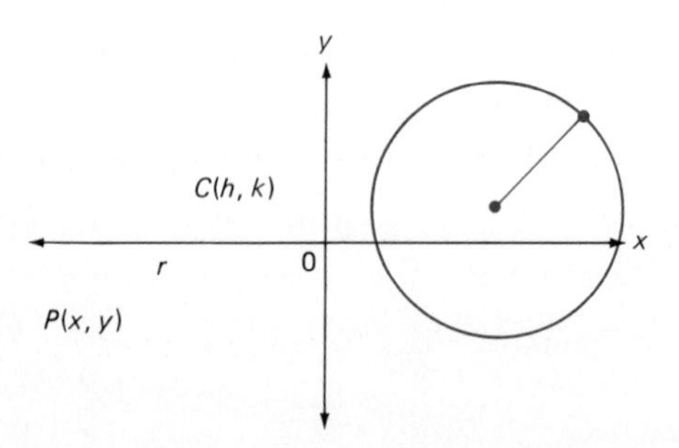

By squaring CP, we obtain the following:

STANDARD FORM FOR THE EQUATION OF A CIRCLE WITH CENTER AT (h, k) AND RADIUS r

$$(x - h)^2 + (y - k)^2 = r^2$$

EXAMPLE 3 Find the center and radius of the circle with this equation: $(x-2)^2 + (y+3)^2 = 4$.

Solution Using $y + 3 = y - (-3)$, rewrite the equation in this form:

$$(x-2)^2 + [(y-(-3)]^2 = 2^2$$

By comparing to the standard form, we find the radius $r = 2$ and the center at $(2, -3)$.

Even though circles are not functions, we include them in this chapter since they tie in nicely with some of our earlier work in this chapter and will prove to be useful in later work as well. Also, their equations are quadratic in two variables.

Circles are not graphs of functions. When $x = 2$ is substituted into the equation of the circle in Example 3, we get two y-values, -1 and -5. This is contrary to the definition of a function that calls for just *one* range value y for each domain value x.

EXAMPLE 4 Write the equation of the circle with center at $(-3, 5)$ and radius 2.

Solution Use $h = -3$, $k = 5$, and $r = 2$ in the standard form to obtain

$$[x-(-3)]^2 + (y-5)^2 = 2^2$$

or

$$(x+3)^2 + (y-5)^2 = 4$$

TEST YOUR UNDERSTANDING

Find the length of the line segment determined by the two points.

1 $(4, 0)$; $(-8, -5)$ **2** $(9, -1)$; $(2, 3)$ **3** $(-7, -5)$; $(3, -13)$

Find the center and radius of each circle.

4 $x^2 + y^2 = 100$ **5** $x^2 + y^2 = 10$

6 $(x-1)^2 + (y+1)^2 = 25$ **7** $(x+\frac{1}{2})^2 + y^2 = 256$

8 $(x+4)^2 + (y+4)^2 = 50$

Write the equation of the circle with the given center and radius in standard form.

9 Center at $(0, 4)$; $r = 5$ **10** Center at $(1, -2)$; $r = \sqrt{3}$

The equation in Example 3 can be written in another form.

$$(x-2)^2 + (y+3)^2 = 4$$
$$x^2 - 4x + 4 + y^2 + 6y + 9 = 4$$
$$x^2 - 4x + y^2 + 6y = -9$$

This last equation no longer looks like the equation of a circle. Starting with such an equation we can convert it back into the standard form of a circle by completing the square in both variables, if necessary. For example, let us begin with

$$x^2 - 4x + y^2 + 6y = -9$$

Note that the major reason for writing the equation of a circle in standard form is that this form enables us to identify the center and the radius of the circle. This information is sufficient to allow us to draw the circle.

Then complete the squares in x and y:

$$(x^2 - 4x + 4) + (y^2 + 6y + 9) = -9 + 4 + 9$$
$$(x - 2)^2 + (y + 3)^2 = 4$$

EXAMPLE 5 Find the center and radius of the circle with equation $9x^2 + 12x + 9y^2 = 77$.

Solution First divide by 9 so that the x^2 and y^2 terms each has a coefficient of 1.

$$x^2 + \tfrac{4}{3}x + y^2 = \tfrac{77}{9}$$

Complete the square in x; add $\frac{4}{9}$ to both sides of the equation.

$$(x^2 + \tfrac{4}{3}x + \tfrac{4}{9}) + y^2 = \tfrac{77}{9} + \tfrac{4}{9}$$
$$(x + \tfrac{2}{3})^2 + y^2 = 9$$

In standard form:

$$[x - (-\tfrac{2}{3})]^2 + (y - 0)^2 = 3^2$$

The center is at $(-\frac{2}{3}, 0)$ and $r = 3$.

TEST YOUR UNDERSTANDING

Find the center and radius of each circle.

1 $x^2 - 6x + y^2 - 10y = 2$

2 $x^2 + y^2 + y = \frac{19}{4}$

3 $x^2 - x + y^2 + 2y = \frac{23}{4}$

4 $16x^2 + 16y^2 - 8x + 32y = 127$

A line that is tangent to a circle touches the circle at only one point and is perpendicular to the radius drawn to the point of tangency.

When the equation of a circle is given and the coordinates of a point P on the circle are known, then the equation of the tangent line to the circle at P can be found. For example, the circle $(x + 3)^2 + (y + 1)^2 = 25$ has center $(-3, -1)$ and $r = 5$. The point $P(1, 2)$ is on this circle because its coordinates satisfy the equation of the circle.

$$(1 + 3)^2 + (2 + 1)^2 = 4^2 + 3^2 = 25$$

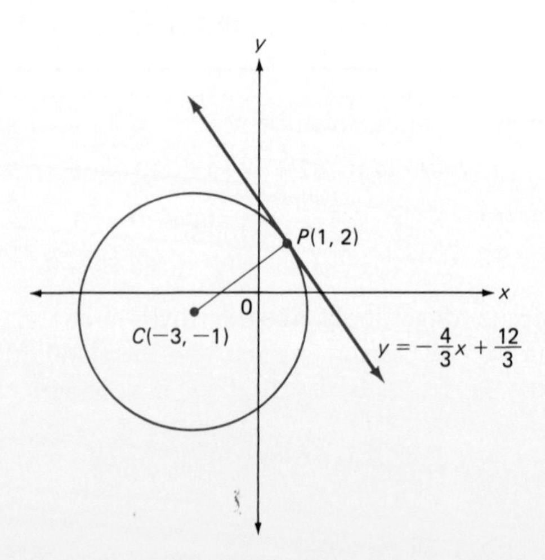

The slope of radius CP is $\frac{2-(-1)}{1-(-3)} = \frac{3}{4}$. Then since the tangent at P is perpendicular to the radius CP, its slope is the negative reciprocal of $\frac{3}{4}$, namely $-\frac{4}{3}$. Now using the point-slope form we get this equation of the tangent at P:

Recall that perpendicular lines have slopes that are negative reciprocals of one another.

$$y - 2 = -\tfrac{4}{3}(x - 1)$$

In slope-intercept form this becomes

$$y = -\tfrac{4}{3}x + \tfrac{10}{3}$$

CAUTION: LEARN TO AVOID MISTAKES LIKE THESE

WRONG	RIGHT
The circle $(x+3)^2 + (y-2)^2 = 7$ has center $(3, -2)$ and radius 7.	The circle has center $(-3, 2)$ and radius $\sqrt{7}$.
The equation of the circle with center $(-1, 0)$ and radius 5 has equation $x^2 + (y+1)^2 = 5$.	The circle has equation $(x+1)^2 + y^2 = 25$.

The center of a circle is the midpoint of each of its diameters. So if the endpoints of a diameter are given, the coordinates of the center can be found using the *midpoint formula* shown below. In the figure PQ is a line segment with midpoint M having coordinates (x', y'). Since x_1, x', and x_2 are on the x-axis, $x' = \frac{x_1 + x_2}{2}$. Similarly, $y' = \frac{y_1 + y_2}{2}$.

The midpoint of a segment on a number line is discussed on page 28.

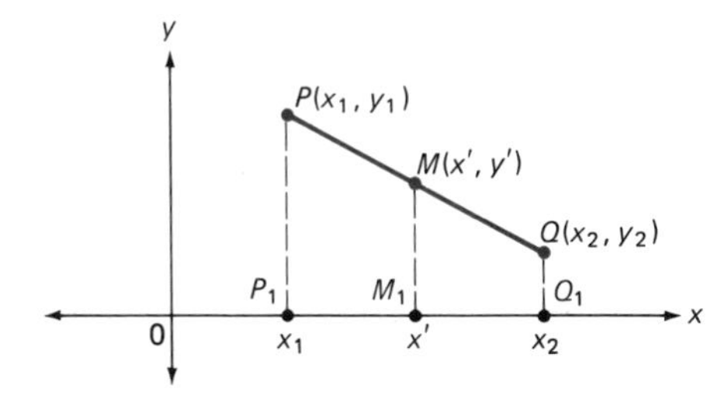

MIDPOINT FORMULA

If (x_1, x_2) and (y_1, y_2) are the endpoints of a line segment, the midpoint of the segment has coordinates

$$\left(\frac{x_1 + x_2}{2}, \frac{y_1 + y_2}{2}\right)$$

EXAMPLE 6 Points $P(2, 5)$ and $Q(-4, -3)$ are the endpoints of a diameter of a circle. Find the center, radius, and equation of the circle.

Solution The center is the midpoint of PQ whose coordinates (x', y') are given by

$$x' = \frac{2 + (-4)}{2} = -1, \qquad y' = \frac{5 + (-3)}{2} = 1$$

The center is located at $C(-1, 1)$. To find the radius, use the distance formula between $C(-1, 1)$ and $P(2, 5)$.

$$r = \sqrt{(-1 - 2)^2 + (1 - 5)^2} = \sqrt{25} = 5$$

The equation of the circle is

$$(x + 1)^2 + (y - 1)^2 = 25$$

EXERCISES 4.5

1 Graph these circles in the same coordinate system.
(a) $x^2 + y^2 = 25$ (b) $x^2 + y^2 = 16$
(c) $x^2 + y^2 = 4$ (d) $x^2 + y^2 = 2$
(e) $x^2 + y^2 = 1$

2 Graph these circles in the same coordinate system.
(a) $(x - 3)^2 + (y - 3)^2 = 9$
(b) $(x + 3)^2 + (y - 3)^2 = 9$
(c) $(x + 3)^2 + (y + 3)^2 = 9$
(d) $(x - 3)^2 + (y + 3)^2 = 9$

Write the equations of each circle in standard form. Find the center and radius for each.

3 $x^2 - 4x + y^2 - 10y = -28$
4 $x^2 - 10x + y^2 - 14y = -25$
5 $x^2 - 8x + y^2 = -14$
6 $x^2 + y^2 + 2y = 7$
7 $x^2 - 20x + y^2 + 20y = -100$
8 $4x^2 - 4x + 4y^2 = 15$
9 $16x^2 + 24x + 16y^2 - 32y = 119$
10 $36x^2 - 48x + 36y^2 + 180y = -160$

Write the equation of each circle in standard form.

11 Center at $(2, 0)$; $r = 2$
12 Center at $(\frac{1}{2}, 1)$; $r = 10$
13 Center at $(-3, 3)$; $r = \sqrt{7}$
14 Center at $(-1, -4)$; $r = 2\sqrt{2}$
15 Draw the circle $x^2 + y^2 = 25$ and the tangent lines at the points $(3, 4)$, $(-3, 4)$, $(3, -4)$, and $(-3, -4)$. Write the equations of these tangent lines.
16 Where are the tangents to the circle $x^2 + y^2 = 4$ whose slopes equal 0? Write their equations.
17 Write the equations of the tangents to the circle $x^2 + y^2 = 4$ that have no slope.

Draw the given circle and the tangent line at the indicated point for each of the following. Write the equation of the tangent line.

18 $x^2 + y^2 = 80$; $(-8, 4)$
19 $x^2 + y^2 = 9$; $(-2, \sqrt{5})$
20 $(x - 4)^2 + (y + 5)^2 = 45$; $(1, 1)$
21 $x^2 + 4x + y^2 - 6y = 60$; $(6, 0)$
22 $x^2 + 14x + y^2 + 18y = 39$; $(5, -4)$
23 $x^2 - 2x + y^2 - 2y = 8$; $(4, 2)$

Find the coordinates of the midpoint of a line segment with endpoints as given.

24 $P(3, 2)$ and $Q(-2, 1)$
25 $P(-2, 4)$ and $Q(3, -8)$
26 $P(-1, 0)$ and $Q(0, 5)$
27 $P(-8, 7)$ and $Q(3, -6)$
28 Points $P(3, -5)$ and $Q(-1, 3)$ are the endpoints of a diameter of a circle. Find the center, radius, and equation of the circle.
*29 Write the equation of the tangent line to the circle $x^2 + y^2 = 80$ at the point in the first quadrant where $x = 4$.

*30 Write the equation of the tangent line to the circle $x^2 + y^2 = 9$ at the point in the third quadrant where $y = -1$.

*31 Write the equation of the tangent line to the circle $x^2 + 14x + y^2 + 18y = 39$ at the point in the second quadrant where $x = -2$.

▬ 32 Suppose a kite is flying at a height of 300 feet above a point *P* on the ground. The kite string is anchored in the ground x feet from *P*.

(a) If s is the length of the string (assume the string forms a straight line) write s as a function of x.

(b) Find s when $x = 400$ feet.

▬ 33 A 13-foot-long board is leaning against the wall of a house so that its base is 5 feet from the wall. When the base of the board is pulled y feet further from the wall, the top of the board drops x feet.

(a) Express y as a function of x.

(b) Find y when $x = 7$.

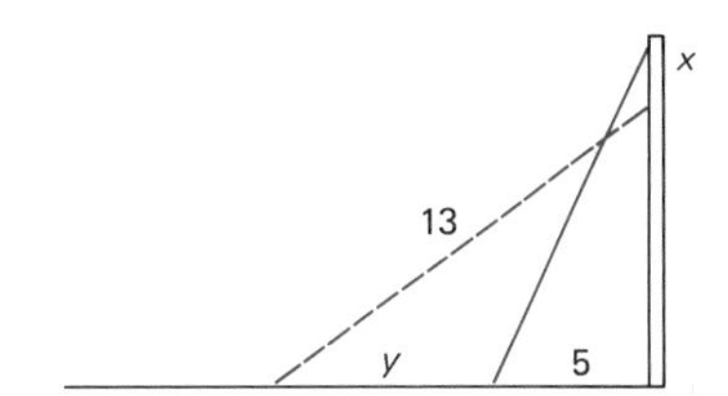

4.6 Solving nonlinear systems

A straight line will intersect a parabola or a circle twice, or once, or not at all. Two parabolas of the form $y = ax^2 + bx + c$ can intersect at most two times; the same is true for two circles. A circle and a parabola can intersect at most four times. These diagrams illustrate some of these possibilities.

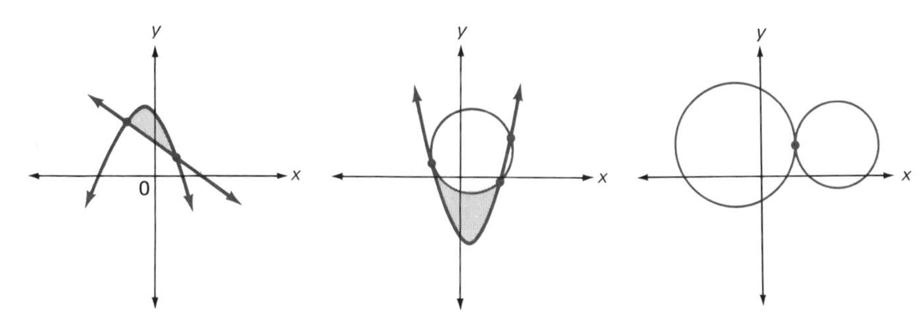

When you study calculus you will learn how to find the areas of the regions between curves. For example, the areas of the shaded regions in the first two diagrams can be found once the coordinates of the points of intersection are known. Here we will address ourselves only to this part of the problem: finding the points of intersection.

In each of the examples that follow, at least one of the two equations will not be linear. Thus we will be learning how to solve certain types of *nonlinear systems*. The underlying strategy in solving such systems will be the same as it was for linear systems. We first eliminate one of the two variables to obtain an equation in one unknown.

EXAMPLE 1 Solve the system and graph:

$$y = x^2$$
$$y = -2x + 8$$

(A parabola and a line.)

Solution Let (x, y) represent the points of intersection. Since these x- and y-values are the same in both equations, we may set the two values for y equal to each other and solve for x.

Various possible cases for nonlinear systems are illustrated by the examples. All the illustrative examples as well as the exercises have been designed so that the solutions are manageable.

$$x^2 = -2x + 8$$
$$x^2 + 2x - 8 = 0$$
$$(x + 4)(x - 2) = 0$$
$$x = -4 \quad \text{or} \quad x = 2$$

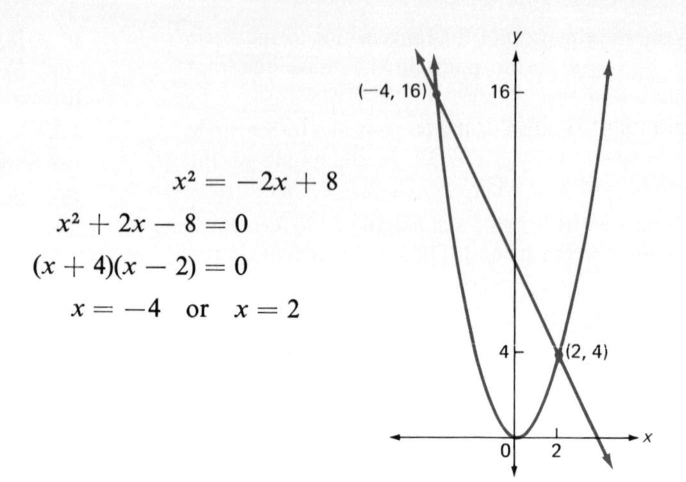

To find the corresponding y-values, either of the original equations may be used. Using $y = -2x + 8$, we have:

For $x = -4$, $y = -2(-4) + 8 = 16$.
For $x = 2$, $y = -2(2) + 8 = 4$.

The solution of the system consists of the two ordered pairs $(-4, 16)$ and $(2, 4)$. The other equation can be used as a check of these results.

(Two parabolas.)

EXAMPLE 2 Solve the system and graph:

$$y = x^2 - 2$$
$$y = -2x^2 + 6x + 7$$

Solution Set the two values for y equal to each other and solve for x.

$$x^2 - 2 = -2x^2 + 6x + 7$$
$$3x^2 - 6x - 9 = 0$$
$$x^2 - 2x - 3 = 0$$
$$(x + 1)(x - 3) = 0$$
$$x = -1 \quad \text{or} \quad x = 3$$

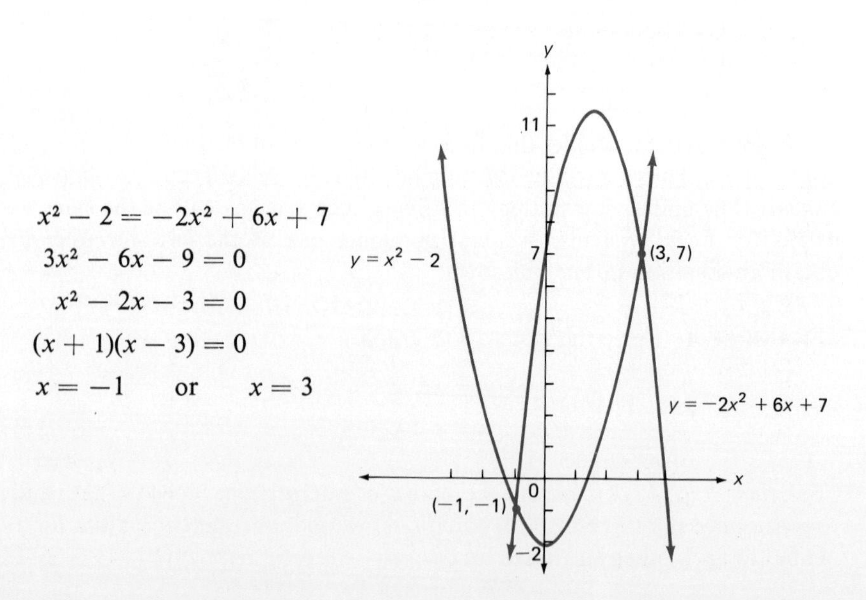

Use $y = x^2 - 2$ to solve for y.

$$y = (-1)^2 - 2 = -1 \qquad y = 3^2 - 2 = 7$$

The points of intersection are $(-1, -1)$ and $(3, 7)$.

EXAMPLE 3 Solve the system and graph.

(A circle and a parabola.)

$$x^2 + y^2 - 8y = -7$$
$$y - x^2 = 1$$

Solution Solve the second equation for x^2.

$$x^2 = y - 1$$

Substitute into the first equation and solve for y.

Note: Alternative methods can be used to solve this example (as well as others). Another easy way begins by adding the given equations. Try it. We may also solve $y - x^2 = 1$ for y and substitute into the first equation. You will find that the latter method is more difficult. With practice you will learn how to find the easier methods.

$$(y - 1) + y^2 - 8y = -7$$
$$y^2 - 7y + 6 = 0$$
$$(y - 1)(y - 6) = 0$$
$$y = 1 \quad \text{or} \quad y = 6$$

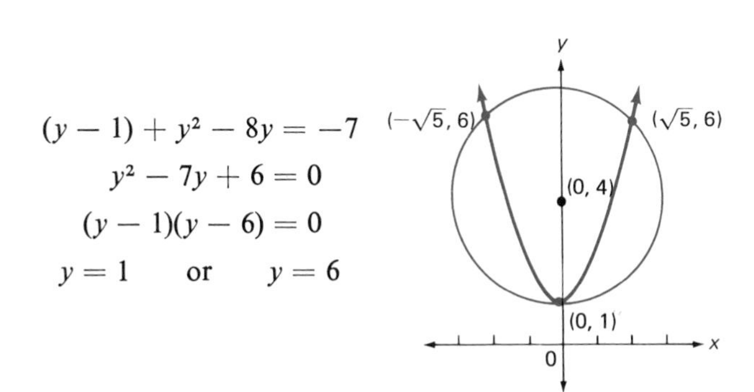

Use $x^2 = y - 1$ to solve for x.

For $y = 1$: $\quad x^2 = 1 - 1 = 0 \qquad x = 0$

For $y = 6$: $\quad x^2 = 6 - 1 = 5 \qquad x = \pm\sqrt{5}$

The points of intersection are $(0, 1)$, $(\sqrt{5}, 6)$, and $(-\sqrt{5}, 6)$.

EXERCISES 4.6

Solve each system and graph.

1. $y = -x^2 - 4x + 1$
   $y - 2x = 10$
2. $3x - 4y = -5$
   $(x + 3)^2 + (y + 1)^2 = 25$
3. $(x + 4)^2 + (y - 1)^2 = 16$
   $(x + 4)^2 + (y - 3)^2 = 4$
4. $y = x^2 - 6x + 9$
   $(x - 3)^2 + (y - 9)^2 = 9$
5. $y = x^2 + 6x + 6$
   $y = -x^2 - 6x + 6$
6. $y = \frac{1}{3}(x - 3)^2 - 3$
   $(x - 3)^2 + (y + 2)^2 = 1$

Solve each system.

7. $y = (x + 1)^2$
   $y = (x - 1)^2$
8. $x^2 + y^2 = 9$
   $y = x^2 - 3$
9. $y = x^2$
   $y = -x^2 + 8x - 16$
10. $y = x^2$
    $y = x^2 - 8x + 24$
11. $7x + 3y = 42$
    $y = -3x^2 - 12x - 15$
12. $y + 2x = 1$
    $x^2 + 4x = 6 - 2y$

13 $y - x = 0$
$(x-2)^2 + (y+5)^2 = 25$

14 $y - 2x = 0$
$(x-2)^2 + (y+5)^2 = 25$

15 $y = -x^2 + 2x$
$x^2 - 2x + y^2 - 2y = 0$

16 $x^2 + 4x + y^2 - 4y = -4$
$(x-2)^2 + (y-2)^2 = 4$

*17 $(x-1)^2 + y^2 = 1$
$x^2 + (y-1)^2 = 1$

*18 $4x + 3y = 25$
$x^2 + y^2 = 25$

*19 $y = \frac{1}{3}(x-3)^2 - 3$
$x^2 - 6x + y^2 + 2y = -6$

*20 $y = x^2 - 6x + 9$
$(x-3)^2 + (y-2)^2 = 58$

4.7

The ellipse and the hyperbola

A **conic section** is a curve formed by the intersection of a plane with a double right-circular cone. These curves, also called **conics**, are known as the **circle, ellipse, parabola,** and **hyperbola**.

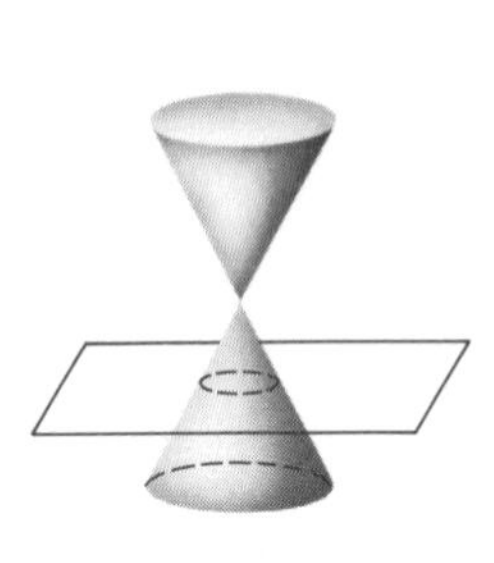
Circle

Parabola

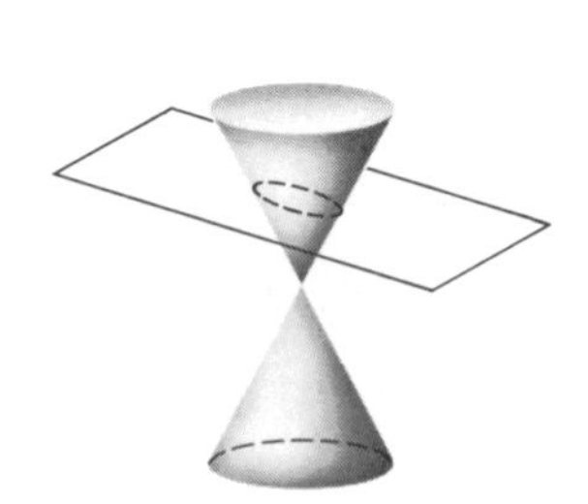
Ellipse

Hyperbola

The figure indicates that the inclination of the plane in relation to the axis of the cone determines the nature of the curve. These four curves have played a vital role in mathematics and its applications from the time of the ancient Greeks until the present day.

Since we have already done some work in this chapter with the parabola and the circle, we now confine our efforts to the study of the ellipse and the hyperbola. Although it is possible to study these curves by using the given geometric interpretations, we will use definitions similar in style to that of a circle as the set of points in a plane equidistant from a fixed point. In the exercises you will also find some work on the parabola along these lines.

By definition, *an ellipse is the set of all points in a plane such that the sum of the distances from two fixed points is a constant.* The two fixed points, F_1 and F_2, are called the **foci** of the ellipse.

We first consider an ellipse whose foci are symmetric about the origin along the x-axis. Thus we let F_1 have coordinates $(-c, 0)$ and F_2 have coordinates $(c, 0)$, where c is some positive number.

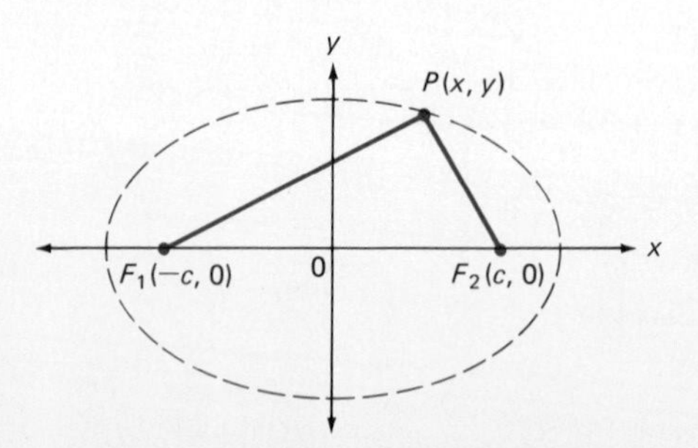

Since the sum of the distances PF_1 and PF_2 must be constant, we choose some positive number a and let this constant equal $2a$. (The form $2a$ will prove to be useful to simplify the algebraic computation.) Thus a point $P(x, y)$ is on the ellipse if and only if $PF_1 + PF_2 = 2a$. (Note $a > c$. Why?)

Using the distance formula gives

$$PF_1 = \sqrt{(x + c)^2 + y^2} \quad \text{and} \quad PF_2 = \sqrt{(x - c)^2 + y^2}$$

Thus

$$PF_1 + PF_2 = \sqrt{(x + c)^2 + y^2} + \sqrt{(x - c)^2 + y^2} = 2a$$

which implies the following:

$$\sqrt{(x + c)^2 + y^2} = 2a - \sqrt{(x - c)^2 + y^2}$$

Squaring both sides and collecting terms, we have

$$a\sqrt{(x - c)^2 + y^2} = a^2 - cx$$

Square both sides again and simplify:

$$(a^2 - c^2)x^2 + a^2y^2 = a^2(a^2 - c^2)$$

Since $a^2 - c^2 > 0$ we may let $b = \sqrt{a^2 - c^2}$ so that $b^2 = a^2 - c^2$. Therefore,

$$b^2x^2 + a^2y^2 = a^2b^2$$

Now divide through by a^2b^2 to obtain the following standard form:

ELLIPSE WITH FOCI AT $(-c, 0)$ AND $(c, 0)$

$$\frac{x^2}{a^2} + \frac{y^2}{b^2} = 1, \qquad \text{where } b^2 = a^2 - c^2$$

The geometric interpretations of a and b can be found from this last equation. Letting $y = 0$ produces the x-intercepts, $x = \pm a$. The points $V_1(-a, 0)$ and $V_2(a, 0)$ are called the **vertices** of the ellipse. The **major axis** of the ellipse is the chord V_1V_2, which has length $2a$. Letting $x = 0$ produces the y-intercepts, $y = \pm b$. The points $(0, -b)$ and $(0, b)$ are the endpoints of the **minor axis**. The intersection of the major and minor axes is the **center** of the ellipse; in this case the center is the origin.

Note that the minor axis has length $2b$, and $2b < 2a$ since $b = \sqrt{a^2 - c^2} < a$.

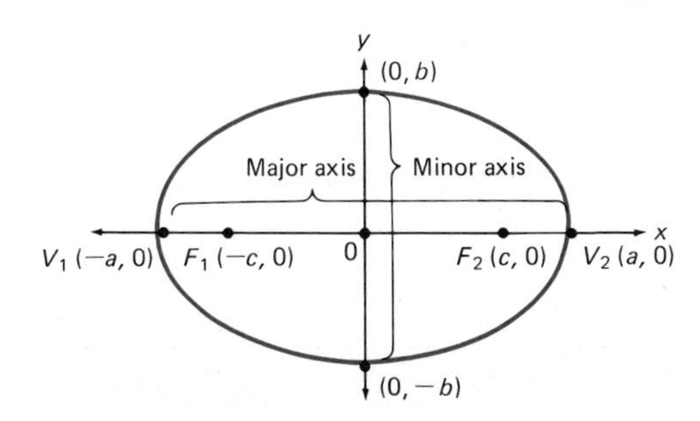

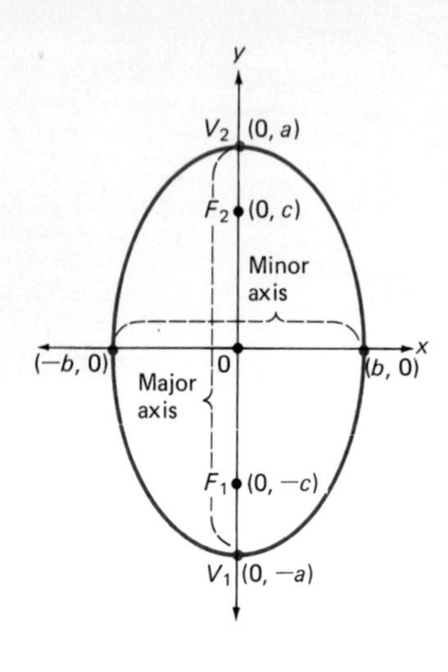

When the foci of the ellipse are on the y-axis, a similar development produces this standard form:

ELLIPSE WITH FOCI AT $(0, -c)$ AND $(0, c)$

$$\frac{x^2}{b^2} + \frac{y^2}{a^2} = 1, \qquad \text{where } b^2 = a^2 - c^2$$

The major axis is on the y-axis, and its endpoints are the vertices $(0, \pm a)$. The minor axis is on the x-axis and has endpoints $(\pm b, 0)$. Center: $(0, 0)$.

EXAMPLE 1 Change $25x^2 + 16y^2 = 400$ into standard form and graph.

Solution Divide both sides of $25x^2 + 16y^2 = 400$ by 400 to obtain $\frac{x^2}{16} + \frac{y^2}{25} = 1$. Since $a^2 = 25$, $a = 5$ and the major axis is on the y-axis with length $2a = 10$. Similarly, $b^2 = 16$ gives $b = 4$, and the minor axis has length $2b = 8$. Also, $c^2 = a^2 - b^2 = 25 - 16 = 9$, so that $c = 3$, which locates the foci at $(0, \pm 3)$.

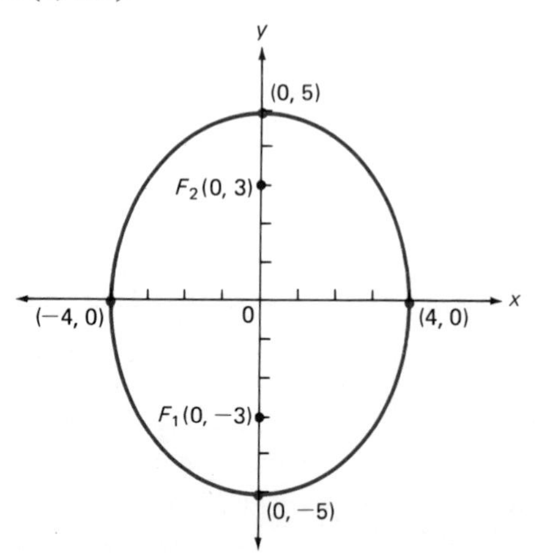

The preceding results can be generalized by allowing the center of the ellipse to be at some point (h, k). If the major axis is horizontal, then the foci have coordinates $(h - c, k)$ and $(h + c, k)$, and it can be shown that the equation has this standard form:

Note: When $a = b$, the standard form produces $(x - h)^2 + (y - k)^2 = a^2$, which is the equation of a circle with center (h, k) and radius a. Thus a circle may be regarded as a special kind of ellipse, one for which the foci and center coincide.

ELLIPSE WITH CENTER (h, k)

$$\frac{(x-h)^2}{a^2} + \frac{(y-k^2)}{b^2} = 1, \qquad \text{where } b^2 = a^2 - c^2$$

What is the equation when the major axis is vertical?

EXAMPLE 2 Write in standard form and graph:

$$4x^2 - 16x + 9y^2 + 18y = 11$$

Solution We follow a procedure much like that used in Section 4.5 for circles, that is, complete the square in both variables.

$$4x^2 - 16x + 9y^2 + 18y = 11$$

$$4(x^2 - 4x) + 9(y^2 + 2y) = 11$$

$$4(x^2 - 4x + 4) + 9(y^2 + 2y + 1) = 11 + 16 + 9$$

$$4(x - 2)^2 + 9(y + 1)^2 = 36$$

Divide both sides by 36:

$$\frac{(x-2)^2}{9} + \frac{(y+1)^2}{4} = 1$$

This is the equation of an ellipse having center at $(2, -1)$, with major axis $2a = 6$ and minor axis $2b = 4$. Since $c^2 = a^2 - b^2 = 5$, $c = \sqrt{5}$, which gives the foci $(2 \pm \sqrt{5}, -1)$.

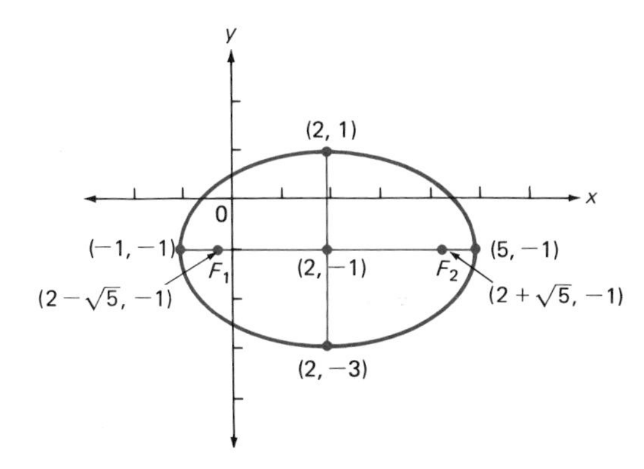

A hyperbola is defined as the set of all points in a plane such that the difference of the distances from two fixed points is constant. The two fixed points, F_1 and F_2, are called the **foci** of the hyperbola, and its **center** is the midpoint of the **transverse axis** F_1F_2. It turns out that a hyperbola consists of two congruent branches which open in opposite directions.

We begin with a hyperbola with foci on the x-axis at $F_1(-c, 0)$ and $F_2(c, 0)$, where $c > 0$. Select a number $a > 0$ so that for any point P on the right branch of the hyperbola we have $PF_1 - PF_2 = 2a$. For any point P on the left branch $PF_2 - PF_1 = 2a$, as in the figure at the top of the following page.

The form $2a$ is used to simplify computations.

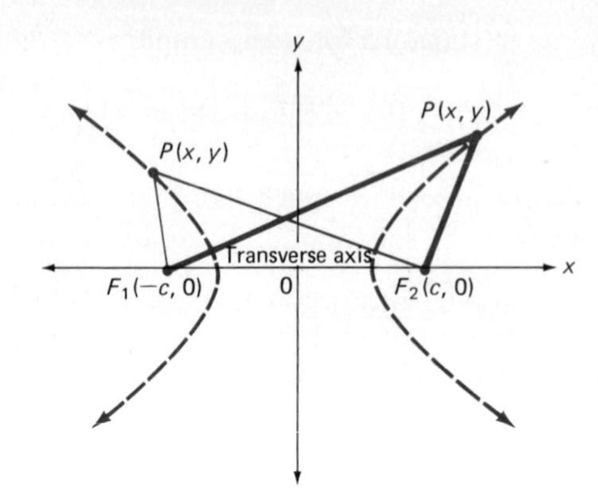

Note that $a < c$, since from triangle F_1PF_2 (P on the right branch) we have $PF_1 < PF_2 + F_1F_2$, which gives $PF_1 - PF_2 < F_1F_2$, or $2a < 2c$.

If we now follow the same type of analysis used to derive the equation of an ellipse, it can be shown that the equation of a hyperbola may be written in this standard form:

HYPERBOLA WITH FOCI AT $(-c, 0)$ AND $(c, 0)$

$$\frac{x^2}{a^2} - \frac{y^2}{b^2} = 1, \qquad \text{where } b^2 = c^2 - a^2$$

Letting $y = 0$ gives $x = \pm a$; the points $V_1(-a, 0)$ and $V_2(a, 0)$ are the **vertices** of the hyperbola.

EXAMPLE 3 Write in standard form and identify the foci and vertices: $16x^2 - 25y^2 = 400$.

Solution Divide through by 400 to place in standard form.

$$\frac{16x^2}{400} - \frac{25y^2}{400} = \frac{400}{400}$$

$$\frac{x^2}{25} - \frac{y^2}{16} = 1$$

Note that $a^2 = 25$ and $b^2 = 16$, so that $a = 5$ and $b = 4$. Then $c^2 = a^2 + b^2 = 25 + 16 = 41$ and $c = \pm\sqrt{41}$. The vertices of the hyperbola are located at $(-5, 0)$ and $(5, 0)$; the foci are at $(-\sqrt{41}, 0)$ and $(\sqrt{41}, 0)$.

Let us return to the standard form for the equation of a hyperbola and solve for y:

$$\frac{x^2}{a^2} - \frac{y^2}{b^2} = 1$$

$$y^2 = \frac{b^2}{a^2}(x^2 - a^2)$$

$$y = \pm\frac{b}{a}\sqrt{x^2 - a^2}$$

Consequently, $|x| \geq a$, which means that there are no points of the hyperbola for $-a < x < a$.

An efficient way to sketch a hyperbola is first to draw the rectangle that is $2a$ units wide and $2b$ units high, as shown in the following figure. Note that the center of the hyperbola is also the center of this rectangle. Draw the diagonals of the rectangle and extend them in both directions; these are the **asymptotes**. Now sketch the hyperbola by beginning at the vertices $(\pm a, 0)$ so that the lines are asymptotes to the curve and the branches are between the asymptotes whose equations are $y = \pm\frac{b}{a}x$.

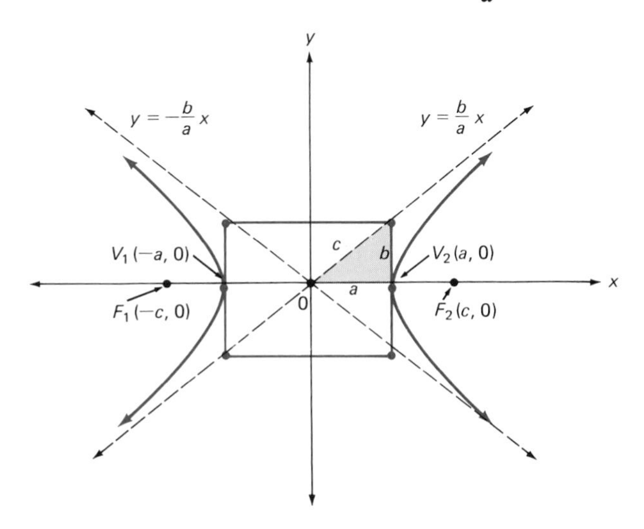

Since $b^2 = c^2 - a^2$, or $a^2 + b^2 = c^2$, it follows that b can be used as a side of a right triangle having hypotenuse c. Thus if we construct a perpendicular at V_2 of length b, the resulting right triangle has hypotenuse c. We also see that this hypotenuse lies on the line $y = \frac{b}{a}x$.

When the foci of a hyperbola are on the y-axis the equation has this standard form:

HYPERBOLA WITH FOCI AT $(0, -c)$ AND $(0, c)$

$$\frac{y^2}{a^2} - \frac{x^2}{b^2} = 1, \qquad \text{where } b^2 = c^2 - a^2$$

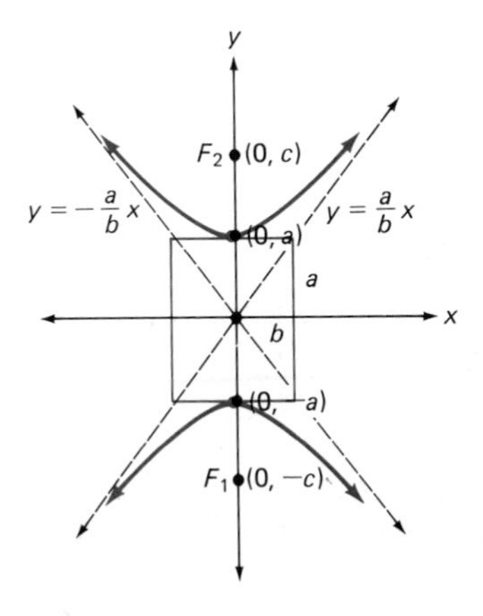

The vertices are $(0, \pm a)$, the transverse axis has length $2a$, and the asymptotes are the lines $y = \pm\frac{a}{b}x$. The branches of this hyperbola open upward and downward.

When the center of the hyperbola is at some point (h, k) and the transverse axis is horizontal, we have the following standard form:

What is the equation when the transverse axis is vertical?

HYPERBOLA WITH CENTER AT (h, k)

$$\frac{(x-h)^2}{a^2} - \frac{(y-k)^2}{b^2} = 1, \qquad \text{where } b^2 = c^2 - a^2$$

EXAMPLE 4 Write in standard form and graph:

$$4x^2 + 16x - 9y^2 + 18y = 29$$

Solution Complete the square in x and y.

$$4(x^2 + 4x) - 9(y^2 - 2y) = 29$$
$$4(x^2 + 4x + 4) - 9(y^2 - 2y + 1) = 29 + 16 - 9$$
$$4(x+2)^2 - 9(y-1)^2 = 36$$

Divide both sides by 36:

$$\frac{(x+2)^2}{9} - \frac{(y-1)^2}{4} = 1$$

This is the standard form for a hyperbola with center at $(-2, 1)$. Since $a^2 = 9$, we have $a = 3$ and the vertices are located 3 units from the center at $(-5, 1)$ and $(1, 1)$. Since $c^2 = a^2 + b^2 = 13$, the foci are located at $\sqrt{13}$ units from the center, namely at $(-2 \pm \sqrt{13}, 1)$.

To sketch the hyperbola first draw the 6-by-4 rectangle with center at $(-2, 1)$ as shown. Draw the asymptotes by extending the diagonals and sketch the branches.

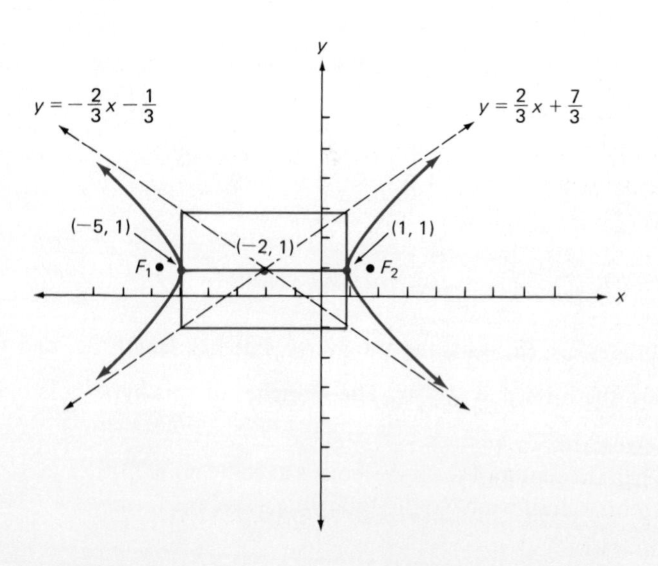

Name the conic section and state the coordinates of the center, vertices, and foci. Give the equations of the asymptotes where applicable.

1 $\frac{x^2}{25}+\frac{y^2}{16}=1$

2 $\frac{x^2}{16}+\frac{y^2}{25}=1$

3 $\frac{x^2}{36}-\frac{y^2}{25}=1$

4 $\frac{y^2}{36}-\frac{x^2}{25}=1$

5 $25x^2+4y^2=100$

6 $x^2-2y^2=6$

7 $x^2+\frac{(y-1)^2}{4}=1$

8 $\frac{(x+2)^2}{16}+\frac{(y+5)^2}{25}=1$

Name the conic and graph.

9 $\frac{x^2}{9}+\frac{y^2}{16}=1$

10 $\frac{x^2}{16}+\frac{y^2}{9}=1$

11 $16y^2-9x^2=144$

12 $9x^2-16y^2=144$

13 $\frac{(x-1)^2}{64}+\frac{(y-2)^2}{36}=1$

14 $\frac{(y-1)^2}{64}-\frac{(x-3)^2}{36}=1$

15 $16(y-3)^2-9(x+2)^2=-144$

16 $16(y-2)^2-9(x+3)^2=144$

Write in standard form and identify.

17 $x^2+y^2-2x+4y+1=0$

18 $x^2+y^2+6x-4y+4=0$

19 $x^2+4y^2+2x-3=0$

20 $x^2-9y^2-2x-8=0$

21 $9x^2+18x-16y^2+96y=279$

22 $4x^2-16x+y^2+8y=-28$

Write the equation of the ellipse in standard form having the given properties.

23 Center (0, 0); horizontal major axis of length 10; minor axis of length 6.

24 Center (0, 0); foci (±2, 0); vertices (±5, 0).

25 Vertices (0, ±5); foci (0, ±3).

26 Center (2, 3); foci (−2, 3) and (6, 3); minor axis of length 8.

27 Center (2, −3); vertical major axis of length 12; minor axis of length 8.

28 Center (−5, 0); foci (−5, ±2); $b=3$.

Write the equation of the hyperbola in standard form having the given properties.

29 Center (0, 0); foci (±6, 0); vertices (±4, 0).

30 Center (0, 0); foci (0, ±4); vertices (0, ±1).

31 Center (−2, 3); vertical transverse axis of length 6; $c=4$.

32 Center (4, 4); vertex (4, 7); $b=2$.

*33 Center (0, 0); asymptotes $y=\pm\frac{1}{2}x$; vertices (±4, 0).

34 The figure describes an instrument that can be used to draw an ellipse. Put thumbtacks at the points on the paper and place a loop of string around them. Pull the string taut by using a pencil point as indicated. Keeping the string taut, move the pencil around the loop. Why does this motion trace out an ellipse? Draw an ellipse using the construction suggested.

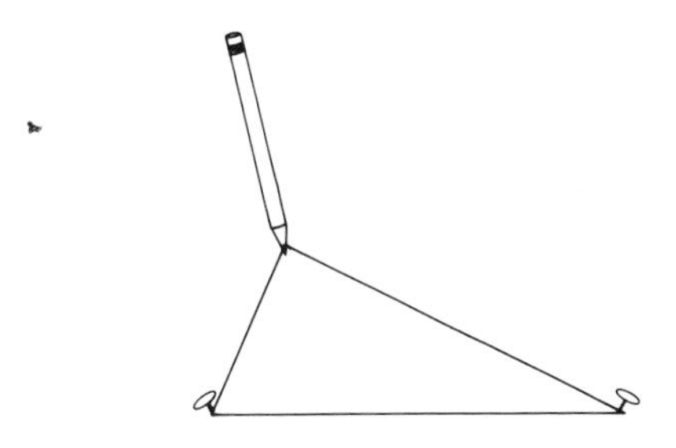

35 A parabola can be defined as the set of all points in a plane equidistant from a given fixed line called the **directrix** and a given fixed point called the **focus**. Let focus F have coordinates $(0, p)$, and let the directrix have equation $y=-p$ as indicated.

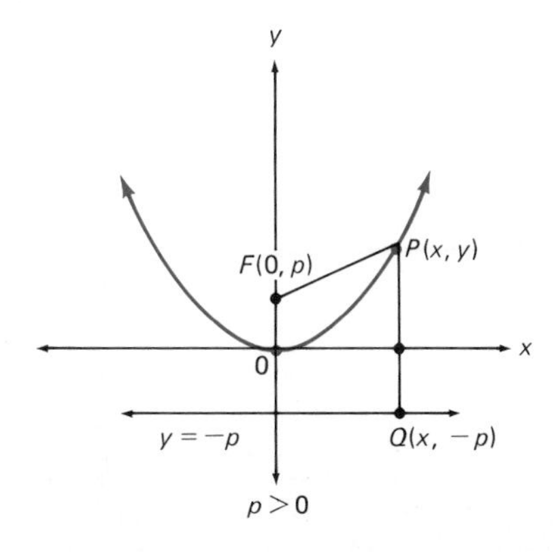

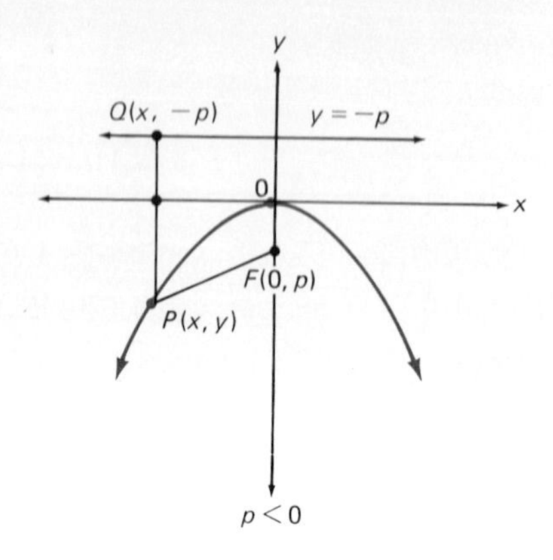

(a) Write the length PF in terms of the coordinates of P and F.

(b) Express the length PQ in terms of y and p.

(c) Equate the results in parts (a) and (b) and derive the form $x^2 = 4py$.

36 Write the equation of the parabola with focus (0, 3) and directrix $y = -3$. Sketch.

37 Write the equation of the parabola with focus (0, −3) and directrix $y = 3$. Sketch.

38 Find the coordinates of the focus and the equation of the directrix for the parabola $y = 2x^2$.

39 Find the coordinates of the focus and the equation of the directrix for the parabola $y = -\frac{1}{4}x^2$.

***40** The origin is the vertex of each parabola having equation $x^2 = 4py$. What is the equation of the parabola with center (h, k) having focus $(h, k + p)$ and directrix $y = k - p$?

***41** A parabola has vertex (2, −5), focus at (2, −3), and directrix $y = -7$. Write its equation in the form of Exercise 40. Then write its equation in the form $y = ax^2 + bx + c$.

***42** Let P be on the right branch of the hyperbola with foci $F_1(-c, 0)$ and $F_2(c, 0)$, and let $PF_1 - PF_2 = 2a$ for $a < c$. Derive the equation $\frac{x^2}{a^2} - \frac{y^2}{b^2} = 1$, where $b^2 = c^2 - a^2$.

4.8

Complex numbers

See Section 2.2 for a review of radicals.

In the definition of a radical, care was taken to avoid the even root of a negative number, such as $\sqrt{-4}$. This was necessary because there is no real number x whose square is −4. Consequently, there can be no real number that satisfies the equation $x^2 + 4 = 0$. Suppose, for the moment, that we could solve $x^2 + 4 = 0$ using our algebraic methods. Then we might write the following:

$$\begin{aligned} x^2 + 4 &= 0 \\ x^2 &= -4 \\ x &= \pm\sqrt{-4} \\ &= \pm\sqrt{4(-1)} \\ &= \pm\sqrt{4}\sqrt{-1} \\ &= \pm 2\sqrt{-1} \end{aligned}$$

It could now be claimed that $2\sqrt{-1}$ is a solution of $x^2 + 4 = 0$. But it is certainly not a *real* solution. What we now do is to introduce formally $\sqrt{-1}$ as a new kind of number; it will be the unit for a new set of numbers, the *imaginary numbers*. The symbol i is used to stand for this number and is defined as follows.

DEFINITION OF i

$$i = \sqrt{-1} \quad \text{and} \quad i^2 = -1$$

Using i, the square root of a negative real number is now defined:

$$\sqrt{-x} = \sqrt{-1}\sqrt{x} = i\sqrt{x} \qquad \text{for } x > 0$$

EXAMPLE 1 Simplify: **(a)** $\sqrt{-16} + \sqrt{-25}$ **(b)** $\sqrt{-16} \cdot \sqrt{-25}$

Solution

(a) $\sqrt{-16} = \sqrt{-1} \cdot \sqrt{16} = i \cdot 4 = 4i$

$\sqrt{-25} = \sqrt{-1} \cdot \sqrt{25} = i \cdot 5 = 5i$

Thus

$$\sqrt{-16} + \sqrt{-25} = 4i + 5i = 9i$$

(b) $\sqrt{-16} \cdot \sqrt{-25} = (4i)(5i) = 20i^2 = 20(-1) = -20$

In the example, $4i$ and $5i$ are combined by using the usual rules of algebra. You will see later that such procedures apply for this new kind of number.

TEST YOUR UNDERSTANDING

Express as the product of a real number and i.

1 $\sqrt{-9}$ **2** $\sqrt{-49}$ **3** $\sqrt{-5}$ **4** $-2\sqrt{-1}$ **5** $\sqrt{-\frac{4}{9}}$

Simplify.

6 $\sqrt{-16} \cdot \sqrt{-25}$

7 $\sqrt{9} \cdot \sqrt{-49}$

8 $\sqrt{-50} + \sqrt{-32} - \sqrt{-8}$

9 $3\sqrt{-20} + 2\sqrt{-45}$

An indicated product of a real number times the imaginary unit i, such as $7i$ or $\sqrt{2}\,i$, is called a **pure imaginary number**. The sum of a real number and a pure imaginary number is called a **complex number**.

A complex number has the form $a + bi$, where a and b are real numbers and $i = \sqrt{-1}$.

We say that the real number a is the **real part** of $a + bi$ and the real number b is called the **imaginary part** of $a + bi$. In general, two complex numbers are equal only when both their real parts and their imaginary parts are equal. Thus

$$a + bi = c + di \quad \text{if and only if} \quad a = c \text{ and } b = d$$

The collection of complex numbers contains all the real numbers, since any real number a can also be written as $a = a + 0i$. Similarly, if b is real, $bi = 0 + bi$, so that the complex numbers also contain the pure imaginaries.

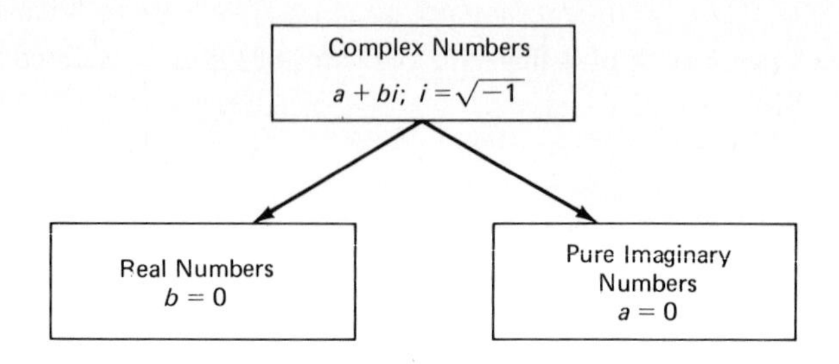

Various sets of numbers are described in Section 1.2.

EXAMPLE 2 Name the sets of numbers to which each of the following belongs: **(a)** $\sqrt{\frac{4}{9}}$ **(b)** $\sqrt{-10}$ **(c)** $-4 + \sqrt{-12}$

Solution

(a) $\sqrt{\frac{4}{9}} = \frac{2}{3} = \frac{2}{3} + 0i$ is complex, real, and rational.

(b) $\sqrt{-10} = i\sqrt{10}$ is complex and pure imaginary.

(c) $-4 + \sqrt{-12} = -4 + i\sqrt{12} = -4 + 2i\sqrt{3}$ is complex.

How should complex numbers be added, subtracted, multiplied, or divided? In answering this question, it must be kept in mind that the real numbers are included in the collection of complex numbers and the definitions we construct for the complex numbers must preserve the established operations for the reals. We do not want, for example, to add complex numbers in such a way that would produce inconsistent results for the addition of real numbers.

We add and subtract complex numbers by combining their real and their imaginary parts separately, according to these definitions.

SUM AND DIFFERENCE OF COMPLEX NUMBERS

$$(a + bi) + (c + di) = (a + c) + (b + d)i$$
$$(a + bi) - (c + di) = (a - c) + (b - d)i$$

Actually, these procedures are quite similar to those used for combining polynomials. For example, compare these two sums:

$$(2 + 3x) + (5 + 7x) = (2 + 5) + (3 + 7)x = 7 + 10x$$
$$(2 + 3i) + (5 + 7i) = (2 + 5) + (3 + 7)i = 7 + 10i$$

Similarly, compare these subtraction problems:

$$(8 + 5x) - (3 + 2x) = (8 - 3) + (5 - 2)x = 5 + 3x$$
$$(8 + 5i) - (3 + 2i) = (8 - 3) + (5 - 2)i = 5 + 3i$$

To decide how to multiply two complex numbers we go back to the set of real numbers and extend the multiplication process. Recall the procedure for multiplying two binomials:

$$(a + b)(c + d) = ac + bc + ad + bd$$

Now compare these two products:

$$(3 + 2x)(5 + 3x) = 15 + 10x + 9x + 6x^2$$
$$(3 + 2i)(5 + 3i) = 15 + 10i + 9i + 6i^2$$

This last expression can be simplified by noting that $10i + 9i = 19i$ and $6i^2 = -6$. The final result is $9 + 19i$.

In general, we can develop a rule for multiplication of two complex numbers by finding the product of $a + bi$ and $c + di$ as follows:

$$\begin{aligned}(a + bi)(c + di) &= ac + adi + bci + bdi^2 \\ &= ac + (ad + bc)i + bd(-1) \\ &= (ac - bd) + (ad + bc)i\end{aligned}$$

PRODUCT OF COMPLEX NUMBERS

$$(a + bi)(c + di) = (ac - bd) + (ad + bc)i$$

Note: In practice, it is easier to find the product by using the procedure for multiplying binomials rather than by memorizing the formal definition.

EXAMPLE 3 Multiply: $(5 - 2i)(3 + 4i)$.

Solution

$$\begin{aligned}(5 - 2i)(3 + 4i) &= 15 - 6i + 20i - 8i^2 \\ &= 15 - 6i + 20i + 8 \\ &= 23 + 14i\end{aligned}$$

Now consider the quotient of two complex numbers such as

$$\frac{2 + 3i}{3 + i}$$

Our objective is to express this quotient in the form $a + bi$. To do so, we use a method similar to rationalizing the denominator. Note what happens when $3 + i$ is multiplied by its **conjugate**, $3 - i$:

The term imaginary is said to have been first applied to these numbers precisely because they seem mysteriously to vanish under certain multiplications.

$$(3 + i)(3 - i) = 9 + 3i - 3i - i^2 = 9 - i^2 = 9 + 1 = 10$$

We are now ready to complete the division problem.

$$\begin{aligned}\frac{2+3i}{3+i} &= \frac{2+3i}{3+i}\cdot\frac{3-i}{3-i}\\ &= \frac{6+9i-2i-3i^2}{9-i^2}\\ &= \frac{6+9i-2i+3}{9+1}\\ &= \frac{9+7i}{10}\\ &= \frac{9}{10}+\frac{7}{10}i\end{aligned}$$

In general, multiplying the numerator and denominator of $\frac{a+bi}{c+di}$ by the conjugate of $c+di$ leads to the following definition for division (see Exercise 71).

Rather than memorize this definition, simply find quotients as in the preceding illustration.

QUOTIENT OF COMPLEX NUMBERS

$$\frac{a+bi}{c+di} = \frac{ac+bd}{c^2+d^2} + \frac{bc-ad}{c^2+d^2}i; \qquad c+di \neq 0$$

The conjugate of a complex number $z = a + bi$ is sometimes denoted as $\bar{z}$, so that $\bar{z} = \overline{a+bi} = a - bi$. In the exercises you will be asked to prove that the conjugate of a sum is equal to the sum of the conjugates; $\overline{z+w} = \bar{z} + \bar{w}$; and similarly for differences, products, and quotients.

EXAMPLE 4 For $z = -2 + 5i$ and $w = 4 - 7i$, verify that **(a)** $\overline{z+w} = \bar{z} + \bar{w}$ and **(b)** $\overline{zw} = \bar{z} \cdot \bar{w}$.

Solution

(a)

$$\begin{aligned}\overline{z+w} &= \overline{(-2+5i)+(4-7i)}\\ &= \overline{2-2i}\\ &= 2+2i\end{aligned}$$

$$\begin{aligned}\bar{z}+\bar{w} &= \overline{-2+5i}+\overline{4-7i}\\ &= (-2-5i)+(4+7i)\\ &= 2+2i\end{aligned}$$

(b)

$$\begin{aligned}\overline{zw} &= \overline{(-2+5i)(4-7i)}\\ &= \overline{27+34i}\\ &= 27-34i\end{aligned}$$

$$\begin{aligned}\bar{z}\bar{w} &= \overline{(-2+5i)}\,\overline{(4-7i)}\\ &= (-2-5i)(4+7i)\\ &= 27-34i\end{aligned}$$

Although we will not go into the details here, it can be shown that some of the basic algebraic rules apply for the complex numbers. For example, the commutative, associative, and distributive laws hold, whereas the rules of order do not apply.

It is also true that the rules for integer exponents apply for complex numbers. For example, $(2 - 3i)^0 = 1$ and $(2 - 3i)^{-1} = \frac{1}{2 - 3i}$. In particular, the integral powers of i are easily evaluated. Following are the first four powers of i.

$$
\begin{aligned}
i &= \sqrt{-1} \\
i^2 &= -1 \\
i^3 &= i^2 \cdot i = -1 \cdot i = -i \\
i^4 &= i^2 \cdot i^2 = (-1)(-1) = 1
\end{aligned}
$$

After the first four powers of i a repeating pattern exists, as can be seen from the next four powers of i:

$$
\begin{aligned}
i^5 &= i^4 \cdot i = 1 \cdot i = i \\
i^6 &= i^4 \cdot i^2 = 1 \cdot i^2 = -1 \\
i^7 &= i^4 \cdot i^3 = 1 \cdot i^3 = -i \\
i^8 &= i^4 \cdot i^4 = 1 \cdot i^4 = 1
\end{aligned}
$$

The cycle for consecutive powers of i is: $i, -1, -i, 1$.

To simplify i^n when $n > 4$, begin by finding the largest multiple of 4 in the integer n as in the next example.

EXAMPLE 5 Simplify (a) i^{22} (b) i^{39}

Solution

(a) $i^{22} = i^{20} \cdot i^2 = (i^4)^5 \cdot i^2 = 1^5 \cdot i^2 = i^2 = -1$

(b) $i^{39} = i^{36} \cdot i^3 = (i^4)^9 \cdot i^3 = 1^9 \cdot i^3 = i^3 = -i$

$20 = 5(4)$ is the largest multiple of 4 in 22
$36 = 9(4)$ is the largest multiple of 4 in 39

The following example illustrates how to operate with a negative integral power of i.

EXAMPLE 6 Express $2i^{-3}$ as the indicated product of a real number and i.

Solution First note that $2i^{-3} = \frac{2}{i^3}$. Next multiply numerator and denominator by i to obtain a real number in the denominator.

$$2i^{-3} = \frac{2}{i^3} \cdot \frac{i}{i} = \frac{2i}{i^4} = \frac{2i}{1} = 2i$$

Alternative Solution

$$
\begin{aligned}
2i^{-3} &= \frac{2}{i^3} = \frac{2}{-i} \\
&= \frac{2}{-i} \cdot \frac{i}{i} \qquad (i \text{ is the conjugate of } -i) \\
&= \frac{2i}{-i^2} = \frac{2i}{-(-1)} \\
&= 2i
\end{aligned}
$$

Many algebraic procedures studied earlier can be extended to complex numbers. For example, the factoring formula for the difference of squares,

$$a^2 - b^2 = (a + b)(a - b)$$

can now be used to factor the sum of squares such as $x^2 + 1$ that was not factorable using real coefficients. This becomes possible after first writing $x^2 + 1$ as $x^2 - (-1)$ and letting $-1 = i^2$. Thus

$$\begin{aligned} x^2 + 1 &= x^2 - (-1) \\ &= x^2 - i^2 \\ &= (x + i)(x - i) \end{aligned}$$

This is a good place to review the quadratic formula developed in Section 4.3.

The quadratic formula used to find the roots of $ax^2 + bx + c = 0$ is also applicable when complex numbers are involved. In fact, the situation in which there are no real roots, because the discriminant is negative, now produces complex roots.

Assume $b^2 - 4ac < 0$:

$$\begin{aligned} x &= \frac{-b \pm \sqrt{b^2 - 4ac}}{2a} = \frac{-b \pm \sqrt{-(4ac - b^2)}}{2a} \\ &= \frac{-b \pm i\sqrt{4ac - b^2}}{2a} \\ &= -\frac{b}{2a} \pm \frac{i\sqrt{4ac - b^2}}{2a} \end{aligned}$$

EXAMPLE 7 Solve $x^2 + x + 1 = 0$.

Solution $b^2 - 4ac = 1 - 4 = -3 < 0$ shows that there are no real roots. However, there are these complex roots:

$$\begin{aligned} x &= \frac{-1 \pm \sqrt{-3}}{2} = \frac{-1 \pm i\sqrt{3}}{2} \\ &= -\frac{1}{2} \pm \frac{\sqrt{3}}{2}i \end{aligned}$$

EXERCISES 4.8

Classify each statement as true or false.

1 Every real number is a complex number.

2 Every complex number is a real number.

3 Every irrational number is a complex number.

4 Every integer can be written in the form $a + bi$.

5 Every complex number may be expressed as an irrational number.

6 Every negative integer may be written as a pure imaginary number.

Express each of the following numbers in the form $a + bi$.

7 $5 + \sqrt{-4}$

8 $7 - \sqrt{-7}$

9 -5

10 $\sqrt{25}$

Express in the form bi.

11 $\sqrt{-16}$

12 $\sqrt{-81}$

13 $\sqrt{-144}$

14 $-\sqrt{-9}$

15 $\sqrt{-\frac{9}{16}}$

16 $\sqrt{-3}$

17 $-\sqrt{-5}$

18 $-\sqrt{-8}$

Simplify.

19 $\sqrt{-9} \cdot \sqrt{-81}$

20 $\sqrt{4} \cdot \sqrt{-25}$

21 $\sqrt{-3} \cdot \sqrt{-2}$

22 $(2i)(3i)$

23 $(-3i^2)(5i)$

24 $(i^2)(i^2)$

25 $\sqrt{-9} + \sqrt{-81}$

26 $\sqrt{-12} + \sqrt{-75}$

27 $\sqrt{-8} + \sqrt{-18}$

28 $2\sqrt{-72} - 3\sqrt{-32}$

29 $\sqrt{-9} - \sqrt{-3}$

30 $3\sqrt{-80} - 2\sqrt{-20}$

Complete the indicated operation. Express all answers in the form $a + bi$.

31 $(7 + 5i) + (3 + 2i)$

32 $(8 + 7i) + (9 - i)$

33 $(8 + 2i) - (3 + 5i)$

34 $(7 + 2i) - (4 - 3i)$

35 $(7 + \sqrt{-16}) + (3 - \sqrt{-4})$

36 $(8 + \sqrt{-49}) - (2 - \sqrt{-25})$

37 $2i(3 + 5i)$

38 $3i(5i - 2)$

39 $(3 + 2i)(2 + 3i)$

40 $(\sqrt{5} + 3i)(\sqrt{5} - 3i)$

41 $(5 - 2i)(3 + 4i)$

42 $(\sqrt{3} + 2i)^2$

43 $\dfrac{3 + 5i}{i}$

44 $\dfrac{5 - i}{i}$

45 $\dfrac{5 + 3i}{2 + i}$

46 $\dfrac{7 - 2i}{2 - i}$

47 $\dfrac{3 - i}{3 + i}$

48 $\dfrac{8 + 3i}{3 - 2i}$

$z = -6 + 8i$ *and* $w = \frac{1}{2} + \frac{1}{2}i$ *to verify the following.*

49 $\overline{z + w} = \bar{z} + \bar{w}$

50 $\overline{z - w} = \bar{z} - \bar{w}$

51 $\overline{zw} = \bar{z} \cdot \bar{w}$

52 $\overline{\left(\dfrac{z}{w}\right)} = \dfrac{\bar{z}}{\bar{w}}$

Express in the form bi.

53 $3i^3$

54 $-5i^5$

55 $2i^7$

56 $3i^{-3}$

57 $-4i^{18}$

58 i^{-32}

Simplify and express each answer in the form $a + bi$.

59 $(3 + 2i)^{-1}$

60 $(3 + 2i)^{-2}$

61 One of the basic rules for operating with radicals is that $\sqrt{ab} = \sqrt{a} \cdot \sqrt{b}$, where a and b are nonnegative real numbers. Prove that this rule does not work when both a and b are negative by showing that $\sqrt{(-4)(-9)} \neq \sqrt{-4} \cdot \sqrt{-9}$.

62 Use the definition $\sqrt{-x} = i\sqrt{x}$ $(x \geq 0)$ to prove that $\sqrt{ab} = \sqrt{a} \cdot \sqrt{b}$ when $a < 0$ and $b \geq 0$.

63 Use complex numbers to factor each binomial.
(a) $x^2 + 4$
(b) $3x^2 + 27$
(c) $x^2 + 2$
(d) $5x^2 + 15$

Solve each quadratic equation for real or complex roots using the quadratic formula or by factoring when possible.

64 $x^2 + 7 = 0$

65 $x^2 + 6x + 25 = 0$

66 $x^3 + 9x = 0$

67 $2x^2 + x = -5$

68 $6x^2 = 7x - 3$

69 $\frac{1}{2}x^2 - 3x + 7 = 1$

70 Find the value of $x^2 + 3x + 5$ when $x = \dfrac{-3 + \sqrt{11}i}{2}$.

71 Write $\dfrac{a + bi}{c + di}$ in the form $x + yi$. (*Hint:* Multiply the numerator and denominator by the conjugate of $c + di$.)

72 The set of complex numbers satisfies the associative property for addition. Verify this by completing this problem in two different ways.

$$(3 + 5i) + (2 + 3i) + (7 + 4i)$$

73 Repeat Exercise 72 for multiplication, using $(3 + i)(3 - i)(4 + 3i)$.

74 We say that $0 = 0 + 0i$ is the additive identity for the complex numbers since $0 + z = z$ for any $z = a + bi$. Find the additive inverse (negative) of z.

Perform the indicated operations and express the answers in the form $a + bi$.

75 $(5 + 4i) + 2(2 - 3i) - i(1 - 5i)$

76 $2i(3 - 4i)(3 - 6i) - 7i$

77 $\dfrac{(2 + i)^2(3 - i)}{2 + 3i}$

78 $\dfrac{1 - 2i}{3 + 4i} - \dfrac{2i - 3}{4 - 2i}$

79 $x^4 - 81 = 0$

80 $x^3 - 1 = 0$

*81 $4x^4 - 35x^2 - 9 = 0$

*82 $3x^3 - 7x^2 + 6x - 14 = 0$

Use $z = a + bi$ and $w = c + di$ to prove the following.

83 $\overline{z + w} = \bar{z} + \bar{w}$

84 $\overline{z - w} = \bar{z} - \bar{w}$

*85 $\overline{\left(\dfrac{z}{w}\right)} = \dfrac{\bar{z}}{\bar{w}}$

*86 $\overline{zw} = \bar{z}\bar{w}$

REVIEW EXERCISES FOR CHAPTER 4

The solutions to the following exercises can be found within the text of Chapter 4. Try to answer each question without referring to the text.

Section 4.1

Graph each of the following.

1 $y = -x^2$

2 $y = \frac{1}{2}x^2$

3 $y = (x - 1)^2$

4 $y = (x + 1)^2$

5 $y = (x + 2)^2 - 2$

6 $y = -2(x - 3)^2 + 4$

7 For Exercise 6, state the coordinates of the vertex and the equation of the axis of symmetry.

Section 4.2

Write in standard form.

8 $y = x^2 + 4x + 3$

9 $y = 2x^2 - 12x + 11$

10 $y = -\frac{1}{3}x^2 - 2x + 1$

11 Find the vertex and axis of symmetry of the parabola $y = 3x^2 - 4x - 2$.

12 Graph the function f where $f(x) = |x^2 - 4|$.

13 State the conditions on the values a and k so that the parabola $y = a(x - h)^2 + k$ opens downward and intersects the x-axis in two points.

14 What are the domain and range of the function in Exercise 13?

Section 4.3

15 Find the x-intercepts of $f(x) = 25x^2 + 30x + 9$.

16 Find the roots of $2x^2 - 9x - 18 = 0$.

17 State the quadratic formula.

18 Solve for x: $2x^2 - 5x + 1 = 0$.

19 Solve for x: $2x^2 = x - 1$.

20 Find the value of the discriminant and decide how many x-intercepts there are:

(a) $y = x^2 - x - 6$

(b) $y = x^2 - 4x + 4$

(c) $y = x^2 - 4x + 5$

21 Solve: $x^2 - x - 6 < 0$.

Section 4.4

22 Find the maximum value or minimum value of the function $f(x) = 2(x + 3)^2 + 5$.

23 Find the maximum value of the quadratic function $f(x) = -\frac{1}{3}x^2 + x + 2$. At which value x does f achieve this maximum?

24 Suppose that 60 meters of fencing is available to enclose a rectangular garden, one side of which will be against the side of a house. What dimensions of the garden will guarantee a maximum area?

25 A ball is thrown straight upward from ground level with an initial velocity of 32 feet per second. The formula $s = 32t - 16t^2$ gives its height in feet, s, after t seconds. (a) What is the maximum height reached by the ball? (b) When does the ball return to the ground?

Section 4.5

26 Find the length of the line segment determined by the points $(-2, 2)$ and $(6, -4)$.

27 Find the center and radius of the circle
$$(x - 2)^2 + (y + 3)^2 = 4$$

28 Find the center and radius of the circle
$$x^2 - 4x + y^2 + 6y = -9$$

29 Find the center and radius of the circle
$$9x^2 + 12x + 9y^2 = 77$$

30 Find the equation of the tangent line to the circle $(x + 3)^2 + (y + 1)^2 = 25$ at the point $(1, 2)$. Sketch.

31 Points $P(2, 5)$ and $Q(-4, -3)$ are the endpoints of a diameter of a circle. Find the center, radius, and equation of the circle.

Section 4.6

Solve each system and graph.

32 $y = x^2$
$y = -2x + 8$

33 $y = x^2 - 2$
$y = -2x^2 + 6x + 7$

34 $x^2 + y^2 - 8y = -7$
$y - x^2 = 1$

Section 4.7

Write in standard form and graph.

35 $4x^2 - 16x + 9y^2 + 18y = 11$

36 $4x^2 + 16x - 9y^2 + 18y = 29$

37 Write in standard form and identify the foci and vertices of $16x^2 - 25y^2 = 400$.

38 Write the standard form for the equation of a hyperbola with center at (h, k) such that the transverse axis is horizontal.

Section 4.8

39 Multiply: $(5 - 2i)(3 + 4i)$.

40 Divide: $\frac{2 + 3i}{3 + i}$.

41 Simplify: (a) i^{22}; (b) i^{39}.

42 For $z = -2 + 5i$ and $w = 4 - 7i$, evaluate $\overline{z + w}$ and $\overline{z \cdot w}$.

43 Write $2i^{-3}$ in the form bi.

44 Solve $x^2 + x + 1 = 0$.

SAMPLE TEST QUESTIONS FOR CHAPTER 4

Use these questions to test your knowledge of the basic skills and concepts of Chapter 4. Then check your answers with those given at the end of the book.

1 Graph $y = (x - 2)^2 + 3$.

2 Graph the function $y = f(x) = x^2 - 9$.

3 Let $y = -5x^2 + 20x - 1$.

(a) Write the quadratic in the standard form $y = a(x - h)^2 + k$.

(b) Give the coordinates of the vertex.

(c) Write the equation of the axis of symmetry.

(d) State the domain and range of the quadratic function.

4 Solve for x: $3x^2 - 8x - 3 = 0$.

5 Find the x-intercepts of $y = -x^2 + 4x + 7$.

Give the value of the discriminant and use this result to describe the x-intercepts, if any.

6 $y = x^2 + 3x + 1$

7 $y = 6x^2 + 5x - 6$

8 Find the maximum or minimum value of the quadratic function and state the x-value at which this occurs: $f(x) = -\frac{1}{2}x^2 - 6x + 2$.

9 In the figure the altitude BC of triangle ABC is 4 feet. The part of the perimeter $PQRCB$ is to be a total of 28 feet. How long should x be so that the area of rectangle $PCRQ$ is a maximum?

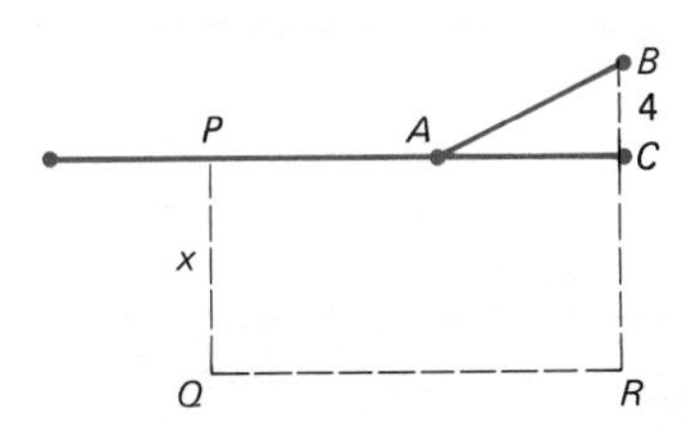

10 The formula $h = 64t - 16t^2$ gives the distance in feet above the ground, h, reached by an object in t seconds. What is the maximum height reached by the object?

11 (a) Draw the circle $(x - 3)^2 + (y + 4)^2 = 25$.

(b) Write the equation of the tangent line to this circle at the point $(6, 0)$.

12 Find the center and radius of the circle

$$4x^2 + 4x + 4y^2 - 56y = -97$$

13 Find the length of the line segment determined by points $A(-2, 5)$ and $B(3, -4)$.

14 Points $P(1, -5)$ and $Q(-3, 3)$ are the endpoints of a diameter of a circle. Find the center, radius, and equation of the circle.

15 Solve the system

$$y = (x - 2)^2 - 2$$
$$(x - 2)^2 + (y - 2)^2 = 4$$

16 Write in standard form and identify the curve

$$4x^2 + 16x - y^2 + 6y = -3$$

Find the coordinates of the center and the vertices.

17 Write the equation of the ellipse in standard form with center at $(0, 0)$, horizontal major axis of length 8, and minor axis of length 6.

18 Write the equation of the hyperbola in standard form with center at $(0, 0)$, foci at $(\pm 8, 0)$, and vertices at $(\pm 6, 0)$.

19 Simplify and express the answer in the form $a + bi$.

$$2(3 - 4i) - 4i^7 + (-2 + 5i)i$$

20 Multiply the complex numbers $3 + 7i$ and $5 - 4i$ and express the result in the form $a + bi$.

21 Divide $3 + 7i$ by $5 - 4i$ and express the result in the form $a + bi$.

22 Simplify $(2 - i)^{-2}$ and express the result in the form $a + bi$.

23 Solve for x: $2x^2 - 5x + 4 = 0$.

ANSWERS TO THE TEST YOUR UNDERSTANDING EXERCISES

Page 124

1 (b) **2** (c) **3** (e) **4** (f) **5** (d) **6** (a)

Page 129

1 16 **2** 25 **3** 9
4 36 **5** $\frac{9}{4}$ **6** $\frac{25}{4}$
7 $y = (x + 2)^2 - 7$ **8** $y = (x - 3)^2 - 2$ **9** $y = (x - 1)^2 + 8$
10 $y = 2(x + 2)^2 - 9$ **11** $y = -(x - \frac{1}{2})^2 - \frac{7}{4}$ **12** $y = \frac{1}{2}(x - 3)^2 - \frac{5}{2}$

Page 134

1 $-8, \frac{3}{2}$ **2** $0, \frac{3}{5}$ **3** $\pm\frac{1}{2}$
4 $\frac{1}{2}$ **5** $\pm 2\sqrt{2}$ **6** None.
7 $-1 + \sqrt{3}, -1 - \sqrt{3}$ **8** $\frac{1}{3}(2 + \sqrt{7}), \frac{1}{3}(2 - \sqrt{7})$ **9** None.
10 $\frac{3}{2}$

Page 135

1 $x = -5$ or $x = 2$ **2** $x = -3$ or $x = 4$ **3** $x = -3$ or $x = 3$ **4** $x = 1 \pm \sqrt{3}$
5 $x = -3 \pm \sqrt{3}$ **6** No real solutions. **7** No real solutions. **8** $x = \dfrac{-1 \pm \sqrt{33}}{4}$

Page 139

1 Minimum value $= -4$ at $x = 5$.
2 Minimum value $= -\frac{5}{6}$ at $x = -\frac{2}{3}$.
3 Minimum value $= \frac{1}{2}$ at $x = 1$.
4 Maximum value $= 131$ at $x = -4$.
5 Maximum value $= -1$ at $x = 0$.
6 Minimum value $= 0$ at $x = 3$.
7 Minimum value $= 0$ at $x = -\frac{7}{5}$.
8 Minimum value $= -\frac{49}{4}$ at $x = -\frac{1}{2}$.

Page 145

1 13 **2** $\sqrt{65}$ **3** $2\sqrt{41}$
4 (0, 0); 10 **5** (0, 0); $\sqrt{10}$ **6** (1, −1); 5
7 $(-\frac{1}{2}, 0)$; 16 **8** (−4, −4); $5\sqrt{2}$ **9** $x^2 + (y - 4)^2 = 25$
10 $(x - 1)^2 + (y + 2)^2 = 3$

Page 146

1 (3, 5); 6 **2** $(0, -\frac{1}{2})$; $\sqrt{5}$ **3** $(\frac{1}{2}, -1)$; $\sqrt{7}$ **4** $(\frac{1}{4}, -1)$; 3

Page 161

1 $3i$ **2** $7i$ **3** $\sqrt{5}i$ **4** $-2i$ **5** $\frac{2}{3}i$ **6** -20
7 $21i$ **8** $7\sqrt{2}i$ **9** $12\sqrt{5}i$

POLYNOMIAL AND RATIONAL FUNCTIONS

5.1 Hints for graphing

The concept of symmetry was used in Chapter 4 when graphing parabolas. Recall that the graph of the function given by $y = f(x) = x^2$ is said to be symmetric about the y-axis. (The y-axis is the axis of symmetry.)

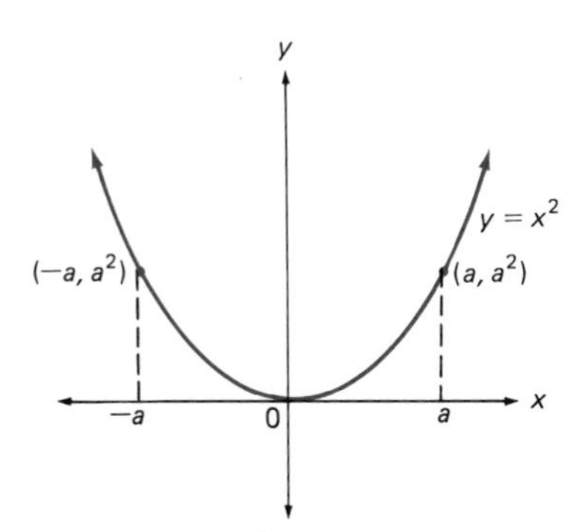

Observe that points such as $(-a, a^2)$ and (a, a^2) are symmetric points about the axis of symmetry. Now we turn our attention to a curve that is symmetric with respect to a point.

As an illustration of a curve that has symmetry through a point we consider the function given by $y = f(x) = x^3$. This function may be referred to as the *cubing function* because for each domain value x, the corresponding range value is the cube of x.

A table of values is a very helpful aid for drawing the graph of a function. Several specific points are located and a smooth curve is drawn through them to show the graph of $y = x^3$.

$y = f(x) = x^3$

x	y
-2	-8
-1	-1
$-\frac{1}{2}$	$-\frac{1}{8}$
0	0
$\frac{1}{2}$	$\frac{1}{8}$
1	1
2	8

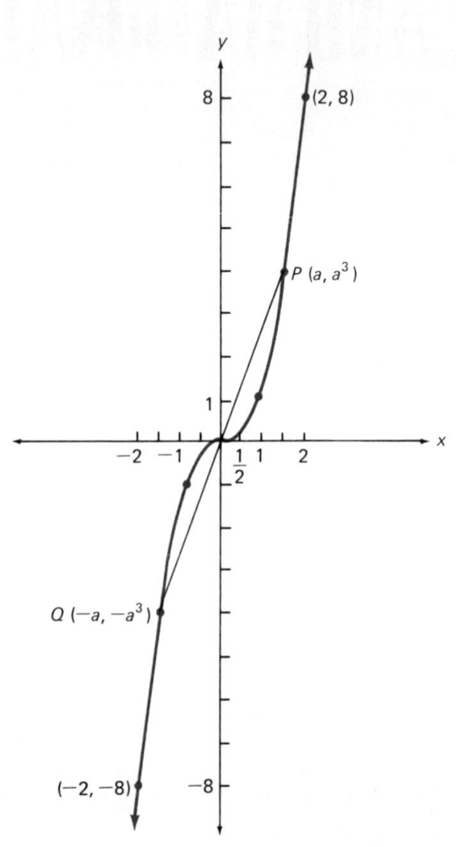

$f(x) = x^3$ is increasing for all x. The curve is concave down on $(-\infty, 0]$ and concave up on $[0, \infty)$.

Domain: all real numbers x
Range: all real numbers y

The table and the graph reveal that the curve is symmetric through the origin. Geometrically this means that whenever a line through the origin intersects the curve at a point P, this line will also intersect the curve in another point Q (on the opposite side of the origin) so that the lengths of OP and OQ are equal. This means that both points (a, a^3) and $(-a, -a^3)$ are on the curve for each value $x = a$. These are said to be **symmetric points** through the origin. In particular, since $(2, 8)$ is on the curve, then $(-2, -8)$ is also on the curve.

In general, the graph of a function $y = f(x)$ is said to be *symmetric through the origin* if for all x in the domain of f, we have

$$f(-x) = -f(x)$$

For the function $f(x) = x^3$, we have $f(-x) = (-x)^3 = -x^3 = -f(x)$. Thus $f(-x) = -f(x)$ and we have symmetry with respect to the origin.

The techniques used for graphing quadratic functions in Chapter 4 can be used for other functions as well. For example, the graph of $y = 2x^3$ can be obtained from the graph of $y = x^3$ by multiplying each of its ordinates by 2. We can also use translations (shifting) as illustrated in the examples that follow.

EXAMPLE 1 Graph $y = g(x) = (x - 3)^3$.

Explain how you can use the graph of g to draw the graph of $y = h(x) = (x - 3)^3 + 2$.

Solution The graph of g is obtained by translating $y = x^3$ by 3 units to the right, as shown below in the left figure.

EXAMPLE 2 Graph $y = f(x) = |(x - 3)^3|$.

Note: $|(x - 3)^3| = |x - 3|^3$.

Solution First graph $y = (x - 3)^3$ as in Example 1. Then take the part of this curve that is below the x-axis $[(x - 3)^3 < 0]$ and reflect it through the x-axis. This is shown below in the figure on the right.

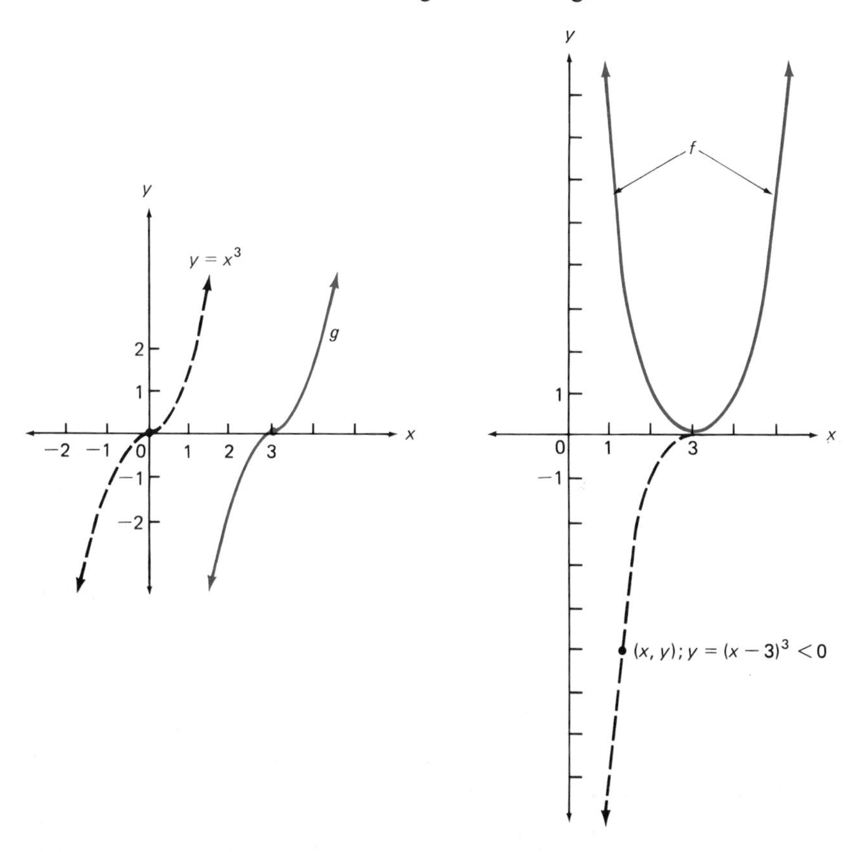

It will be helpful to collect some of the general observations that are useful in graphing functions.

HELP WITH GRAPHING

1 If $f(x) = f(-x)$, the curve is symmetric about the y-axis.

2 If $f(-x) = -f(x)$, the curve is symmetric through the origin.

3 The graph of $y = af(x)$ can be obtained by multiplying the ordinates of the curve $y = f(x)$ by the value a. The case $a = -1$ gives $y = -f(x)$, which is the reflection of $y = f(x)$ through the x-axis.

4 The graph of $y = |f(x)|$ can be obtained from the graph of $y = f(x)$ by taking the part of $y = f(x)$ that is below the x-axis and reflecting it through the x-axis.

Try to supply a specific example that illustrates each of these items.

In each of the following, h is positive.

5 The graph of $y = f(x - h)$ can be obtained by shifting $y = f(x)$ h units to the right.

6 The graph of $y = f(x + h)$ can be obtained by shifting $y = f(x)$ h units to the left.

7 The graph of $y = f(x) + h$ can be obtained by shifting $y = f(x)$ h units upward.

8 The graph of $y = f(x) - h$ can be obtained by shifting $y = f(x)$ h units downward.

EXAMPLE 3 In the figure the curve C_1 is obtained by shifting the curve C with equation $y = x^3$ horizontally, and C_2 is obtained by shifting C_1 vertically. What are the equations of C_1 and C_2?

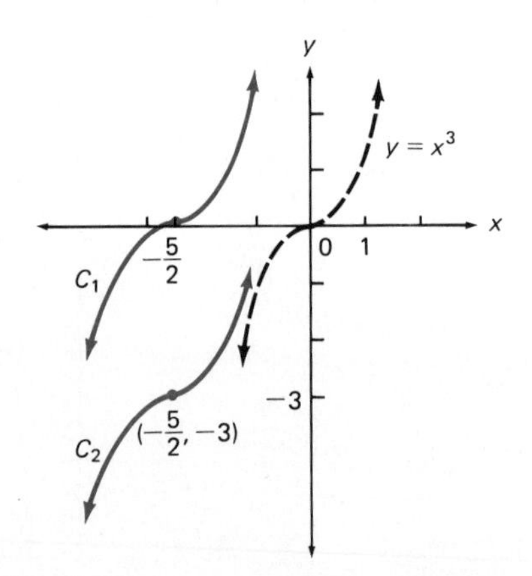

Solution

C_1: $y = (x + \frac{5}{2})^3$ ⟵ C is shifted $\frac{5}{2}$ units left.

C_2: $y = (x + \frac{5}{2})^3 - 3$ ⟵ C_1 is shifted 3 units down.

EXAMPLE 4 Graph: (a) $y = f(x) = x^4$; (b) $y = 2x^4$.

Solution

(a) Since $f(-x) = (-x)^4 = x^4 = f(x)$, the graph is symmetric about the y-axis. Use the table of values to locate the right half of the curve; the symmetry gives the rest.

(b) Multiply each of the ordinates of $y = x^4$ by 2 to obtain the graph of $y = 2x^4$.

x	y
0	0
$\frac{1}{2}$	$\frac{1}{16}$
1	1
$\frac{3}{2}$	$\frac{81}{16}$
2	16

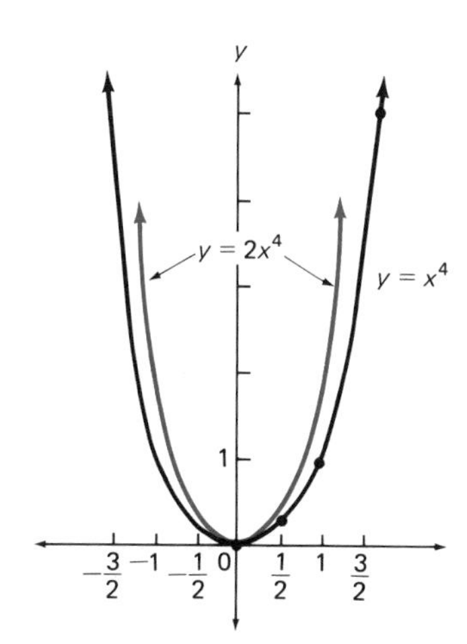

EXERCISES 5.1

In Exercises 1–8, graph each set of curves in the same coordinate system. For each exercise use a dashed curve for the first equation and a solid curve for each of the others.

1 $y = x^2$, $y = (x - 3)^2$

2 $y = x^3$, $y = (x + 2)^3$

3 $y = x^3$, $y = -x^3$

4 $y = x^4$, $y = (x - 4)^4$

5 $f(x) = x^3$, $g(x) = \frac{1}{2}x^3$, $h(x) = \frac{1}{4}x^3$

6 $f(x) = x^3$, $g(x) = (x - 3)^3 - 3$, $h(x) = (x + 3)^3 + 3$

7 $f(x) = x^4$, $g(x) = (x - 1)^4 - 1$, $h(x) = (x - 2)^4 - 2$

8 $f(x) = x^4$, $g(x) = -\frac{1}{8}x^4$, $h(x) = -x^4 - 4$

9 Graph $y = |(x + 1)^3|$.

10 Graph $y = 2(x - 1)^3 + 3$.

11 Graph $y = f(x) = |x|$, $y = g(x) = |x - 3|$, and $y = h(x) = |x - 3| + 2$, on the same axes.

Find the equation of the curve C which is obtained from the dashed curve by a horizontal or vertical shift.

12

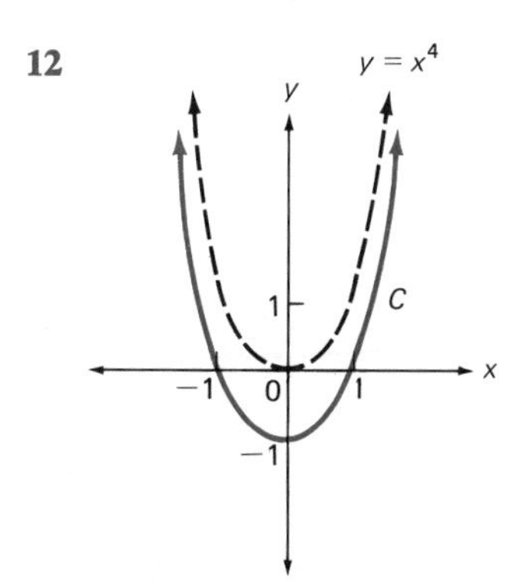

13

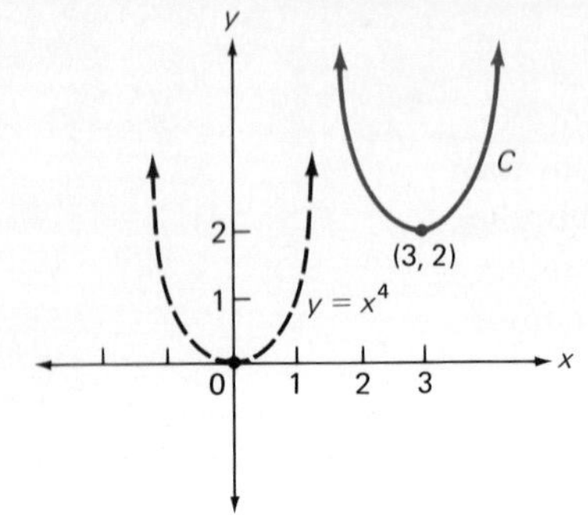

15

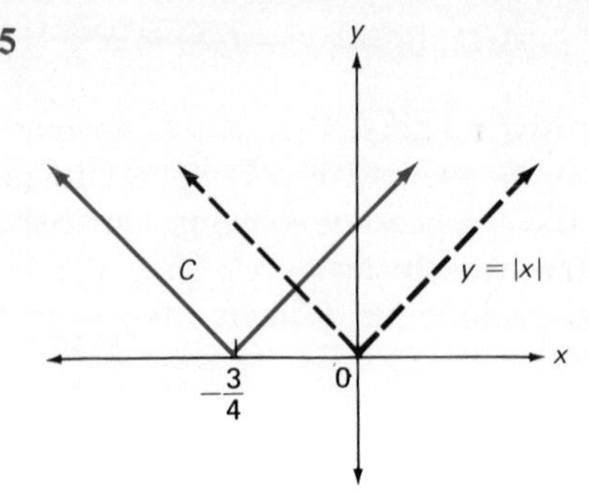

14

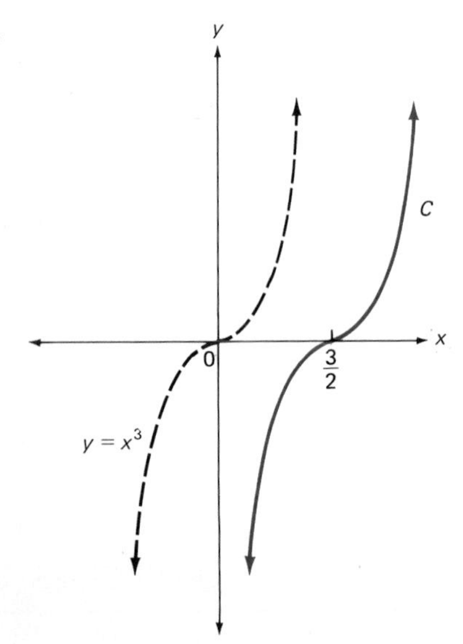

Graph each of the following.

16 $y = |x^4 - 16|$ **17** $y = |x^3 - 1|$

18 $y = |-1 - x^3|$

Graph each of the following. (Hint: Consider the expansion of $(a \pm b)^n$ for appropriate n.)

***19** $y = x^3 + 3x^2 + 3x + 1$

***20** $y = x^3 - 6x^2 + 12x - 8$

***21** $y = -x^3 + 3x^2 - 3x + 1$

***22** $y = x^4 - 4x^3 + 6x^2 - 4x + 1$

■ *Evaluate and simplify the difference quotients.*

23 $\dfrac{f(x) - f(3)}{x - 3}$ where $f(x) = x^3$.

24 $\dfrac{f(-2 + h) - f(-2)}{h}$ where $f(x) = x^3$.

25 $\dfrac{f(1 + h) - f(1)}{h}$ where $f(x) = x^4$.

26 $\dfrac{f(x) - f(-1)}{x + 1}$ where $f(x) = x^4 + 1$.

5.2

Graphing some special rational functions

A *rational expression* is a ratio of polynomials. Such expressions may be used to define rational functions and are considered in more detail in later sections. At this point we introduce the topic by exploring several special rational functions.

Consider the function $y = f(x) = \dfrac{1}{x}$, a rational function whose domain consists of all numbers except zero. The denominator x is a polynomial of degree 1, and the numerator $1 = 1x^0$ is a (constant) polynomial of degree zero.

To draw the graph of this function, first observe that it is symmetric through the origin because

$$f(-x) = \frac{1}{-x} = -\frac{1}{x} = -f(x)$$

Moreover, in $y = \frac{1}{x}$ both variables must have the same sign since $xy = 1$, a positive number. That is, x and y must both be positive or both be negative. Thus the graph will only appear in quadrants I and III. Next, we use a table of values to obtain points for the curve in the first quadrant. Finally, use the symmetry with respect to the origin to obtain the remaining portion of the graph in the third quadrant.

The function is decreasing on $(-\infty, 0)$ and on $(0, \infty)$. The curve is concave down on $(-\infty, 0)$ and concave up on $(0, \infty)$.

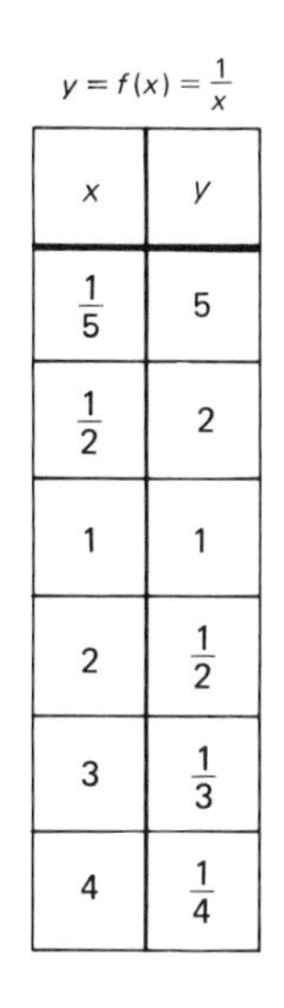

$y = f(x) = \frac{1}{x}$

x	y
$\frac{1}{5}$	5
$\frac{1}{2}$	2
1	1
2	$\frac{1}{2}$
3	$\frac{1}{3}$
4	$\frac{1}{4}$

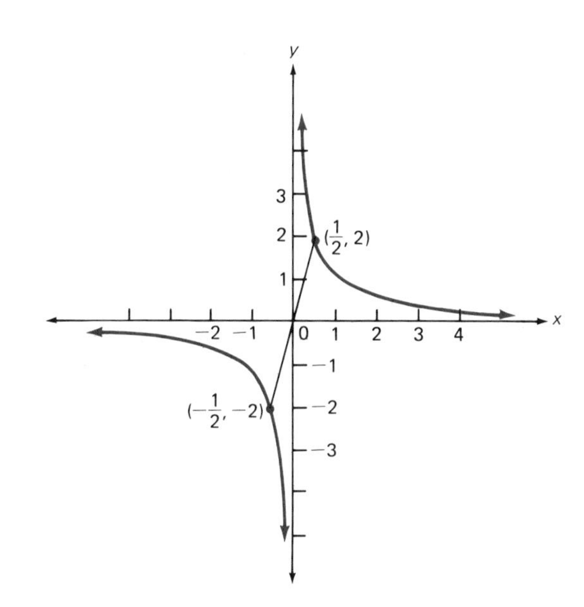

Verify these table entries for $y = \frac{1}{x}$.

x	y
10	.1
100	.01
1000	.001
10,000	.0001
⋮	⋮
Getting very large	Getting close to 0

Domain: all $x \neq 0$
Range: all $y \neq 0$

Observe that the curve approaches the x-axis in quadrant I. That is, as the values for x become large, the values for y approach zero. Also, as the values for x approach zero in the first quadrant, the y-values become very large. A similar observation can be made about the curve in the third quadrant. We say that the axes are **asymptotes** for the curve; the curve is *asymptotic* to the axes.

EXAMPLE 1 Sketch the graph of $g(x) = \frac{1}{x - 3}$. Find the asymptotes.

Solution Using $f(x) = \frac{1}{x}$, we have $f(x - 3) = \frac{1}{x - 3} = g(x)$. Therefore, the graph of g can be drawn by shifting the graph of $f(x) = \frac{1}{x}$ by 3 units to the right. See the figure at the top of page 178.

The x-axis and the vertical line $x = 3$ are the asymptotes for the graph. The domain is $x \neq 3$, and the range is $y \neq 0$.

Describe the graph of $y = h(x) = \dfrac{1}{x+3}$.

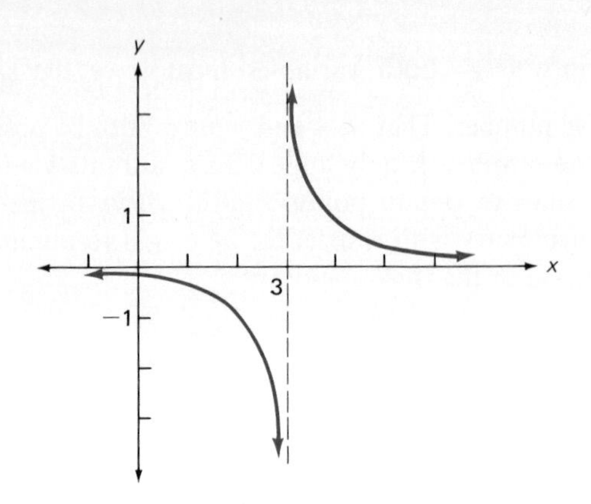

EXAMPLE 2 Graph $y = \dfrac{1}{x^2}$.

In symbols, we write "As $x \longrightarrow \infty$, $y \longrightarrow 0$." That is, as x becomes larger and larger in value, the value of y becomes very small. Similarly, as $x \longrightarrow 0$, $y \longrightarrow \infty$.

Solution First note that $x \neq 0$. For all other values of x we have $x^2 > 0$, so that the curve will appear in quadrants I and II only. As the values for x become large, the values for $y = \dfrac{1}{x^2}$ become very small. Moreover, as x approaches zero, y becomes very large. Thus the axes are asymptotes to the curve. Note that the curve is symmetric about the y-axis. That is,

$$f(x) = \frac{1}{x^2} = \frac{1}{(-x)^2} = f(-x)$$

$y = \dfrac{1}{x^2}$

x	y
$\pm\frac{1}{2}$	4
$\pm\frac{1}{10}$	100
$\pm\frac{1}{100}$	10,000

As x is getting close to 0, y is getting very large.

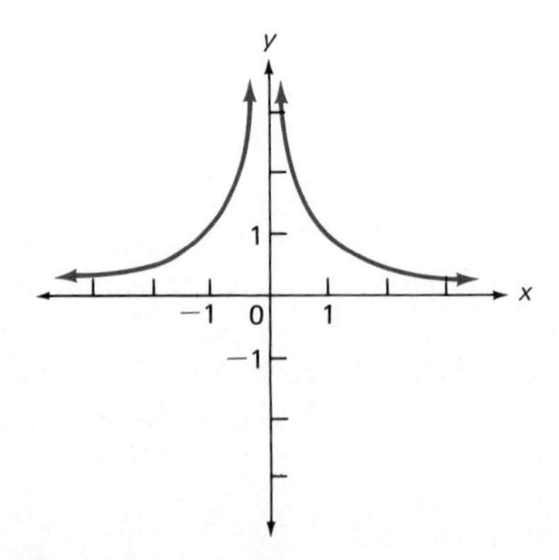

The domain is $x \neq 0$, and the range is $y > 0$.

EXAMPLE 3 Graph $y = \dfrac{1}{(x+2)^2} - 3$. What are the asymptotes? Find the domain and the range.

Solution Shift the graph of $y = \dfrac{1}{x^2}$ in Example 2 by 2 units left and then 3 units down.

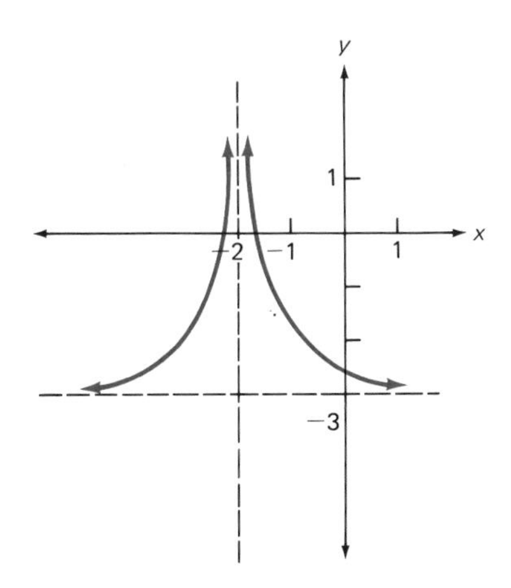

Asymptotes: $x = -2$, $y = -3$ Domain: $x \neq -2$
Range: $y > -3$

The curves studied in this chapter, as well as the parabolas and straight lines discussed in earlier chapters, are very useful in the study of calculus. Having an almost instant recall of the graphs of the following functions will be helpful in future work:

$$y = x \qquad y = |x| \qquad y = x^2 \qquad y = x^3 \qquad y = x^4 \qquad y = \frac{1}{x} \qquad y = \frac{1}{x^2}$$

Not only have you learned what these curves look like, but just as important, you have also learned how to obtain other curves from them by appropriate translations and reflections.

EXERCISES 5.2

In Exercises 1–8, graph each set of curves in the same coordinate system. For each exercise use a dashed curve for the first equation and a solid curve for the other.

1 $y = \frac{1}{x}, y = \frac{2}{x}$

2 $y = \frac{1}{x}, y = -\frac{1}{x}$

3 $y = \frac{1}{x}, y = \frac{1}{x+2}$

4 $y = \frac{1}{x}, y = \frac{1}{x} + 5$

5 $y = \frac{1}{x}, y = \frac{1}{2x}$

6 $y = \frac{1}{x}, y = \frac{1}{x} - 5$

7 $y = \frac{1}{x^2}, y = \frac{1}{(x-3)^2}$

8 $y = \frac{1}{x^2}, y = -\frac{1}{x^2}$

Graph each of the following. Find all asymptotes, if any.

9 $y = \frac{1}{x+4} - 2$

10 $y = \frac{1}{(x-2)^2} + 3$

11 $y = \frac{1}{|x-2|}$

12 $y = \frac{1}{x^3}$

13 $xy = 3$

14 $xy = -2$

15 $xy - y = 1$

16 $xy - 2x = 1$

***17** Graph $y = \left|\frac{1}{x} - 1\right|$.

***18** Graph $x = \frac{1}{y^2}$. Why is y not a function of x?

Find the difference quotients and simplify.

19 $\frac{f(x) - f(3)}{x - 3}$ where $f(x) = \frac{1}{x}$.

20 $\frac{f(1+h) - f(1)}{h}$ where $f(x) = \frac{1}{x^2}$.

5.3

Polynomial and rational functions

Factoring methods were studied in Chapter 2. We will now use factored forms to determine the signs of polynomial and rational functions, which, in turn, will aid in sketching their graphs.

As a first illustration, consider the polynomial function given by $f(x) = x^3 + 4x^2 - x - 4$. To obtain the factored form, use factoring by grouping.

STEP 1 for graphing polynomial functions. Factor the polynomial and find the x-intercepts.

$$\begin{aligned} f(x) &= x^3 + 4x^2 - x - 4 \\ &= (x+4)x^2 - (x+4) \\ &= (x+4)(x^2-1) \\ &= (x+4)(x+1)(x-1) \end{aligned}$$

Set each factor equal to 0 to get the roots $-4, -1, 1$ of $f(x) = 0$, which are also the x-intercepts of $y = f(x)$. Since there are no other x-intercepts, the curve must stay above or below the x-axis for each of the intervals determined by $-4, -1, 1$.

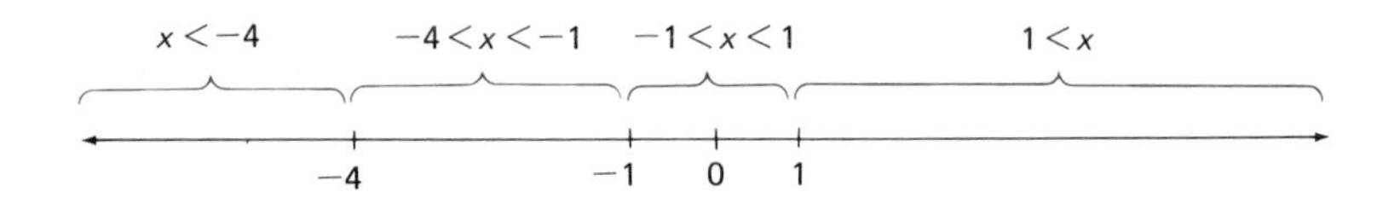

When $f(x) > 0$ the curve is above the x-axis, and it is below for $f(x) < 0$. The following table of signs contains this information, and is used to sketch the graph. (See pages 22–24 for the details in forming such a table.)

STEP 2. Form a table of signs for $f(x)$.

Interval	$(-\infty, -4)$	$(-4, -1)$	$(-1, 1)$	$(1, \infty)$
Sign of $x + 4$	−	+	+	+
Sign of $x + 1$	−	−	+	+
Sign of $x - 1$	−	−	−	+
Sign of $f(x)$	−	+	−	+
Position of curve relative to x-axis	below	above	below	above

In general, a polynomial of degree n has at most $n - 1$ turning points and at most n x-intercepts.

Notice that there are two turning points and three x-intercepts. It turns out that for a third-degree polynomial there are at most two turning points and at most three x-intercepts. There may be less of either; for example $y = x^3$ has no turning points and one x-intercept.

$f(x) = x^3 + 4x^2 - x - 4$

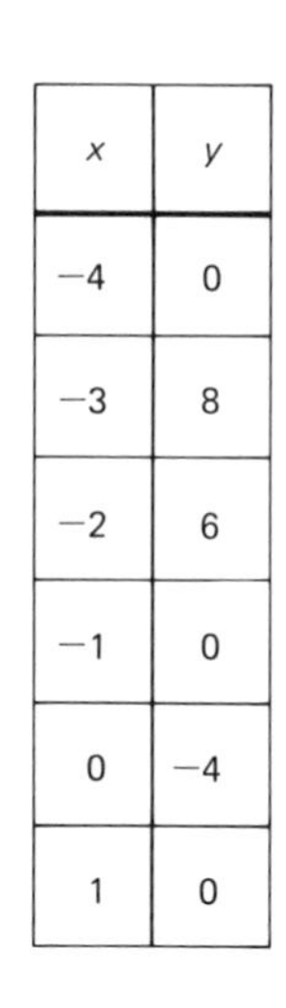

x	y
−4	0
−3	8
−2	6
−1	0
0	−4
1	0

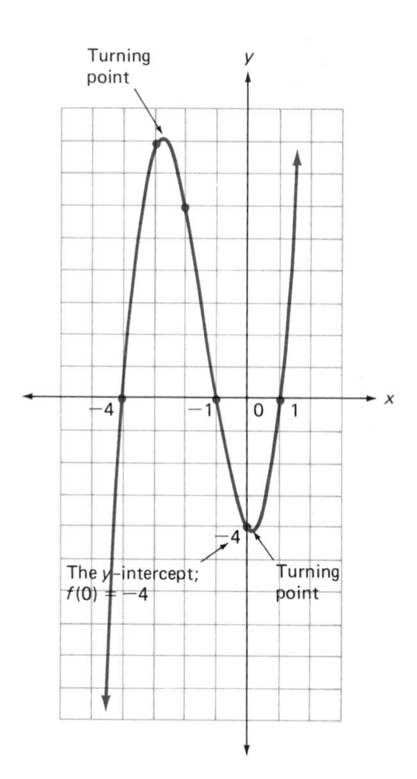

STEP 3. Sketch the graph using the x-intercepts, the signs of $f(x)$, and some additional points including the y-intercept.

To graph rational functions, first locate the asymptotes if there are any. For example, note that $y = \dfrac{1}{x - 2}$ has a vertical asymptote at $x = 2$, since as x is taken close to 2, y gets very large in absolute value. It has the horizontal asymptote $y = 0$, because as $|x|$ is taken very large, the y-values get close to 0.

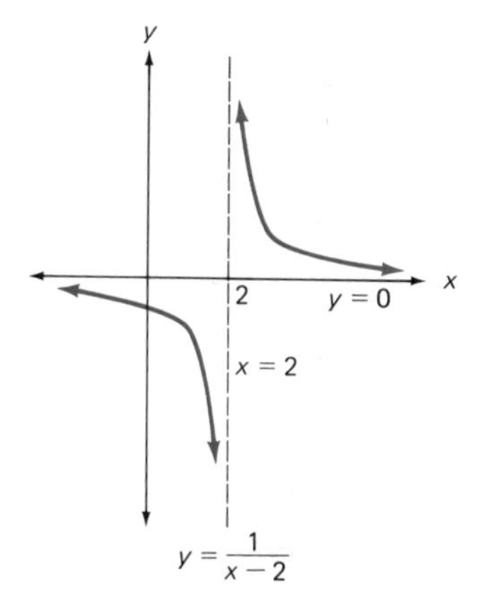

A graph of a rational function has a vertical asymptote at each value a where the denominator is 0, and the numerator is not 0.

The method for finding the horizontal asymptote (if any) for a rational function f can be summarized by referring to the general form of a rational function.

In general, a rational function may have numerous vertical asymptotes, but at most one horizontal asymptote.

$$f(x) = \frac{a_n x^n + a_{n-1}x^{n-1} + \cdots + a_0}{b_m x^m + b_{m-1}x^{m-1} + \cdots + b_0}, \quad a_n \neq 0 \neq b_m$$

The numerator and denominator are polynomials in x of degree n and m respectively (see page 52).

1 If $n < m$, then $y = 0$ is the horizontal asymptote.

2 If $n = m$, then $y = \dfrac{a_n}{b_m}$ is the horizontal asymptote.

3 If $n > m$, there is no horizontal asymptote.

TEST YOUR UNDERSTANDING

Find the vertical and horizontal asymptotes for each rational function.

1 $y = \dfrac{5}{(x-2)(x-3)}$ 2 $y = \dfrac{x-2}{x+3}$ 3 $y = \dfrac{x^2-1}{x-2}$

4 $y = \dfrac{x}{(x-4)^2}$ 5 $y = \dfrac{2-x}{x^2-x-12}$ 6 $y = \dfrac{2x^3+x}{x^2-x}$

EXAMPLE 1 Graph $y = f(x) = \dfrac{x-2}{x^2-3x-4}$.

Solution

STEP 1 for graphing rational functions. Factor the numerator and denominator.

$$f(x) = \frac{x-2}{(x+1)(x-4)}$$

STEP 2. Find the x-intercepts.

$f(x) = 0$ when the numerator $x - 2 = 0$, so the x-intercept is 2.

STEP 3. Find the vertical asymptotes.

Setting the denominator $(x+1)(x-4)$ equal to 0 gives the vertical asymptotes $x = -1$ and $x = 4$.

STEP 4. Find the horizontal asymptote.

The horizontal asymptote is $y = 0$ since the degree of the numerator is 1, which is less than 2, the degree of the denominator.

The numbers for which the numerator or denominator equals 0 are -1, 2, and 4. They determine the four intervals in the table of signs for f.

STEP 5. Form a table of signs for f in the intervals determined by the numbers for which the numerator or denominator equals 0.

Interval	$(-\infty, -1)$	$(-1, 2)$	$(2, 4)$	$(4, \infty)$
Sign of $x+1$	$-$	$+$	$+$	$+$
Sign of $x-2$	$-$	$-$	$+$	$+$
Sign of $x-4$	$-$	$-$	$-$	$+$
Sign of $f(x)$	$-$	$+$	$-$	$+$
Position of curve relative to x-axis	below	above	below	above

STEP 6. Graph the curve. First draw the asymptotes and locate the intercepts. Use the table of signs and some selected points to complete the graph.

When graphing, remember that the closer the curve is to a vertical asymptote, the steeper it gets. And the larger $|x|$ gets, the closer the curve is to the horizontal asymptote; the curve gets "flatter." Selected points on the curve will help to draw the correct shape.

x	y
-2	$-\frac{2}{3}$
$-\frac{5}{4}$	-2.5
$-.9$	5.9
0	$\frac{1}{2}$
2	0
3	$-\frac{1}{4}$
3.9	-3.9
4.2	3.1
5	$\frac{1}{2}$

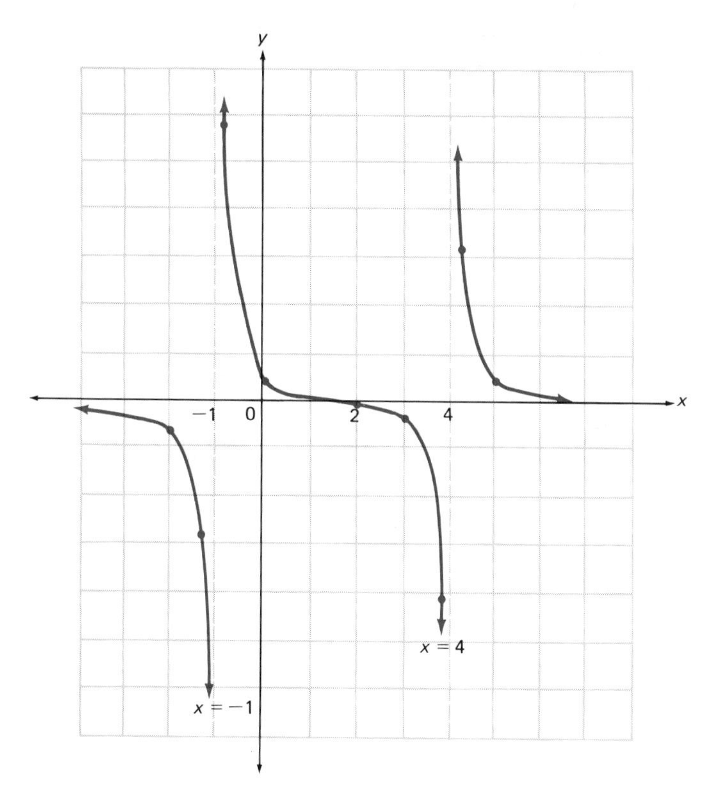

This is the graph of

$$f(x) = \frac{x-2}{x^2-3x-4}.$$

Part of our curve sketching is being done on "faith." After all, what is the shape of a curve connecting two points? Here are some possibilities.

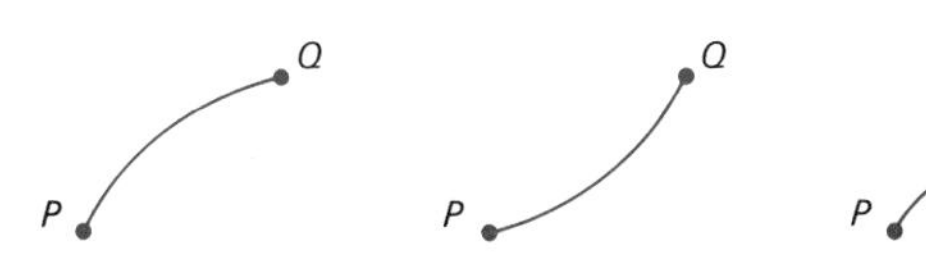

Locating more points between P and Q will not really answer the question no matter how close the chosen points are, since the question as to how to connect them still remains. More advanced methods studied in calculus will help answer such questions, and curve sketching can then be done with less ambiguity.

EXERCISES 5.3

Use the table method to determine the signs of f.

1 $f(x) = (x-1)(x-2)(x-3)$

2 $f(x) = x(x+2)(x-2)$

3 $f(x) = \dfrac{(3x-1)(x+4)}{x^2(x-2)}$

4 $f(x) = \dfrac{x(2x+3)}{(x+2)(x-5)}$

Use the table method to solve each inequality.

5 $(x^2 + 2)(x - 4)(x + 1) > 0$

6 $x(x + 1)(x + 2) < 0$

7 $\dfrac{x - 10}{3(x + 1)(5x - 1)} < 0$

8 $\dfrac{3}{(x + 2)(4x + 3)} > 0$

In Exercises 9–22, sketch the graph of the polynomial functions. Indicate all intercepts.

9 $f(x) = (x + 3)(x + 1)(x - 2)$

10 $f(x) = (x - 1)(x - 3)(x - 5)$

11 $f(x) = (x + 3)(x + 1)(x - 1)(x - 3)$

12 $f(x) = -x(x + 4)(x^2 - 4)$

13 $f(x) = x^3 - 4x$

14 $f(x) = 9x - x^3$

15 $f(x) = x^3 + 3x$

16 $f(x) = x^3 + x^2 - 6x$

17 $f(x) = -x^3 - x^2 + 6x$

18 $f(x) = x^3 + 2x^2 - 9x - 18$

19 $f(x) = x^4 - 4x^2$

20 $f(x) = (x - 1)^2(3 + 2x - x^2)$

21 $f(x) = x^4 - 6x^3 + 8x^2$

22 $f(x) = x(x^2 - 1)(x^2 - 4)$

In Exercises 23–36, sketch the graph of the rational function. Indicate the asymptotes and the intercepts.

23 $f(x) = \dfrac{1}{(x - 1)(x + 2)}$

24 $f(x) = \dfrac{1}{x^2 - 4}$

25 $f(x) = \dfrac{1}{4 - x^2}$

26 $f(x) = \dfrac{2}{x^2 - x - 6}$

27 $f(x) = \dfrac{x}{x^2 - 1}$

28 $f(x) = \dfrac{x + 1}{x^2 + x - 2}$

29 $f(x) = \dfrac{3}{x^2 + 1}$

30 $f(x) = \dfrac{2}{x + x^2}$

31 $f(x) = \dfrac{x + 2}{x - 2}$

32 $f(x) = \dfrac{x}{1 - x}$

33 $f(x) = \dfrac{x - 1}{x + 3}$

34 $f(x) = \dfrac{x^2 + x}{x^2 - 4}$

35 $f(x) = \dfrac{x^2 + x - 2}{x^2 + x - 12}$

36 $f(x) = \dfrac{1 - x^2}{x^2 - 9}$

▬ *In Exercises 37–42, simplify, obtain factored forms, and determine the signs of f. (The combination of such skills will be needed in the study of calculus.)*

37 $g(x) = 2(2x - 1)(x - 2) + 2(x - 2)^2$

38 $g(x) = 2(x - 3)(x + 2)^2 + 2(x + 2)(x - 3)^2$

39 $g(x) = \dfrac{x^2 - 2x(x + 2)}{x^4}$

40 $g(x) = \dfrac{(x - 3)(2x) - (x^2 - 5)}{(x - 3)^2}$

41 $g(x) = 2\left(\dfrac{x - 5}{x + 2}\right)\dfrac{(x + 2) - (x - 5)}{(x + 2)^2}$

42 $g(x) = \dfrac{5(x + 10)^2 - 10x(x + 10)}{(x + 10)^4}$

5.4

Equations and inequalities with fractions

You already have used properties of equality and inequality to solve equations and inequalities. Now we are ready to extend these ideas to include statements that involve fractions. Consider, for example, this equation:

$$\frac{x - 2}{x} = \frac{3}{4}$$

The first step in finding the solution is to find the least common denominator (LCD) of the two fractions. In this case, the LCD is $4x$. We then use the multiplication property of equality and multiply each member of the equation by $4x$.

Notice that multiplication by the LCD transforms the original equation into one that does not involve any fractions. Thereafter we are able to finish the solution using methods we have previously studied.

$$4x \cdot \left(\frac{x - 2}{x}\right) = 4x \cdot \left(\frac{3}{4}\right)$$

$$4(x - 2) = 3x$$

$$4x - 8 = 3x$$

$$x = 8$$

Check: $\frac{8-2}{8} = \frac{6}{8} = \frac{3}{4}$

The equation that we have just discussed is also called a *proportion.* A proportion is a statement that two ratios are equal, such as the following:

$$\frac{a}{b} = \frac{c}{d}$$

This is often read "*a* is to *b* as *c* is to *d*," and may be written in this form:

$$a : b = c : d$$

The proportion is in the form of a fractional equation and can be simplified by multiplying both sides by bd.

$$\frac{a}{b} = \frac{c}{d}$$

$$(bd)\frac{a}{b} = (bd)\frac{c}{d}$$

$$ad = bc$$

PROPORTION PROPERTY

If $\frac{a}{b} = \frac{c}{d}$, then $ad = bc$.

Using this property, the solution to the earlier equation begins in this way:

If $\frac{x-2}{x} = \frac{3}{4}$, then $4(x-2) = 3x$.

EXAMPLE 1 Solve for x: $\frac{2x-5}{x+1} - \frac{3}{x^2+x} = 0.$

At times, multiplication by the LCD can give rise to a quadratic equation, as in Example 1. Try to explain each step in the solution.

Solution Factor the denominator in the second fraction.

$$\frac{2x-5}{x+1} - \frac{3}{x(x+1)} = 0$$

Multiply by the LCD which is $x(x+1)$.

$$x(2x-5) - 3 = 0$$

$$2x^2 - 5x - 3 = 0$$

$$(2x+1)(x-3) = 0$$

$$2x + 1 = 0 \quad \text{or} \quad x - 3 = 0$$

$$x = -\tfrac{1}{2} \quad \text{or} \quad x = 3$$

You can check these answers in the original equation.

It is especially important to check each solution of a fractional equation. The reason for this can best be explained through the use of another example.

EXAMPLE 2 Solve for x: $\dfrac{3}{x-3} + 2 = \dfrac{x}{x-3}$.

Solution The LCD here is $x - 3$; thus we multiply each member of the equation by this term.

$$(x-3)\left(\frac{3}{x-3}\right) + (x-3)(2) = (x-3)\left(\frac{x}{x-3}\right)$$

$$3 + 2(x-3) = x$$

$$3 + 2x - 6 = x$$

$$2x - 3 = x$$

$$x = 3$$

In this example we could have noticed at the outset that $x - 3 \neq 0$, and thus $x \neq 3$. In other words, it is wise to notice such restrictions on the variable before starting the solution. These are the values of the variable that would cause division by zero.

The procedure we have just followed *assumed* that there was a solution to the given equation. However, if we replace x by 3 in the original equation, we obtain the following statement:

$$\frac{3}{0} + 2 = \frac{3}{0}$$

Since division by zero is meaningless, we have an impossible equation, and therefore conclude that there are no solutions. That is, there is no replacement for x that will make the original equation true.

TEST YOUR UNDERSTANDING

Solve for x and check your results. Use the proportion property where appropriate.

1 $\dfrac{x}{3} + \dfrac{x}{2} = 10$

2 $\dfrac{3}{x} + \dfrac{2}{x} = 10$

3 $\dfrac{x-3}{8} = 4$

4 $\dfrac{8}{x-3} = 4$

5 $\dfrac{x+3}{x} = \dfrac{2}{3}$

6 $\dfrac{2}{x+3} = \dfrac{3}{x+3}$

7 $\dfrac{3x}{2} - \dfrac{2x}{3} = \dfrac{1}{4}$

8 $\dfrac{2}{3x} + \dfrac{3}{2x} = 4$

9 $\dfrac{3x-1}{4} - \dfrac{x-1}{2} = 1$

10 $\dfrac{x}{x-6} + \dfrac{2}{x} = \dfrac{1}{x-6}$

11 $\dfrac{x}{x-1} - \dfrac{3}{4x} = \dfrac{3}{4} - \dfrac{1}{x-1}$

12 $\dfrac{2}{x} - \dfrac{3}{2x} = \dfrac{11}{3} - \dfrac{4}{3x}$

Some statements of inequality that involve fractions can be solved very much like equations. First, find the LCD of the fractions. Next, multiply each member of the inequality by this quantity in order to eliminate all fractions. Then solve, using the addition and multiplication properties of inequality.

EXAMPLE 3 Solve: $\frac{x}{3} - \frac{2x+1}{4} > 1.$

Solution

$$12\left(\frac{x}{3}\right) - 12\left(\frac{2x+1}{4}\right) > 12(1)$$

$$4x - 3(2x+1) > 12$$

$$4x - 6x - 3 > 12$$

$$-2x > 15$$

$$x < -\frac{15}{2} \quad \text{(Why?)}$$

EXAMPLE 4 Solve for x: $\frac{x+1}{x} < 3.$

An alternative method begins by multiplying through by x. However this is more difficult because it calls for two cases, $x > 0$ and $x < 0$.

Solution First rewrite as follows:

$$\frac{x+1}{x} - 3 < 0$$

$$\frac{1-2x}{x} < 0$$

Now use a table of signs to obtain the solution $x < 0$ or $x > \frac{1}{2}$.

The general procedures for solving problems outlined in Section 1.3 apply to problems that involve fractions as well. The reader is advised to review that material at this time. Then study the following illustrative example.

EXAMPLE 5 Working alone Harry can mow a lawn in 3 hours. Elliot can complete the same job in 2 hours. How long will it take them working together?

Working together should enable the boys to do the job in less time than it would take either of them to do the job alone. If Elliot can do the job alone in 2 hours, together the two boys should take less than this time.

Solution To solve work problems of this type, we first consider the part of the job that can be done in 1 hour.

Let x = time (in hours) to do the job together

Then $\frac{1}{x}$ = portion of job done in 1 hour working together

Also: $\frac{1}{2}$ = portion of job done by Elliot in 1 hour

$\frac{1}{3}$ = portion of job done by Harry in 1 hour

Working together, the part of the job they can do in one hour is $\frac{1}{2} + \frac{1}{3}$. Thus

$$\frac{1}{2} + \frac{1}{3} = \frac{1}{x}$$

$$3x + 2x = 6$$

$$5x = 6$$

$$x = 1\tfrac{1}{5}$$

Together they need $1\frac{1}{5}$ hours, or 1 hours, 12 minutes.

The final example of this section illustrates how the solution of a rational equation is used in solving a nonlinear system.

EXAMPLE 6 Solve the system

$$y = \frac{2}{x}$$

$$y = -2x + 5$$

Solution Set both values for y equal to each other.

$$\frac{2}{x} = -2x + 5$$

Multiply by x.

$$2 = -2x^2 + 5x$$

Solve for x.

$$2x^2 - 5x + 2 = 0$$

$$(2x - 1)(x - 2) = 0$$

$$x = \tfrac{1}{2} \quad \text{or} \quad x = 2$$

For $x = \frac{1}{2}$, $y = -2(\frac{1}{2}) + 5 = 4$. For $x = 2$, $y = -2(2) + 5 = 1$. The points of intersection are $(\frac{1}{2}, 4)$ and $(2, 1)$. Check these points in the original system.

EXERCISES 5.4

Solve for x and check your results.

1 $\frac{x}{2} - \frac{x}{5} = 6$

2 $\frac{2}{x} - \frac{5}{x} = 6$

3 $\frac{x-1}{2} = \frac{x+2}{4}$

4 $\frac{2}{x-1} = \frac{4}{x+2}$

5 $\frac{5}{x} - \frac{3}{4x} = 1$

6 $\frac{3x}{4} - \frac{3}{2} = \frac{x}{6} + \frac{4x}{3}$

7 $\frac{2x+1}{5} - \frac{x-2}{3} = 1$

8 $\frac{3x-2}{4} - \frac{x-1}{3} = \frac{1}{2}$

9 $\frac{x+4}{2x-10} = \frac{8}{7}$

10 $\frac{10}{x} - \frac{1}{2} = \frac{15}{2x}$

11 $\frac{2}{x+6} - \frac{2}{x-6} = 0$

12 $\frac{1}{3x-1} + \frac{1}{3x+1} = 0$

13 $\frac{x+1}{x+10} = \frac{1}{2x}$

14 $\frac{3}{x+1} = \frac{9}{x^2 - 3x - 4}$

15 $\frac{x^2}{2} - \frac{3x}{2} + 1 = 0$

16 $\frac{x+1}{x-1} - \frac{2}{x(x-1)} = \frac{4}{x}$

17 $\frac{5}{x^2 - 9} = \frac{3}{x+3} - \frac{2}{x-3}$

18 $\frac{3}{x^2 - 25} = \frac{5}{x-5} - \frac{5}{x+5}$

19 $\frac{1}{x^2 + 4} + \frac{1}{x^2 - 4} = \frac{18}{x^4 - 16}$

20 $\frac{3}{x^3 - 8} = \frac{1}{x-2}$

21 $\frac{2x-5}{x+1} - \frac{3}{x^2 + x} = 0$

22 $\frac{2}{x} = -2x + 5$

23 $\frac{10 - 5x}{3x} - \frac{2}{x+5} = \frac{8 - 4x}{x+5}$

24 $\frac{2-x}{x+1} + \frac{x+8}{x-2} = \frac{4-x}{x^2 - x - 2}$

25 $\frac{3}{2x^2 - 3x - 2} - \frac{x+2}{2x+1} = \frac{2x}{10 - 5x}$

26 $\frac{2x-2}{x^2 + 2x} - \frac{5x-6}{6x} = \frac{1-x}{x+5}$

In Exercises 27–34 solve for the indicated variable.

27 $\frac{v^2}{K} = \frac{2g}{m}$, for m

28 $A = \frac{h}{2}(b + B)$, for B

29 $S = \pi(r_1 + r_2)s$, for r_1

30 $V = \frac{1}{3}\pi r^2 h$, for h

31 $d = \frac{s-a}{n-1}$, for s

32 $S = \frac{n}{2}[2a + (n-1)d]$, for d

33 $\frac{1}{f} = \frac{1}{m} + \frac{1}{p}$, for m

34 $c = \frac{2ab}{a+b}$, for b

Solve the inequalities.

35 $\frac{x}{2} - \frac{x}{3} \leq 5$

36 $\frac{x}{3} - \frac{x}{2} \leq 5$

37 $\frac{x+3}{4} - \frac{x}{2} > 1$

38 $\frac{x}{3} - \frac{x-1}{2} < 2$

39 $\frac{1}{2}(x+1) - \frac{2}{3}(x-2) < \frac{1}{6}$

40 $\frac{1}{2}(x-2) - \frac{1}{5}(x-1) > 1$

41 $\frac{2x+3}{x} < 1$

42 $\frac{x}{x-1} > 2$

43 What number must be subtracted from both the numerator and denominator of the fraction $\frac{11}{23}$ to give a fraction whose value is $\frac{2}{5}$?

44 The denominator of a fraction is 3 more than the numerator. If 5 is added to the numerator and 4 is subtracted from the denominator, the value of the new fraction is 2. Find the original fraction.

45 One pipe can empty a tank in 3 hours. A second pipe takes 4 hours to complete the same job. How long will it take to empty the tank if both pipes are used?

46 Working together Wendy and Julie can paint their room in 3 hours. If it takes Wendy 5 hours to do the job alone, how long would it take Julie to paint the room working by herself?

47 Find two fractions whose sum is $\frac{2}{3}$ if the smaller fraction is $\frac{1}{2}$ the larger one. (*Hint:* Let x represent the larger fraction.)

48 A rope that is 20 feet long is cut into two pieces. The ratio of the smaller piece to the larger piece is $\frac{3}{5}$. Find the length of the shorter piece.

49 The shadow of a tree is 20 feet long at the same time that a 1-foot-high flower casts a 4-inch shadow. How high is the tree?

50 The area A of a triangle is given by $A = \frac{bh}{2}$, where b is the length of the base and h is the length of the altitude.

(a) Solve for h in terms of A and b.
(b) Find h when $A = 100$ and $b = 25$.

51 If 561 is divided by a certain number, the quotient is 29 and the remainder is 10. Find the number. (*Hint:* For $N \div D$ we have $\frac{N}{D} = Q + \frac{R}{D}$, where Q is the quotient and R is the remainder.)

52 A student received grades of 72, 75, and 78 on three tests. What must her score on the next test be for her to have an average grade of 80 for all four tests?

53 The denominator of a certain fraction is 1 more than the numerator. If the numerator is increased by $2\frac{1}{2}$, the value will be equal to the reciprocal of the original fraction. Find the original fraction.

*54 Dan takes twice as long as George to complete a

certain job. Working together they can complete the job in 6 hours. How long will it take Dan to complete the job by himself?

***55** Prove:

$$\text{If } \frac{a}{b} = \frac{c}{d}, \text{ then } \frac{a+b}{b} = \frac{c+d}{d}$$

***56** Prove:

$$\text{If } \frac{a}{b} = \frac{c}{d}, \text{ then } \frac{a}{a+b} = \frac{c}{c+d}$$

***57** A bookstore has a stock of 30 paperback copies of *Algebra*, as well as 50 hardcover copies of the same book. They wish to increase their stock for the new semester. Based on past experience, they want their final numbers of paperback and hardcover copies to be in the ratio 4 to 3. However, the publisher stipulates that they will only sell the store 2 copies of the paperback edition for each copy of the hardcover edition ordered. Under these conditions, how many of each edition should the store order to achieve the 4:3 ratio?

58 To find out how wide a certain river is, a pole 20 feet high is set straight up on one of the banks. Another pole 4 feet long is also set straight up, on the same side, some distance away from the embankment. The observer waits until the shadow of the 20-foot pole just reaches the other side of the river. At that time he measures the shadow of the 4-foot pole and finds it to be 34 feet. Use this information to determine the width of the river.

***59** A certain college gives 4 points per credit for a grade of *A*, 3 points per credit for a *B*, 2 points per credit for a *C*, 1 point per credit for a *D*, and 0 for an *F*. A student is taking 15 credits for the semester. She expects *A*'s in a 4 credit course and in a 3 credit course. She also expects a *B* in another 3 credit course and a *D* in a 2 credit course. What grade must she get in the fifth course in order to earn a 3.4 grade point average for the term? (*Hint:* Grade point average $= \dfrac{\text{total points}}{\text{total credits}}$.)

***60** Solve $f(x) < 0$, where $f(x) = \dfrac{1}{x^3 + 6x^2 + 12x + 8}$, and graph $y = f(x)$.

Solve each system and graph.

61 $2x + 3y = 7$
$y = \dfrac{1}{x}$

***62** $y = x^2 - 2x - 4$
$y = -\dfrac{8}{x}$

5.5

Variation

If a car is traveling at a constant rate of 40 miles per hour, then the distance d traveled in t hours is given by $d = 40t$. The change in the distance is "directly" affected by the change in the time; as t increases, so does d. We say that *d is directly proportional to t.* This is because $d = 40t$ converts to the proportion $\dfrac{d}{t} = 40$. We also say that *d varies directly as t* and that 40 is the *constant of variation.*

Direct variation is a functional relationship in the sense that $y = kx$ defines y to be a function of x.

DIRECT VARIATION

y varies directly as x if

$$y = kx$$

for some constant of variation k.

EXAMPLE 1

(a) Write the equation that expresses this direct variation: y varies directly as x, and y is 8 when x is 12.

(b) Find y for $x = 30$.

Solution

(a) Since y varies directly as x, we have $y = kx$ for some constant k. To find k, substitute the given values for the variables and solve.

$$8 = k(12)$$

$$\tfrac{2}{3} = k$$

Thus $y = \frac{2}{3}x$.

(b) For $x = 30$, $y = \frac{2}{3}(30) = 20$.

Numerous examples of direct variation can be found in geometry. Here are some illustrations.

Circumference of a circle of radius r:

$C = 2\pi r$; C varies directly as the radius r;
2π is the constant of variation.

Area of a circle of radius r:

$A = \pi r^2$; A varies directly as the square of r;
π is the constant of variation.

Area of an equilateral triangle of side s:

$A = \dfrac{\sqrt{3}}{4}s^2$; A varies directly as s^2;

$\dfrac{\sqrt{3}}{4}$ is the constant of proportionality.

Volume of a cube of side e:

$V = e^3$; V varies directly as the cube of e (as e^3);
1 is the constant of variation.

EXAMPLE 2 According to Hooke's law, the force F required to stretch a spring x units is directly proportional to the length x. If a force of 20 pounds is needed to stretch a spring 3 inches, how far is the spring stretched by a force of 60 pounds?

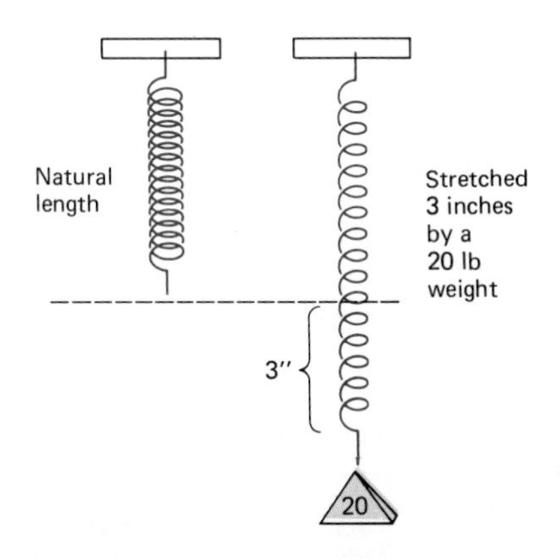

Throughout this section the constant of variation k will be a positive number.

Solution Since F varies directly as x, we have $F = kx$. Solve for k by substituting the known values for F and x.

$$20 = k(3)$$
$$\frac{20}{3} = k$$

Thus $F = \frac{20}{3}x$. Now let $F = 60$ and solve for x.

$$60 = \frac{20}{3}x$$
$$60\left(\frac{3}{20}\right) = x$$
$$9 = x$$

Thus 60 pounds stretches the spring 9 inches.

When y varies directly as x, the variables x and y increase or decrease together. There are other situations where one variable increases as the other decreases. We refer to this as *inverse variation*.

INVERSE VARIATION

y varies inversely as x if

$$y = \frac{k}{x}$$

for some constant of variation k.

EXAMPLE 3 According to Boyle's law, the pressure P of a compressed gas is inversely proportional to the volume V. Suppose that there is a pressure of 25 pounds per square inch when the volume of gas is 400 cubic inches. Find the pressure when the gas is compressed to 200 cubic inches.

Solution Since P varies inversely as V, we have

$$P = \frac{k}{V}$$

Substitute the known values for P and V and solve for k.

$$25 = \frac{k}{400}$$
$$10{,}000 = k$$

Thus $P = \dfrac{10{,}000}{V}$, and when $V = 200$ we have

Note that the pressure increases as the volume decreases.

$$P = \frac{10{,}000}{200} = 50$$

The pressure is 50 pounds per square inch.

The variation of a variable may depend on more than one other variable. Here are some illustrations:

$z = kxy$; z varies *jointly* as x and y.
$z = kx^2y$; z varies *jointly* as x^2 and y.
$z = \frac{k}{xy}$; z varies *inversely* as x and y.
$z = \frac{kx}{y}$; z varies *directly* as x and *inversely* as y.
$w = \frac{kxy^3}{z}$; w varies *jointly* as x and y^3 and *inversely* as z.

EXAMPLE 4 Describe the variation given by these equations:

(a) $z = kx^2y^3$ **(b)** $z = \frac{kx^2}{y}$ **(c)** $V = \pi r^2 h$

Solution

(a) z varies jointly as x^2 and y^3.
(b) z varies directly as x^2 and inversely as y.
(c) V varies jointly as r^2 and h.

EXAMPLE 5 Suppose that z varies directly as x and inversely as the square of y. If $z = \frac{1}{3}$ when $x = 4$ and $y = 6$, find z when $x = 12$ and $y = 4$.

Solution

$$z = \frac{kx}{y^2}$$

$$\frac{1}{3} = \frac{k(4)}{6^2}$$

$$\frac{1}{3} = \frac{k}{9}$$

$$3 = k$$

Thus $z = \frac{3x}{y^2}$. When $x = 12$ and $y = 4$ we have

$$z = \frac{3(12)}{4^2} = \frac{9}{4}$$

EXERCISES 5.5

In Exercises 1–4, write the equation for the given variation and identify the constant of variation.

1 The perimeter P of a square varies directly as the side s.

2 The area of a circle varies directly as the square of the radius.

3 The area of a rectangle 5 centimeters wide varies directly as its length.

4 The volume of a rectangular-shaped box 10 centimeters high varies jointly as the length and width.

In Exercises 5–9, write the equation for the given variation using k as the constant of variation.

5 z varies directly as x and y^3.

6 z varies inversely as x and y^3.

7 z varies directly as x and inversely as y^3.

8 w varies directly as x and y and z.

9 w varies directly as x^2 and inversely as y and z.

In Exercises 10–14, find the constant of variation.

10 y varies directly as x; $y = 4$ when $x = \frac{2}{3}$.

11 s varies directly as t^2; $s = 50$ when $t = 10$.

12 y varies inversely as x; $y = 15$ when $x = \frac{1}{3}$.

13 u varies directly as v and w; $u = 2$ when $v = 15$ and $w = \frac{2}{3}$.

14 z varies directly as x and inversely as the square of y; $z = \frac{7}{2}$ when $x = 14$ and $y = 6$.

15 a varies inversely as the square of b and $a = 10$ when $b = 5$. Find a when $b = 25$.

16 z varies jointly as x and y; $z = \frac{3}{2}$ when $x = \frac{5}{6}$ and $y = \frac{9}{20}$. Find z when $x = 2$ and $y = 7$.

17 s varies jointly as l and the square of w; $s = \frac{10}{3}$ when $l = 12$ and $w = \frac{5}{6}$. Find s when $l = 15$ and $w = \frac{9}{4}$.

18 The cost C of producing x number of articles varies directly as x. If it costs \$560 to produce 70 articles, what is C when $x = 400$?

19 If a ball rolls down an inclined plane, the distance traveled varies directly as the square of the time. If the ball rolls 12 feet in 2 seconds, how far will it roll in 3 seconds?

20 The volume V of a right circular cone varies jointly as the square of the radius r of the base, and the altitude h. If $V = 8\pi$ cubic centimeters when $r = 2$ centimeters and $h = 6$ centimeters, find the formula for the volume V.

▬ 21 A force of 2.4 kilograms is needed to stretch a spring 1.8 centimeters. Use Hooke's law to determine the force required to stretch the spring 3.0 centimeters.

▬ 22 The force required to compress a metal spring from its natural length is directly proportional to the change in the length of the spring. If 235 pounds is required to compress the spring from its natural length of 18 inches to a length of 15 inches, how much force is required to compress it from 18 inches to 12 inches?

23 Fifty kilograms per square centimeter is the pressure exerted by 150 cubic centimeters of a gas. Use Boyle's law to find the pressure if the gas is compressed to 100 cubic centimeters.

24 The gas in Exercise 23 expands to 500 cubic centimeters. What is the pressure?

▬ 25 If we neglect air resistance, the distance that an object will fall from a height near the surface of the earth is directly proportional to the square of the time it falls. If the object falls 256 feet in 4 seconds, how far will it fall in 7 seconds?

26 The volume of a right circular cylinder varies jointly as its height and the square of the radius of the base. The volume is 360π cubic centimeters when the height is 10 centimeters and the radius is 6 centimeters. Find V when $h = 18$ centimeters and $r = 5$ centimeters.

27 If the volume of a sphere varies directly as the cube of its radius and $V = 288\pi$ cubic inches when $r = 6$ inches, find V when $r = 2$ inches.

*28 The resistance to the flow of electricity through a wire depends on the length and thickness of the wire. The resistance R is measured in *ohms* and varies directly as the length l and inversely as the square of the diameter d. If a wire 200 feet long with diameter 0.16 inch has a resistance of 64 ohms, how much resistance will there be if only 50 feet of wire is used?

*29 A wire made of the same material as the wire in Exercise 28 is 100 feet long. Find R if $d = 0.4$ inch. Find R if $d = 0.04$ inch.

*30 If z varies jointly as x and y, how does x vary with respect to y and z?

▬ 31 A rectangular shaped beam is to be cut from a round log with a 2.5 foot diameter.

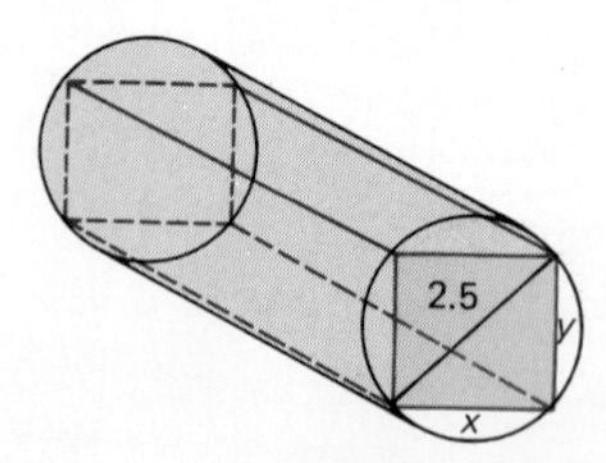

The strength s of the beam varies jointly as its height y and the square of its width x.

(a) Write s as a function of y.

(b) Write s as a function of x.

(c) Use a calculator to find s to two decimal places when $x = y$. Express the answer in terms of the constant of proportionality.

5.6 Synthetic division

In this and in the following two sections we will learn some methods for finding the x-intercepts of polynomial functions and rational functions. First let us briefly review the process used in division of polynomials; it is very much like the division algorithm in arithmetic. Consider the quotient

$$(x^3 - x^2 - 5x + 6) \div (x - 2)$$

Quotient
Divisor | Dividend

STEP 1. Divide x^3 by x. Place the result in the quotient.

$$\begin{array}{r|l} & x^2 \\ \hline x - 2 & x^3 - x^2 - 5x + 6 \end{array}$$

STEP 2. Multiply $x - 2$ by x^2 and subtract.

$$\begin{array}{r|l} & x^2 \\ \hline x - 2 & x^3 - x^2 - 5x + 6 \\ & \underline{x^3 - 2x^2} \\ & \qquad x^2 \end{array}$$

STEP 3. Bring down the next term, $-5x$, and divide again: $x^2 \div x = x$.

$$\begin{array}{r|l} & x^2 + x \\ \hline x - 2 & x^3 - x^2 - 5x + 6 \\ & \underline{x^3 - 2x^2} \\ & \qquad x^2 - 5x \end{array}$$

STEP 4. Repeat this process until you have used all terms in the dividend. The completed division follows:

$$\begin{array}{r|l} & x^2 + x - 3 \\ \hline x - 2 & x^3 - x^2 - 5x + 6 \\ & \underline{x^3 - 2x^2} \\ & \qquad x^2 - 5x \\ & \qquad \underline{x^2 - 2x} \\ & \qquad\quad -3x + 6 \\ & \qquad\quad \underline{-3x + 6} \\ & \qquad\qquad\quad 0 \end{array}$$

The remainder is 0. To check, multiply $x^2 + x - 3$ by $x - 2$. The result should be $x^3 - x^2 - 5x + 6$.

When dividing two polynomials, be sure that both the dividend and the divisor are in descending order of powers of the variable. Also, missing powers of the variable in the dividend should be denoted by using a form of zero as in Example 1.

EXAMPLE 1 Divide: $(5x + 3x^3 - 8) \div (x + 3)$.

Solution First write the dividend in the form $3x^3 + 0x^2 + 5x - 8$; then divide.

$$
\begin{array}{r|l}
 & 3x^2 - 9x + 32 \\
\hline
x + 3 & 3x^3 + 0x^2 + 5x - 8 \\
 & 3x^3 + 9x^2 \\
\cline{2-2}
 & \quad - 9x^2 + 5x \\
 & \quad - 9x^2 - 27x \\
\cline{2-2}
 & \qquad\qquad 32x - 8 \\
 & \qquad\qquad 32x + 96 \\
\cline{2-2}
 & \qquad\qquad\quad - 104
\end{array}
$$

The quotient is $3x^3 - 9x + 32$ and the remainder is -104. To check, note that the dividend should be equal to the product of the quotient and the divisor, plus the remainder. Verify that the following is correct.

$$(x + 3)(3x^2 - 9x + 32) + (-104) = 3x^3 + 5x - 8$$

We now develop a special procedure for handling long-division problems with polynomials where the divisor is of the form $x \pm c$. To discover this procedure let us first examine the following long-division problem.

$$
\begin{array}{r|l}
 & x^2 + 3x - 5 \\
\hline
x - 2 & x^3 + x^2 - 11x + 12 \\
 & x^3 - 2x^2 \\
\cline{2-2}
 & \quad + 3x^2 - 11x \\
 & \quad + 3x^2 - 6x \\
\cline{2-2}
 & \qquad\qquad - 5x + 12 \\
 & \qquad\qquad - 5x + 10 \\
\cline{2-2}
 & \qquad\qquad\qquad +2
\end{array}
$$

Now it should be clear that all of the work done involved the coefficients of the variables and the constants. Thus we could just as easily complete the division by omitting the variables, as long as we write the coefficients in the proper places. The division problem would then look like this:

$$
\begin{array}{r|l}
 & 1 + 3 - 5 \\
\hline
1 - 2 & 1 + 1 - 11 + 12 \\
 & \textcircled{1} - 2 \\
\cline{2-2}
 & \quad + 3 - 11 \\
 & \quad + \textcircled{3} - 6 \\
\cline{2-2}
 & \qquad\quad - 5 + 12 \\
 & \qquad\quad - \textcircled{5} + 10 \\
\cline{2-2}
 & \qquad\qquad\quad +2
\end{array}
$$

Since the circled numerals are repetitions of those immediately above them, this process can be further shortened by deleting them. Moreover, since these circled numbers are the products of the numbers in the quotient by the 1 in the divisor, we may also eliminate this 1. Thus we have the following:

$$
\begin{array}{rrrrr}
 & 1 & +3 & -5 & \\
\cline{2-5}
-2 \big| & 1 & +1 & -11 & +12 \\
 & & -2 & & \\
\cline{3-3}
 & & +3 & -11 & \\
 & & & -6 & \\
\cline{4-4}
 & & & -5 & +12 \\
 & & & & +10 \\
\cline{5-5}
 & & & & +2
\end{array}
$$

$\longrightarrow$ It is not necessary to bring down the -11 and 12.

$$
\begin{array}{rrrrr}
 & 1 & +3 & -5 & \\
\cline{2-5}
-2 \big| & 1 & +1 & -11 & +12 \\
 & & -2 & & \\
\cline{3-3}
 & & +3 & & \\
 & & & -6 & \\
\cline{4-4}
 & & & -5 & \\
 & & & & +10 \\
\cline{5-5}
 & & & & +2
\end{array}
$$

$\longrightarrow$ Move the numerals upward.

$$
\begin{array}{rrrrr}
 & 1 & +3 & -5 & \\
\cline{2-5}
-2 \big| & 1 & +1 & -11 & +12 \\
 & & -2 & -6 & +10 \\
\cline{3-5}
 & & +3 & -5 & +2
\end{array}
$$

When the top numeral 1 is brought down, then the last line contains the coefficients of the quotient and the remainder. So eliminate the line above the dividend.

$$
\begin{array}{r|rrr|r}
-2 & 1 & +1 & -11 & +12 \\
 & & -2 & -6 & +10 \\
\hline
 & 1 & +3 & -5 & +2 \quad \text{remainder} \\
\text{Quotient:} & \multicolumn{3}{l}{x^2 + 3x - 5} &
\end{array}
$$

We can further simplify this process by changing the sign of the divisor, making it $+2$ instead of -2. This change allows us to add throughout rather than subtract, as follows.

$$
\begin{array}{r|rrr|r}
+2 & 1 & +1 & -11 & +12\} \ \text{coefficients of dividend} \\
 & & +2 & +6 & -10 \\
\hline
\text{coefficients of quotient:} & 1 & +3 & -5 & +2 \quad \text{remainder} \\
\text{Quotient:} & \multicolumn{3}{l}{x^2 + 3x - 5} &
\end{array}
$$

The long-division process has now been condensed to this short form. Doing a division problem by this short form is called **synthetic division**, as illustrated in the examples that follow.

EXAMPLE 2 Use synthetic division to find $(2x^3 - 9x^2 + 10x - 7) \div (x - 3)$.

Solution Write the coefficients of the dividend in descending order. Change the sign of the divisor (change -3 to $+3$).

$$
\begin{array}{r|rrrr}
+3 & 2 & -9 & +10 & -7 \\
\hline
 & & & &
\end{array}
$$

Now bring down the first term, 2, and multiply by $+3$.

$$
\begin{array}{r|rrrr}
+3 & 2 & -9 & +10 & -7 \\
 & & +6 & & \\
\hline
 & 2 & & &
\end{array}
$$

Add -9 and $+6$ to obtain the sum -3. Multiply this sum by $+3$ and repeat the process to the end. The completed example should look like this:

$$
\begin{array}{r|rrr|r}
+3 & 2 & -9 & +10 & -7 \\
 & & +6 & -9 & +3 \\
\hline
 & 2 & -3 & +1 & -4
\end{array}
$$

Since the original dividend began with x^3 (third degree), the quotient will begin with x^2 (second degree). Thus we read the last line as implying a quotient of $2x^2 - 3x + 1$ and a remainder of -4. Check this result by using the long-division process.

The synthetic division process has been developed for divisors of the form $x - c$. (Thus, in Example 2, $c = 3$.) A minor adjustment also permits divisors by polynomials of the form $x + c$. For example, a divisor of $x + 2$ may be written as $x - (-2)$; $c = -2$.

Note that the quotient in a synthetic division problem is always a polynomial of degree one less than that of the dividend. This is so because the divisor has degree 1. The bottom line in the synthetic division process, except for the last entry on the right, gives the coefficients of the quotient: a polynomial in standard form.

EXAMPLE 3 Use synthetic division to find the quotient and the remainder.

$$\left(-\tfrac{1}{3}x^4 + \tfrac{1}{6}x^2 - 7x - 4\right) \div (x + 3)$$

Solution Write $x + 3$ as $x - (-3)$. Since there is no x^3 term in the dividend, use $0x^3$.

$$\begin{array}{r|rrrrr} -3 & -\frac{1}{3} & +\,0 & +\,\frac{1}{6} & -\,7 & -\,4 \\ & & +\,1 & -\,3 & +\,\frac{17}{2} & -\,\frac{9}{2} \\ \hline & -\frac{1}{3} & +\,1 & -\,\frac{17}{6} & +\,\frac{3}{2} & \big|\,-\frac{17}{2} \end{array} = \text{remainder}$$

Quotient: $-\frac{1}{3}x^3 + x^2 - \frac{17}{6}x + \frac{3}{2}$ Remainder: $-\frac{17}{2}$

Check this result.

EXERCISES 5.6

Complete Exercises 1–6 by long division and by synthetic division.

1 $(x^3 - 2x^2 - 5x + 6) \div (x - 3)$
2 $(x^3 - x^2 - 5x + 2) \div (x + 2)$
3 $(2x^3 + x^2 - 3x + 7) \div (x + 1)$
4 $(3x^3 - 2x^2 + x - 1) \div (x - 1)$
5 $(x^3 + 5x^2 - 7x + 8) \div (x - 2)$
6 $(x^3 - 3x^2 + x - 5) \div (x + 3)$

Use synthetic division to find each quotient and check.

7 $(x^4 - 3x^3 + 7x^2 - 2x + 1) \div (x + 2)$
8 $(x^4 + x^3 - 2x^2 + 3x - 1) \div (x - 2)$
9 $(2x^4 - 3x^2 + 4x - 2) \div (x - 1)$
10 $(3x^4 + x^3 - 2x + 3) \div (x + 1)$
11 $(x^3 - 27) \div (x - 3)$
12 $(x^3 - 27) \div (x + 3)$
13 $(x^3 + 27) \div (x + 3)$
14 $(x^3 + 27) \div (x - 3)$
15 $(x^4 - 16) \div (x - 2)$
16 $(x^4 - 16) \div (x + 2)$
17 $(x^4 + 16) \div (x + 2)$
18 $(4x^5 - x^3 + 5x^2 + \frac{3}{2}x - \frac{1}{2}) \div (x + \frac{1}{2})$
19 $(x^4 - \frac{1}{2}x^3 + \frac{1}{3}x^2 - \frac{1}{4}x + \frac{1}{5}) \div (x - 1)$

Use long division to find each quotient.

*20 $(2x^3 + 9x^2 - 3x - 1) \div (2x - 1)$
*21 $(x^3 - x^2 - x + 10) \div (x^2 - 3x + 5)$
*22 $(3x^5 + 4x^3 - 12x^2) \div (x^2 - x)$

23 A division problem can be represented in this way:

$$\frac{\text{dividend}}{\text{divisor}} = \text{quotient} + \frac{\text{remainder}}{\text{divisor}}$$

Using N for dividend, D for divisor, Q for quotient, and R for remainder, we can write this relationship as follows:

$$\frac{N}{D} = Q + \frac{R}{D}$$

Multiply each term of this equation by D and explain how this justifies the usual method of checking a division problem by multiplication.

5.7 The remainder and factor theorems

When the polynomial $p(x) = 2x^3 - 9x^2 + 10x - 7$ is divided by $x - 3$, the quotient is the polynomial $q(x) = 2x^2 - 3x + 1$ and the remainder $r = -4$. (See Example 2, page 197.) As a check we see that

$$\underbrace{2x^3 - 9x^2 + 10x - 7}_{p(x)} = \underbrace{(2x^2 - 3x + 1)}_{q(x)} \cdot (x - 3) + \underbrace{(-4)}_{r}$$

In general, whenever a polynomial $p(x)$ is divided by $x - c$ we have

$$p(x) = q(x)(x - c) + r$$

Another form of this is $\frac{p(x)}{x-c} = q(x) + \frac{r}{x-c}$.

where $q(x)$ is the quotient and r is the (constant) remainder. Since this equation holds for all x, we may let $x = c$ and obtain

$$\begin{aligned} p(c) &= q(c)(c - c) + r \\ &= q(c) \cdot 0 + r \\ &= r \end{aligned}$$

This result may be summarized as follows:

REMAINDER THEOREM

If a polynomial $p(x)$ is divided by $x - c$, the remainder is equal to $p(c)$.

EXAMPLE 1 Find the remainder when $p(x) = 3x^3 - 5x^2 + 7x + 5$ is divided by $x - 2$.

Solution By the remainder theorem, the answer is $p(2)$.

$$\begin{aligned} p(x) &= 3x^3 - 5x^2 + 7x + 5 \\ p(2) &= 3(2)^3 - 5(2)^2 + 7(2) + 5 \\ &= 23 \end{aligned}$$

EXAMPLE 2 Let $f(x) = x^3 - 2x^2 + 3x - 1$. Use synthetic division and the remainder theorem to find $f(3)$.

Solution

$$\begin{array}{r|rrrr} 3 & 1 & -2 & +3 & -1 \\ & & +3 & +3 & +18 \\ \hline & 1 & +1 & +6 & \boxed{+17} \end{array} = \text{remainder} = f(3)$$

TEST YOUR UNDERSTANDING

Use synthetic division to find the remainder r when p(x) is divided by x − c. Verify that r = p(c) by substituting x = c into p(x).

1 $p(x) = x^5 - 7x^4 + 4x^3 + 10x^2 - x - 5;\ \ x - 1$

2 $p(x) = x^4 + 11x^3 + 11x^2 + 11x + 10;\ \ x + 10$

3 $p(x) = x^4 + 11x^3 + 11x^2 + 11x + 10;\ \ x - 10$

4 $p(x) = 6x^3 - 40x^2 + 25;\ \ x - 6$

Once again, we are going to consider the division of a polynomial $p(x)$ by a divisor of the form $x - c$. First note that

$$p(x) = q(x)(x - c) + r$$

where $q(x)$ is the quotient and r is the (constant) remainder. Now suppose that $r = 0$. Then the remainder theorem gives $p(c) = r = 0$, and the preceding equation becomes

$$p(x) = q(x)(x - c)$$

It follows that $x - c$ is a *factor* of $p(x)$. Conversely, suppose that $x - c$ is a factor of $p(x)$. This means there is another polynomial, say $q(x)$, so that

$$p(x) = q(x)(x - c)$$

or

$$p(x) = q(x)(x - c) + 0$$

which tells us that when $p(x)$ is divided by $x - c$ the remainder is zero. These observations comprise the following result:

If $p(c) = 0$, then c is a *zero* of the polynomial and is a root of $p(x) = 0$.

FACTOR THEOREM

A polynomial $p(x)$ has a factor $x - c$ if and only if $p(c) = 0$.

EXAMPLE 3 Show that $x - 2$ is a factor of $p(x) = x^3 - 3x^2 + 7x - 10$.

Solution By the factor theorem we can state that $x - 2$ is a factor of $p(x)$ if $p(2) = 0$.

$$\begin{aligned} p(2) &= 2^3 - 3(2)^2 + 7(2) - 10 \\ &= 0 \end{aligned}$$

EXAMPLE 4 **(a)** Use the factor theorem to show that $x + 3$ is a factor of $p(x) = x^3 - x^2 - 8x + 12$. **(b)** Factor $p(x)$ completely.

Solution

(a) First write $x + 3 = x - (-3)$, so that $c = -3$. Then use synthetic division.

$$\begin{array}{r|rrrr} -3 & 1 & -1 & -8 & +12 \\ & & -3 & +12 & -12 \\ \hline & 1 & -4 & +4 & \boxed{+0} \end{array}$$

Since $p(-3) = 0$, the factor theorem tells us that $x + 3$ is a factor of $p(x)$.

(b) Synthetic division has produced the quotient $x^2 - 4x + 4$. Therefore, since $x + 3$ is a factor of $p(x)$, we get

$$x^3 - x^2 - 8x + 12 = (x^2 - 4x + 4)(x + 3)$$

To get the complete factored form, observe that $x^2 - 4x + 4 = (x - 2)^2$. Thus

$$x^3 - x^2 - 8x + 12 = (x - 2)^2(x + 3)$$

EXERCISES 5.7

In Exercises 1–6, use synthetic division and the remainder theorem.

1 $f(x) = x^3 - x^2 + 3x - 2$; find $f(2)$.

2 $f(x) = 2x^3 + 3x^2 - x - 5$; find $f(-1)$.

3 $f(x) = x^4 - 3x^2 + x + 2$; find $f(3)$.

4 $f(x) = x^4 + 2x^3 - 3x - 1$; find $f(-2)$.

5 $f(x) = x^5 - x^3 + 2x^2 + x - 3$; find $f(1)$.

6 $f(x) = 3x^4 + 2x^3 - 3x^2 - x + 7$; find $f(-3)$.

Find the remainder for each division by substitution, using the remainder theorem. That is, in Exercise 7 (for example) let $f(x) = x^3 - 2x^2 + 3x - 5$ and find $f(2) = r$.

7 $(x^3 - 2x^2 + 3x - 5) \div (x - 2)$

8 $(x^3 - 2x^2 + 3x - 5) \div (x + 2)$

9 $(2x^3 + 3x^2 - 5x + 1) \div (x - 3)$

10 $(3x^4 - x^3 + 2x^2 - x + 1) \div (x + 3)$

11 $(4x^5 - x^3 - 3x^2 + 2) \div (x + 1)$

12 $(3x^5 - 2x^4 + x^3 - 7x + 1) \div (x - 1)$

In Exercises 13–23, show that the given binomial $x - c$ is a factor of $p(x)$, and then factor $p(x)$ completely.

13 $p(x) = x^3 + 6x^2 + 11x + 6$; $x + 1$

14 $p(x) = x^3 - 6x^2 + 11x - 6$; $x - 1$

15 $p(x) = x^3 + 5x^2 - 2x - 24$; $x - 2$

16 $p(x) = -x^3 + 11x^2 - 23x - 35$; $x - 7$

17 $p(x) = -x^3 + 7x + 6$; $x + 2$

18 $p(x) = x^3 + 2x^2 - 13x + 10$; $x + 5$

19 $p(x) = 6x^3 - 25x^2 - 29x + 20$; $x - 5$

20 $p(x) = 12x^3 - 22x^2 - 100x - 16$; $x + 2$

21 $p(x) = x^4 + 4x^3 + 3x^2 - 4x - 4$; $x + 2$

22 $p(x) = x^4 - 8x^3 + 7x^2 + 72x - 144$; $x - 4$

23 $p(x) = x^6 + 6x^5 + 8x^4 - 6x^3 - 9x^2$; $x + 3$

*24 When $x^2 + 5x - 2$ is divided by $x + n$, the remainder is -8. Find all possible values of n and check by division.

*25 Find d so that $x + 6$ is a factor of

$$x^4 + 4x^3 - 21x^2 + dx + 108$$

*26 Find b so that $x - 2$ is a factor of

$$x^3 + bx^2 - 13x + 10$$

*27 Find a so that $x - 10$ is a factor of

$$ax^3 - 25x^2 + 47x + 30$$

5.8 The rational root theorem

Consider the polynomial equation

$$(3x + 2)(5x - 4)(2x - 3) = 0$$

To find the roots, set each factor equal to zero.

$$3x + 2 = 0 \qquad 5x - 4 = 0 \qquad 2x - 3 = 0$$
$$x = -\tfrac{2}{3} \qquad x = \tfrac{4}{5} \qquad x = \tfrac{3}{2}$$

Now multiply the original three factors and keep careful note of the details of this multiplication. Your result should be

$$30x^3 - 49x^2 - 10x + 24 = 0$$

which must have the same three rational roots.

As you analyze this multiplication it becomes clear that the constant 24 is the product of the three constants in the binomials, 2, -4, and -3. Also, the leading coefficient, 30, is the product of the three original coefficients of x in the binomials, namely 3, 5, and 2.

Furthermore, 3, 5, and 2 are also the denominators of the roots $-\frac{2}{3}$, $\frac{4}{5}$, and $\frac{3}{2}$. Therefore, the denominators of the rational roots are all factors of 30, and their numerators are all factors of 24.

These results are not accidental. It turns out that we have been discussing the following general result:

RATIONAL ROOT THEOREM

Let $f(x) = a_nx^n + a_{n-1}x^{n-1} + \cdots + a_1x + a_0$ ($a_n \neq 0$) be an nth-degree polynomial with integer coefficients. If $\dfrac{p}{q}$ is a rational root of $f(x) = 0$, where $\dfrac{p}{q}$ is in lowest terms, then p is a factor of a_0 and q is a factor of a_n.

Let us see how this theorem can be applied to find the rational roots of

$$f(x) = 4x^3 - 16x^2 + 11x + 10 = 0$$

Begin by listing all factors of the constant 10 and of the leading coefficient 4.

Factors of 10: $\pm 1, \pm 2, \pm 5, \pm 10$ (possible numerators)
Factors of 4: $\pm 1, \pm 2, \pm 4$ (possible denominators)

Possible rational roots (take each number in the first row and divide by each number in the second row): $\pm 1, \pm\frac{1}{2}, \pm\frac{1}{4}, \pm 2, \pm 5, \pm\frac{5}{2}, \pm\frac{5}{4}, \pm 10$

If $f(c) = 0$, then c is a root; if $f(c) \neq 0$, then c is not a root.

To decide which (if any) of these are roots of $f(x) = 0$, we could substitute the values directly into $f(x)$. However, it is easier to use synthetic division because in most cases it leads to easier computations and also makes quotients available. Therefore, we proceed by using synthetic division with divisors c, where c is a possible rational root.

$$\begin{array}{r|rrrr} 1 & 4 & -16 & +11 & +10 \\ & & +4 & -12 & -1 \\ \hline & 4 & -12 & -1 & +9 \end{array} \qquad \begin{array}{r|rrrr} -1 & 4 & -16 & +11 & +10 \\ & & -4 & +20 & -31 \\ \hline & 4 & -20 & +31 & -21 \end{array}$$

Since $f(1) = 9 \neq 0$, 1 is *not* a root.

Since $f(-1) = -21 \neq 0$, -1 is *not* a root.

$$\begin{array}{r|l} \frac{1}{2} & 4 - 16 + 11 + 10 \\ & \quad + 2 - 7 + 2 \\ \hline & 4 - 14 + 4 \mid +12 \end{array} \qquad \begin{array}{r|l} -\frac{1}{2} & 4 - 16 + 11 + 10 \\ & \quad - 2 + 9 - 10 \\ \hline & 4 - 18 + 20 \mid + 0 \end{array}$$

Since $f(\frac{1}{2}) = 12 \neq 0$, $\frac{1}{2}$ is *not* a root.

Since $f(-\frac{1}{2}) = 0$, $-\frac{1}{2}$ *is* a root.

By the factor theorem it follows that $x - (-\frac{1}{2}) = x + \frac{1}{2}$ is a factor of $f(x)$, and synthetic division gives the other factor, $4x^2 - 18x + 20$.

$$f(x) = (x + \tfrac{1}{2})(4x^2 - 18x + 20)$$

To find other roots of $f(x) = 0$ we could proceed by using the rational root theorem for $4x^2 - 18x + 20 = 0$. But this is unnecessary because the quadratic expression is factorable.

$$\begin{aligned} f(x) &= (x + \tfrac{1}{2})(4x^2 - 18x + 20) \\ &= (x + \tfrac{1}{2})(2)(2x^2 - 9x + 10) \\ &= 2(x + \tfrac{1}{2})(x - 2)(2x - 5) \end{aligned}$$

The solution of $f(x) = 0$ can now be found by setting each factor equal to zero. The solutions are $x = -\frac{1}{2}$, $x = 2$, and $x = \frac{5}{2}$.

EXAMPLE 1 Factor $f(x) = x^3 + 6x^2 + 11x + 6$.

Solution Since the leading coefficient is 1, whose only factors are ± 1, the possible denominators of a rational root of $f(x) = 0$ can only be ± 1. Hence the possible rational roots must all be factors of $+6$, namely ± 1, ± 2, ± 3, and ± 6. Use synthetic division to test these cases.

Whenever a polynomial has 1 as the leading coefficient, then any rational root will be an integer that is a factor of the constant term of the polynomial.

$$\begin{array}{r|l} 1 & 1 + 6 + 11 + 6 \\ & \quad + 1 + 7 + 18 \\ \hline & 1 + 7 + 18 \mid +24 = r \end{array}$$

Since $r = f(1) \neq 0$, $x - 1$ is *not* a factor of $f(x)$.

$$\begin{array}{r|l} -1 & 1 + 6 + 11 + 6 \\ & \quad - 1 - 5 - 6 \\ \hline & 1 + 5 + 6 \mid +0 = r \end{array}$$

Since $r = f(-1) = 0$, $x - (-1) = x + 1$ *is* a factor of $f(x)$.

$$x^3 + 6x^2 + 11x + 6 = (x + 1)(x^2 + 5x + 6)$$

Now factor the trinomial:

$$x^3 + 6x^2 + 11x + 6 = (x + 1)(x + 2)(x + 3)$$

TEST YOUR UNDERSTANDING

For each p(x), find (a) the possible rational roots of $p(x) = 0$, (b) the factored form of $p(x)$, and (c) the roots of $p(x) = 0$.

1 $p(x) = x^3 - 3x^2 - 10x + 24$

2 $p(x) = x^4 + 6x^3 + x^2 - 24x + 16$

3 $p(x) = 4x^3 + 20x^2 - 23x + 6$

4 $p(x) = 3x^4 - 13x^3 + 7x^2 - 13x + 4$

EXAMPLE 2 Solve for x: $p(x) = 2x^5 + 7x^4 - 18x^2 - 8x + 8 = 0$.

Solution The possible rational roots are $\pm 1, \pm\frac{1}{2}, \pm 2, \pm 4$, and ± 8. Testing these possibilities (left to right), the first root we find is $\frac{1}{2}$.

$$\begin{array}{r|rrrrrr} \frac{1}{2} & 2 & +7 & +0 & -18 & -8 & +8 \\ & & +1 & +4 & +2 & -8 & -8 \\ \hline & 2 & +8 & +4 & -16 & -16 & +0 \end{array}$$

Therefore, $x - \frac{1}{2}$ is a factor of $p(x)$.

$$\begin{aligned} p(x) &= (x - \tfrac{1}{2})(2x^4 + 8x^3 + 4x^2 - 16x - 16) \\ &= 2(x - \tfrac{1}{2})(x^4 + 4x^3 + 2x^2 - 8x - 8) \end{aligned}$$

To find other roots of $p(x) = 0$ it now becomes necessary to solve

$$x^4 + 4x^3 + 2x^2 - 8x - 8 = 0$$

The possible rational roots for this equation are $\pm 1, \pm 2, \pm 4$, and ± 8. However, values like ± 1 that were tried before need not be tried again. Why? We find that $x = -2$ is a root:

$$\begin{array}{r|rrrrr} -2 & 1 & +4 & +2 & -8 & -8 \\ & & -2 & -4 & +4 & +8 \\ \hline & 1 & +2 & -2 & -4 & +0 \end{array}$$

$$\begin{aligned} x^4 + 4x^3 + 2x^2 - 8x - 8 &= (x + 2)(x^3 + 2x^2 - 2x - 4) \\ &= (x + 2)[x^2(x + 2) - 2(x + 2)] \\ &= (x + 2)(x^2 - 2)(x + 2) \\ &= (x + 2)^2(x^2 - 2) \end{aligned}$$

This gives

$$\begin{aligned} p(x) &= 2(x - \tfrac{1}{2})(x^4 + 4x^3 + 2x^2 - 8x - 8) \\ &= 2(x - \tfrac{1}{2})(x + 2)^2(x^2 - 2) \\ &= 2(x - \tfrac{1}{2})(x + 2)^2(x + \sqrt{2})(x - \sqrt{2}) \end{aligned}$$

Setting each factor equal to zero produces the solutions of $p(x) = 0$:

$$x = \tfrac{1}{2}, \quad x = -2, \quad x = -\sqrt{2}, \quad x = \sqrt{2}$$

Solve for x.

1 $x^3 + 2x^2 - 29x + 42 = 0$

2 $x^3 + x^2 - 21x - 45 = 0$

3 $2x^3 - 15x^2 + 24x + 16 = 0$

4 $3x^3 + 2x^2 - 75x - 50 = 0$

5 $x^3 + 3x^2 + 3x + 1 = 0$

6 $x^4 + 3x^3 + 3x^2 + x = 0$

7 $x^4 + 6x^3 + 7x^2 - 12x - 18 = 0$

8 $x^4 + 6x^3 + 2x^2 - 18x - 15 = 0$

9 $x^4 - x^3 - 5x^2 - x - 6 = 0$

10 $x^4 + 2x^3 - 7x^2 - 18x - 18 = 0$

11 $x^4 - 5x^3 + 3x^2 + 15x - 18 = 0$

12 $-x^5 + 5x^4 - 3x^3 - 15x^2 + 18x = 0$

13 $x^4 + 4x^3 - 7x^2 - 36x - 18 = 0$

14 $2x^3 - 5x - 3 = 0$

15 $6x^3 - 25x^2 + 21x + 10 = 0$

16 $3x^4 - 11x^3 - 3x^2 - 6x + 8 = 0$

Factor.

17 $x^3 - 3x^2 - 10x + 24$

18 $-x^3 - 3x^2 + 24x + 80$

19 $x^3 - 28x - 48$

20 $6x^4 + 9x^3 + 9x - 6$

Solve the system.

21 $y = x^3 - 3x^2 + 3x - 1$
$y = 7x - 13$

22 $y = -x^3$
$y = -3x^2 + 4$

*23 $y = 4x^3 - 7x^2 + 10$
$y = x^3 + 43x - 5$

*24 $y = x^2 + 4x$
$(x - 1)^2 + (y - 6)^2 = 37$

*25 Show that $2x^3 - 5x^2 - x + 8 = 0$ has no rational roots.

*26 Solve the system and graph:

$$y = 2x - 5$$

$$y = \frac{1}{x^2 - 2x + 1}$$

5.9 Decomposing rational functions

In Chapter 2 we learned how to combine rational expressions. For example, combining the fractions in

$$(1)\quad \frac{6}{x-4} + \frac{3}{x-2} \qquad \text{produces} \qquad (2)\quad \frac{9x-24}{(x-4)(x-2)}$$

It is now our goal to start with a rational expression such as (2) and *decompose* it into the form (1). When this is accomplished we say that $\frac{9x-24}{(x-4)(x-2)}$ has been decomposed into (simpler) **partial fractions.**

$$\frac{9x-24}{(x-4)(x-2)} = \frac{6}{x-4} + \frac{3}{x-2}$$

We will only consider examples that involve linear factors in the denominator.

First observe that each factor in the denominator on the left serves as a denominator of a partial fraction on the right. Let us assume, for the moment, that the numerators 6 and 3 are not known. Then it is reasonable to begin by writing

$$\frac{9x-24}{(x-4)(x-2)} = \frac{A}{x-4} + \frac{B}{x-2}$$

where A and B are the constants to be found. To find these values, first clear fractions by multiplying both sides by $(x-4)(x-2)$.

$$(x-4)(x-2)\cdot\frac{9x-24}{(x-4)(x-2)}=(x-4)(x-2)\left[\frac{A}{x-4}+\frac{B}{x-2}\right]$$

$$9x-24=A(x-2)+B(x-4)$$

Since we want this equation to hold for all values of x, we may select specific values for x that will produce the constants A and B. Observe that when $x=4$ the term $B(x-4)$ will become zero.

$$\begin{aligned}9(4)-24&=A(4-2)+B(4-4)\\12&=2A+0\\6&=A\end{aligned}$$

Similarly, B can be found by letting $x=2$.

$$\begin{aligned}9(2)-24&=A(2-2)+B(2-4)\\-6&=0-2B\\3&=B\end{aligned}$$

EXAMPLE 1 Decompose $\dfrac{6x^2+x-37}{(x-3)(x+2)(x-1)}$ into partial fractions.

Solution Since there are three linear factors in the denominator we begin with the form

$$\frac{6x^2+x-37}{(x-3)(x+2)(x-1)}=\frac{A}{x-3}+\frac{B}{x+2}+\frac{C}{x-1}$$

Multiply by $(x-3)(x+2)(x-1)$ to clear fractions.

$$6x^2+x-37=A(x+2)(x-1)+B(x-3)(x-1)+C(x-3)(x+2)$$

Since the second and third terms on the right have the factor $x-3$, the value $x=3$ will make these two terms zero.

$$\begin{aligned}6(3)^2+3-37&=A(3+2)(3-1)+B(3-3)(3-1)+C(3-3)(3+2)\\54+3-37&=A(5)(2)+0+0\\20&=10A\\2&=A\end{aligned}$$

To find B, use $x=-2$.

$$\begin{aligned}6(-2)^2+(-2)-37&=A(-2+2)(-2-1)+B(-2-3)(-2-1)\\&\quad+C(-2-3)(-2+2)\\24-2-37&=0+B(-5)(-3)+0\\-15&=15B\\-1&=B\end{aligned}$$

To find C, let $x=1$.

$$6(1)^2 + 1 - 37 = 0 + 0 + C(1 - 3)(1 + 2)$$
$$-30 = -6C$$
$$5 = C$$

Substituting the values for A, B, and C into the original form produces the desired decomposition.

$$\frac{6x^2 + x - 37}{(x - 3)(x + 2)(x - 1)} = \frac{2}{x - 3} - \frac{1}{x + 2} + \frac{5}{x - 1}$$

Check this result by combining the fractions on the right side.

TEST YOUR UNDERSTANDING

Decompose into partial fractions.

1 $\dfrac{8x - 19}{(x - 2)(x - 3)}$

2 $\dfrac{1}{(x + 2)(x - 4)}$

3 $\dfrac{6x^2 - 22x + 18}{(x - 1)(x - 2)(x - 3)}$

Factor the denominator and decompose into partial fractions.

4 $\dfrac{4x + 6}{x^2 + 5x + 6}$

5 $\dfrac{23x - 1}{6x^2 + x - 1}$

6 $\dfrac{5x^2 - 24x - 173}{x^3 + 4x^2 - 31x - 70}$

Let us look at a somewhat different example.

$$\frac{7}{x + 3} - \frac{4}{(x + 3)^2} = \frac{7x + 17}{(x + 3)^2}$$

Note that the least common denominator is the highest power of the linear factor in either denominator.

Now assume that the specific numerators are not known, and begin the decomposition process in this way:

$$\frac{7x + 17}{(x + 3)^2} = \frac{A}{x + 3} + \frac{B}{(x + 3)^2}$$

Clear fractions:

(1) $7x + 17 = A(x + 3) + B$

To find B, let $x = -3$.

$$7(-3) + 17 = A(0) + B$$
$$-4 = B$$

Substitute this value for B into Equation (1).

(2) $7x + 17 = A(x + 3) - 4$

Now find A by substituting some easy value for x, say $x = 0$, into (2).

$$7(0) + 17 = A(0 + 3) - 4$$
$$17 = 3A - 4$$
$$7 = A$$

Note: If the original denominator had been $(x+3)^3$, then we would have used the additional fraction $\frac{C}{(x+3)^3}$ to start with.

Substituting these values for A and B into our original form produces the decomposition.

$$\frac{7x+17}{(x+3)^2} = \frac{7}{x+3} + \frac{-4}{(x+3)^2} = \frac{7}{x+3} - \frac{4}{(x+3)^2}$$

EXAMPLE 2 Decompose $\frac{6+26x-x^2}{(2x-1)(x+2)^2}$ into partial fractions.

Solution Begin with this form:

$$\frac{6+26x-x^2}{(2x-1)(x+2)^2} = \frac{A}{2x-1} + \frac{B}{x+2} + \frac{C}{(x+2)^2}$$

Clear fractions.

$$(1)\quad 6+26x-x^2 = A(x+2)^2 + B(2x-1)(x+2) + C(2x-1)$$

Find A by substituting $x = \frac{1}{2}$.

$$6 + 13 - \tfrac{1}{4} = A(\tfrac{5}{2})^2 + 0 + 0$$

$$\tfrac{75}{4} = \tfrac{25}{4}A$$

$$3 = A$$

Find C by letting $x = -2$.

$$6 - 52 - 4 = 0 + 0 + C(-5)$$

$$-50 = -5C$$

$$10 = C$$

Substitute $A = 3$ and $C = 10$ into Equation (1).

$$(2)\quad 6+26x-x^2 = 3(x+2)^2 + B(2x-1)(x+2) + 10(2x-1)$$

To find B, use a simple value like $x = 1$ in (2).

$$6 + 26 - 1 = 3(9) + B(1)(3) + 10(1)$$

$$-6 = 3B$$

$$-2 = B$$

Then the decomposition is

$$\frac{6+26x-x^2}{(2x-1)(x+2)^2} = \frac{3}{2x-1} - \frac{2}{x+2} + \frac{10}{(x+2)^2}$$

You can check this by combining the right side.

An alternative method for finding the unknown constants in a decomposition problem makes use of the techniques of solving linear systems studied in Section 3.6. This method begins the same as before. In Example 2, for instance, we again reach this equation after clearing fractions:

$$6+26x-x^2 = A(x+2)^2 + B(2x-1)(x+2) + C(2x-1)$$

Convert each side to a polynomial in standard form (using decreasing exponents). Thus

$$-x^2 + 26x + 6 = Ax^2 + 4Ax + 4A + 2Bx^2 + 3Bx - 2B + 2Cx - C$$
$$= (A + 2B)x^2 + (4A + 3B + 2C)x + (4A - 2B - C)$$

Now call on the fact that *when two polynomials written in standard form are equal, their coefficients are the same*. According to this criterion we may write:

Coefficients of x^2:	$-1 = A + 2B$
Coefficients of x:	$26 = 4A + 3B + 2C$
Constants:	$6 = 4A - 2B - C$

As a working rule *always try to find as many of the unknown constants by the substitution method before you equate coefficients.*

We now have a linear system of three equations in A, B, and C. The solution to this system is $A = 3$, $B = -2$, and $C = 10$.

Thus far, in every decomposition problem the degree of the polynomial in the numerator has been less than the degree in the denominator. Here is an example where this is not the case.

$$\frac{2x^3 + 3x^2 - x + 16}{x^2 + 2x - 3}$$

In such cases the *first* step is to divide.

$$\begin{array}{r|l}
 & 2x - 1 \\
x^2 + 2x - 3 & 2x^3 + 3x^2 - x + 16 \\
 & 2x^3 + 4x^2 - 6x \\ \hline
 & -x^2 + 5x \\
 & -x^2 - 2x + 3 \\ \hline
 & 7x + 13
\end{array}$$

(remainder has degree *less* than degree of divisor)

Now write

$$\frac{2x^3 + 3x^2 - x + 16}{x^2 + 2x - 3} = \text{quotient} + \frac{\text{remainder}}{\text{divisor}}$$
$$= 2x - 1 + \frac{7x + 13}{x^2 + 2x - 3}$$

The problem will be completed by decomposing $\frac{7x + 13}{x^2 + 2x - 3}$. For illustrative purposes we will use the alternative procedure described.

$$\frac{7x + 13}{x^2 + 2x - 3} = \frac{7x + 13}{(x - 1)(x + 3)} = \frac{A}{x - 1} + \frac{B}{x + 3}$$

Clear fractions.

$$7x + 13 = A(x + 3) + B(x - 1)$$
$$= (A + B)x + (3A - B)$$

Equate coefficients and solve for A and B.

$$\begin{array}{rl}
A + B &= 7 \\
3A - B &= 13 \quad \text{(add)} \\ \hline
4A &= 20 \\
A &= 5
\end{array}$$

Substitute into $A + B = 7$ to get $B = 7 - 5 = 2$. Therefore, the final decomposition is

$$\frac{2x^3 + 3x^2 - x + 16}{x^2 + 2x - 3} = 2x - 1 + \frac{5}{x - 1} + \frac{2}{x + 3}$$

> *Caution:* When the degree of the numerator is *not* less than the degree in the denominator, you *must* divide first. If this step is ignored, the resulting decomposition will be wrong. For example, suppose that you started *incorrectly* in this way:
>
> $$\frac{2x^3 + 3x^2 - x + 16}{(x - 1)(x + 2)} = \frac{A}{x - 1} + \frac{B}{x + 3}$$
>
> This approach will produce the following *incorrect* answer:
>
> $$\frac{2x^3 + 3x^2 - x + 16}{(x - 1)(x + 3)} = \frac{5}{x - 1} + \frac{2}{x + 3}$$

EXERCISES 5.9

Decompose into partial fractions.

1 $\dfrac{2x}{(x + 1)(x - 1)}$

2 $\dfrac{x}{x^2 - 4}$

3 $\dfrac{x + 7}{x^2 - x - 6}$

4 $\dfrac{4x^2 + 16x + 4}{(x + 3)(x^2 - 1)}$

5 $\dfrac{5x^2 + 9x - 56}{(x - 4)(x - 2)(x + 1)}$

6 $\dfrac{x}{(x - 3)^2}$

7 $\dfrac{3x - 3}{(x - 2)^2}$

8 $\dfrac{2 - 3x}{x^2 + x}$

9 $\dfrac{3x - 30}{15x^2 - 14x - 8}$

10 $\dfrac{2x + 1}{(2x + 3)^2}$

11 $\dfrac{x^2 - x - 4}{(x - 2)^3}$

12 $\dfrac{x^2 + 5x + 8}{(x - 3)(x + 1)^2}$

In Exercises 13–15, first divide and then complete the decomposition into partial fractions.

13 $\dfrac{x^3 - x + 2}{x^2 - 1}$

14 $\dfrac{4x^2 - 14x + 2}{4x^2 - 1}$

15 $\dfrac{12x^4 - 12x^3 + 7x^2 - 2x - 3}{4x^2 - 4x + 1}$

Decompose into partial fractions.

16 $\dfrac{10x^2 - 16}{x^4 - 5x^2 + 4}$

17 $\dfrac{10x^3 - 15x^2 - 35x}{x^2 - x - 6}$

*18 $\dfrac{25x^3 + 10x^2 + 31x + 5}{25x^2 + 10x + 1}$

*19 $\dfrac{x^5 - 3x^4 + 2x^3 + x^2 + x + 4}{x^3 - 3x^2 + 3x - 1}$

REVIEW EXERCISES FOR CHAPTER 5

The solutions to the following exercises can be found within the text of Chapter 5. Try to answer each question without referring to the text.

Section 5.1

raph each of the following.

$y = f(x) = x^3$

$y = f(x) = |(x - 3)^3|$

$= 2x^4$

nder what conditions is the graph of a function $= f(x)$ said to be symmetric through the origin?

5 Under what conditions is the graph of $y = f(x)$ symmetric about the y-axis?

6 Explain how the graph of $y = f(x + h)$ can be obtained from the graph of $y = f(x)$ when $h > 0$, and also when $h < 0$.

7 Explain how the graph of $y = f(x) + h$ can be obtained from the graph of $y = f(x)$ when $h > 0$, and also when $h < 0$.

Section 5.2

8 Draw the graph of the function $y = f(x) = \dfrac{1}{x}$.

On what interval is the curve concave down? Where is it concave up?

Sketch each graph. Find the asymptotes.

9 $g(x) = \dfrac{1}{x - 3}$

10 $y = \dfrac{1}{(x + 2)^2} - 3$

11 Draw the graph of $y = \dfrac{1}{x^2}$ and describe the symmetry of the curve.

Section 5.3

12 Construct a table of signs for the function $f(x) = x^3 + 4x^2 - x - 4$.

13 Graph the function in Exercise 12.

14 What are the conditions needed for the graph of a rational function to have a vertical asymptote?

15 Graph: $y = \dfrac{x - 2}{x^2 - 3x - 4}$.

Section 5.4

Solve for x.

16 $\dfrac{x - 2}{x} = \dfrac{3}{4}$

17 $\dfrac{2x - 5}{x + 1} - \dfrac{3}{x^2 + x} = 0$

18 $\dfrac{3}{x - 3} + 2 = \dfrac{x}{x - 3}$

19 $\dfrac{x}{3} - \dfrac{2x + 1}{4} > 1$

20 State the proportion property.

21 Working alone Harry can mow a lawn in 3 hours. Elliot can complete the same job in 2 hours. How long will it take them working together?

22 Solve the system

$$y = \frac{2}{x}$$

$$y = -2x + 5$$

Section 5.5

23 Write the equation that expresses this direct variation: y varies directly as x, and y is 8 when x is 12. Find y for $x = 30$.

24 According to Hooke's law, the force F required to stretch a spring x units is directly proportional to the length x. If a force of 20 pounds is needed to stretch a spring 3 inches, how far is the spring stretched by a force of 60 pounds?

25 According to Boyle's law, the pressure of a compressed gas is inversely proportional to the volume V. Suppose that there is a pressure of 25 pounds per square inch when the volume of gas is 400 cubic inches. Find the pressure when the gas is compressed to 200 cubic inches.

26 Suppose that z varies directly as x and inversely as the square of y. If $z = \frac{1}{3}$ when $x = 4$ and $y = 6$, find z when $x = 12$ and $y = 4$.

Section 5.6

27 Use long division to find the quotient

$$(x^3 - x^2 - 5x + 6) \div (x - 2)$$

28 Use synthetic division to find

$$(2x^3 - 9x^2 + 10x - 7) \div (x - 3)$$

29 Use synthetic division to find the quotient and the remainder:
$(-\frac{1}{3}x^4 + \frac{1}{6}x^2 - 7x - 4) \div (x + 3)$

Section 5.7

30 State the remainder theorem.

31 State the factor theorem.

32 Find the remainder if $p(x) = 3x^3 - 5x^2 + 7x + 5$ is divided by $x - 2$.

33 Let $f(x) = x^3 - 2x^2 + 3x - 1$. Use synthetic division and the remainder theorem to find $f(3)$.

34 Show that $x - 2$ is a factor of $p(x) = x^3 - 3x^2 + 7x - 10$.

35 Use the factor theorem to show that $x + 3$ is a factor of $p(x) = x^3 - x^2 + 8x + 12$. Then factor $p(x)$ completely.

Section 5.8

36 State the rational root theorem.

37 What are the possible rational roots of $p(x) = 4x^3 - 16x^2 + 11x + 10 = 0$?

38 Factor $f(x) = x^3 + 6x^2 + 11x + 6$.

39 Factor $p(x) = 2x^5 + 7x^4 - 18x^2 - 8x + 8$ and find the roots of $p(x) = 0$.

Section 5.9

40 Decompose $\dfrac{6x^2 + x - 37}{(x - 3)(x + 2)(x - 1)}$ into partial fractions.

41 Decompose $\dfrac{6 + 26x - x^2}{(2x - 1)(x + 2)^2}$ into partial fractions.

42 Decompose $\dfrac{2x^3 + 3x^2 - x + 16}{x^2 + 2x - 3}$ into partial fractions.

SAMPLE TEST QUESTIONS FOR CHAPTER 5

Use these questions to test your knowledge of the basic skills and concepts of Chapter 5. Then check your answers with those given at the end of the book.

Graph each function and write the equation of the asymptotes if there are any.

1 $f(x) = (x + 2)^3 - \frac{3}{2}$

2 $y = f(x) = x^3$

3 $y = f(x) = -\dfrac{1}{x - 2}$

4 $y = f(x) = \dfrac{x - 1}{x^2 - x - 2}$

5 Graph: $y = x^3 - x^2 - 4x + 4$.

6 Determine the signs of $f(x) = \dfrac{x^2 - 2x}{x + 3}$.

Solve for x.

7 $\dfrac{6}{x} = 2 + \dfrac{3}{x + 1}$ **8** $\dfrac{x}{2} - \dfrac{3x + 1}{3} > 2$

9 A piece of wire that is 10 feet long is cut into two pieces. The ratio of the smaller piece to the larger piece is $\frac{3}{4}$. Find the length of the larger piece.

10 Working alone, Dave can wash his car in 45 minutes. If Ellen helps him, they can do the job together in 30 minutes. How long would it take Ellen to wash the car by herself?

11 z varies directly as x and inversely as y. If $z = \frac{2}{3}$ when $x = 2$ and $y = 15$, find z when $x = 4$ and $y = 10$.

12 If the volume of a sphere varies directly as the cube of its radius and $V = 36\pi$ cubic inches when $r = 3$ inches, find V when $r = 6$ inches.

13 Use synthetic division to divide

$$2x^5 + 5x^4 - x^2 - 21x + 7 \text{ by } x + 3$$

14 **(a)** Let $p(x) = 27x^4 - 36x^3 + 18x^2 - 4x + 1$. Use the remainder theorem to evaluate $p(\frac{1}{3})$.
(b) Use the result of part (a) and the factor theorem to determine whether or not $x - \frac{1}{3}$ is a factor of $p(x)$.

15 Show that $x - 2$ is a factor of $p(x) = x^4 - 4x^3 + 7x^2 - 12x + 12$, and factor $p(x)$ completely.

16 Make use of the rational root theorem to factor $f(x) = x^4 + 5x^3 + 4x^2 - 3x + 9$.

17 Solve the system:

$$y = x^3$$
$$y - 19x = -30$$

18 Decompose $\dfrac{x - 15}{x^2 - 25}$ into partial fractions.

ANSWERS TO THE TEST YOUR UNDERSTANDING EXERCISES

Page 182

1 Vertical asymptotes: $x = 2$, $x = 3$; horizontal asymptote: $y = 0$
2 Vertical asymptote: $x = -3$; horizontal asymptote: $y = 1$
3 Vertical asymptote: $x = 2$; no horizontal asymptote
4 Vertical asymptote: $x = 4$; horizontal asymptote: $y = 0$
5 Vertical asymptotes: $x = -3$, $x = 4$; horizontal asymptote: $y = 0$
6 Vertical asmptotes: $x = 1$; no horizontal asymptote

Page 186

1 12 **2** $\frac{1}{2}$ **3** 35 **4** 5 **5** -9 **6** No solution.
7 $\frac{3}{10}$ **8** $\frac{13}{24}$ **9** 3 **10** $-4, 3$ **11** $-3, -1$ **12** $\frac{1}{2}$

Page 200

1
$$\begin{array}{r|rrrrrl} 1 & 1 & -7 & +4 & +10 & -1 & -5 \\ & & +1 & -6 & -2 & +8 & +7 \\ \hline & 1 & -6 & -2 & +8 & +7 & \lvert +2 = r \end{array}$$
$p(1) = 1 - 7 + 4 + 10 - 1 - 5 = 2$

2
$$\begin{array}{r|rrrrl} -10 & 1 & +11 & +11 & +11 & +10 \\ & & -10 & -10 & -10 & -10 \\ \hline & 1 & +1 & +1 & +1 & \lvert +0 = r \end{array}$$
$p(-10) = 10{,}000 - 11{,}000 + 1100 - 110 + 10 = 0$

3 $$\begin{array}{r|l} 10 & 1 + 11 + 11 + 11 + 10 \\ & \quad + 10 + 210 + 2210 + 22210 \\ \hline & 1 + 21 + 221 + 2221 \,|\, + 22220 = r \end{array}$$

$$p(10) = 10{,}000 + 11{,}000 + 1100 + 110 + 10 = 22{,}220$$

4 $$\begin{array}{r|l} 6 & 6 - 40 + 0 + 25 \\ & \quad + 36 - 24 - 144 \\ \hline & 6 - 4 - 24 \,|\, - 119 = r \end{array}$$

$$p(6) = 6(6)^3 - 40(6)^2 + 25 = -119$$

Page 204

1 **(a)** $\pm 1, \pm 2, \pm 3, \pm 4, \pm 6, \pm 8, \pm 12, \pm 24$

(b) $(x + 3)(x - 2)(x - 4)$

(c) $-3, 2, 4$

2 **(a)** $\pm 1, \pm 2, \pm 4, \pm 8, \pm 16$

(b) $(x + 4)^2(x - 1)^2$

(c) $-4, 1$

3 **(a)** $\pm 1, \pm \frac{1}{2}, \pm \frac{1}{4}, \pm 2, \pm 3, \pm \frac{3}{2}, \pm \frac{3}{4}, \pm 6$

(b) $(x + 6)(2x - 1)^2$

(c) $-6, \frac{1}{2}$

4 **(a)** $\pm 1, \pm \frac{1}{3}, \pm 2, \pm \frac{2}{3}, \pm 4, \pm \frac{4}{3}$

(b) $(3x - 1)(x - 4)(x^2 + 1)$

(c) $\frac{1}{3}, 4$

Page 207

1 $\dfrac{3}{x - 2} + \dfrac{5}{x - 3}$

2 $-\dfrac{\frac{1}{6}}{x + 2} + \dfrac{\frac{1}{6}}{x - 4}$

3 $\dfrac{1}{x - 1} + \dfrac{2}{x - 2} + \dfrac{3}{x - 3}$

4 $\dfrac{6}{x + 3} - \dfrac{2}{x + 2}$

5 $\dfrac{4}{3x - 1} + \dfrac{5}{2x + 1}$

6 $\dfrac{4}{x + 7} - \dfrac{2}{x - 5} + \dfrac{3}{x + 2}$

RADICAL FUNCTIONS AND EQUATIONS

6.1 Graphing some special radical functions

A radical expression in x, such as $\sqrt{x}$, may be used to define a function f, where $f(x) = \sqrt{x}$, the *square root function.* The domain of f consists of all real numbers $x \geq 0$ since the square root of a negative number is not a real number.

To graph $y = \sqrt{x}$, it is helpful to first square both sides to obtain $y^2 = x$; that is, $x = y^2$. Recall the graph of $y = x^2$ and obtain the graph of $x = y^2$ by reversing the role of the variables.

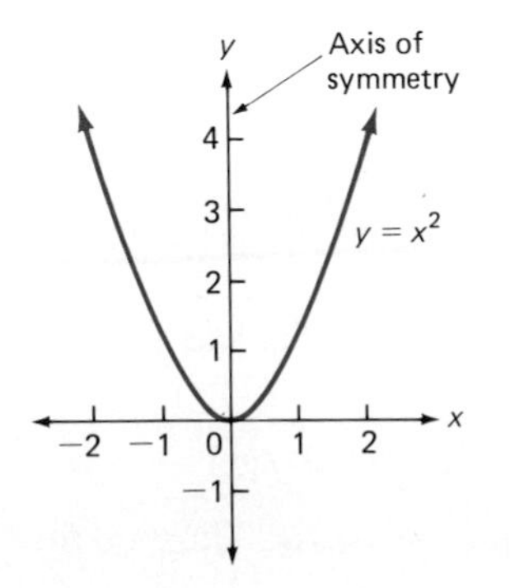

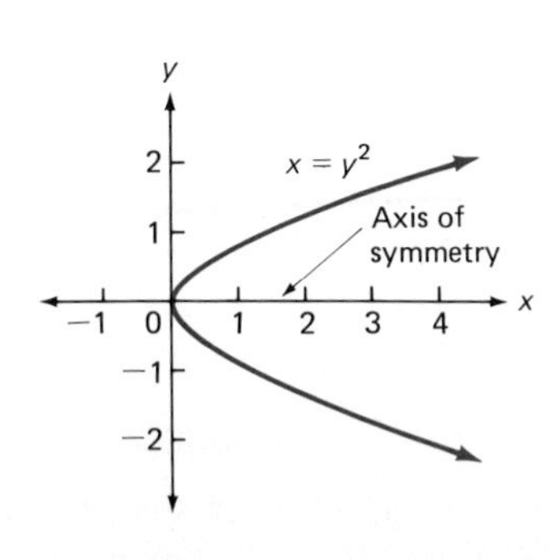

y	$x = y^2$
−2	4
−1	1
0	0
1	1
2	4

The preceding "sideways" parabola is *not* the graph of a function having x as the independent variable because for $x > 0$ there are two corresponding y-values. But if the bottom branch is removed, we have the correct graph of $y = \sqrt{x}$, for $x \geq 0$. (What is the equation for the bottom branch?)

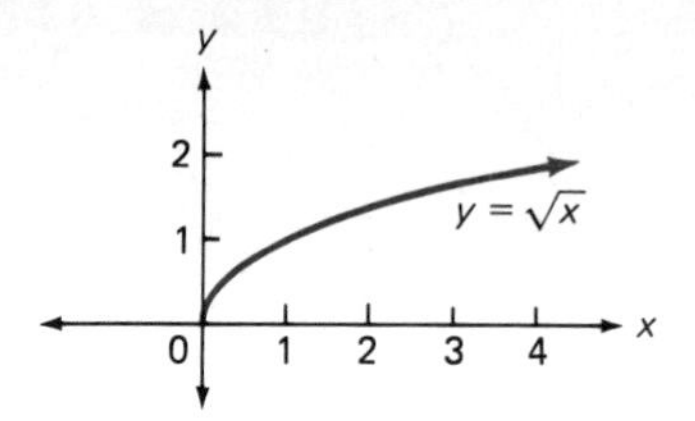

Except for the origin, this graph is in the first quadrant, as could have been predicted by observing that for all $x > 0$, $\sqrt{x} > 0$. You can also verify that the specific points given in the following table are on the graph.

Such a table of values presents us with an alternate method for sketching the graph of $y = \sqrt{x}$; plot the points in the table and connect with a smooth curve.

x	0	$\frac{1}{4}$	$\frac{9}{16}$	1	$\frac{9}{4}$	2	3	4
$y = \sqrt{x}$	0	$\frac{1}{2}$	$\frac{3}{4}$	1	$\frac{3}{2}$	$\sqrt{2}$	$\sqrt{3}$	2

EXAMPLE 1 Find the domain of $y = g(x) = \sqrt{x - 2}$ and graph.

Solution Since the square root of a negative number is not a real number, the expression $x - 2$ must be nonnegative; therefore, the domain of g consists of all $x \geq 2$. The graph of g may be found by shifting the graph of $y = \sqrt{x}$ by 2 units to the right.

Note: $x - 2 \geq 0$, thus $x \geq 2$.

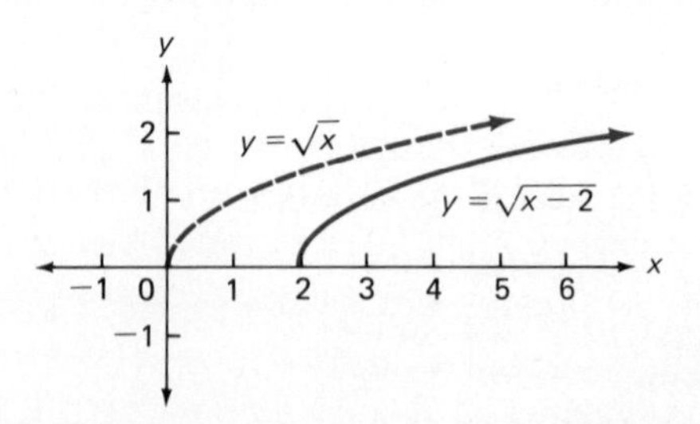

EXAMPLE 2 Find the domain of $y = f(x) = \sqrt{|x|}$ and graph.

Solution Since $|x| \geq 0$ for all x, the domain of f consists of all real numbers. To graph f, first note that

It is also helpful to locate a few specific points on the curve as an aid to graphing.

$$f(-x) = \sqrt{|-x|} = \sqrt{|x|} = f(x)$$

Therefore, the graph is symmetric about the y-axis. Thus we first find the graph for $x \geq 0$ and use symmetry to obtain the rest. For $x \geq 0$, we get $|x| = x$ and $f(x) = \sqrt{|x|} = \sqrt{x}$.

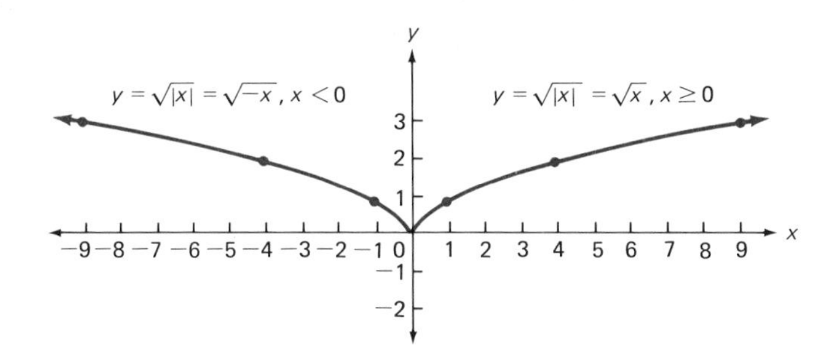

x	y
0	0
1	1
4	2
9	3
−1	1
−4	2
−9	3

EXAMPLE 3 Find the domain of $y = h(x) = x^{-1/2}$ and graph by using a table of values.

Solution Note that $h(x) = x^{-1/2} = \dfrac{1}{x^{1/2}} = \dfrac{1}{\sqrt{x}}$. Thus the domain consists of all $x > 0$. Furthermore, $\dfrac{1}{\sqrt{x}} > 0$ for all x, so we know that the graph must be in the first quadrant only. Plot the points in the table and connect with a smooth curve.

Explain why $x \neq 0$ in Example 3.

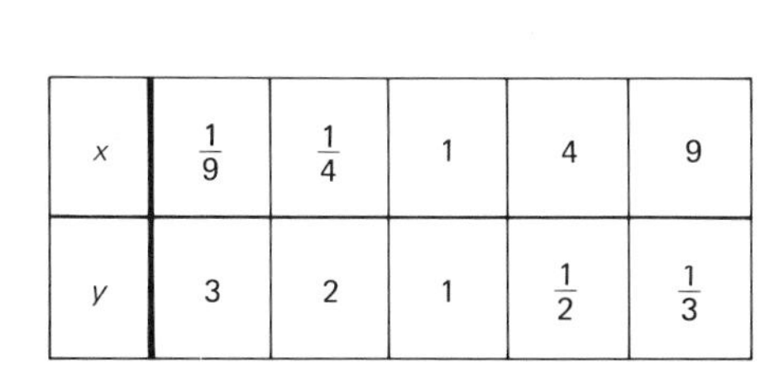

x	$\frac{1}{9}$	$\frac{1}{4}$	1	4	9
y	3	2	1	$\frac{1}{2}$	$\frac{1}{3}$

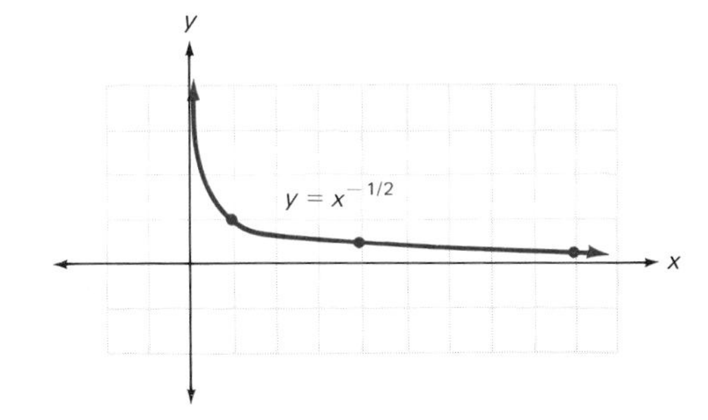

Note that the closer x is to zero, the larger are the corresponding y-values. Also, as the values of x get larger, the corresponding y-values get closer to 0. These observations suggest that the coordinate axes are asymptotes to the curve $y = x^{-1/2}$.

TEST YOUR UNDERSTANDING

Graph each pair of functions on the same set of axes and state the domain of g.

1 $f(x) = \sqrt{x}$, $g(x) = \sqrt{x} - 2$

2 $f(x) = -\sqrt{x}$, $g(x) = -\sqrt{x - 1}$

3 $f(x) = \sqrt{x}$, $g(x) = 2\sqrt{x}$

4 $f(x) = \dfrac{1}{\sqrt{x}}$, $g(x) = \dfrac{1}{\sqrt{x + 2}}$

The graph of the *cube root function* $y = \sqrt[3]{x}$ can be found by a process similar to that used for $y = \sqrt{x}$. First take $y = \sqrt[3]{x}$ and cube both sides to get $y^3 = x$. Now recall the graph of $y = x^3$ and obtain the graph of $x = y^3$ by reversing the role of the variables.

y	$x = y^3$
−2	−8
−1	−1
0	0
1	1
2	8

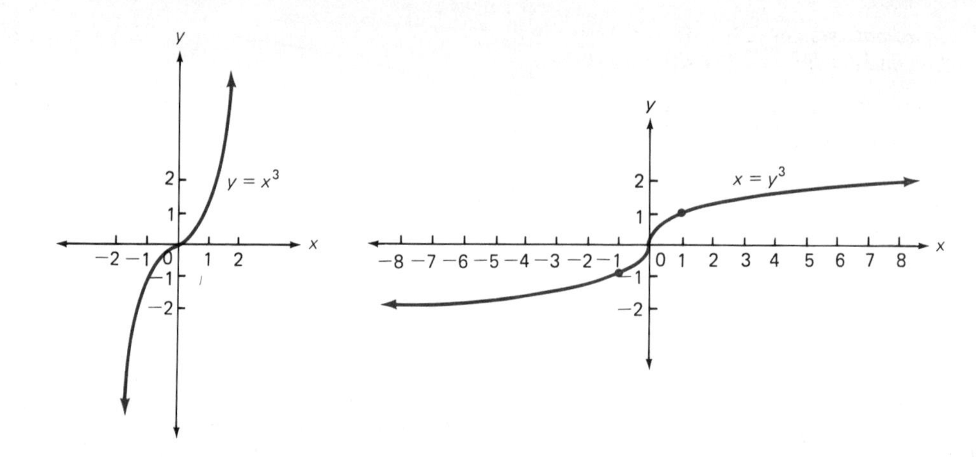

The next example illustrates that a function can be defined by using more than one expression.

EXAMPLE 4 Graph f defined on the domain $-2 \leq x < 5$ as follows:

$$f(x) = \begin{cases} x^2 - 1 & \text{for } -2 \leq x < 2 \\ \sqrt{x - 2} & \text{for } 2 \leq x < 5 \end{cases}$$

Solution The first part of f is given by $f(x) = x^2 - 1$ for $2 \leq x < 2$. This is an arc of a parabola obtained by shifting the graph of $y = x^2$ downward 1 unit.

The second part of f is given by $f(x) = \sqrt{x - 2}$ for $2 \leq x < 5$. This is an arc of the square root curve obtained by shifting the graph of $y = \sqrt{x}$ two units to the right.

When $x = 2$, the radical part of f is used; $f(2) = \sqrt{2 - 2} = 0$. So there is a solid dot at (2, 0) and an open dot at (2, 3). Also, there is an open dot for $x = 5$ since 5 is not a domain value of f.

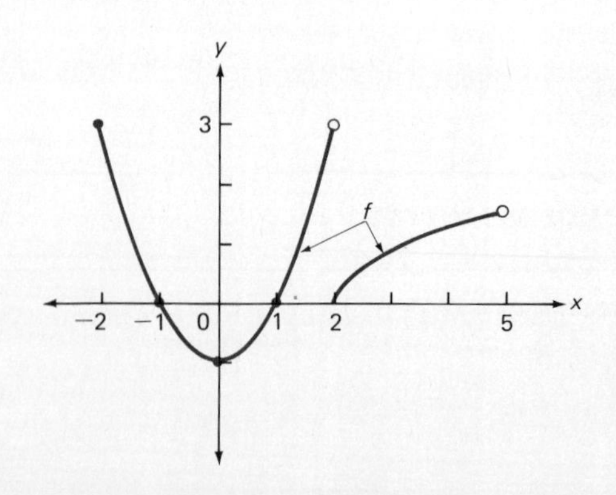

In Exercises 1–4, graph each set of curves on the same coordinate system. Use a dashed curve for the first equation and a solid curve for the second.

1 $f(x) = \sqrt{x}$, $g(x) = \sqrt{x-1}$

2 $f(x) = -\sqrt{x}$, $g(x) = -2\sqrt{x}$

3 $f(x) = \sqrt[3]{x}$, $g(x) = \sqrt[3]{x+2}$

4 $f(x) = -\sqrt[3]{x}$, $g(x) = -\sqrt[3]{x-3}$

In Exercises 5–16, find the domain of f, sketch the graph, and give the equations of the asymptotic lines if there are any.

5 $f(x) = \sqrt{x+2}$

6 $f(x) = x^{1/2} + 2$

7 $f(x) = \sqrt{x-3} - 1$

8 $f(x) = -\sqrt{x}$

9 $f(x) = \sqrt{-x}$

10 $f(x) = \sqrt{(x-2)^2}$

11 $f(x) = 2\sqrt[3]{x}$

12 $f(x) = |\sqrt[3]{x}|$

13 $f(x) = -x^{1/3}$

14 $f(x) = \sqrt[3]{-x}$

15 $f(x) = \dfrac{1}{\sqrt{x}} - 1$

16 $f(x) = \dfrac{1}{\sqrt{x-2}}$

17 (a) Explain why the graph of $f(x) = \dfrac{1}{\sqrt[3]{x}}$ is symmetric through the origin.
(b) What is the domain of f?
(c) Use a table of values to graph f.
(d) What are the equations of the asymptotes?

18 Find the domain of $f(x) = \dfrac{1}{\sqrt[3]{x+1}}$, sketch the graph, and give the equations of the asymptotes.

*19 Find the graph of the function $y = \sqrt[4]{x}$ by raising both sides of the equation to the fourth power and comparing to the graph of $y = x^4$.

*20 Reflect the graph of $y = x^2$, for $x \geq 0$, through the line $y = x$. Obtain the equation of this new curve by interchanging variables in $y = x^2$ and solving for y.

*21 Follow the instructions of Exercise 20 with $y = x^3$ for all values x.

In Exercises 22 and 23, the function f is defined by using more than one expression. Graph f on its given domain.

22 $f(x) = \begin{cases} \sqrt{x} & \text{for } 0 \leq x \leq 4 \\ 10 - \frac{1}{2}x^2 & \text{for } 4 < x < 6 \end{cases}$

23 $f(x) = \begin{cases} -2x - 1 & \text{for } -3 \leq x < 0 \\ \sqrt[3]{x-1} & \text{for } 0 \leq x \leq 2 \end{cases}$

■ *In Exercises 24–26, verify the equation involving the difference quotient for the given radical function (see Section 3.1).*

24 $\dfrac{f(x) - f(25)}{x - 25} = \dfrac{1}{\sqrt{x} + 5}$; $f(x) = \sqrt{x}$
(factor $x - 25$ as the difference of squares)

25 $\dfrac{f(4+h) - f(4)}{h} = \dfrac{1}{\sqrt{4+h} + 2}$; $f(x) = \sqrt{x}$
(rationalize the numerator)

26 $\dfrac{f(x) - f(9)}{x - 9} = -\dfrac{1}{\sqrt{x} + 3}$; $f(x) = -\sqrt{x}$

■ *Exercises 27–30 call for the extraction of radical functions from geometric situations (see Section 3.8).*

27 A runner starts at point A, goes to point P that is x miles from B, and then runs to D.

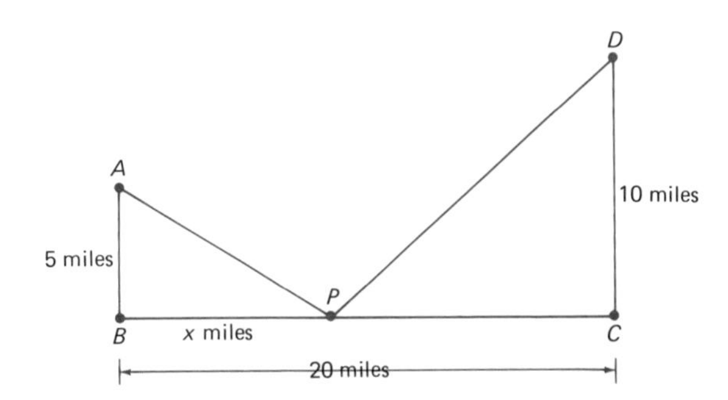

(a) Write the total distance d traveled as a function of x.
(b) The runner averages 12 miles per hour from A to P and 10 miles per hour from P to D. Write the time t for the trip as a function of x. (*Hint:* Use time = distance/rate.)

28 In the figure on the next page, AC is along the shoreline of a lake, and the distance from A to C is 12 miles. P represents the starting point of a swimmer who swims at 3 miles per hour along the hypotenuse PB. P is 5 miles from point A on the shore-line. After reaching B he walks at 6 miles per hour to C. Express the total time t of the trip as a function of x, where x is the distance from A to B. (*Hint:* Use time = distance/rate.)

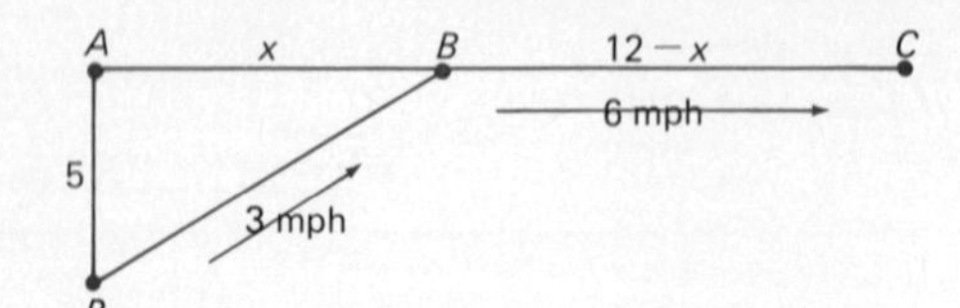

29 Express the distance d from the origin to a point (x, y) on the line $3x + y = 6$ as a function of x.

30 Express the distance d from the point (2, 5) to a point (x, y) on the line $3x + y = 6$ as a function of x.

6.2

Radical equations

To graph $y = f(x) = \sqrt[3]{x+1} - 2$, it is helpful to know the intercepts. The y-intercept is easy to find:

$$f(0) = \sqrt[3]{0+1} - 2 = 1 - 2 = -1$$

Finding the x-intercepts for this graph, as well as for many others, calls for more involved techniques. Since the x-intercepts are values for which $y = f(x) = 0$, we need to solve this equation:

$$\sqrt[3]{x+1} - 2 = 0$$

First isolate the radical by adding 2 to each side.

$$\sqrt[3]{x+1} = 2$$

Cube both sides and solve for x.

$$(\sqrt[3]{x+1})^3 = 2^3$$
$$x + 1 = 8$$
$$x = 7$$

Check this result by substituting into the original equation.

$$\sqrt[3]{7+1} - 2 = \sqrt[3]{8} - 2 = 2 - 2 = 0$$

The solution is $x = 7$ and the graph of $y = \sqrt[3]{x+1} - 2$ crosses the x-axis at (7, 0).

EXAMPLE 1 Find the x-intercepts and the domain of $f(x) = \sqrt{x^2 + x - 6}$.

Solution The x-intercepts occur when $f(x) = 0$. Let $\sqrt{x^2 + x - 6} = 0$ and solve for x.

Check: for $x = -3$, $\sqrt{(-3)^2 + (-3) - 6} = \sqrt{9 - 3 - 6} = \sqrt{0} = 0$; for $x = 2$, $\sqrt{2^2 + 2 - 6} = \sqrt{0} = 0$.

$$(\sqrt{x^2 + x - 6})^2 = (0)^2 \longleftarrow \text{square both sides}$$
$$x^2 + x - 6 = 0$$
$$(x + 3)(x - 2) = 0$$
$$x = -3 \quad \text{or} \quad x = 2$$

Therefore, the x-intercepts are -3 and 2.

The domain of f is the solution of the inequality $(x + 3)(x - 2) \geq 0$, which consists of the intervals $(-\infty, -3]$ and $[2, \infty)$.

You can solve this inequality by forming a table of signs or by noting where the parabola $y = x^2 + x - 6$ is above the x-axis.

For the remainder of this section we focus primarily on the solutions of radical equations in one variable. Keep in mind, however, that the solutions of $f(x) = 0$ can be regarded as the x-intercepts of the graph of the related equation in two variables, $y = f(x)$.

The solution in Example 1 made use of this principle:

If $a = b$, then $a^n = b^n$

This statement says that every solution of $a = b$ will also be a solution of $a^n = b^n$. Sometimes it is convenient to apply this principle after other changes have been made. For instance, to solve the equation

$$\sqrt{x+4} + 2 = x$$

first isolate the radical on one side of the equation.

$$\sqrt{x+4} = x - 2$$

Note: If the radical in $\sqrt{x+4} + 2 = x$ is not first isolated, it is still possible to solve the equation, but the work will be more involved. Try it. Also observe that $x = 5$ is the x-intercept of the curve $y = \sqrt{x+4} + 2 - x$.

Now square and solve for x.

$$(\sqrt{x+4})^2 = (x-2)^2$$
$$x + 4 = x^2 - 4x + 4$$
$$0 = x^2 - 5x$$
$$0 = x(x-5)$$

Thus we get $x = 0$ or $x = 5$. Let us check these values in the given equation.

For $x = 0$: $\sqrt{0+4} + 2 = \sqrt{4} + 2 = 2 + 2 = 4 \neq 0$;
0 is *not* a solution.

For $x = 5$: $\sqrt{5+4} + 2 = \sqrt{9} + 2 = 3 + 2 = 5$;
5 is the *only* solution.

How did the *extraneous* solution zero arise? In going from

$$\sqrt{x+4} = x - 2 \qquad \text{to} \qquad (\sqrt{x+4})^2 = (x-2)^2$$

we used the basic principle: If $a = b$, then $a^n = b^n$. Therefore, every solution of the first equation is also a solution for the second. But this principle is not always reversible. In particular, both 0 and 5 are solutions of the second equation, but only 5 is a solution of the first. In summary, the process of raising both sides of an equation in x to the nth power can introduce false solutions. It is therefore vital to check all possible solutions in the original equation.

TEST YOUR UNDERSTANDING

Find the x-intercepts and the domain of f.

1 $f(x) = \sqrt{x} - 5$

2 $f(x) = \dfrac{1}{\sqrt{x+1}} - \dfrac{1}{3}$

3 $f(x) = \sqrt{x^2 - 3x}$

Solve each equation.

4 $\sqrt{x+2} = \sqrt{2x-5}$

5 $\sqrt{x^2+9} = -5$

6 $\dfrac{1}{\sqrt{x}} = 3$

7 $\dfrac{4}{\sqrt{x-1}} = 2$

8 $\sqrt{x^2-5} = 2$

9 $\sqrt[3]{x+3} - 2 = 0$

10 $\sqrt[4]{2x-1} = 3$

11 $\sqrt{x+16} - x = 4$

12 $2x = 1 + \sqrt{1-2x}$

Radical equations may contain more than one radical. For such cases it is usually best to transform the equation first into one with as few radicals on each side as possible. For example:

$$\sqrt{x-7} - \sqrt{x} = 1$$

$$\sqrt{x-7} = \sqrt{x} + 1$$

CAUTION: $(\sqrt{x}+1)^2 \neq x^2+1$. Use $(a+b)^2 = a^2+2ab+b^2$ with $a = \sqrt{x}$ and $b = 1$.

Now square each side.

$$(\sqrt{x-7})^2 = (\sqrt{x}+1)^2$$

$$x - 7 = x + 2\sqrt{x} + 1$$

$$-8 = 2\sqrt{x}$$

$$-4 = \sqrt{x}$$

Square again:

$$16 = x$$

The radical signs may be replaced by appropriate fractional exponents and solved by the same methods; see Example 2.

Check: $\sqrt{16-7} - \sqrt{16} = \sqrt{9} - \sqrt{16} = 3 - 4 = -1 \neq 1$. Therefore, this equation has no solution, which could have been observed at an earlier stage as well. For example, $-4 = \sqrt{x}$ has no solution. (Why not?)

EXAMPLE 2 Solve for x:

$$\frac{3(2x-1)^{1/2}}{x-3} - \frac{(2x-1)^{3/2}}{(x-3)^2} = 0$$

Solution Multiply through by $(x-3)^2$ to clear fractions.

$$(x-3)^2\left[\frac{3(2x-1)^{1/2}}{x-3}\right] - (x-3)^2\left[\frac{(2x-1)^{3/2}}{(x-3)^2}\right] = (x-3)^2(0)$$

$$3(x-3)(2x-1)^{1/2} - (2x-1)^{3/2} = 0$$

Factor out $(2x-1)^{1/2}$. This is easier to see if we let $a = (2x-1)^{1/2}$ and factor a out of $3(x-3)a - a^3$.

$$[3(x-3) - (2x-1)](2x-1)^{1/2} = 0$$

$$(x-8)(2x-1)^{1/2} = 0$$

Set each factor equal to zero.

$$x - 8 = 0 \quad \text{or} \quad (2x-1)^{1/2} = 0$$

$$x = 8 \quad \text{or} \quad 2x - 1 = 0$$

$$x = 8 \quad \text{or} \quad x = \tfrac{1}{2}$$

Check in the original equation to show that both $x = \frac{1}{2}$ and $x = 8$ are solutions for Example 2.

As you may have noticed, the algebraic techniques developed earlier involving other kinds of expressions often carry over into this work. Here is an example that uses our knowledge of quadratics in conjunction with radicals.

EXAMPLE 3 Solve for x: $\sqrt[3]{x^2} + \sqrt[3]{x} - 20 = 0$.

Solution First write the equation by using rational exponents.

$$x^{2/3} + x^{1/3} - 20 = 0$$

Then think of $x^{2/3}$ as the square of $x^{1/3}$, $x^{2/3} = (x^{1/3})^2$, and use the substitution $u = x^{1/3}$ as follows:

$$x^{2/3} + x^{1/3} - 20 = 0$$
$$(x^{1/3})^2 + x^{1/3} - 20 = 0$$
$$u^2 + u - 20 = 0$$
$$(u + 5)(u - 4) = 0 \quad \text{(factoring the quadratic)}$$
$$u + 5 = 0 \quad \text{or} \quad u - 4 = 0$$
$$u = -5 \quad \text{or} \quad u = 4$$

Alternatively, we may keep the radical sign and proceed with $u = \sqrt[3]{x}$.

Now replace u by $x^{1/3}$.

$$x^{1/3} = -5 \quad \text{or} \quad x^{1/3} = 4$$

Cubing gives

$$x = -125 \quad \text{or} \quad x = 64$$

Check to show that both values are solutions of the given equation.

EXAMPLE 4 Solve the system and graph:

$$y = \sqrt[3]{x}$$
$$y = \tfrac{1}{4}x$$

Solution For the points of intersection the x- and y-values are the same in both equations. Thus

$$\tfrac{1}{4}x = \sqrt[3]{x} \longleftarrow x^{1/3} \text{ can be used in place of } \sqrt[3]{x}$$

Cube both sides and solve for x.

$$\tfrac{1}{64}x^3 = x$$
$$x^3 = 64x$$
$$x^3 - 64x = 0$$
$$x(x^2 - 64) = 0 \quad \text{(factoring out } x\text{)}$$
$$x(x + 8)(x - 8) = 0$$
$$x = 0 \quad \text{or} \quad x = -8 \quad \text{or} \quad x = 8$$

CAUTION: A common error is to take $x^3 - 64x = 0$ and divide through by x to get $x^2 - 64 = 0$. This step produces the roots ± 8. The root 0 has been lost because we divided by x, and 0 is the number for which the factor x in $x(x^2 - 64)$ is zero. You may always divide by a nonzero constant and get an equivalent form of the equation. But when you divide by a variable quantity there is the danger of losing some roots, those for which the divisor is 0.

Substitute these values into either of the given equations to obtain the corresponding y-values. The remaining equation can be used for checking. The solutions are $(-8, -2)$, $(0, 0)$, and $(8, 2)$.

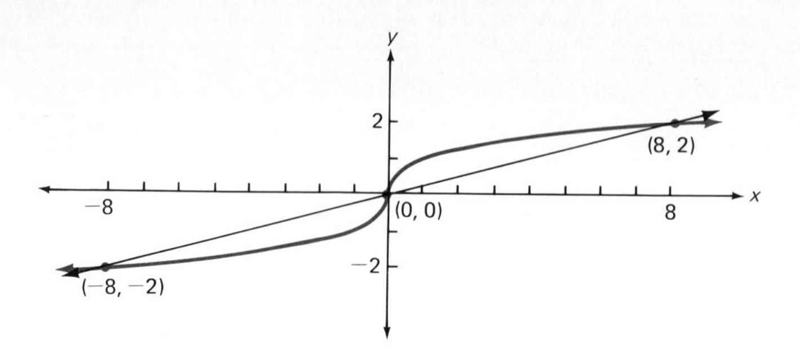

EXERCISES 6.2

In Exercises 1–6, find the x-intercepts of each curve and find the domain of f.

1 $f(x) = \sqrt{x-1} - 4$

2 $f(x) = \sqrt{3x-2} - 5$

3 $f(x) = \sqrt{x^2 + 2x}$

4 $f(x) = \sqrt{x - x^2}$

5 $f(x) = \sqrt{x^2 - 5x - 6}$

6 $f(x) = \sqrt{2x^2 - x - 2} - 1$

Solve each equation in Exercises 7–30.

7 $\sqrt{4x+9} - 7 = 0$

8 $2 + \sqrt{7x-3} = 7$

9 $(3x+1)^{1/2} = (2x+6)^{1/2}$

10 $\sqrt{x-1} - \sqrt{2x-11} = 0$

11 $\sqrt{x^2 - 36} = 8$

12 $3x = \sqrt{3 - 5x - 3x^2}$

13 $\sqrt{x^2 + \frac{1}{2}} = \frac{1}{\sqrt{3}}$

14 $3\sqrt{x} = 2\sqrt{3}$

15 $\frac{8}{\sqrt{x+2}} = 4$

16 $\left(\frac{5x+4}{2}\right)^{1/3} = 3$

17 $\frac{1}{\sqrt{2x-1}} = \frac{3}{\sqrt{5-3x}}$

18 $\sqrt[3]{2x+7} = 3$

19 $\sqrt[4]{1-3x} = \frac{1}{2}$

20 $2 + \sqrt[5]{7x-4} = 0$

21 $\sqrt{x} + \sqrt{x-5} = 5$

22 $\sqrt{x-7} = 7 - \sqrt{x}$

23 $x\sqrt{4-x} - \sqrt{9x-36} = 0$

24 $2x - 5\sqrt{x} - 3 = 0$

25 $x = 8 - 2\sqrt{x}$

26 $\frac{(2x+2)^{3/2}}{(x+9)^2} - \frac{(2x+2)^{1/2}}{x+9} = 0$

27 $\sqrt{x^2 - 6x} = x - \sqrt{2x}$

28 $x^{1/3} - 3x^{1/6} + 2 = 0$

29 $4x^{2/3} - 12x^{1/3} + 9 = 0$

30 $\sqrt[4]{3x^2 + 4} = x$

In Exercises 31 and 32 solve the system and then sketch each graph.

31 $y = \sqrt{x}$
$y = \frac{1}{2}x$

32 $y = \frac{2}{\sqrt{x}}$
$x + 3y = 7$

In Exercises 33–36, solve for the indicated variable in terms of the others.

33 $y = \sqrt{16 - x^2}$, for x, $0 < x < 4$

34 $x = \frac{3}{5}\sqrt{25 - y^2}$, for y, $0 < y < 5$

35 $\frac{1}{2\sqrt{x}} + \frac{f}{2\sqrt{y}} = 0$, for f

36 $\frac{1}{\sqrt{xy}}(xf + y) = f$, for f

37 The distance from the point $(3, 0)$ to a point $P(x, y)$ on the curve $y = \sqrt{x}$ is $3\sqrt{5}$. Find the coordinates of P. (*Hint:* Use the formula for the distance between two points.)

38 In the figure, rectangle $ABCD$ is inscribed in a circle of radius 10. Express the area of the rectangle as a function of x.

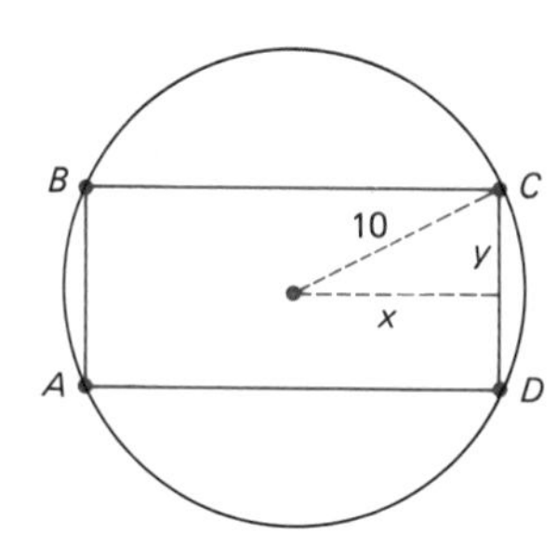

(*Hint:* Solve for y in terms of x.)

39 The volume of a right circular cylinder with altitude h and radius r is 5 cm^3. Solve for r in terms of h and express the surface area of the cylinder as a function of h.

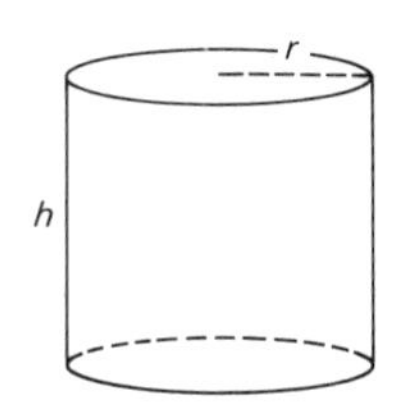

40 A right circular cone with height h and radius r is inscribed in a sphere of radius 1 as shown. Solve for r in terms of h and express the volume of the cone as a function of h.

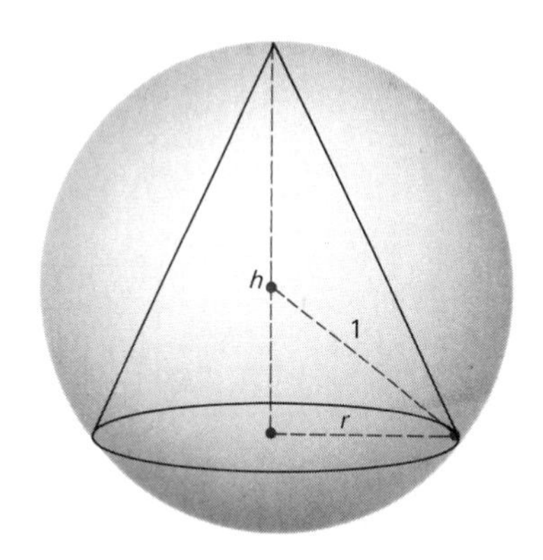

6.3 Determining the signs of radical functions

The methods used for determining the signs of rational functions are easily modified to determine the signs of functions defined in terms of radical expressions. As a first example, consider this function:

$$f(x) = (x - 5)\sqrt{x - 3}$$

Observe that the factor $\sqrt{x - 3}$ indicates $x \geq 3$. (Why?) Therefore, we do not need to consider any values for $x < 3$. The table shows that $f(x)$ is negative on the interval $(3, 5)$ and positive on the interval $(5, \infty)$.

Interval	$(3, 5)$	$(5, \infty)$
Sign of $\sqrt{x - 3}$	+	+
Sign of $x - 5$	−	+
Sign of $f(x)$	−	+

$(3, 5)$ is the interval consisting of all x where $3 < x < 5$.
$(5, \infty)$ is the interval consisting of all x where $x > 5$.

EXAMPLE 1 Find the domain of $f(x) = \dfrac{x}{\sqrt{4 - x^2}}$ and determine its signs.

Solution The domain consists of those x for which $4 - x^2 > 0$. To solve this inequality we note that $y = 4 - x^2$ is a parabola that opens downward and crosses the x-axis at ± 2. Hence $4 - x^2 > 0$ on the interval $(-2, 2)$, which is the domain of f. The sign of $f(x)$ depends only on the numerator since the denominator is always positive. Hence $f(x) > 0$ on $(0, 2)$, and $f(x) < 0$ on $(-2, 0)$.

TEST YOUR UNDERSTANDING

Find the domain of each function and determine its signs.

1 $f(x) = \dfrac{x - 4}{\sqrt{x}}$ **2** $f(x) = \dfrac{\sqrt{x + 2}}{x}$ **3** $f(x) = \dfrac{2}{3\sqrt[3]{x}}$

Algebraic procedures developed earlier can be helpful with this type of work. For example, the signs of

$$f(x) = x^{1/2} + \tfrac{1}{2}x^{-1/2}(x - 3)$$

can be determined after combining and simplifying as follows:

$$\begin{aligned} f(x) &= x^{1/2} + \tfrac{1}{2}x^{-1/2}(x - 3) \\ &= x^{1/2} + \frac{x - 3}{2x^{1/2}} \\ &= \frac{x^{1/2}(2x^{1/2})}{2x^{1/2}} + \frac{x - 3}{2x^{1/2}} \qquad \text{(the common denominator is } 2x^{1/2}) \\ &= \frac{2x + x - 3}{2x^{1/2}} \\ &= \frac{3x - 3}{2x^{1/2}} = \frac{3(x - 1)}{2\sqrt{x}} \end{aligned}$$

$(0, 1)$: $0 < x < 1$
$(1, \infty)$: $x > 1$

Interval	$(0, 1)$	$(1, \infty)$
Sign of $\sqrt{x}$	+	+
Sign of $x - 1$	−	+
Sign of $f(x)$	−	+

Hence $f(x) < 0$ on $(0, 1)$; $f(x) > 0$ on $(1, \infty)$.

EXAMPLE 2 Find the domain of f, the roots of $f(x) = 0$, and determine the signs of f where

$$f(x) = \frac{x^2 - 1}{2\sqrt{x - 1}} + 2x\sqrt{x - 1}$$

Solution The radical in the denominator calls for $x - 1 > 0$. Therefore, the domain consists of all $x > 1$. Now simplify as follows:

$$\begin{aligned} f(x) &= \frac{x^2 - 1}{2\sqrt{x-1}} + 2x\sqrt{x-1} \\ &= \frac{x^2 - 1}{2\sqrt{x-1}} + \frac{(2x\sqrt{x-1})(2\sqrt{x-1})}{2\sqrt{x-1}} \\ &= \frac{x^2 - 1 + 4x(x-1)}{2\sqrt{x-1}} \\ &= \frac{5x^2 - 4x - 1}{2\sqrt{x-1}} \\ &= \frac{(5x+1)(x-1)}{2\sqrt{x-1}} \\ &= \frac{5(x+\frac{1}{5})(x-1)}{2\sqrt{x-1}} \end{aligned}$$

The numerator is zero when $x = -\frac{1}{5}$ or $x = 1$. But these values are *not* in the domain of f. Since a fraction can only be zero when the numerator is zero, it follows that $f(x) = 0$ has no solutions. The signs of $f(x)$ are given in this brief table:

Interval	$(1, \infty)$
Sign of $x + \frac{1}{5}$	+
Sign of $x - 1$	+
Sign of $\sqrt{x-1}$	+
Sign of $f(x)$	+

$(1, \infty)$: $x > 1$

Thus $f(x) > 0$ on its domain $(1, \infty)$.

CAUTION: LEARN TO AVOID MISTAKES LIKE THESE

WRONG	RIGHT
$(9 - x^2)^{1/2}(9 - x^2)^{3/2} = (9 - x^2)^{3/4}$	$(9 - x^2)^{1/2}(9 - x^2)^{3/2} = (9 - x^2)^{(1/2)+(3/2)} = (9 - x^2)^2$
$\frac{2}{\sqrt{9-x^2}} + 3\sqrt{9-x^2} = \frac{2 + 3\sqrt{9-x^2}}{\sqrt{9-x^2}}$	$\frac{2}{\sqrt{9-x^2}} + 3\sqrt{9-x^2} = \frac{2}{\sqrt{9-x^2}} + \frac{3(9-x^2)}{\sqrt{9-x^2}} = \frac{2 + 3(9-x^2)}{\sqrt{9-x^2}}$

WRONG	RIGHT
$x^{-1/3} + x^{2/3} = x^{-1/3}(1 + x^{1/3})$	$x^{-1/3} + x^{2/3} = x^{-1/3}(1 + x)$
$x^2\sqrt{1+x} = \sqrt{x^2 + x^3}$	$x^2\sqrt{1+x} = \sqrt{x^4}\sqrt{1+x}$ $= \sqrt{x^4 + x^5}$

EXERCISES 6.3

In Exercises 1–10, (a) find the domain of f, (b) determine the signs of f, and (c) find the roots of f(x) = 0.

1 $f(x) = (x+4)\sqrt{x} - 2$

2 $f(x) = (x+4)\sqrt[3]{x} - 2$

3 $f(x) = (x-2)^{2/3}$

4 $f(x) = \frac{2}{3}(x-2)^{-1/3}$

5 $f(x) = \dfrac{9+x}{9\sqrt{x}}$

6 $f(x) = \dfrac{9-x}{9x^{3/2}}$

7 $f(x) = \dfrac{(x+4)\sqrt[3]{x} - 2}{3\sqrt[3]{x}}$

8 $f(x) = \dfrac{-5x(x-4)}{2\sqrt{5-x}}$

9 $f(x) = \dfrac{\sqrt{9-x^2}}{x}$

10 $f(x) = \dfrac{x}{\sqrt{x^2-4}}$

In Exercises 11–18, take the expression at the left and change it into the equivalent form given at the right.

11 $x^{1/2} + (x-4)\frac{1}{2}x^{-1/2}$; $\dfrac{3x-4}{2\sqrt{x}}$

12 $x^{1/3} + (x-1)\frac{1}{3}x^{-2/3}$; $\dfrac{4x-1}{3\sqrt[3]{x^2}}$

13 $\frac{1}{2}(4-x^2)^{-1/2}(-2x)$; $-\dfrac{x}{\sqrt{4-x^2}}$

14 $\dfrac{-\frac{1}{2}(25-x^2)^{-1/2}(-2x)}{25-x^2}$; $\dfrac{x}{\sqrt{(25-x^2)^3}}$

15 $\dfrac{x}{3(x-1)^{2/3}} + (x-1)^{1/3}$; $\dfrac{4x-3}{3(\sqrt[3]{x-1})^2}$

16 $(x+1)^{1/2}(2) + (2x+1)\frac{1}{2}(x+1)^{-1/2}$; $\dfrac{6x+5}{2\sqrt{x+1}}$

17 $\dfrac{x}{(5x-6)^{4/5}} + (5x-6)^{1/5}$; $\dfrac{6(x-1)}{(5x-6)^{4/5}}$

18 $x^{-3/2} - \dfrac{1}{9}x^{-1/2}$; $\dfrac{9-x}{9x^{3/2}}$

Simplify and determine the signs of f.

19 $f(x) = \sqrt{x} - \dfrac{1}{\sqrt{x}}$

20 $f(x) = \frac{3}{2}x^{1/2} - \frac{3}{2}x^{-1/2}$

21 $f(x) = x^{1/2} + \frac{1}{2}x^{-1/2}(x-9)$

22 $f(x) = x^{2/3} + \frac{2}{3}x^{-1/3}(x-10)$

23 $f(x) = \dfrac{x^2}{2}(x-2)^{-1/2} + 2x(x-2)^{1/2}$

24 $f(x) = \dfrac{\sqrt{x-1} - \dfrac{x}{2\sqrt{x-1}}}{x-1}$

6.4

Combining functions

What does it mean to add two functions f and g? Remember, a function is basically a correspondence; so how do we add correspondences? The answer turns out to be relatively simple.

To illustrate, let us use f and g, where

$$f(x) = \frac{1}{x^3-1} \quad \text{and} \quad g(x) = \sqrt{x}$$

The domain of f consists of all $x \neq 1$, and g has domain all $x \geq 0$. The sum of f and g will be symbolized by $f + g$. We will take as the domain of $f + g$ all values x that are in *each* of the domains of f and g simultaneously. Therefore $f + g$ has domain $x \geq 0$ and $x \neq 1$.

It might be helpful if you momentarily used $f \oplus g$ to emphasize that this plus is not our old familiar addition.

Now consider any x in this new domain. What will be the corresponding range value for $f + g$? That is, how should $(f + g)(x)$ be defined? The answer is almost instinctive. For this x, take the two range values $f(x) = \frac{1}{x^3 - 1}$ and $g(x) = \sqrt{x}$, and add them to get $f(x) + g(x)$. Thus

$$(f + g)(x) = f(x) + g(x) = \frac{1}{x^3 - 1} + \sqrt{x}$$

Note that the plus sign in $f(x) + g(x)$ is addition of real numbers.

Specifically, for $x = 4$ we have

$$(f + g)(4) = f(4) + g(4) = \frac{1}{64 - 1} + \sqrt{4} = \frac{1}{63} + 2 = \frac{127}{63}$$

The difference, product, and quotient are found in a similar manner. Using f and g above, we get

$$(f - g)(x) = f(x) - g(x) = \frac{1}{x^3 - 1} - \sqrt{x}$$

$$(f \cdot g)(x) = f(x)g(x) = \frac{1}{x^3 - 1} \cdot \sqrt{x} = \frac{\sqrt{x}}{x^3 - 1}$$

$$\frac{f}{g}(x) = \frac{f(x)}{g(x)} = \frac{\frac{1}{x^3 - 1}}{\sqrt{x}} = \frac{1}{(x^3 - 1)\sqrt{x}}$$

The domains of $f - g$ and $f \cdot g$ are the same as for $f + g$: namely, $x \geq 0$ and $x \neq 1$. The domain of $\frac{f}{g}$ also has those x common to the domains of f and g except those for which $g(x) = 0$; $x > 0$ and $x \neq 1$.

EXAMPLE 1 Let $f(x) = \frac{1}{x^3 - 1}$ and $g(x) = \sqrt{x}$.

(a) Evaluate $f(4)$, and $g(4)$, and compute $f(4) \cdot g(4)$.

(b) Use the expression for $(f \cdot g)(x)$, as in the preceding discussion, to evaluate $(f \cdot g)(4)$.

Solution

(a) $f(4) = \frac{1}{4^3 - 1} = \frac{1}{63}$, $g(4) = \sqrt{4} = 2$, and $f(4) \cdot g(4) = \frac{1}{63} \cdot 2 = \frac{2}{63}$

(b) $(f \cdot g)(x) = \frac{\sqrt{x}}{x^3 - 1}$

$(f \cdot g)(4) = \frac{\sqrt{4}}{4^3 - 1} = \frac{2}{63}$

We are ready to state the general definition for forming the sum, difference, product, and quotient of two functions.

For functions f and g, the functions $f+g$, $f-g$, $f \cdot g$, and $\frac{f}{g}$ have range values given by

$$(f+g)(x) = f(x) + g(x)$$
$$(f-g)(x) = f(x) - g(x)$$
$$(f \cdot g)(x) = f(x)g(x)$$
$$\frac{f}{g}(x) = \frac{f(x)}{g(x)}$$

The domains of $f+g$, $f-g$, and $f \cdot g$ are all the same and consist of all x common to the domains of f and g. The domain of $\frac{f}{g}$ has all x common to the domains of f and g except for those x where $g(x) = 0$.

We will study one more way of forming new functions from given functions. It is referred to as the *composition* of functions. Let f and g be given by

$$f(x) = \frac{1}{x-2} \quad \text{and} \quad g(x) = \sqrt{x}$$

The output of f becomes the input of g: $f(6) = \frac{1}{4}$ comes out of f and goes into g.

Take a specific value in the domain of f, say $x = 6$. Then the corresponding range value is $f(6) = \frac{1}{4}$. Take this range value and use it as a domain value for g to produce $g(\frac{1}{4}) = \sqrt{\frac{1}{4}} = \frac{1}{2}$. This work may be condensed in this way.

$$g(f(6)) = g(\tfrac{1}{4}) = \sqrt{\tfrac{1}{4}} = \tfrac{1}{2}$$

Here are two more illustrations using the condensed notation.

read this as "g of f of 10" $\longrightarrow$ $g(f(10)) = g\left(\frac{1}{10-2}\right) = g\left(\frac{1}{8}\right) = \sqrt{\frac{1}{8}} = \frac{1}{2\sqrt{2}}$

$$g\left(f\left(\frac{9}{4}\right)\right) = g\left(\frac{1}{\frac{9}{4}-2}\right) = g\left(\frac{1}{\frac{1}{4}}\right) = g(4) = \sqrt{4} = 2$$

The roles of f and g may be interchanged. Thus

read this as "f of g of 10" $\longrightarrow$ $f(g(10)) = f(\sqrt{10}) = \frac{1}{\sqrt{10}-2}$

and

$$f\left(g\left(\frac{9}{4}\right)\right) = f\left(\sqrt{\frac{9}{4}}\right) = f\left(\frac{3}{2}\right) = \frac{1}{\frac{3}{2}-2} = -2$$

In some cases this process does not work. For instance, if $x = -3$, then $f(-3) = -\frac{1}{5}$; $g(-\frac{1}{5}) = \sqrt{-\frac{1}{5}}$ is not a real number. We therefore say that $g(f(-3))$ is undefined or that it does not exist.

TEST YOUR UNDERSTANDING

For each pair of functions f and g, evaluate (if possible) each of the following:

(a) $g(f(1))$ **(b)** $f(g(1))$ **(c)** $f(g(0))$ **(d)** $g(f(-2))$

1 $f(x) = 3x - 1$; $g(x) = x^2 + 4$

2 $f(x) = \sqrt{x}$; $g(x) = x^2$

3 $f(x) = \sqrt[3]{3x - 1}$; $g(x) = 5x$

4 $f(x) = \dfrac{x + 2}{x - 1}$; $g(x) = x^3$

The preceding computations for specific values of x can be stated in terms of any allowable x. For instance, using $f(x) = \dfrac{1}{x - 2}$ and $g(x) = \sqrt{x}$, we have

$$g(f(x)) = g\left(\frac{1}{x - 2}\right) = \sqrt{\frac{1}{x - 2}} = \frac{1}{\sqrt{x - 2}}$$

This new correspondence between a domain value x and the range value $\dfrac{1}{\sqrt{x - 2}}$ is referred to as the **composite function of g by f** (or the *composition of g by f*). This composite function is denoted by $g \circ f$. That is, for the given functions f and g, we form the composite of g by f, whose range values $(g \circ f)(x)$ are defined by

$$(g \circ f)(x) = g(f(x)) = \frac{1}{\sqrt{x - 2}}$$

$g(f(x))$ is read as "g of f of x."

The domain of $g \circ f$ will consist of all values x in the domain of f such that $f(x)$ is in the domain of g. Since $f(x) = \dfrac{1}{x - 2}$ has domain all $x \neq 2$ and the domain of $g(x) = \sqrt{x}$ is all $x \geq 0$, the domain of $g \circ f$ is all $x \neq 2$ for which $\dfrac{1}{x - 2}$ is positive, that is, all $x > 2$.

Reversing the role of the two functions gives the composite of f by g, where

$$(f \circ g)(x) = f(g(x)) = f(\sqrt{x}) = \frac{1}{\sqrt{x} - 2}$$

$f(g(x))$ is read as "f of g of x."

The domain of $f \circ g$ consists of all $x \geq 0$ except $x \neq 4$.

Here is the definition of composite functions.

For functions f and g the composite function g by f, denoted $g \circ f$, has range values defined by

$$(g \circ f)(x) = g(f(x))$$

and domain consisting of all x in the domain of f for which $f(x)$ is in the domain of g.

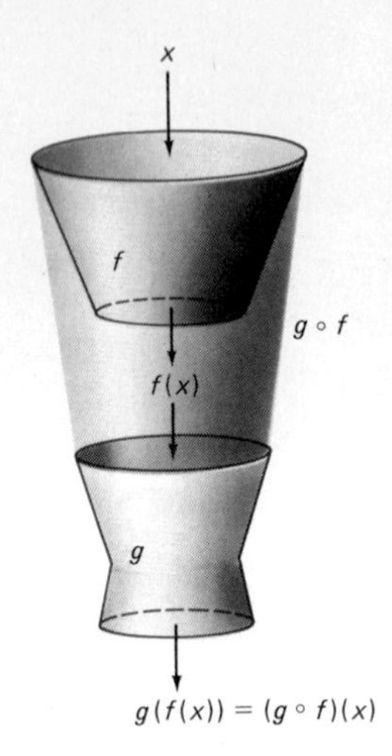

It may help you to remember the construction of composites by looking at the following schematic diagram.

$$x \xrightarrow{f} f(x) \xrightarrow{g} g(f(x)) = (g \circ f)(x)$$

$$g \circ f$$

It is also helpful to view the composition $(g \circ f)(x) = g(f(x))$ as consisting of an "inner" function f and an "outer" function g.

EXAMPLE 2 Form the composite functions $f \circ g$ and $g \circ f$ and give their domains, where $f(x) = \dfrac{1}{x^2 - 1}$ and $g(x) = \sqrt{x}$.

Solution We find that $f \circ g$ is given by

$$(f \circ g)(x) = f(g(x)) = f(\sqrt{x}) = \frac{1}{(\sqrt{x})^2 - 1} = \frac{1}{x - 1}$$

The domain of $f \circ g$ excludes $x < 0$ since such x are not in the domain of g. And it excludes $x = 1$ since $g(1) = 1$, which is not in the domain of f. Therefore, the domain of $f \circ g$ is $x \geq 0$ and $x \neq 1$. Moreover, $g \circ f$ is given by

$$(g \circ f)(x) = g(f(x)) = g\left(\frac{1}{x^2 - 1}\right) = \sqrt{\frac{1}{x^2 - 1}} = \frac{1}{\sqrt{x^2 - 1}}$$

The domain of $g \circ f$ excludes $x = \pm 1$ since these are not in the domain of f. And it excludes $-1 < x < 1$, since negative values of $f(x)$ are not in the domain of g. Therefore, the domain of $g \circ f$ is seen to be $x < -1$ or $x > 1$.

In general, $f(g(x)) \neq g(f(x))$.

From Example 2 it follows that $(f \circ g)(x) \neq (g \circ f)(x)$. We conclude that, in general, the composite of f by g is not equal to the composite of g by f.

The composition of functions may be extended to include more than two functions. For example, if $f(x) = \sqrt{x}$, $g(x) = x^2 + 1$, and $h(x) = \dfrac{1}{x}$, then the composition of f by g by h, denoted $f \circ g \circ h$, is given by

$$(f \circ g \circ h)(x) = f(g(h(x))) \qquad h \text{ is the "inner" function.}$$

$$= f\left(g\left(\frac{1}{x}\right)\right) \qquad g \text{ is the "middle" function.}$$

$$= f\left(\frac{1}{x^2} + 1\right) \qquad f \text{ is the "outer" function.}$$

$$= \sqrt{\frac{1}{x^2} + 1}$$

In Exercises 1–6, let $f(x) = 2x - 3$ *and* $g(x) = 3x + 2$.

1 (a) Find $f(1)$, $g(1)$, and $f(1) + g(1)$.

(b) Find $(f + g)(x)$ and state the domain of $f + g$.

(c) Use the result in part (b) to evaluate $(f + g)(1)$.

2 (a) Find $g(2)$, $f(2)$, and $g(2) - f(2)$.

(b) Find $(g - f)(x)$ and state the domain of $g - f$.

(c) Use the result in part (b) to evaluate $(g - f)(2)$.

3 (a) Find $f(\frac{1}{2})$, $g(\frac{1}{2})$, and $f(\frac{1}{2}) \cdot g(\frac{1}{2})$.

(b) Find $(f \cdot g)(x)$ and state the domain of $f \cdot g$.

(c) Use the result in part (b) to evaluate $(f \cdot g)(\frac{1}{2})$.

4 (a) Find $g(-2)$, $f(-2)$, and $\frac{g(-2)}{f(-2)}$.

(b) Find $\frac{g}{f}(x)$ and state the domain of $\frac{g}{f}$.

(c) Use the result in part (b) to evaluate $\frac{g}{f}(-2)$.

5 (a) Find $g(0)$ and $f(g(0))$.

(b) Find $(f \circ g)(x)$ and state the domain of $f \circ g$.

(c) Use the result in part (b) to evaluate $(f \circ g)(0)$.

6 (a) Find $f(0)$ and $g(f(0))$.

(b) Find $(g \circ f)(x)$ and state the domain of $g \circ f$.

(c) Use the result in part (b) to evaluate $(g \circ f)(0)$.

For each of the functions in Exercises 7–12, find the following:

(a) $(f + g)(x)$; domain of $f + g$.

(b) $\left(\frac{f}{g}\right)(x)$; domain $\frac{f}{g}$.

(c) $(f \circ g)(x)$; domain $f \circ g$.

7 $f(x) = x^2$, $g(x) = \sqrt{x}$

8 $f(x) = 5x - 1$, $g(x) = \frac{5}{1 + 3x}$

9 $f(x) = x^3 - 1$, $g(x) = \frac{1}{x}$

10 $f(x) = 3x - 1$, $g(x) = \frac{1}{3}x + \frac{1}{3}$

11 $f(x) = x^2 + 6x + 8$, $g(x) = \sqrt{x - 2}$

12 $f(x) = \sqrt[3]{x}$, $g(x) = x^2$

For each pair of functions in Exercises 13–16, find the following:

(a) $(g - f)(x)$; domain of $g - f$.

(b) $(g \cdot f)(x)$; domain of $g \cdot f$.

(c) $(g \circ f)(x)$; domain of $g \circ f$.

13 $f(x) = -2x + 5$, $g(x) = 4x - 1$

14 $f(x) = |x|$, $g(x) = 3|x|$

15 $f(x) = 2x^2 - 1$, $g(x) = \frac{1}{2x}$

16 $f(x) = \sqrt{2x + 3}$, $g(x) = x^2 - 1$

In Exercises 17–24, find $(f \circ g)(x)$ and $(g \circ f)(x)$.

17 $f(x) = x^2$, $g(x) = x - 1$

18 $f(x) = |x - 3|$, $g(x) = 2x + 3$

19 $f(x) = \frac{x}{x - 2}$, $g(x) = \frac{x + 3}{x}$

20 $f(x) = x^3 - 1$, $g(x) = \frac{1}{x^3 + 1}$

21 $f(x) = \sqrt{x + 1}$, $g(x) = x^4 - 1$

22 $f(x) = 2x^3 - 1$, $g(x) = \sqrt[3]{\frac{x + 1}{2}}$

23 $f(x) = \sqrt{x}$, $g(x) = 4$

24 $f(x) = \sqrt[3]{1 - x}$, $g(x) = 1 - x^3$

25 Let $f(x) = \frac{1}{x}$, $g(x) = 2x - 1$, and $h(x) = x^{1/3}$. Find the following:

(a) $(f \circ g \circ h)(x)$ (b) $(g \circ f \circ h)(x)$

(c) $(h \circ f \circ g)(x)$

26 Let $f(x) = x + 2$, $g(x) = \sqrt{x}$, and $h(x) = x^3$. Find the following:

(a) $(f \circ g \circ h)(x)$ (b) $(f \circ h \circ g)(x)$

(c) $(g \circ f \circ h)(x)$ (d) $(g \circ h \circ f)(x)$

(e) $(h \circ f \circ g)(x)$ (f) $(h \circ g \circ f)(x)$

27 Let $f(x) = \frac{1}{x}$. Find $(f \circ f)(x)$ and $(f \circ f \circ f)(x)$.

28 Let $f(x) = x^2$, $g(x) = \frac{1}{x - 1}$, and $h(x) = 1 + \frac{1}{x}$. Find the following:

(a) $(f \circ h \circ g)(x)$ (b) $(g \circ h \circ f)(x)$

(c) $(h \circ g \circ f)(x)$

***29** Let $f(x) = x^3$. Find a function g so that $(f \circ g)(x) = x$ and $(g \circ f)(x) = x$.

***30** Let $f(x) = x$. Find $(f \circ g)(x)$ and $(g \circ f)(x)$ for any function g.

***31** If $f(x) = 2x + 1$, find $g(x)$ so that $(f \circ g)(x) = 2x^2 - 4x + 1$.

6.5

Decomposition of composite functions

One of the most useful skills needed in the study of calculus is the ability to recognize that a given function may be viewed as the composition of two or more functions. For instance, let h be given by

$$h(x) = \sqrt{x^2 + 2x + 2}$$

If we let $f(x) = x^2 + 2x + 2$ and $g(x) = \sqrt{x}$, then the composite g by f is

$$\begin{aligned}(g \circ f)(x) &= g(f(x)) \\ &= g(x^2 + 2x + 2) \\ &= \sqrt{x^2 + 2x + 2} = h(x)\end{aligned}$$

Thus the given function h has been *decomposed* into the composition of the two functions f and g. Such decompositions are not unique. More than one decomposition is possible. For h we may also use

$$t(x) = x^2 + 2x \qquad s(x) = \sqrt{x + 2}$$

Then

$$\begin{aligned}(s \circ t)(x) &= s(t(x)) \\ &= s(x^2 + 2x) \\ &= \sqrt{x^2 + 2x + 2} = h(x)\end{aligned}$$

You would most likely agree that the first of these decompositions is more "natural." Just which decomposition one is to choose will, in later work, depend on the situation. For our purposes some additional examples will help to demonstrate the decompositions that are desirable. Other answers for each of these examples are possible.

EXAMPLE 1 Find f and g such that $h = f \circ g$, where $h(x) = \left(\frac{1}{3x - 1}\right)^5$ and the inner function g is rational.

Solution Let $g(x) = \frac{1}{3x - 1}$ and $f(x) = x^5$.

$$\begin{aligned}(f \circ g)(x) &= f(g(x)) \\ &= f\left(\frac{1}{3x - 1}\right) \\ &= \left(\frac{1}{3x - 1}\right)^5 = h(x)\end{aligned}$$

EXAMPLE 2 Write $h(x) = \frac{1}{(3x - 1)^5}$ as the composite of two functions so that the inner function is a binomial.

Solution Let $g(x) = 3x - 1$ and $f(x) = \frac{1}{x^5}$.

$$\begin{aligned}(f \circ g)(x) &= f(g(x)) \\ &= f(3x - 1) \\ &= \frac{1}{(3x - 1)^5} = h(x)\end{aligned}$$

EXAMPLE 3 Decompose $h(x) = \sqrt{(x^2 - 3x)^5}$ into two functions so that the inner function is a binomial.

Solution First write $h(x) = (x^2 - 3x)^{5/2}$. Now let $f(x) = x^2 - 3x$ and let $g(x) = x^{5/2}$.

$$\begin{aligned}(g \circ f)(x) &= g(f(x)) \\ &= g(x^2 - 3x) \\ &= (x^2 - 3x)^{5/2} \\ &= h(x)\end{aligned}$$

EXAMPLE 4 Decompose h in Example 3 so that the outer function is a monomial.

Solution Write $h(x) = (\sqrt{x^2 - 3x})^5$. Let $f(x) = \sqrt{x^2 - 3x}$ and $g(x) = x^5$.

$$\begin{aligned}(g \circ f)(x) &= g(f(x)) \\ &= g(\sqrt{x^2 - 3x}\,) \\ &= (\sqrt{x^2 - 3x}\,)^5 \\ &= h(x)\end{aligned}$$

TEST YOUR UNDERSTANDING

For each function h, find functions f and g so that $h = g \circ f$.

1 $h(x) = (8x - 3)^5$

2 $h(x) = \sqrt[5]{8x - 3}$

3 $h(x) = \sqrt{\dfrac{1}{8x - 3}}$

4 $h(x) = \left(\dfrac{5}{7 + 4x^2}\right)^3$

5 $h(x) = \dfrac{(2x + 1)^4}{(2x - 1)^4}$

6 $h(x) = \sqrt{(x^4 - 2x^2 + 1)^3}$

Examples 3 and 4 showed two different ways of decomposing $h(x) = \sqrt{(x^2 - 3x)^5}$ into the composition of two functions. It is also possible to express h as the composition of three functions. For example, we may use these functions:

$$f(x) = x^2 - 3x \qquad g(x) = x^5 \qquad t(x) = \sqrt{x}$$

Then

$$\begin{aligned}(t \circ g \circ f)(x) &= t(g(f(x))) && f \text{ is the "inner" function.} \\ &= t(g(x^2 - 3x)) && g \text{ is the "middle" function.} \\ &= t((x^2 - 3x)^5) && t \text{ is the "outer" function.} \\ &= \sqrt{(x^2 - 3x)^5} \\ &= h(x)\end{aligned}$$

EXERCISES 6.5

In Exercises 1–14, find functions f and g so that $h(x) = (f \circ g)(x)$. In each case let the inner function g be a polynomial or a rational function.

1 $h(x) = (3x + 1)^2$

2 $h(x) = (x^2 - 2x)^3$

3 $h(x) = \sqrt{1 - 4x}$

4 $h(x) = \sqrt[3]{x^2 - 1}$

5 $h(x) = \left(\frac{x + 1}{x - 1}\right)^2$

6 $h(x) = \left(\frac{1 - 2x}{1 + 2x}\right)^3$

7 $h(x) = (3x^2 - 1)^{-3}$

8 $h(x) = \left(1 + \frac{1}{x}\right)^{-2}$

9 $h(x) = \sqrt{\frac{x}{x - 1}}$

10 $h(x) = \sqrt[3]{\frac{x - 1}{x}}$

11 $h(x) = \sqrt{(x^2 - x - 1)^3}$

12 $h(x) = \sqrt[3]{(1 - x^4)^2}$

13 $h(x) = \frac{2}{\sqrt{4 - x^2}}$

14 $h(x) = -\left(\frac{3}{x - 1}\right)^5$

In Exercises 15–24, find three functions f, g, and h such that $k(x) = (h \circ g \circ f)(x)$.

15 $k(x) = (\sqrt{2x + 1})^3$

16 $k(x) = \sqrt[3]{(2x - 1)^2}$

17 $k(x) = \sqrt{\left(\frac{x}{x + 1}\right)^5}$

18 $k(x) = \left(\sqrt[7]{\frac{x^2 - 1}{x^2 + 1}}\right)^4$

19 $k(x) = (x^2 - 9)^{2/3}$

20 $k(x) = (5 - 3x)^{5/2}$

21 $k(x) = -\sqrt{(x^2 - 4x + 7)^3}$

22 $k(x) = -\left(\frac{2x}{1 - x}\right)^{2/5}$

23 $k(x) = (1 + \sqrt{2x - 11})^2$

24 $k(x) = \sqrt[3]{(x^2 - 4)^5 - 1}$

25 Find f so that $((x + 1)^2 + 1)^2 = (f \circ f)(x)$.

26 Find f so that

$$\sqrt{1 + \sqrt{1 + \sqrt{1 + x}}} = (f \circ f \circ f)(x).$$

REVIEW EXERCISES FOR CHAPTER 6

The solutions to the following exercises can be found within the text of Chapter 6. Try to answer each question without referring to the text.

Section 6.1

In Exercises 1–3, find the domain and graph.

1 $y = \sqrt{x - 2}$

2 $y = \sqrt{|x|}$

3 $y = x^{-1/2}$

4 Graph: $f(x) = \begin{cases} x^2 - 1 & \text{if } -2 \leq x < 2 \\ \sqrt{x - 2} & \text{if } 2 \leq x < 5 \end{cases}$

Section 6.2

5 Find the x-intercepts of $y = \sqrt[3]{x + 1} - 2$.

6 Find the x-intercepts and the domain of $f(x) = \sqrt{x^2 + x - 6}$.

Solve the equations in Exercises 7–9.

7 $\sqrt{x + 4} + 2 = x$

8 $\sqrt{x - 7} - \sqrt{x} = 1$

9 $\sqrt[3]{x^2} + \sqrt[3]{x} - 20 = 0$

10 Solve the system and graph:

$$y = \sqrt[3]{x}$$
$$y = \tfrac{1}{4}x$$

Section 6.3

11 Find the domain of $f(x) = (x - 5)\sqrt{x - 3}$ and determine the signs of f.

12 Find the domain of $f(x) = \frac{x}{\sqrt{4 - x^2}}$ and determine its signs.

13 Determine the signs of $f(x) = x^{1/2} + \frac{1}{2}x^{-1/2}(x - 3)$.

14 Find the domain of f, solve $f(x) = 0$, and determine the signs of f, where

$$f(x) = \frac{x^2 - 1}{2\sqrt{x - 1}} + 2x\sqrt{x - 1}$$

Section 6.4

15 Determine $(f + g)(x)$, $(f - g)(x)$, $(f \cdot g)(x)$, and $\frac{g}{f}(x)$, where $f(x) = \frac{1}{x^3 - 1}$ and $g(x) = \sqrt{x}$.

16 What are the domains of $f + g$, $f - g$, and $f \cdot g$ given in Exercise 15?

17 What is the domain of $\frac{f}{g}$ given in Exercise 15?

18 Evaluate $g\left(f\left(\frac{9}{4}\right)\right)$ and $f\left(g\left(\frac{9}{4}\right)\right)$, where $f(x) = \frac{1}{x - 2}$ and $g(x) = \sqrt{x}$.

19 Find $(g \circ f)(x)$ for f and g in Exercise 18 and state the domain.

20 Find $(f \circ g)(x)$ for f and g in Exercise 18 and state the domain.

21 Form the composites $g \circ f$ and $f \circ g$ and give their domains, where $f(x) = \frac{1}{x^2 - 1}$ and $g(x) = \sqrt{x}$.

22 Find $(f \circ g \circ h)(x)$, where $f(x) = \sqrt{x}$, $g(x) = x^2 + 1$, and $h(x) = \dfrac{1}{x}$.

Section 6.5

23 Find functions f and g so that $h(x) = \sqrt{x^2 + 2x + 2} = (g \circ f)(x)$.

24 Find functions f and g so that $h = f \circ g$, where $h(x) = \left(\dfrac{1}{3x - 1}\right)^5$.

25 Decompose $h(x) = \sqrt{(x^2 - 3x)^5}$ into two functions so that the inner function is a binomial.

26 Decompose h in Exercise 25 so that the outer function is a polynomial.

27 Show that $h(x) = \sqrt{(x^2 - 3x)^5}$ can be written as the composition of three functions.

SAMPLE TEST QUESTIONS FOR CHAPTER 6

Use these questions to test your knowledge of the basic skills and concepts of Chapter 6. Then check your answers with those given at the end of the book.

Graph each function, state its domain, and give the equations of the asymptotes if there are any.

1 $f(x) = -\sqrt[3]{x - 2}$

2 $g(x) = \dfrac{1}{\sqrt{x}} + 2$

Solve each equation.

3 $\sqrt{18x + 5} - 9x = 1$

4 $6x^{2/3} + 5x^{1/3} - 4 = 0$

5 $\sqrt{x - 7} + \sqrt{x + 9} = 8$

6 Convert $(x + 1)^{1/2}(2x) + (x + 1)^{-1/2}\left(\dfrac{x^2}{2}\right)$ into the equivalent form $\dfrac{x(5x + 4)}{2\sqrt{x + 1}}$. Show your work.

In Questions 7 and 8, determine the signs of f.

7 $f(x) = \dfrac{\sqrt[3]{x - 4}}{x - 2}$

8 $f(x) = x^{1/2} - \frac{1}{2}x^{-1/2}(x + 4)$

9 Let $f(x) = \dfrac{1}{x^2 - 1}$ and $g(x) = \sqrt{x + 2}$. Find $(f - g)(x)$ and $\dfrac{f}{g}(x)$ and state their domains.

10 For $f(x) = \dfrac{1}{1 - x^2}$ and $g(x) = \sqrt{x}$ find the composites $(f \circ g)(x)$ and $(g \circ f)(x)$ and state their domains.

11 Find functions f and g so that $h(x) = \sqrt[3]{(x - 2)^2} = (f \circ g)(x)$, where g is a binomial.

12 Let $F(x) = \dfrac{1}{(2x - 1)^{3/2}}$. Find functions f, g, and h so that $(f \circ g \circ h)(x) = F(x)$.

13 If y is positive in $\dfrac{x^2}{16} - \dfrac{y^2}{9} = 1$, then $y =$

(a) $\frac{3}{4}x - 3$ **(b)** $\sqrt{\frac{3}{4}x - 3}$

(c) $\frac{3}{4}x - 1$ **(d)** $\frac{3}{4}\sqrt{x^2 - 16}$

(e) $\sqrt{\frac{9}{16}x^2 + 1}$

14 State the domain of $f(x) = \sqrt[3]{3x + 1} - x - 1$ and find the x-intercepts.

15 Solve the system:

$$y = 2\sqrt{x - 1}$$
$$2y - x = 2$$

ANSWERS TO THE TEST YOUR UNDERSTANDING EXERCISES

Page 217

1 Domain of g: $x \geq 0$

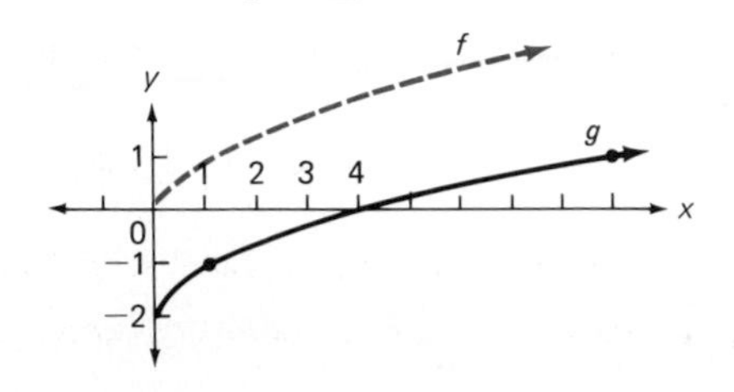

2 Domain of g: $x \geq 1$

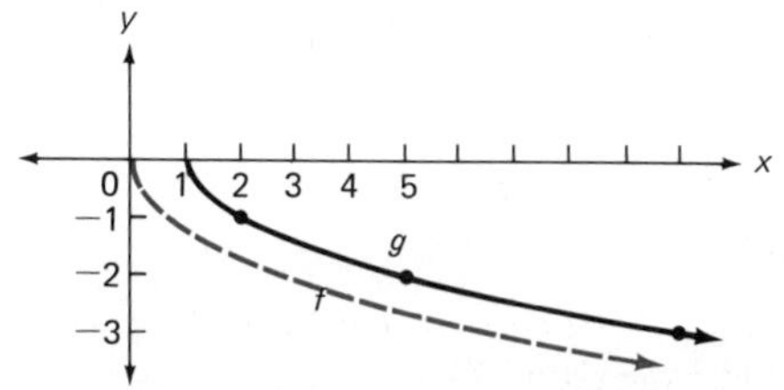

3 Domain of g: $x \geq 0$

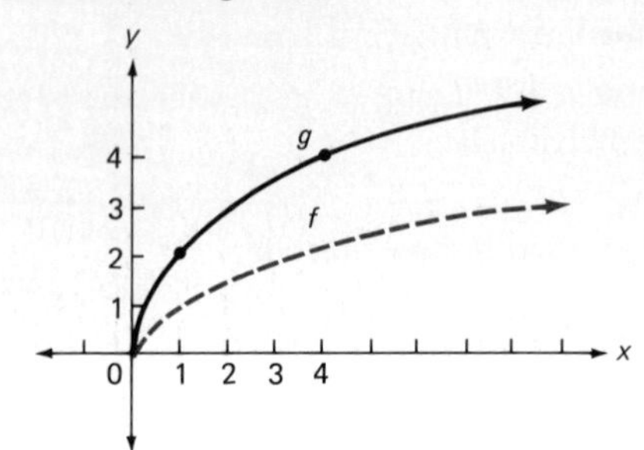

4 Domain of g: $x > -2$

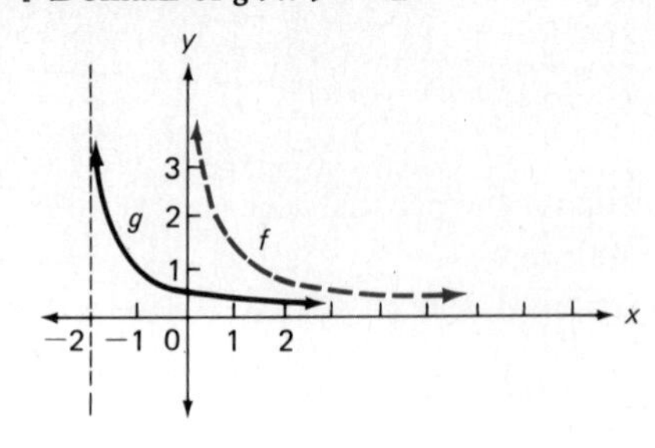

Page 221

1 25; all $x \geq 0$
2 8; all $x > -1$
3 0, 3; all $x \leq 0$ or all $x \geq 3$
4 7
5 No solutions.
6 $\frac{1}{9}$
7 5
8 $-3, 3$
9 5
10 41
11 0
12 $\frac{1}{2}$

Page 226

1 Domain: $x > 0$
$f(x) > 0$ on $(4, \infty)$
$f(x) < 0$ on $(0, 4)$

2 Domain: $x \geq -2$ and $x \neq 0$
$f(x) > 0$ on $(0, \infty)$
$f(x) < 0$ on $(-2, 0)$

3 Domain: all $x \neq 0$
$f(x) > 0$ on $(0, \infty)$
$f(x) < 0$ on $(-\infty, 0)$

Page 231

1 **(a)** $g(f(1)) = g(2) = 8$
(b) $f(g(1)) = f(5) = 14$
(c) $f(g(0)) = f(4) = 11$
(d) $g(f(-2)) = g(-7) = 53$

2 **(a)** $g(f(1)) = g(1) = 1$
(b) $f(g(1)) = f(1) = 1$
(c) $f(g(0)) = f(0) = 0$
(d) $g(f(-2))$ is undefined.

3 **(a)** $g(f(1)) = g(\sqrt[3]{2}) = 5\sqrt[3]{2}$
(b) $f(g(1)) = f(5) = \sqrt[3]{14}$
(c) $f(g(0)) = f(0) = -1$
(d) $g(f(-2)) = g(\sqrt[3]{-7}) = -5\sqrt[3]{7}$

4 **(a)** $g(f(1))$ is undefined.
(b) $f(g(1))$ is undefined.
(c) $f(g(0)) = f(0) = -2$
(d) $g(f(-2)) = g(0) = 0$

Page 235

(Other answers are possible.)

1 $f(x) = 8x - 3$; $g(x) = x^5$
2 $f(x) = 8x - 3$; $g(x) = \sqrt[5]{x}$
3 $f(x) = \dfrac{1}{8x - 3}$; $g(x) = \sqrt{x}$
4 $f(x) = \dfrac{5}{7 + 4x^2}$; $g(x) = x^3$
5 $f(x) = \dfrac{2x + 1}{2x - 1}$; $g(x) = x^4$
6 $f(x) = x^4 - 2x^2 + 1$; $g(x) = \sqrt{x^3}$

EXPONENTIAL AND LOGARITHMIC FUNCTIONS

7.1 Inverse functions

By definition, a function has each domain value x corresponding to exactly one range value y. Some (but not all) functions have the additional property that to every range value y there corresponds exactly one domain value x. Such functions are said to be **one-to-one functions**. To understand this concept, consider the functions $y = x^2$ and $y = x^3$.

For a one-to-one function you can start at a range value y and trace it back to exactly one domain value x.

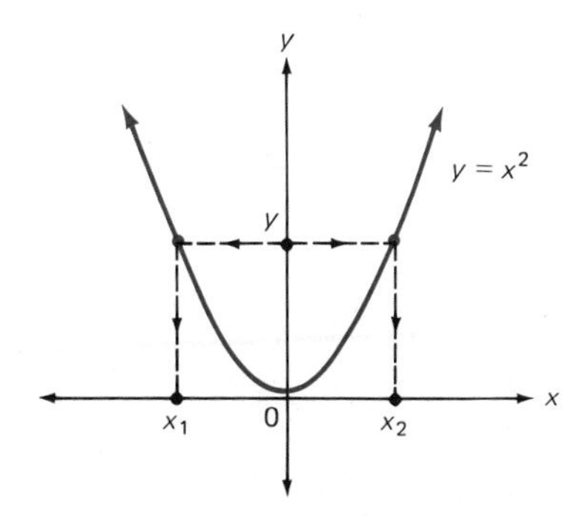

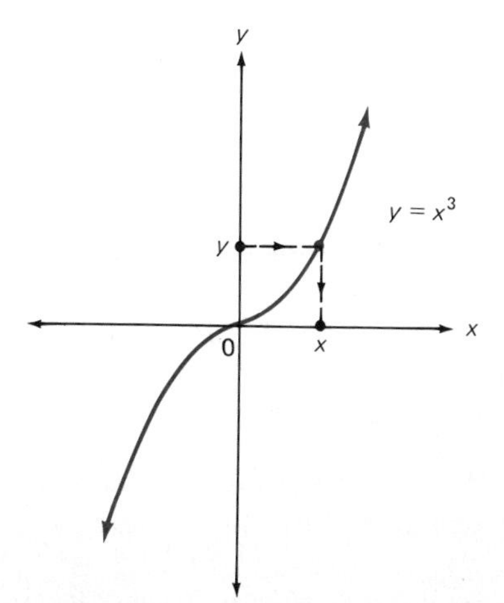

$y = x^2$ *is not* a one-to-one function. There are two domain values for a range value $y > 0$.

$y = x^3$ *is* a one-to-one function. There is exactly one domain value x for each range value y.

DEFINITION

A function f is a one-to-one function if and only if for each range value there corresponds exactly one domain value.

Once the graph of a function is known, there is a simple visual test for determining the one-to-one property. Consider a horizontal line through each range value y, as in the preceding figures. If the line meets the curve exactly once, then we have a one-to-one function; otherwise, it is not one-to-one.

TEST YOUR UNDERSTANDING

Use the horizontal line test described above to determine which of the following are one-to-one functions.

1 $y = x^2 - 2x + 1$	**2** $y = \sqrt{x}$	**3** $y = \frac{1}{x}$
4 $y = \lvert x \rvert$	**5** $y = 2x + 1$	**6** $y = \sqrt[3]{x}$
7 $y = -x$	**8** $y = [x]$	**9** $y = \frac{1}{x^2}$

If the variables in $y = x^3$ are interchanged, we obtain $x = y^3$. Here are the graphs for these equations, together with the two graphs on the same axes.

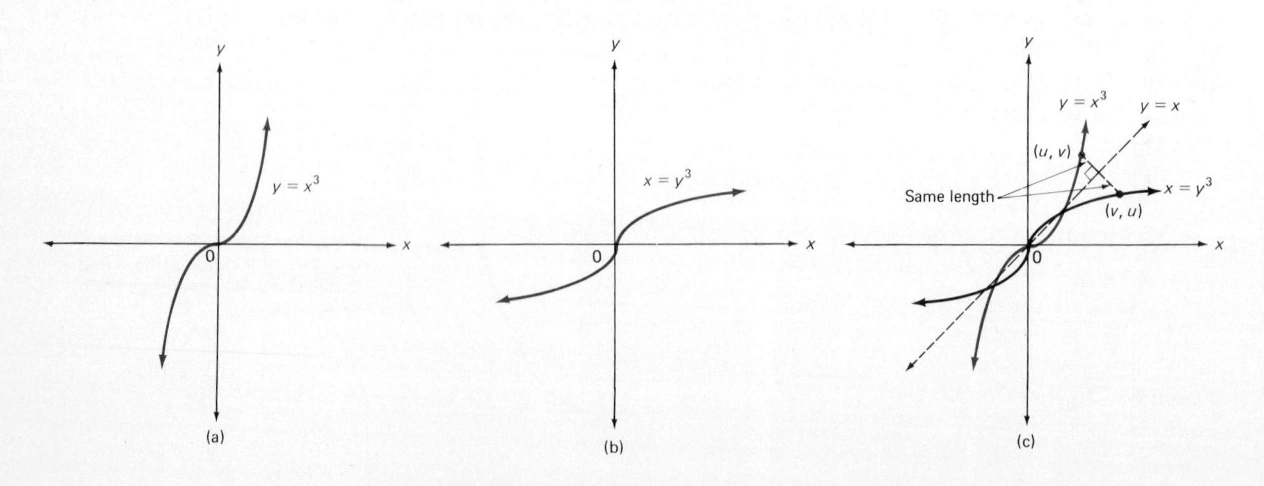

Because of this interchange of coordinates, the two curves are reflections of each other through the line $y = x$, as in part (c) of the figure. Another way to describe this relationship is to say that they are mirror images of one another through the "mirror line" $y = x$.

If the paper were folded along the line $y = x$, the two curves would coincide.

The equation $x = y^3$ may be solved for y by taking the cube root of both sides: $y = \sqrt[3]{x}$. The two equations $x = y^3$ and $y = \sqrt[3]{x}$ are equivalent and therefore define the same function of x. However, since $y = \sqrt[3]{x}$ shows *explicitly* how y depends on x, it is the preferred form.

We began with the one-to-one function $y = f(x) = x^3$ and, by interchanging coordinates, arrived at the new function $y = g(x) = \sqrt[3]{x}$. If the composites $f \circ g$ and $g \circ f$ are formed, something surprising happens.

$$(f \circ g)(x) = f(g(x)) = f(\sqrt[3]{x}) = (\sqrt[3]{x})^3 = x$$

$$(g \circ f)(x) = g(f(x)) = g(x^3) = \sqrt[3]{x^3} = x$$

Function f cubes and function g "uncubes"; f and g are inverse functions.

In each case we obtained the same value x that we started with; whatever one of the functions does to a value x, the other function undoes. Whenever two functions act on each other in such a manner, we say that they are *inverse functions* or that either function is the inverse of the other.

DEFINITION OF INVERSE FUNCTIONS

Two functions f and g are said to be *inverse functions* if and only if:

1 For each x in the domain of g, $g(x)$ is in the domain of f and

$$(f \circ g)(x) = f(g(x)) = x$$

2 For each x in the domain of f, $f(x)$ is in the domain of g and

$$(g \circ f)(x) = g(f(x)) = x$$

The notation f^{-1} is also used to represent the inverse of f. Thus $f(f^{-1}(x)) = x$ and $f^{-1}(f(x)) = x$.

It turns out (as suggested by our work with $y = x^3$) that every one-to-one function f has an inverse g and that their graphs are reflections of each other through the line $y = x$. The technique of interchanging variables, used to obtain $y = \sqrt[3]{x}$ from $y = x^3$, can be applied to many situations, as illustrated in the following examples.

$y = x^2$ is not one-to-one. Its reflection in the line $y = x$ has equation $x = y^2$, or, $y = \pm\sqrt{x}$ which is not a function.

EXAMPLE 1 Find the inverse g of $y = f(x) = 2x + 3$. Then show that $(f \circ g)(x) = x = (g \circ f)(x)$, and graph both on the same axes.

Solution Interchange variables in $y = 2x + 3$ and solve for y.

$$x = 2y + 3$$

$$2y = x - 3$$

$$y = \tfrac{1}{2}x - \tfrac{3}{2}$$

Using $y = g(x) = \frac{1}{2}x - \frac{3}{2}$, we have

$$\begin{aligned}(f \circ g)(x) = f(g(x)) &= f(\tfrac{1}{2}x - \tfrac{3}{2}) \\ &= 2(\tfrac{1}{2}x - \tfrac{3}{2}) + 3 \\ &= x - 3 + 3 \\ &= x\end{aligned}$$

To find the inverse g of f, begin with

$$y = f(x).$$

Then interchange variables

$$x = f(y)$$

and solve for y in terms of x, producing

$$y = g(x).$$

$$(g \circ f)(x) = g(f(x)) = g(2x + 3)$$
$$= \tfrac{1}{2}(2x + 3) - \tfrac{3}{2}$$
$$= x + \tfrac{3}{2} - \tfrac{3}{2}$$
$$= x$$

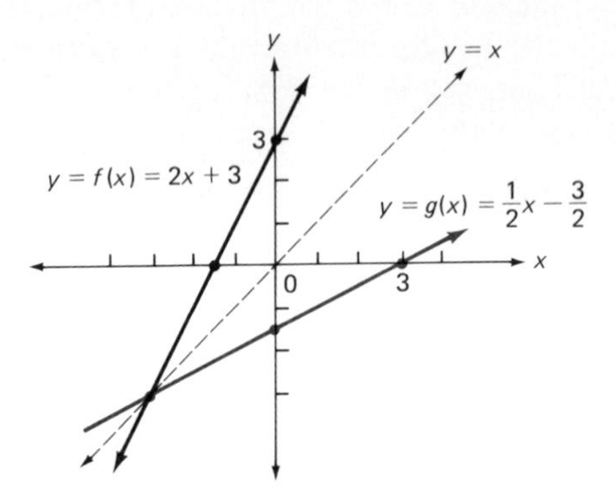

EXAMPLE 2 Follow the instructions in Example 1 for $y = f(x) = \sqrt{x}$.

Solution Interchange variables in $y = \sqrt{x}$ to get $x = \sqrt{y}$. At this point we see that x cannot be negative: $x \geq 0$. Solving for y by squaring produces $y = x^2$. Using the letter g for this inverse function, we have $y = g(x) = x^2$ with domain $x \geq 0$.

$$(f \circ g)(x) = f(g(x)) = f(x^2) = \sqrt{x^2} = |x| = x \qquad (\text{since } x \geq 0)$$
$$(g \circ f)(x) = g(f(x)) = g(\sqrt{x}) = (\sqrt{x})^2 = x$$

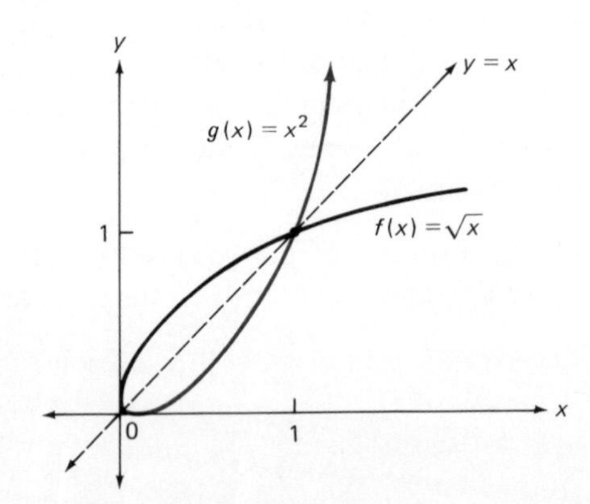

The process of interchanging variables to find the inverse of a function can also be explained in terms of the criterion $f(g(x)) = x$. Begin with a function, say $f(x) = 3x - 2$, and let g be its inverse. Thus $f(g(x)) = x$. Substituting $g(x)$ into the equation for f gives

$$x = f(g(x)) = 3g(x) - 2$$

Solving this last equation for $g(x)$ produces

$$g(x) = \tfrac{1}{3}x + \tfrac{2}{3}$$

Now you can see that in taking $y = f(x) = 3x - 2$ and interchanging the variables to get $x = 3y - 2$, we are really letting y stand for the inverse function g. Thus

$$x = 3y - 2 = 3g(x) - 2 = f(g(x))$$

In many cases you will find it more efficient to use the form $x = 3y - 2$ and solve for y. This procedure is simpler than using $x = 3g(x) - 2$ and solving for $g(x)$. But regardless of which notation is used, this procedure has its limitations. Thus if the defining expression $y = f(x)$ is complicated, it may be algebraically difficult, or even impossible, to do what we did in the special cases above. We will avoid such situations by limiting our work in this section to functions for which the procedure can be applied.

EXERCISES 7.1

In Exercises 1–6, use the horizontal line test to decide if the function is one-to-one.

1 $f(x) = (x - 1)^2$

2 $f(x) = x^3 - 3x^2$

3 $f(x) = \dfrac{1 - x}{x}$

4 $f(x) = \dfrac{1}{x - 2}$

5 $f(x) = \sqrt{x} + 3$

6 $f(x) = \dfrac{1}{x^2 - x - 2}$

In Exercises 7–12, show that f and g are inverse functions according to the criteria $(f \circ g)(x) = x$ and $(g \circ f)(x) = x$. Then graph both functions and the line $y = x$ on the same axes.

7 $f(x) = \tfrac{1}{3}x - 3$; $g(x) = 3x + 9$

8 $f(x) = 2x - 6$; $g(x) = \tfrac{1}{2}x + 3$

9 $f(x) = (x + 1)^3$; $g(x) = \sqrt[3]{x} - 1$

10 $f(x) = -(x + 2)^3$; $g(x) = -\sqrt[3]{x} - 2$

11 $(x) = \dfrac{1}{x - 1}$; $g(x) = \dfrac{1}{x} + 1$

12 $f(x) = x^2 + 2$ for $x \geq 2$; $g(x) = \sqrt{x - 2}$

In Exercises 13–24, find the inverse function g of the given function f.

13 $y = f(x) = (x - 5)^3$

14 $y = f(x) = x^{1/3} - 3$

15 $y = f(x) = \tfrac{2}{3}x - 1$

16 $y = f(x) = -4x + \tfrac{2}{5}$

17 $y = f(x) = (x - 1)^5$

18 $y = f(x) = -x^5$

19 $y = f(x) = x^{3/5}$

20 $y = f(x) = x^{5/3} + 1$

21 $y = f(x) = \dfrac{2}{x - 2}$

22 $y = f(x) = -\dfrac{1}{x} - 1$

23 $y = f(x) = x^{-5}$

24 $y = f(x) = \dfrac{1}{\sqrt[3]{x - 2}}$

In Exercises 25–28, verify that the function is its own inverse by showing that $(f \circ f)(x) = x$.

25 $f(x) = \dfrac{1}{x}$

26 $f(x) = \sqrt{4 - x^2}$, $0 \leq x \leq 2$

27 $f(x) = \dfrac{x}{x - 1}$

28 $f(x) = \dfrac{3x - 8}{x - 3}$

In Exercises 29–34, find the inverse g of the given function f, and graph both in the same coordinate system.

29 $y = f(x) = (x + 1)^2$; $x \geq -1$

30 $y = f(x) = x^2 - 4x + 4$; $x \geq 2$

31 $y = f(x) = \dfrac{1}{\sqrt{x}}$

32 $y = f(x) = -\sqrt{x}$

33 $y = f(x) = x^2 - 4$; $x \geq 0$

34 $y = f(x) = 4 - x^2$; $0 \leq x \leq 2$

*35 Aside from the linear function $y = x$, what other

linear functions are their own inverses? (*Hint:* Inverse functions are reflections of one another through the line $y = x$.)

*36 In the figure curve C_1 is the graph of a one-to-one function $y = f(x)$ and curve C_2 is the graph of $y = g(x)$. The points on C_2 have been obtained by interchanging the first and second coordinates of the points on curve C_1. For a specific value x in the domain of f, the point $P(x, f(x))$ on C_1 produces $Q(f(x), x)$ on C_2. How does this figure demonstrate that $(g \circ f)(x) = x$?

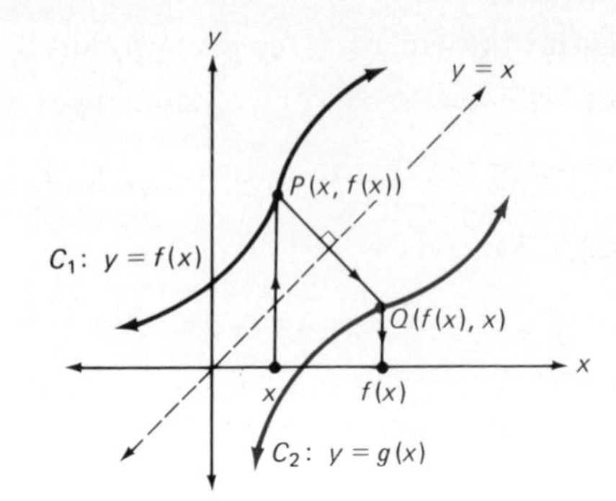

7.2

Exponential functions

$20{,}000 = (10{,}000)2^1$
$40{,}000 = (10{,}000)2^2$
$80{,}000 = (10{,}000)2^3$

Imagine that a bacterial culture is growing at such a rate that after each hour the number of bacteria has doubled. Thus if there were 10,000 bacteria when the culture started to grow, then after 1 hour the number would have grown to 20,000, after 2 hours there would be 40,000, and so on. It becomes reasonable to say that

$$y = (10{,}000)2^x$$

gives the number y of bacteria present after x hours. This equation defines an **exponential function** with independent variable x and dependent variable y. We call such a function exponential because its exponent is a variable. In this section we study exponential functions of the form $y = b^x$ for $b > 0$. The number b is often referred to as the *base* number.

What does the graph of $y = f(x) = 2^x$ look like? We can get a good idea by forming a table of values, plotting the points, and connecting them with a smooth curve.

The function is increasing and the curve is concave up. The x-axis is a horizontal asymptote.

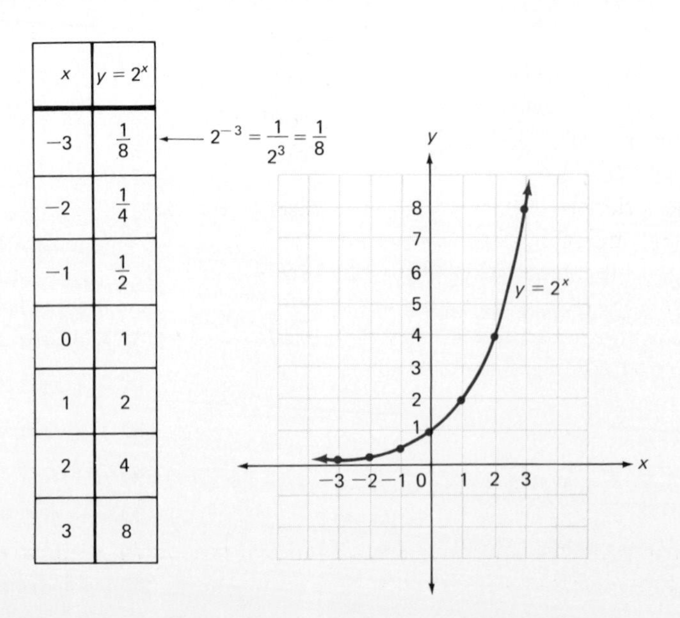

x	$y = 2^x$
-3	$\frac{1}{8}$
-2	$\frac{1}{4}$
-1	$\frac{1}{2}$
0	1
1	2
2	4
3	8

The accuracy of this graph can be improved by using more points. Consider rational values of x, such as $\frac{1}{2}$ or $\frac{3}{2}$:

$2^{1/2} = \sqrt{2} = 1.4$ (correct to one decimal place from Appendix Table I)

$2^{3/2} = (\sqrt{2})^3 = (1.4)^3 = 2.7$

Using irrational values for x, such as $\sqrt{2}$ or π, is another matter entirely. (Remember that our development of exponents stopped with the rationals.) To give a precise meaning of such numbers is beyond the scope of this course. It does turn out, however, that the indicated shape of the curve $y = 2^x$ is correct and that the formal definitions of values like $2^{\sqrt{2}}$ are made so that they "fit" the curve.

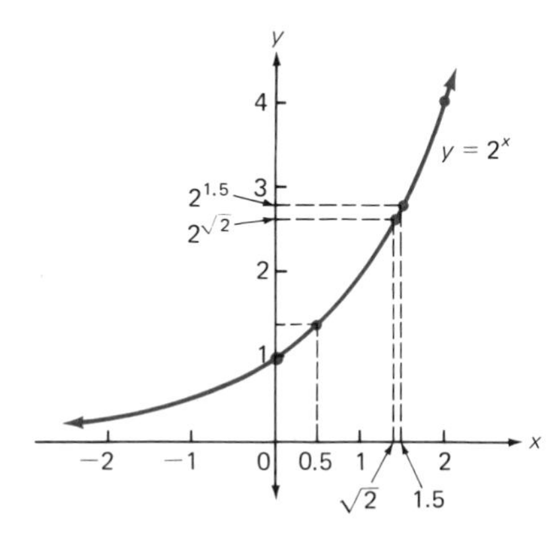

A calculator can be used to better understand such numbers as $2^{\sqrt{2}}$. For example, verify these rational powers of 2 to four decimal places.

$$2^{1.4} = 2.6390$$
$$2^{1.41} = 2.6574$$
$$2^{1.414} = 2.6647$$
$$2^{1.4142} = 2.6651$$

See Ex. 34–35 for similar computations involving irrational powers.

Since the rational exponents are getting closer to the irrational number $\sqrt{2}$, the corresponding powers are getting closer to $2^{\sqrt{2}}$. So the exponential approximations suggest that $2^{\sqrt{2}} = 2.67$ to two decimal places. Now find $2^{\sqrt{2}}$ directly on a calculator and compare.

In advanced work it can be shown that for any positive bases a and b, the following hold for all real numbers r and s.

$$b^r b^s = b^{r+s} \qquad \frac{b^r}{b^s} = b^{r-s} \qquad (b^r)^s = b^{rs}$$

$$a^r b^r = (ab)^r \qquad b^0 = 1 \qquad b^{-r} = \frac{1}{b^r}$$

Our earlier work with the same rules, for rational exponents, can now serve as the basis for accepting these results.

EXAMPLE 1 Graph the curve $y = 8^x$ on the interval $[-1, 1]$ by using a table of values.

Solution

x	y
-1	$\frac{1}{8}$
$-\frac{2}{3}$	$\frac{1}{4}$
$-\frac{1}{3}$	$\frac{1}{2}$
0	1
$\frac{1}{3}$	2
$\frac{2}{3}$	4
1	8

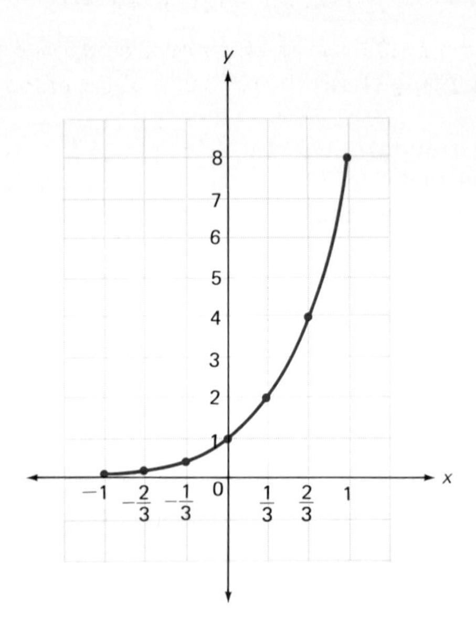

> Can you think of reasons why values of $b \leq 0$ are also excluded as bases for exponential functions?

Our attention has been restricted to the exponential functions $y = b^x$, where $b > 1$, all of which have the same general shape as $y = 2^x$. For $b = 1$, we get $y = 1^x = 1$ for all x. Since this is a constant function, we do not use base $b = 1$ in the classification of exponential functions.

The remaining base values for our purposes are those for which $0 < b < 1$. In particular, if $b = \frac{1}{2}$, we get $y = \left(\frac{1}{2}\right)^x = \frac{1}{2^x}$.

x	$y = 2^{-x}$
-3	8
-2	4
-1	2
0	1
1	$\frac{1}{2}$
2	$\frac{1}{4}$
3	$\frac{1}{8}$

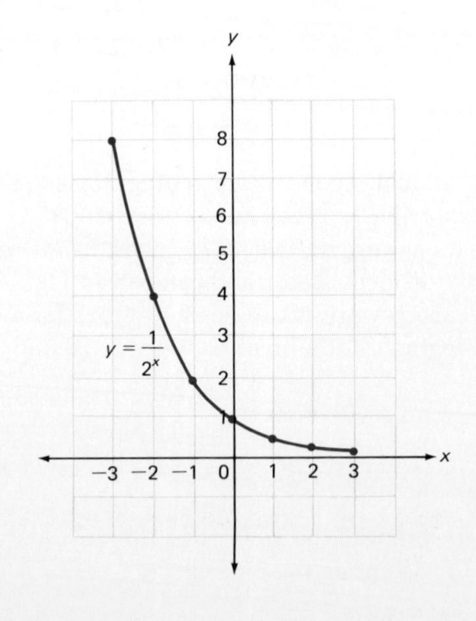

The graph of $y = g(x) = \frac{1}{2^x}$ can also be found by comparing it to the graph of $y = f(x) = 2^x$. Since $g(x) = \frac{1}{2^x} = 2^{-x} = f(-x)$, the y-values for g are the same as the y-values for f on the opposite side of the y-axis. In other words, the graph of g is the reflection of the graph of f through the y-axis.

All the curves $y = b^x$ for $0 < b < 1$ have this same basic shape. The curve is concave up, the function is decreasing, and $y = 0$ is a horizontal asymptote.

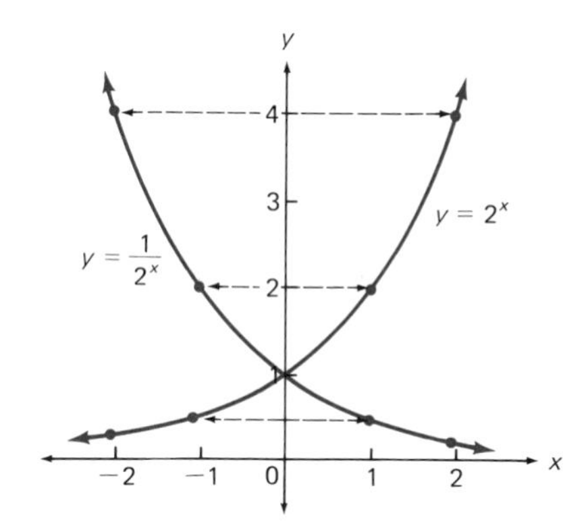

EXAMPLE 2 Use the graph of $y = f(x) = 2^x$ to sketch the curves $y = g(x) = 2^{x-3}$ and $y = h(x) = 2^x - 1$.

Solution Since $g(x) = f(x - 3)$, the graph of g can be obtained by shifting $y = 2^x$ by 3 units to the right. Moreover, since $h(x) = f(x) - 1$ the graph of h can be found by shifting $y = 2^x$ down 1 unit.

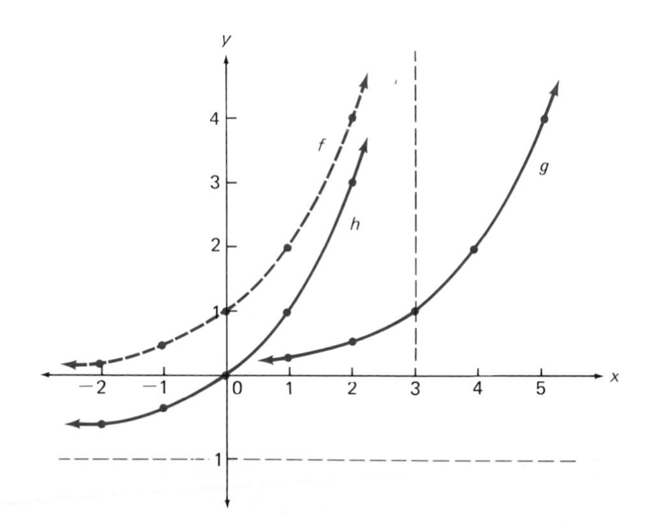

Following are some important properties of the function $y = b^x$, for $b > 0$ and $b \neq 1$.

PROPERTIES OF $y = f(x) = b^x$

1 The domain consists of all real numbers x.
2 The range consists of all positive numbers y.
3 The function is increasing (the curve is rising) when $b > 1$, and it is decreasing (the curve is falling) when $0 < b < 1$.
4 The curve is concave up for $b > 1$ and for $0 < b < 1$.
5 It is a one-to-one function.
6 The point (0, 1) is on the curve. There is no x-intercept.
7 The x-axis is a horizontal asymptote to the curve, toward the left for $b > 1$ and toward the right for $0 < b < 1$.
8 $b^{x_1}b^{x_2} = b^{x_1+x_2}$; $b^{x_1}/b^{x_2} = b^{x_1-x_2}$; $(b^{x_1})^{x_2} = b^{x_1x_2}$.

This form of the one-to-one property can sometimes be applied to the solutions of equations.

The one-to-one property of a function f may be stated in this way:

$$\text{If } f(x_1) = f(x_2), \text{ then } x_1 = x_2$$

That is, since $f(x_1)$ and $f(x_2)$ represent the same range value there can only be one corresponding domain value; consequently, $x_1 = x_2$. Using $f(x) = b^x$, this statement means that if $b^{x_1} = b^{x_2}$, then $x_1 = x_2$.

EXAMPLE 3 Use the one-to-one property to solve for t in the exponential equation $5^{t^2} = 625$.

Solution Since $625 = 5^4$ we get

$$5^{t^2} = 5^4$$

Now use the one-to-one property of $f(x) = 5^x$ to equate exponents. Thus

$$t^2 = 4$$

$$t = \pm 2$$

Check: $5^{(\pm 2)^2} = 5^4 = 625$.

EXERCISES 7.2

In Exercises 1–10, graph the exponential function f by making use of a brief table of values. Then use this curve to sketch the graph of g. Indicate the horizontal asymptotes.

1 $f(x) = 2^x$; $g(x) = 2^{x+3}$
2 $f(x) = 3^x$; $g(x) = 3^x - 2$
3 $f(x) = 4^x$; $g(x) = -(4^x)$
4 $f(x) = 5^x$; $g(x) = (\frac{1}{5})^x$
5 $f(x) = (\frac{3}{2})^x$; $g(x) = (\frac{3}{2})^{-x}$
6 $f(x) = 8^x$; $g(x) = 8^{x-2} + 3$
7 $f(x) = 3^x$; $g(x) = 2(3^x)$
8 $f(x) = 3^x$; $g(x) = \frac{1}{2}(3^x)$
9 $f(x) = 2^{x/2}$; $g(x) = 2^{x/2} - 3$
10 $f(x) = 4^x$; $g(x) = 4^{1-x}$

In Exercises 11–14, sketch the curves on the same axes.

11 $y = (\frac{3}{2})^x$, $y = 2^x$, $y = (\frac{5}{2})^x$
12 $y = (\frac{1}{4})^x$, $y = (\frac{1}{3})^x$, $y = (\frac{1}{2})^x$
13 $y = 2^{|x|}$, $y = -(2^{|x|})$
14 $y = 2^x$, $y = 2^{-x}$, $y = 2^x - 2^{-x}$
(*Hint:* Subtract ordinates.)

In Exercises 15–32, use the one-to-one property of an appropriate exponential function to solve the indicated equation.

15 $2^x = 64$

16 $3^x = 81$

17 $2^{x^2} = 512$

18 $3^{x-1} = 27$

19 $5^{2x+1} = 125$

20 $2^{x^3} = 256$

21 $7^{x^2+x} = 49$

22 $b^{x^2+x} = 1$

23 $\frac{1}{2^x} = 32$

24 $\frac{1}{10^x} = 10{,}000$

25 $9^x = 3$

26 $64^x = 8$

27 $9^x = 27$

28 $64^x = 16$

29 $(\frac{1}{49})^x = 7$

30 $5^x = \frac{1}{125}$

31 $(\frac{27}{8})^x = \frac{9}{4}$

32 $(0.01)^x = 1000$

33 Graph the functions $y = 2^x$ and $y = x^2$ in the same coordinate system for the interval [0, 5]. (Use a larger unit on the x-axis than on the y-axis.) What are the points of intersection?

34 Use a calculator to verify that $\sqrt{3} = 1.732050\ldots$. Now fill in the powers of 2 in the table, rounding off each entry to four decimal places.

x	1.7	1.73	1.732	1.7320	1.73205
2^x					

On the basis of the results, what is your estimate of $2^{\sqrt{3}}$ to three decimal places? Now find $2^{\sqrt{3}}$ directly on the calculator and compare.

35 Follow the instructions in Exercise 34 for these numbers:

(a) $3^{\sqrt{2}}$ **(b)** $3^{\sqrt{3}}$

(c) $2^{\sqrt{5}}$ **(d)** 4^{π}

***36** Solve for x: $(6^{2x})(4^x) = 1728$.

***37** Solve for x: $(5^{2x+1})(7^{2x}) = 175$.

7.3 Logarithmic functions

It was pointed out in the preceding section that $y = f(x) = b^x$, for $b > 0$ and $b \neq 1$, is a one-to-one function. Since every one-to-one function has an inverse, it follows that f has an inverse. The graph of the inverse function g is the reflection of $y = f(x)$ through the line $y = x$. Here are two typical cases for $b > 1$ and for $0 < b < 1$.

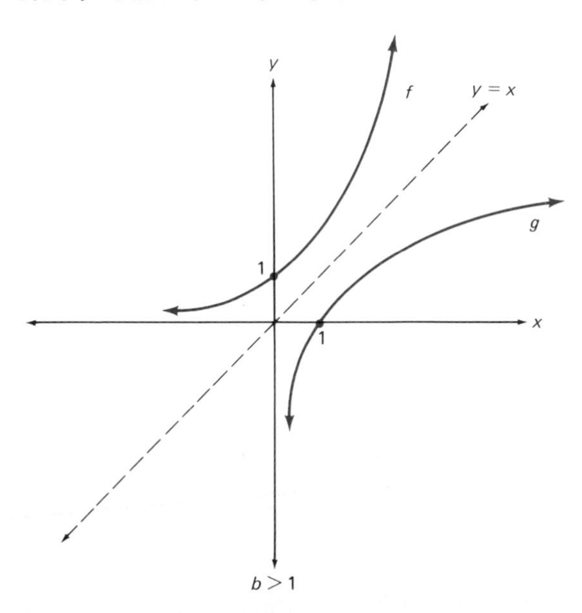

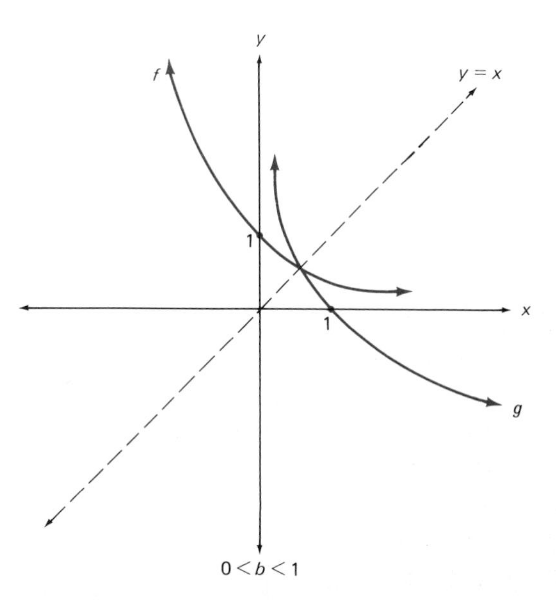

An equation for the inverse function g can be obtained by interchanging the roles of the variables in $y = b^x$. Thus $x = b^y$ is an equation for g. Unfortunately, we have no way of solving $x = b^y$ to get y explicitly in terms of x. To overcome this difficulty we create some new terminology.

The equation $x = b^y$ tells us that *y is the exponent on b that produces x.* In situations like this the word **logarithm** is used in place of *exponent*. A logarithm, then, is an exponent. Now we may say that *y is the logarithm on b that produces x.* This description can be abbreviated to $y = \text{logarithm}_b x$ and abbreviating further we reach the final form

Note: $y = b^x$ and $y = \log_b x$ are inverse functions.

$$y = \log_b x$$

which is read "y equals log x to the base b" or "y equals log x base b."

It is important to realize that we are only defining (not proving) the equation $y = \log_b x$ to have the same meaning as $x = b^y$. In other words, the logarithmic form $y = \log_b x$ and the exponential form $x = b^y$ are equivalent. And since they are equivalent they define the same function g:

$$y = g(x) = \log_b x.$$

Now we know that $y = f(x) = b^x$ and $y = g(x) = \log_b x$ are inverse functions. Consequently,

$$f(g(x)) = f(\log_b x) = b^{\log_b x} = x \quad \text{and} \quad g(f(x)) = g(b^x) = \log_b(b^x) = x$$

EXAMPLE 1 Write the equation of the inverse function g of $y = f(x) = 2^x$ and graph both on the same axes.

Solution The inverse g has equation $y = g(x) = \log_2 x$, and its graph can be obtained by reflecting $y = f(x) = 2^x$ through the line $y = x$.

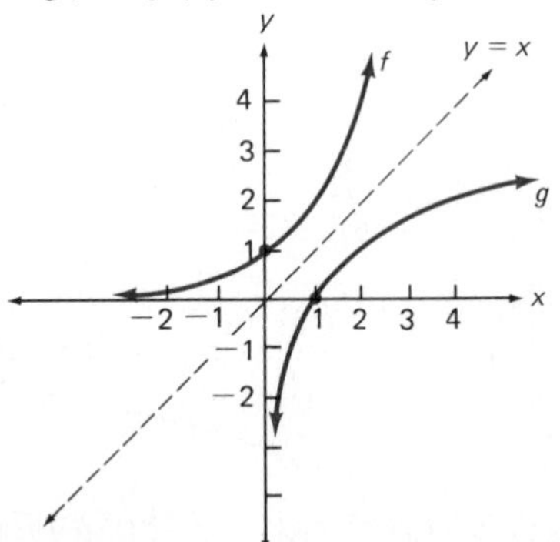

TEST YOUR UNDERSTANDING

1 Find the equation of the inverse of $y = 3^x$ and graph both on the same axes.

2 Find the equation of the inverse of $y = (\frac{1}{3})^x$ and graph both on the same axes.

Let $y = f(x) = \log_5 x$. Describe how the graph of each of the following can be obtained from the graph of f.

3 $g(x) = \log_5(x + 2)$

4 $g(x) = 2 + \log_5 x$

5 $g(x) = -\log_5 x$

6 $g(x) = 2 \log_5 x$

We found $y = \log_b x$ by interchanging the role of the variables in $y = b^x$. As a consequence of this switching, the domains and ranges of the two functions are also interchanged. Thus

Domain of $y = \log_b x$ is the same as the range of $y = b^x$.

Range of $y = \log_b x$ is the same as the domain of $y = b^x$.

These results are incorporated into the following list of important properties of the function $y = \log_b x$, where $b > 0$ and $b \neq 1$.

PROPERTIES OF $y = f(x) = \log_b x$

1 The domain consists of all positive numbers x.
2 The range consists of all real numbers y.
3 The function increases (the curve is rising) for $b > 1$, and it decreases (the curve is falling) for $0 < b < 1$.
4 The curve is concave down for $b > 1$, and it is concave up for $0 < b < 1$.
5 It is a one-to-one function; if $\log_b(x_1) = \log_b(x_2)$, then $x_1 = x_2$.
6 The point (1, 0) is on the graph. There is no y-intercept.
7 The y-axis is a vertical asymptote to the curve in the downward direction for $b > 1$ and in the upward direction for $0 < b < 1$.
8 $\log_b(b^x) = x$ and $b^{\log_b x} = x$.

EXAMPLE 2 Find the domain for $y = \log_2(x - 3)$.

Solution In $y = \log_2(x - 3)$ the quantity $x - 3$ plays the role that x does in $\log_2 x$. Thus $x - 3 > 0$ and the domain consists of all $x > 3$.

The equation $y = \log_b x$ is equivalent to $x = b^y$ (not to be confused with the inverse $y = b^x$). Studying the following list of special cases will help you to understand this equivalence.

Logarithmic form $\log_b x = y$	*Exponential form* $b^y = x$
$\log_5 25 = 2$	$5^2 = 25$
$\log_{27} 9 = \frac{2}{3}$	$27^{2/3} = 9$
$\log_6 \frac{1}{36} = -2$	$6^{-2} = \frac{1}{36}$
$\log_b 1 = 0$	$b^0 = 1$

Of the two forms $y = \log_b x$ and $x = b^y$, the exponential form is usually easier to work with. Consequently, when there is a question concerning

$y = \log_b x$ it is often useful to convert to the exponential form. For instance, to evaluate $\log_9 27$ we write

$$y = \log_9 27$$

and convert to exponential form. Thus

$$9^y = 27$$

If you happen to recognize that $9^{3/2} = 27$, you will see that $y = \frac{3}{2}$. Otherwise, try writing each side of $9^y = 27$ with the same base. Since $27 = 3^3$ and $9^y = (3^2)^y = 3^{2y}$, we have

$$3^{2y} = 3^3$$

$$2y = 3 \qquad (f(t) = 3^t \text{ is one-to-one})$$

$$y = \frac{3}{2}$$

EXAMPLE 3 Solve for b: $\log_b 8 = \frac{3}{4}$.

Solution Convert to exponential form.

$$b^{3/4} = 8$$

Take the $\frac{4}{3}$ power of both sides.

$8^{4/3} = (\sqrt[3]{8})^4 = 2^4$

$$(b^{3/4})^{4/3} = 8^{4/3}$$

$$b = 16$$

EXERCISES 7.3

Sketch the graph of the function f. Reflect this curve through the line y = x to obtain the graph of the inverse function g, and write the equation for g.

1 $y = f(x) = 4^x$
2 $y = f(x) = 5^x$
3 $y = f(x) = (\frac{1}{3})^x$
4 $y = f(x) = (0.2)^x$

Describe how the graph of h can be obtained from the graph of g. Find the domain of h, and write the equation of the vertical asymptote.

5 $g(x) = \log_3 x$; $h(x) = \log_3(x + 2)$
6 $g(x) = \log_5 x$; $h(x) = \log_5(x - 1)$
7 $g(x) = \log_8 x$; $h(x) = 2 + \log_8 x$
8 $g(x) = \log_{10} x$; $h(x) = 2 \log_{10} x$

Sketch the graph of f and state its domain.

9 $f(x) = \log_{10} x$
10 $f(x) = -\log_{10} x$
11 $f(x) = |\log_{10} x|$
12 $f(x) = \log_{10}(-x)$
13 $f(x) = \log_{10}|x|$
14 $f(x) = \log_{1/10}(x + 1)$

Convert from the exponential to the logarithmic form.

15 $2^8 = 256$
16 $5^{-3} = \frac{1}{125}$
17 $(\frac{1}{3})^{-1} = 3$
18 $81^{3/4} = 27$
19 $17^0 = 1$
20 $(\frac{1}{49})^{-1/2} = 7$

Convert from the logarithmic form to the exponential form.

21 $\log_{10} 0.0001 = -4$
22 $\log_{64} 4 = \frac{1}{3}$
23 $\log_{\sqrt{2}} 2 = 2$
24 $\log_{13} 13 = 1$
25 $\log_{12} \frac{1}{1728} = -3$
26 $\log_{27/8} \frac{9}{4} = \frac{2}{3}$

Solve for the indicated quantity: y, x, or b.

27 $\log_2 16 = y$
28 $\log_{1/2} 32 = y$
29 $\log_{1/3} 27 = y$
30 $\log_7 x = -2$
31 $\log_{1/6} x = 3$
32 $\log_8 x = -\frac{2}{3}$
33 $\log_b 125 = 3$
34 $\log_b 8 = \frac{3}{2}$
35 $\log_b \frac{1}{8} = -\frac{3}{2}$
36 $\log_{100} 10 = y$
37 $\log_{27} 3 = y$
38 $\log_{1/16} x = \frac{1}{4}$

39 $\log_b \frac{16}{81} = 4$

40 $\log_8 x = -3$

41 $\log_b \frac{1}{27} = -\frac{3}{2}$

42 $\log_{\sqrt{3}} x = 2$

43 $\log_{\sqrt{8}} (\frac{1}{8}) = y$

44 $\log_b \frac{1}{128} = -7$

45 $\log_{0.001} 10 = y$

46 $\log_{0.2} 5 = y$

47 $\log_9 x = 1$

Evaluate the expressions in Exercises 48–49.

*48 $\log_2(\log_4 256)$

*49 $\log_{3/4}(\log_{1/27} \frac{1}{81})$

By interchanging the roles of the variables, find the inverse function g. Show that $(f \circ g)(x) = x$ and $(g \circ f)(x) = x$.

*50 $y = f(x) = 2^{x+1}$

*51 $y = f(x) = \log_3(x + 3)$

7.4 The laws of logarithms

From the rules of exponents we have

$$2^3 \cdot 2^4 = 2^{3+4} = 2^7$$

Now let us focus on just the exponential part:

$$3 + 4 = 7$$

The three exponents involved here can be expressed as logarithms.

Logarithms are exponents.

$$3 = \log_2 8 \qquad \text{because } 2^3 = 8$$

$$4 = \log_2 16 \qquad \text{because } 2^4 = 16$$

$$7 = \log_2 128 \qquad \text{because } 2^7 = 128$$

Substituting these expressions into $3 + 4 = 7$ gives

$$\log_2 8 + \log_2 16 = \log_2 128$$

Furthermore, since $128 = 8 \cdot 16$, we have

$$\log_2 8 + \log_2 16 = \log_2(8 \cdot 16)$$

This is a special case of the first law of logarithms:

LAWS OF LOGARITHMS

If M and N are positive, $b > 0$ and $b \neq 1$, then

LAW 1. $\log_b MN = \log_b M + \log_b N$

LAW 2. $\log_b \frac{M}{N} = \log_b M - \log_b N$

LAW 3. $\log_b(N^k) = k \log_b N$

Law 1 says that the log of a product is the sum of the logs of the factors. Can you give similar interpretations for Laws 2 and 3?

Since logarithms are exponents it is not surprising that these laws can be proved by using the appropriate rules of exponents. Following is a proof of Law 1; the proofs of Laws 2 and 3 are left as exercises.

Let

$$\log_b M = r \qquad \text{and} \qquad \log_b N = s$$

Convert to exponential form:

$$M = b^r \qquad \text{and} \qquad N = b^s$$

Multiply the two equations:

$$MN = b^r b^s = b^{r+s}$$

Then convert to logarithmic form:

$$\log_b MN = r + s$$

Substitute for r and s to get the final result:

$$\log_b MN = \log_b M + \log_b N$$

EXAMPLE 1 For positive numbers A, B, and C, show that

$$\log_b \frac{AB^2}{C} = \log_b A + 2 \log_b B - \log_b C$$

Solution

$$\log_b \frac{AB^2}{C} = \log_b(AB^2) - \log_b C \qquad \text{(Law 2)}$$

$$= \log_b A + \log_b B^2 - \log_b C \qquad \text{(Law 1)}$$

$$= \log_b A + 2 \log_b B - \log_b C \qquad \text{(Law 3)}$$

EXAMPLE 2 Express $\frac{1}{2} \log_b x - 3 \log_b(x - 1)$ as the logarithm of a single expression in x.

Solution

Identify the laws of logarithms being used in Examples 2 and 3.

$$\frac{1}{2} \log_b x - 3 \log_b(x - 1) = \log_b x^{1/2} - \log_b(x - 1)^3$$

$$= \log_b \frac{x^{1/2}}{(x - 1)^3}$$

$$= \log_b \frac{\sqrt{x}}{(x - 1)^3}$$

EXAMPLE 3 Given: $\log_b 2 = 0.6931$ and $\log_b 3 = 1.0986$; find $\log_b \sqrt{12}$.

Solution

$$\log_b \sqrt{12} = \log_b 12^{1/2} = \tfrac{1}{2} \log_b 12$$

$$= \tfrac{1}{2} \log_b(3 \cdot 4) = \tfrac{1}{2} [\log_b 3 + \log_b 4]$$

$$= \tfrac{1}{2} [\log_b 3 + \log_b 2^2]$$

$$= \tfrac{1}{2} [\log_b 3 + 2 \log_b 2]$$

$$= \tfrac{1}{2} \log_b 3 + \log_b 2$$

$$= \tfrac{1}{2}(1.0986) + 0.6931$$

$$= 1.2424$$

TEST YOUR UNDERSTANDING

Convert the given logarithms into expressions involving $\log_b A$, $\log_b B$, and $\log_b C$.

1 $\log_b ABC$

2 $\log_b \frac{A}{BC}$

3 $\log_b \frac{(AB)^2}{C}$

4 $\log_b AB^2C^3$

5 $\log_b \frac{A\sqrt{B}}{C}$

6 $\log_b \frac{\sqrt[3]{A}}{(BC)^3}$

Change each expression into the logarithm of a single expression in x.

7 $\log_b x + \log_b x + \log_b 3$

8 $2 \log_b(x - 1) + \frac{1}{2} \log_b x$

9 $\log_b(2x - 1) - 3 \log_b(x^2 + 1)$

10 $\log_b x - \log_b(x - 1) - 2 \log_b(x - 2)$

Use the information given in Example 3 to find these logarithms.

11 $\log_b 18$

12 $\log_b \frac{16}{27}$

Examples 4 to 6 illustrate how the laws of logarithms can be used to solve logarithmic equations.

EXAMPLE 4 Solve for x: $\log_8(x - 6) + \log_8(x + 6) = 2$.

Solution First note that in $\log_8(x - 6)$ we must have $x - 6 > 0$, or $x > 6$. Similarly, $\log_8(x + 6)$ calls for $x > -6$. Therefore, the only solutions, if there are any, must satisfy $x > 6$.

$$\log_8(x - 6) + \log_8(x + 6) = 2$$

$$\log_8(x - 6)(x + 6) = 2 \quad \text{(Law 1)}$$

$$\log_8(x^2 - 36) = 2$$

$$x^2 - 36 = 8^2 \quad \text{(converting to exponential form)}$$

$$x^2 - 100 = 0$$

$$(x + 10)(x - 10) = 0$$

$$x = -10 \quad \text{or} \quad x = 10$$

The only possible solutions are -10 and 10. Our initial observation that $x > 6$ automatically eliminates -10. (If that initial observation had not been made, -10 could still have been eliminated by checking in the given equation.) The value $x = 10$ can be checked as follows:

$$\log_8(10 - 6) + \log_8(10 + 6) = \log_8 4 + \log_8 16$$

$$= \tfrac{2}{3} + \tfrac{4}{3} = 2$$

EXAMPLE 5 Solve: $\log_{10}(x^3 - 1) - \log_{10}(x^2 + x + 1) = 1$.

Solution

$$\log_{10}(x^3 - 1) - \log_{10}(x^2 + x + 1) = 1$$

$$\log_{10}\frac{x^3 - 1}{x^2 + x + 1} = 1 \quad \text{(Law 2)}$$

$$\log_{10}\frac{(x-1)(x^2+x+1)}{x^2+x+1} = 1 \qquad \text{(by factoring)}$$

$$\log_{10}(x-1) = 1$$

$$x - 1 = 10^1 \qquad \text{(Why?)}$$

$$x = 11$$

Check: $\log_{10}(11^3 - 1) - \log_{10}(11^2 + 11 + 1) = \log_{10}1330 - \log_{10}133$

$$= \log_{10}\tfrac{1330}{133}$$

$$= \log_{10}10 = 1$$

EXAMPLE 6 Solve: $\log_3 2x - \log_3(x+5) = 0$.

Alternate Solution

$\log_3 2x - \log_3(x+5) = 0$

$\log_3 2x = \log_3(x+5)$

$2x = x + 5$ **(Why?)**

$x = 5$

Solution

$$\log_3 2x - \log_3(x+5) = 0$$

$$\log_3\frac{2x}{x+5} = 0$$

$$\frac{2x}{x+5} = 3^0$$

$$\frac{2x}{x+5} = 1$$

$$2x = x + 5$$

$$x = 5$$

Check: $\log_3 2(5) - \log_3(5+5) = \log_3 10 - \log_3 10 = 0$.

CAUTION: LEARN TO AVOID MISTAKES LIKE THESE

WRONG	RIGHT
$\log_b A + \log_b B = \log_b(A+B)$	$\log_b A + \log_b B = \log_b AB$
$\log_b(x^2-4) = \log_b x^2 - \log_b 4$	$\log_b(x^2-4)$ $= \log_b(x+2)(x-2)$ $= \log_b(x+2) + \log_b(x-2)$
$(\log_b x)^2 = 2\log_b x$	$(\log_b x)^2 = (\log_b x)(\log_b x)$
$\log_b A - \log_b B = \dfrac{\log_b A}{\log_b B}$	$\log_b A - \log_b B = \log_b \dfrac{A}{B}$
If $2\log_b x = \log_b(3x+4)$, then $2x = 3x + 4$	If $2\log_b x = \log_b(3x+4)$, then $\log_b x^2 = \log_b(3x+4)$ and $x^2 = 3x + 4$.

Use the laws of logarithms (as much as possible) to convert the given logarithms into expressions involving sums and differences.

1 $\log_b \frac{3x}{x+1}$

2 $\log_b \frac{x^2}{x-1}$

3 $\log_b \frac{\sqrt{x^2-1}}{x}$

4 $\log_b \frac{1}{x}$

5 $\log_b \frac{1}{x^2}$

6 $\log_b \sqrt{\frac{x+1}{x-1}}$

Convert each expression into the logarithm of a single expression in x.

7 $\log_b(x+1) - \log_b(x+2)$

8 $\log_b x + 2\log_b(x-1)$

9 $\frac{1}{2}\log_b(x^2-1) - \frac{1}{2}\log_b(x^2+1)$

10 $\log_b(x+2) - \log_b(x^2-4)$

11 $3\log_b x - \log_b 2 - \log_b(x+5)$

12 $\frac{1}{3}\log_b(x-1) + \log_b 3 - \frac{1}{3}\log_b(x+1)$

Use the appropriate laws of logarithms to explain why each statement is correct.

13 $\log_b 27 + \log_b 3 = \log_b 243 - \log_b 3$

14 $\log_b 16 + \log_b 4 = \log_b 64$

15 $-2\log_b \frac{4}{9} = \log_b \frac{81}{16}$

16 $\frac{1}{2}\log_b 0.0001 = -\log_b 100$

Find the logarithms in Exercises 17–22 by using the laws of logarithms and the given information that $\log_b 2 = .3010$, $\log_b 3 = 0.4771$, and $\log_b 5 = 0.6990$. Assume that all logs have base b.

17 **(a)** $\log 4$ **(b)** $\log 8$ **(c)** $\log \frac{1}{2}$

18 **(a)** $\log \sqrt{2}$ **(b)** $\log 9$ **(c)** $\log 12$

19 **(a)** $\log 48$ **(b)** $\log \frac{2}{3}$ **(c)** $\log 125$

20 **(a)** $\log 50$ **(b)** $\log 10$ **(c)** $\log \frac{25}{6}$

21 **(a)** $\log \sqrt[3]{5}$ **(b)** $\log \sqrt{20^3}$ **(c)** $\log \sqrt{900}$

22 **(a)** $\log 0.2$ **(b)** $\log 0.25$ **(c)** $\log 2.4$

Solve for x and check.

23 $\log_{10} x + \log_{10} 5 = 2$

24 $\log_{10} x + \log_{10} 5 = 1$

25 $\log_{10} 5 - \log_{10} x = 2$

26 $\log_{10}(x+21) + \log_{10} x = 2$

27 $\log_{12}(x-5) + \log_{12}(x-5) = 2$

28 $\log_3 x + \log_3(2x+51) = 4$

29 $\log_{16} x + \log_{16}(x-4) = \frac{5}{4}$

30 $\log_2(x^2) - \log_2(x-2) = 3$

31 $\log_{10}(3-x) - \log_{10}(12-x) = -1$

32 $\log_{10}(3x^2-5x-2) - \log_{10}(x-2) = 1$

33 $\log_{1/7} x + \log_{1/7}(5x-28) = -2$

34 $\log_{1/3} 12x^2 - \log_{1/3}(20x-9) = -1$

35 $\log_{10}(x^3-1) - \log_{10}(x^2+x+1) = -2$

36 $2\log_{10}(x-2) = 4$

37 $2\log_{25} x - \log_{25}(25-4x) = \frac{1}{2}$

38 $\log_3(8x^3+1) - \log_3(4x^2-2x+1) = 2$

***39** Prove Law 2. (*Hint:* Follow the proof of Law 1 using $\frac{b^r}{b^s} = b^{r-s}$.)

***40** Prove Law 3. (*Hint:* Use $(b^r)^k = b^{rk}$.)

***41** Solve for x: $(x+2)\log_b b^x = x$.

***42** Solve for x: $\log_{N^2} N = x$.

***43** Solve for x: $\log_x(2x)^{3x} = 4x$.

***44** **(a)** Explain why $\log_b b = 1$.
(b) Show that $(\log_b a)(\log_a b) = 1$.
(*Hint:* Use Law 3 and the result $b^{\log_b x} = x$.)

***45** Use $B^{\log_B N} = N$ to derive $\log_B N = \frac{\log_b N}{\log_b B}$. (*Hint:* Begin by taking the log base b of both sides.)

7.5

The base *e*

The graphs of $y = b^x$ for $b > 1$ all have the same basic shape, as shown in the following figure. Notice that the larger the value of b, the faster the curve rises toward the right and the faster it approaches the x-axis toward the left. You can use your imagination to see that as all possible base values $b > 1$ are considered, the corresponding curves will completely fill in the shaded regions.

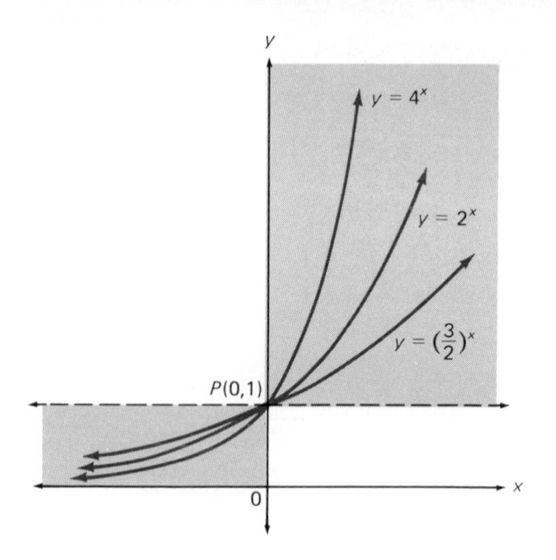

All such curves pass through point $P(0, 1)$. The tangent lines to these curves through P are relatively flat (small positive slope) for values of b close to 1 and very steep for large values of b. The slopes of these tangents consist of all numbers $m > 0$.

These figures show the curves $y = 2^x$ an l $y = 3^x$, including the tangent through point $P(0, 1)$.

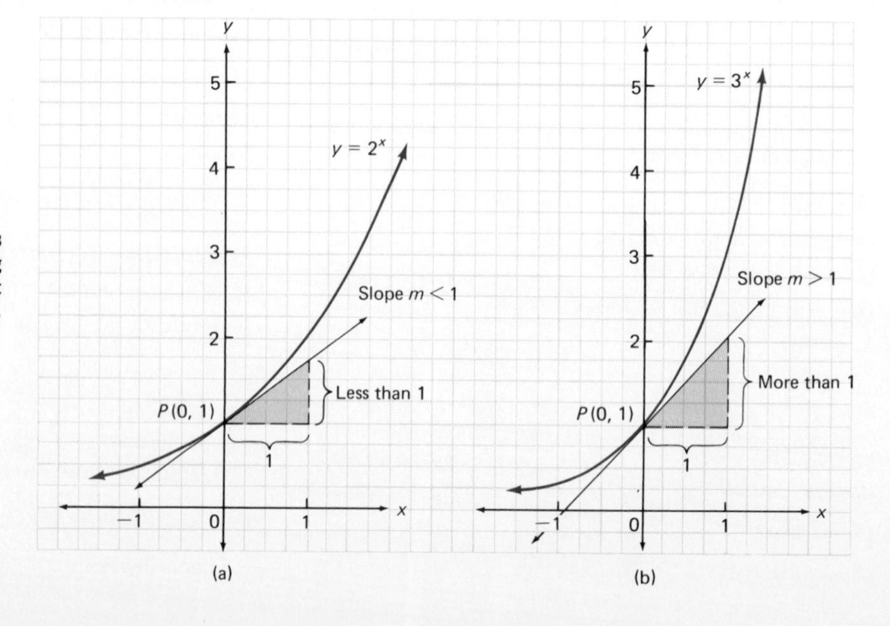

From the grid marks you can observe that the slope of the tangent to $y = 2^x$ is less than 1 because for a horizontal change of 1 unit the vertical change is less than 1 unit. Similarly, you can see that the slope of the tangent to $y = 3^x$ is slightly more than 1. We suspect that there must be a value b so that the slope of the tangent to the corresponding exponential function through point P is exactly equal to 1. In advanced courses it can be shown that such a value does exist. This number plays a very important role in mathematics, and it is designated by the letter e.

e is the real number such that the tangent to the graph of $y = e^x$ at the point $P(0, 1)$ has slope equal to 1.

The number e is also closely related to the expression $\left(1 + \frac{1}{n}\right)^n$. The larger n is taken, the closer $\left(1 + \frac{1}{n}\right)^n$ gets to e. For example.

$(1 + \frac{1}{10})^{10} = 2.59374$

$(1 + \frac{1}{100})^{100} = 2.70481$

$(1 + \frac{1}{1000})^{1000} = 2.71692$

See Exercise 57.

Since the curve $y = e^x$ is between $y = 2^x$ and $y = 3^x$, we expect that e satisfies $2 < e < 3$. This is correct; in fact, it turns out that e is an irrational number that is closer to 3 than to 2. Carried to five decimal places, $e = 2.71828$.

For theoretical purposes e is the most important base number for exponential and logarithmic functions. The inverse of $y = e^x$ is given by $y = \log_e x$. In place of $\log_e x$ we will now write **ln x**, which is called the **natural log of x**. Thus $x = e^y$ and $y = \ln x$ are equivalent.

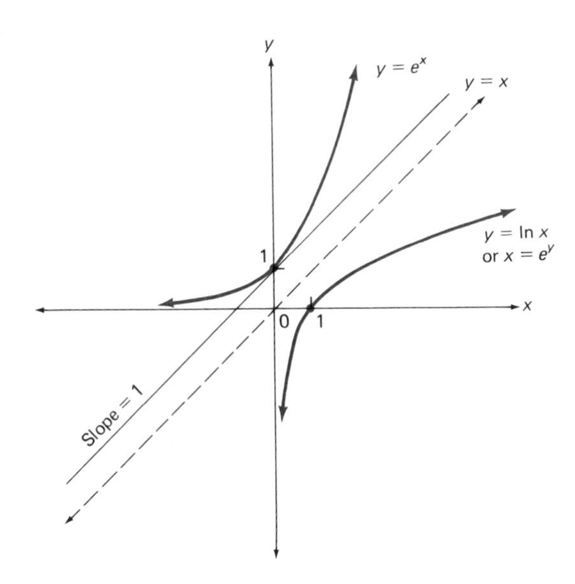

What is the equation of the tangent to $y = e^x$ at (0, 1)?

Since $e > 1$, the properties of $y = b^x$ and $y = \log_b x$ $(b > 1)$ carry over to $y = e^x$ and $y = \ln x$. We collect these properties here for easy reference.

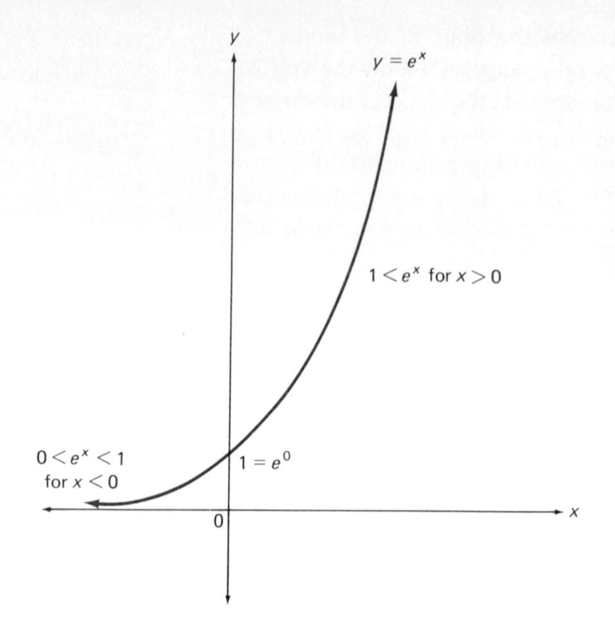

PROPERTIES $y = e^x$

1 Domain: all reals.
2 Range: all $y > 0$.
3 Increasing function.
4 Curve is concave up.
5 One-to-one function;
if $e^{x_1} = e^{x_2}$, then $x_1 = x_2$.
6 $0 < e^x < 1$ for $x < 0$;
$e^0 = 1$; $e^x > 1$ for $x > 0$.
7 $e^{x_1}e^{x_2} = e^{x_1+x_2}$
$\frac{e^{x_1}}{e^{x_2}} = e^{x_1-x_2}$
$(e^{x_1})^{x_2} = e^{x_1x_2}$
8 $e^{\ln x} = x$.
9 Horizontal asymptote: $y = 0$.

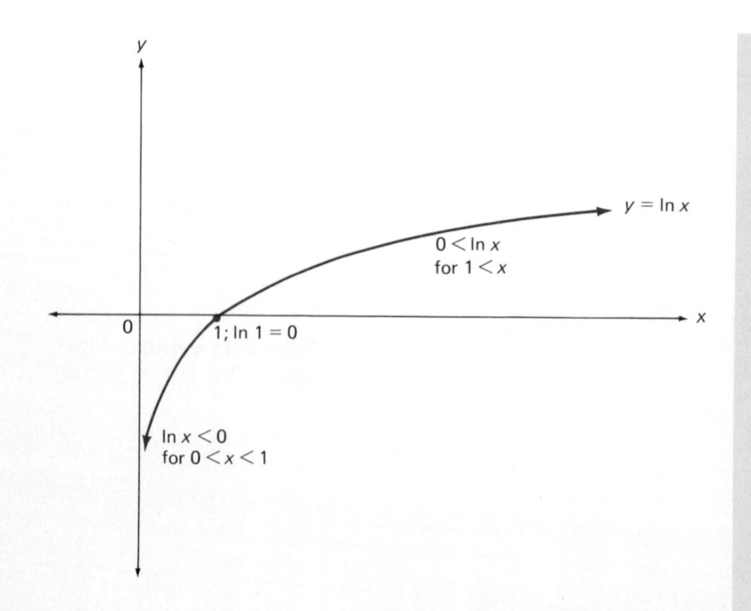

PROPERTIES OF $y = \ln x$

1 Domain: all $x > 0$.
2 Range: all reals.
3 Increasing function.
4 Curve is concave down.
5 One-to-one function;
if $\ln x_1 = \ln x_2$, then $x_1 = x_2$.
6 $\ln x < 0$ for $0 < x < 1$;
$\ln 1 = 0$; $\ln x > 0$ for $x > 1$.
7 $\ln x_1x_2 = \ln x_1 + \ln x_2$
$\ln \frac{x_1}{x_2} = \ln x_1 - \ln x_2$
$\ln x_1^{x_2} = x_2 \ln x_1$
8 $\ln e^x = x$.
9 Vertical asymptote; $x = 0$.

The examples that follow utilize the base e and are solved in a manner similar to those done earlier for other bases.

EXAMPLE 1 (**a**) Find the domain of $y = \ln (x - 2)$; (**b**) Sketch $y = \ln x^2$ for $x > 0$.

Solution

(a) Since the domain of $y = \ln x$ is all $x > 0$, the domain of $y = \ln(x - 2)$ has all x for which $x - 2 > 0$; all $x > 2$.

(b) Since $y = \ln x^2 = 2 \ln x$, we obtain the graph by multiplying the ordinates of $y = \ln x$ by 2.

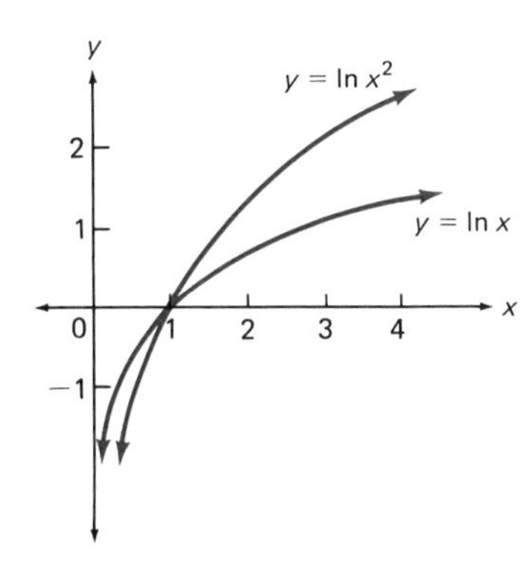

EXAMPLE 2 Solve for t: **(a)** $e^{\ln(2t-1)} = 5$; **(b)** $e^{2t-1} = 5$.

Solution

(a)
$$e^{\ln(2t-1)} = 5$$
$$2t - 1 = 5 \qquad \text{(Property 8 for } y = e^x\text{)}$$
$$2t = 6$$
$$t = 3$$

(b)
$$e^{2t-1} = 5$$
$$2t - 1 = \ln 5 \qquad \text{(Why?)}$$
$$2t = 1 + \ln 5$$
$$t = \tfrac{1}{2}(1 + \ln 5)$$

To three decimal places, $t = \frac{1}{2}(1 + \ln 5) = 1.305$

Check: for (b): $e^{2[1/2(1+\ln 5)]-1} = e^{1+\ln 5-1} = e^{\ln 5} = 5$.

EXAMPLE 3 Solve for x: $\ln(x + 1) = 1 + \ln x$.

Solution

$$\ln(x + 1) - \ln x = 1$$
$$\ln \frac{x + 1}{x} = 1$$

Now convert to exponential form:

$$\frac{x + 1}{x} = e$$
$$ex = x + 1$$
$$(e - 1)x = 1$$
$$x = \frac{1}{e - 1}$$

Check: $\ln\left(\frac{1}{e-1}+1\right)=\ln\frac{e}{e-1}=\ln e-\ln(e-1)=1+\ln(e-1)^{-1}$

$$=1+\ln\frac{1}{(e-1)}.$$

EXAMPLE 4 **(a)** Show $h(x)=\ln(x^2+5)$ as the composite of two functions. **(b)** Show $F(x)=e^{\sqrt{x^2-3x}}$ as the composite of three functions.

Solution

(a) Let $f(x)=\ln x$ and $g(x)=x^2+5$. Then

$$(f\circ g)(x)=f(g(x))=f(x^2+5)=\ln(x^2+5)=h(x)$$

(b) Let $f(x)=e^x$, $g(x)=\sqrt{x}$, and $h(x)=x^2-3x$. Then

$$\begin{aligned}(f\circ g\circ h)(x)&=f(g(h(x))) && h \text{ is the "inner" function.}\\ &=f(g(x^2-3x)) && g \text{ is the "middle" function.}\\ &=f(\sqrt{x^2-3x}) && f \text{ is the "outer" function.}\\ &=e^{\sqrt{x^2-3x}}=F(x)\end{aligned}$$

(Other solutions are possible.)

EXAMPLE 5 Determine the signs of $f(x)=x^2e^x+2xe^x$.

Solution We find that $f(x)=x^2e^x+2xe^x=xe^x(x+2)$ in which $e^x>0$ for all x, and the other factors are zero when $x=0$ or $x=-2$.

Interval	$(-\infty, -2)$	$(-2, 0)$	$(0, \infty)$
Sign of $x+2$	$-$	$+$	$+$
Sign of x	$-$	$-$	$+$
Sign of e^x	$+$	$+$	$+$
Sign of $f(x)$	$+$	$-$	$+$

$$f(x)>0 \text{ on the intervals } (-\infty,-2) \text{ and } (0,\infty)$$
$$f(x)<0 \text{ on the interval } (-2,0)$$

EXERCISES 7.5

Sketch each pair of functions on the same axes.

1 $y=e^x$; $y=e^{x-2}$

2 $y=e^x$; $y=2e^x$

3 $y=\ln x$; $y=\frac{1}{2}\ln x$

4 $y=\ln x$; $y=\ln(x+2)$

5 $y=\ln x$; $y=\ln(-x)$

6 $y = e^x$; $y = e^{-x}$

7 $y = e^x$; $y = e^x + 2$

8 $y = \ln x$; $y = \ln |x|$

9 $f(x) = -e^{-x}$, $g(x) = 1 - e^{-x}$

10 $g(x) = 1 - e^{-x}$, $s(x) = 1 - e^{-2x}$

11 $g(x) = 1 - e^{-x}$, $t(x) = 1 - e^{(-1/2)x}$

12 $u(x) = 1 - e^{-3x}$, $v(x) = 1 - e^{(-1/3)x}$

Explain how the graph of f can be obtained from the curve $y = \ln x$. (Hint: First apply the appropriate rules of logarithms.)

13 $f(x) = \ln ex$

14 $f(x) = \ln \dfrac{x}{e}$

15 $f(x) = \ln \sqrt{x}$

16 $f(x) = \ln \dfrac{1}{x}$

17 $f(x) = \ln(x^2 - 1) - \ln(x + 1)$

18 $f(x) = \ln x^{-3}$

Find the domain.

19 $f(x) = \ln(x + 2)$

20 $f(x) = \ln |x|$

21 $f(x) = \ln(2x - 1)$

22 $f(x) = \dfrac{1}{\ln x}$

23 $f(x) = \dfrac{\ln(x - 1)}{x - 2}$

24 $f(x) = \ln(\ln x)$

Use the laws of logarithms (as much as possible) to write $\ln f(x)$ as an expression involving sums and differences.

25 $f(x) = \dfrac{5x}{x^2 - 4}$

26 $f(x) = x\sqrt{x^2 + 1}$

27 $f(x) = \dfrac{(x - 1)(x + 3)^2}{\sqrt{x^2 + 2}}$

28 $f(x) = \sqrt{\dfrac{x + 7}{x - 7}}$

29 $f(x) = \sqrt{x^3(x + 1)}$

30 $f(x) = \dfrac{x}{\sqrt[3]{x^2 - 1}}$

Convert each expression into the logarithm of a single expression.

31 $\frac{1}{2} \ln x + \ln(x^2 + 5)$

32 $\ln 2 + \ln x - \ln(x - 1)$

33 $3 \ln(x + 1) + 3 \ln(x - 1)$

34 $\ln(x^3 - 1) - \ln(x^2 + x + 1)$

35 $\frac{1}{2} \ln x - 2 \ln(x - 1) - \frac{1}{3} \ln(x^2 + 1)$

Simplify.

36 $\ln(e^{3x})$

37 $e^{\ln \sqrt{x}}$

38 $\ln(x^2 e^3)$

39 $e^{-2 \ln x}$

40 $(e^{\ln x})^2$

41 $\ln\left(\dfrac{e^x}{e^{x-1}}\right)$

Solve for x.

42 $e^{3x+5} = 100$

43 $e^{-0.01x} = 27$

44 $e^{x^2} = e^x e^{3/4}$

45 $e^{\ln(1-x)} = 2x$

46 $\ln x + \ln 2 = 1$

47 $\ln(x + 1) = 0$

48 $\ln x = -2$

49 $\ln e^{\sqrt{x+1}} = 3$

50 $e^{\ln(6x^2-4)} = 5x$

51 $\ln(x^2 - 4) - \ln(x + 2) = 0$

52 $(e^{x+2} - 1) \ln(1 - 2x) = 0$

53 $\ln x = \frac{1}{2} \ln 4 + \frac{2}{3} \ln 8$

54 $\frac{1}{2} \ln(x + 4) = \ln(x + 2)$

55 $\ln x = 2 + \ln(1 - x)$

56 $\ln(x^2 + x - 2) = \ln x + \ln(x - 1)$

57 Use a calculator to complete the table. Round off the entries to four decimal places.

n	2	10	100	500	1000	5000	10,000
$\left(1 + \frac{1}{n}\right)^n$							

■ *Show that each function is the composite of two functions.*

58 $h(x) = e^{2x+3}$

59 $h(x) = e^{-x^2+x}$

60 $h(x) = \ln (1 - 2x)$

61 $h(x) = \ln \dfrac{x}{x + 1}$

62 $h(x) = (e^x + e^{-x})^2$

63 $h(x) = \sqrt[3]{\ln x}$

■ *Show that each function is the composite of three functions.*

64 $F(x) = e^{\sqrt{x+1}}$

65 $F(x) = e^{(3x-1)^2}$

66 $F(x) = [\ln(x^2 + 1)]^3$

67 $F(x) = \ln \sqrt{e^x + 1}$

■ *Determine the signs of each function.*

68 $f(x) = xe^x + e^x$

69 $f(x) = e^{2x} - 2xe^{2x}$

70 $f(x) = -3x^2e^{-3x} + 2xe^{-3x}$

71 $f(x) = 1 + \ln x$

72 Show that $(e^x + e^{-x})^2 - (e^x - e^{-x})^2 = 4$.

73 Show that $\ln\left(\dfrac{x}{4} - \dfrac{\sqrt{x^2 - 4}}{4}\right) = -\ln(x + \sqrt{x^2 - 4})$.

*74 Solve for x: $\dfrac{e^x + e^{-x}}{2} = 1$.

*75 Solve for x in terms of y if $y = \dfrac{e^x}{2} - \dfrac{1}{2e^x}$. (*Hint:* Let $u = e^x$ and solve the resulting quadratic in u.)

7.6

Exponential growth and decay

There is a large variety of applied problems dealing with exponential and logarithmic functions. Before considering some of these applications it will be helpful first to learn how to solve for x in an equation such as $2^x = 35$.

$$2^x = 35$$

$$\ln 2^x = \ln 35 \qquad \text{(taking the natural log of both sides)}$$

$$x \ln 2 = \ln 35 \qquad \text{(Why?)}$$

$$x = \frac{\ln 35}{\ln 2}$$

An approximation for x can be found by using Appendix Table III. The entries in this table give values of $\ln x$ to three decimal places. (In most cases $\ln x$ is irrational.) From this table we have $\ln 2 = 0.693$. Even though $\ln 35$ is not given (directly) in the table, it can still be found by applying the second law of logarithms.

Note that the values found in the tables of logarithms are approximations. For the sake of simplicity, however, we will use the equals sign (=).

$$\ln 35 = \ln(3.5)(10) = \ln 3.5 + \ln 10$$

$$= 1.253 + 2.303 \qquad \text{(Table III)}$$

$$= 3.556$$

Now we have

$$x = \frac{\ln 35}{\ln 2} = \frac{3.556}{0.693} = 5.13$$

As a rough check we see that 5.13 is reasonable since $2^5 = 32$.

TEST YOUR UNDERSTANDING

Solve each equation for x in terms of natural logarithms. Approximate the answer by using Table III.

1 $4^x = 5$ **2** $4^{-x} = 5$ **3** $(\frac{1}{2})^x = 12$

4 $2^{3x} = 10$ **5** $4^x = 15$ **6** $67^x = 4$

At the beginning of Section 7.2, we developed the formula $y = (10{,}000)2^x$, which gives the number of bacteria present after x hours of growth; 10,000 is the initial number of bacteria. How long will it take for this bacterial culture to grow to 100,000? To answer this question we let $y = 100{,}000$ and solve for x.

$$(10{,}000)2^x = 100{,}000$$

$$2^x = 10 \qquad \text{(divide by 10,000)}$$

$$x \ln 2 = \ln 10$$

$$x = \frac{\ln 10}{\ln 2}$$

$$= \frac{2.303}{0.693} = 3.32$$

It will take about 3.3 hours.

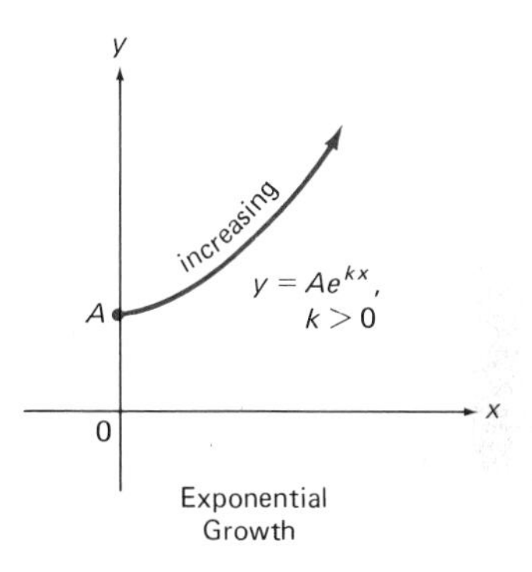

Exponential Growth

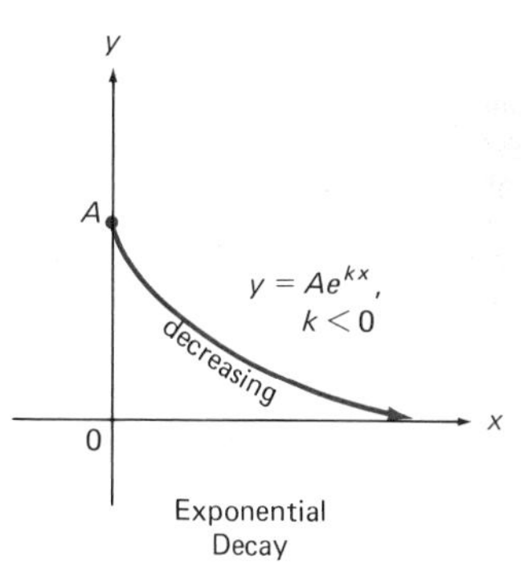

Exponential Decay

In the preceding illustration the exponential and logarithmic functions were used to solve a problem of "exponential growth." Many problems involving exponential growth, or decay, can be solved by using the general formula

$$y = f(x) = Ae^{kx}$$

which shows how the amount of a substance y depends on the time x. Since $f(0) = A$, A represents the initial amount of the substance and k is a constant. In a given situation $k > 0$ signifies that y is growing (increasing) with time. For $k < 0$ the substance is decreasing. (Compare to the graphs of $y = e^x$ and $y = e^{-x}$.)

The preceding bacterial problem also fits this general form. This can be seen by substituting $2 = e^{\ln 2}$ into $y = (10{,}000)2^x$:

$$y = (10{,}000)2^x = (10{,}000)(e^{\ln 2})^x = 10{,}000e^{(\ln 2)x}$$

EXAMPLE 1 A radioactive substance is decaying (it is changing into another element) according to the formula $y = Ae^{-0.2x}$, where y is the amount of material remaining after x years.

(a) If the initial amount $A = 80$ grams, how much is left after 3 years?

(b) The **half-life** of a radioactive substance is the time it takes for half of it to decompose. Find the half-life of this substance in which $A = 80$ grams.

Solution

(a) Since $A = 80$, $y = 80e^{-0.2x}$. We need to solve for the amount y when $x = 3$.

$$y = 80e^{-0.2x}$$
$$= 80e^{-0.2(3)}$$
$$= 80e^{-0.6}$$
$$= 80(0.549) \qquad \text{(Table II)}$$
$$= 43.920$$

There will be about 43.9 grams after 3 years.

(b) This question calls for the time x at which only half of the initial amount is left. Consequently, the half-life x is the solution to $40 = 80e^{-0.2x}$. Divide each side by 80:

$$\tfrac{1}{2} = e^{-0.2x}$$

Take the natural log of both sides, or change to logarithmic form, to obtain $-0.2x = \ln \frac{1}{2}$. Since $\ln \frac{1}{2} = \ln 1 - \ln 2 = -\ln 2$, we solve for x as follows.

$$-0.2x = -\ln 2$$
$$x = \frac{\ln 2}{0.2}$$
$$= 3.465$$

The half-life is approximately 3.465 years.

Carbon-14, often written as ^{14}C, is a radioactive isotope of carbon with a half-life of about 5750 years. By finding how much ^{14}C is contained in the remains of a formerly living organism, it becomes possible to determine what percentage this is of the original amount of ^{14}C at the time of death. Once this information is given, the formula $y = Ae^{kx}$ will enable us to date the age of the remains. The dating will be done after we solve for the constant k. Since the amount of ^{14}C after 5750 years will be $\frac{A}{2}$, we have

Explain each step in this solution.

$$\frac{A}{2} = Ae^{5750k}$$
$$\tfrac{1}{2} = e^{5750k}$$
$$5750k = \ln \tfrac{1}{2}$$
$$k = \frac{\ln 0.5}{5750}$$

Substitute this value for k in the general formula $y = Ae^{kx}$:

$$y = Ae^{(\ln 0.5/5750)x}$$

EXAMPLE 2 An animal skeleton is found to contain one-fourth of its original amount of ^{14}C. How old is the skeleton?

Solution Let x be the age of the skeleton. Then

$$\tfrac{1}{4}A = Ae^{(\ln 0.5/5750)x}$$
$$\tfrac{1}{4} = e^{(\ln 0.5/5750)x}$$
$$\left(\frac{\ln 0.5}{5750}\right)x = \ln \frac{1}{4} = -\ln 4$$
$$x = \frac{(5750)(-\ln 4)}{\ln 0.5}$$
$$= 11{,}500$$

The skeleton is about 11,500 years old.

Estimate the value of y in $y = Ae^{kx}$ for the given values of A, k, and x.

1 $A = 100, k = 0.75, x = 4$

2 $A = 25, k = 0.5, x = 10$

3 $A = 1000, k = -1.8, x = 2$

4 $A = 12.5, k = -0.04, x = 50$

Solve for k. Leave the answer in terms of natural logarithms.

5 $5000 = 50e^{2k}$

6 $75 = 150e^{10k}$

7 $\frac{A}{3} = Ae^{4k}$

8 $\frac{A}{2} = Ae^{100k}$

9 A bacterial culture is growing according to the formula $y = 10{,}000e^{0.6x}$, where x is the time in days. Estimate the number of bacteria after 1 week.

10 Estimate the number of bacteria after the culture in Exercise 9 has grown for 12 hours.

11 How long will it take for the bacterial culture in Exercise 9 to triple in size?

12 How long will it take until the number of bacteria in Exercise 9 reaches 1,000,000?

13 A certain radioactive substance decays according to the exponential formula

$$S = S_0e^{-0.04t}$$

where S_0 is the initial amount of the substance and S is the amount of the substance left after t years. If there were 50 grams of the radioactive substance to begin with, how long will it take for half of it to decay?

14 Show that when the formula in Exercise 13 is solved for t, the result is

$$t = -25 \ln \frac{S}{S_0}.$$

15 A radioactive substance is decaying according to the formula $y = Ae^{kx}$, where x is the time in years. The initial amount $A = 10$ grams, and 8 grams remain after 5 years.

(a) Find k. Leave the answer in terms of natural logs.

(b) Estimate the amount remaining after 10 years.

(c) Find the half-life to the nearest tenth of a year.

16 The half-life of radium is approximately 1690 years. A laboratory has 50 milligrams of radium.

(a) Use the half-life to solve for k in $y = Ae^{kx}$. Leave your answer in terms of logs.

(b) To the nearest 10 years, how long does it take until there are 40 milligrams left?

17 Suppose that 5 grams of a radioactive substance decrease to 4 grams in 30 seconds. What is its half-life to the nearest tenth of a second?

18 How long does it take for two-thirds of the radioactive material in Exercise 17 to decay? Give your answer to the nearest tenth of a second.

19 When the population growth of a certain city was first studied, the population was 22,000. It was found that the population P grows with respect to time t (in years) by the exponential formula

$$P = (22{,}000)(10^{0.0163t})$$

How long will it take for the city to double its population?

20 How long will it take for the population of the city described in Exercise 19 to triple?

21 An Egyptian mummy is found to contain 60% of its ^{14}C. To the nearest 100 years, how old is the mummy? (*Hint:* If A is the original amount of ^{14}C, then $\frac{3}{5}A$ is the amount left.)

22 A skeleton contains one-hundredth of its original amount of ^{14}C. To the nearest 1000 years, how old is the skeleton?

23 Answer the question in Exercise 22 if one-millionth of its ^{14}C is left.

7.7

Scientific notation

To write very large or very small numbers, the scientist often makes use of a form called **scientific notation**. As you will see, scientific notation is helpful in simplifying certain kinds of computations. Here are some illustrations of scientific notation:

$$623{,}000 = 6.23 \times 10^5 \qquad 0.00623 = 6.23 \times 10^{-3}$$

$$6230 = 6.23 \times 10^3 \qquad 0.0000623 = 6.23 \times 10^{-5}$$

It is easy to verify that these are correct. For example:

$$6.23 \times 10^5 = 6.23 \times 100{,}000 = 623{,}000$$

$$6.23 \times 10^{-3} = 6.23 \times \frac{1}{10^3} = \frac{6.23}{1000} = 0.00623$$

The four illustrations above indicate that *a number has been put into scientific notation when it has been expressed as the product of a number between 1 and 10 and an integral power of 10.*

WRITING A NUMBER IN SCIENTIFIC NOTATION

Place the decimal point behind the first nonzero digit. (This produces the number between 1 and 10.) Then determine the power of 10 by counting the number of places you moved the decimal point. If you moved the decimal point to the left, then the power is positive; and if you moved it to the right, it is negative.

Illustrations:

$$2{,}070{,}000. = 2.07 \times 10^6$$

six places left

$$0.00000084 = 8.4 \times 10^{-7}$$

seven places right

To convert a number given in scientific notation back into standard notation, all you need do is move the decimal point as many places as indicated by the exponent of 10. Move the decimal point to the right if the exponent is positive and to the left if it is negative.

Illustrations:

$1.21 \times 10^4 = 12{,}100 \longleftarrow$ decimal point was moved four places right

$1.21 \times 10^{-2} = 0.0121 \longleftarrow$ decimal point was moved two places left

EXAMPLE 1 Use scientific notation to compute $\frac{1}{800{,}000}$.

Solution

$$\frac{1}{800{,}000} = \frac{1}{8 \times 10^5} = \frac{1}{8} \times \frac{1}{10^5} = 0.125 \times 10^{-5}$$
$$= 0.00000125$$

In scientific notation the solution to Example 1 is written 1.25×10^{-6}.

EXAMPLE 2 Use scientific notation to evaluate

$$\frac{(2{,}310{,}000)^2}{(11{,}200{,}000)((0.000825)}$$

Solution

$$\frac{(2{,}310{,}000)^2}{(11{,}200{,}000)(0.000825)} = \frac{(2.31 \times 10^6)^2}{(1.12 \times 10^7)(8.25 \times 10^{-4})}$$
$$= \frac{(2.31)^2 \times (10^6)^2}{(1.12 \times 10^7)(8.25 \times 10^{-4})}$$
$$= \frac{(2.31)^2}{(1.12)(8.25)} \times \frac{10^{12}}{(10^7)(10^{-4})}$$
$$= 0.5775 \times 10^9 = 577{,}500{,}000$$

This is a good place to review the rules of exponents given in Section 2.1.

EXERCISES 7.7

Write each number in scientific notation.

1 4680 **2** 0.0092
3 0.92 **4** 0.9
5 7,583,000 **6** 93,000,000
7 25 **8** 36.09
9 0.000000555 **10** 0.57721
11 202.4 **12** 7.93

Write each number in standard form.

13 7.89×10^4 **14** 7.89×10^{-4}
15 3.0×10^3 **16** 3.0×10^{-3}
17 1.74×10^{-1} **18** 1.74×10^0
19 1.74×10^1 **20** 2.25×10^5
21 9.06×10^{-2}

Express each of the following as a single power of 10.

22 $\frac{10^{-3} \times 10^5}{10}$ **23** $\frac{10^8 \times 10^4 \times 10^{-5}}{10^2 \times 10^3}$

24 $\frac{10^{-3}}{10^{-5}}$ **25** $\frac{10^1 \times 10^2 \times 10^3 \times 10^4}{10^{10}}$

26 $\frac{10^9 \times 10^{-2}}{10^6 \times 10^{-9}}$ **27** $\frac{(10^2)^3 \times 10^{-1}}{(10^{-3})^4}$

Compute, using scientific notation.

28 $\frac{1}{5000}$ **29** $\frac{1}{0.0005}$

30 $\frac{2}{80{,}000}$ **31** $\frac{0.0064}{0.000016}$

32 $\frac{(6000)(720)}{12{,}000}$ **33** $\frac{(0.000025)}{(0.0625)(0.02)}$

34 $\frac{(240)(0.000032)}{(0.008)(12{,}000)}$ **35** $\frac{4{,}860{,}000}{(0.081)(19{,}200)}$

36 $\frac{(0.0111)(66{,}600)(555)}{(22{,}200)(0.000333)}$

Perform the indicated operations in Exercises 37–42, using scientific notation.

37 $\sqrt{1{,}440{,}000}$ **38** $(0.0006)^3$

39 $\frac{\sqrt{0.000625}}{3125}$ **40** $\frac{(40)^4(0.015)^2}{24{,}000}$

41 $\dfrac{(1{,}728{,}000)^{1/3}}{(0.06)(400)^2}$

42 $[(0.002)(0.2)(200)(20{,}000)]^{1/2}$

43 Light travels at a rate of about 186,000 miles per second. The average distance from the sun to the earth is 93,000,000 miles. Use scientific notation to find how long it takes light to reach the earth from the sun.

*44 Based on information given in Exercise 43, use scientific notation to show that 1 light-year (the distance light travels in 1 year) is approximately $5.87 \times 10^{12} = 5{,}870{,}000{,}000{,}000$ miles.

7.8

Common logarithms

Logarithms were developed about 350 years ago. Since then they have been widely used to simplify involved numerical computations. Much of this work can now be done more efficiently with the aid of computers and calculators. However, logarithmic computations will help us to better understand the theory of logarithms, which plays an important role in many parts of mathematics (including calculus) and in its applications.

For scientific and technical work, numbers are often written in scientific notation and we will therefore be using logarithms to the base 10, called **common logarithms**.

Below is an excerpt of Appendix Table IV, page 381. It contains the common logarithms of three-digit numbers from 1.00 to 9.99. To find a logarithm, say $\log_{10}3.47$, first find the entry 3.4 in the left-hand column under the heading x. Now in the row for 3.4 and in the column headed by the digit 7 you will find the entry .5403. This is the common logarithm of 3.47. We write

Note that the values found in the tables of logarithms are approximations. For the sake of simplicity, however, we will use the equals sign (=).

$$\log_{10}3.47 = 0.5403$$

By reversing this process we can begin with $\log_{10}x = 0.5403$ and find x.

x	0	1	2	3	4	5	6	7	8	9
⋮	⋮	⋮	⋮	⋮	⋮	⋮	⋮	⋮	⋮	⋮
3.3	.5185	.5198	.5211	.5224	.5237	.5250	.5263	.5276	.5289	.5302
3.4	.5315	.5328	.5340	.5353	.5366	.5378	.5391	.5403	.5416	.5428
3.5	.5441	.5453	.5465	.5478	.5490	.5502	.5514	.5527	.5539	.5551
⋮	⋮	⋮	⋮	⋮	⋮	⋮	⋮	⋮	⋮	⋮

The common logarithms in Table IV are four-place decimals between 0 and 1. Except for the case $\log_{10}1 = 0$, they are all approximations. The fact that they are between 0 and 1 will be taken up in the exercises.

Verify these illustrations using Table IV:

$$\log_{10}3.07 = 0.4871 \qquad \log_{10}8.88 = 0.9484$$

$$\text{If } \log_{10}x = 0.7945, \text{ then } x = 6.23$$

To find $\log_{10}N$, where N may not be between 1 and 10, we first write N in scientific notation.

$$N = x10^c$$

where c is an integer and $1 \leq x < 10$. For example:

$$62{,}300 = 6.23(10^4) \qquad 0.00623 = 6.23(10^{-3})$$

This form of N, in conjunction with Table IV, will allow us to find $\log_{10}N$. In general,

$$\begin{aligned} \log_{10}N &= \log_{10}(x10^c) \\ &= \log_{10}x + \log_{10}10^c \qquad \text{(Law 1 for logs)} \\ &= \log_{10}x + c \qquad \text{(Why?)} \end{aligned}$$

The integer c is the **characteristic** of $\log_{10}N$, and the four-place decimal fraction $\log_{10}x$ is its **mantissa**. Using $N = 62{,}300$, we have

$$\begin{aligned} \log_{10}62{,}300 = \log_{10}6.23(10^4) &= \log_{10}6.23 + \log_{10}10^4 \\ &= \log_{10}6.23 + 4 \\ &= 0.7945 + 4 \qquad \text{(Table IV)} \\ &= 4.7945 \end{aligned}$$

Note the distinction:

$\log N \longleftarrow$ common logarithm, base 10

$\ln N \longleftarrow$ natural logarithm, base e

Since all logarithms considered here are to the base 10, we will simplify the notation and drop the subscript 10 from the logarithmic statements. Thus we write $\log N$ instead of $\log_{10}N$.

EXAMPLE 1 Find log 0.0419.

Solution

$$\begin{aligned} \log 0.0419 &= \log 4.19(10^{-2}) \\ &= \log 4.19 + \log 10^{-2} \\ &= 0.6222 + (-2) \end{aligned}$$

Suppose that in Example 1 the mantissa 0.6222 and the negative characteristic are combined:

$$\begin{aligned} 0.6222 + (-2) = -1.3778 &= -(1 + 0.3778) \\ &= -1 + (-0.3778) \end{aligned}$$

Since Table IV does not have negative mantissas, like -0.3778, we avoid such combining and preserve the form of log 0.0419 so that its mantissa is positive. For computational purposes there are other useful forms of $0.6222 + (-2)$ in which the mantissa 0.6222 is preserved. Note that $-2 = 8 - 10$, $18 - 20$, and so forth. Thus

$$0.6222 + (-2) = 0.6222 + 8 - 10 = 8.6222 - 10 = 18.6222 - 20$$

Similarly,

$$\log 0.00569 = 7.7551 - 10 = 17.7551 - 20$$
$$\log 0.427 = 9.6304 - 10 = 29.6304 - 30$$

An efficient way to find N, if $\log N = 6.1239$, is to find the three-digit number x from Table IV corresponding to the mantissa 0.1239. Then multiply

x by 10^6. Thus, since $\log 1.33 = 0.1239$, we have

$$N = 1.33(10^6) = 1{,}330{,}000$$

In the following explanation you can discover why this technique works.

$$\begin{aligned}\log N &= 6.1239\\ &= 6 + 0.1239\\ &= 6 + \log 1.33\\ &= \log 10^6 + \log 1.33\\ &= \log 10^6(1.33)\\ &= \log 1{,}330{,}000\end{aligned}$$

Therefore, $\log N = \log 1{,}330{,}000$, and we conclude that $N = 1{,}330{,}000$.

TEST YOUR UNDERSTANDING

Find the common logarithm.

1 $\log 267$ **2** $\log 26.7$ **3** $\log 2.67$
4 $\log 0.267$ **5** $\log 0.0267$ **6** $\log 42{,}000$
7 $\log 0.000813$ **8** $\log 7990$ **9** $\log 0.00111$

Find N.

10 $\log N = 2.8248$ **11** $\log N = 0.8248$
12 $\log N = 9.8248 - 10$ **13** $\log N = 0.8248 - 3$
14 $\log N = 7.7126$ **15** $\log N = 18.9987 - 20$

EXAMPLE 2 Estimate $P = (936)(0.00847)$ by using (common) logarithms.

For easy reference:

Law 1. $\log MN = \log M + \log N$
Law 2. $\log \frac{M}{N} = \log M - \log N$
Law 3. $\log N^k = k \log N$

Solution

$$\begin{aligned}\log P &= \log (963)(0.00847)\\ &= \log 963 + \log 0.00847 \qquad \text{(Law 1)}\end{aligned}$$

Now use Table IV.

$$\left.\begin{aligned}\log 963 &= \ \ 2.9836\\ \log 0.00847 &= \ \ 7.9279 - 10\end{aligned}\right\} \text{(add)}$$

$$\log P = 10.9115 - 10 = 0.9115$$

$$P = 8.16(10^0) = 8.16$$

For a more accurate procedure, see Exercise 21. Exercise 20 shows how to find $\log x$ when $0 \le x < 1$ and x has more than three digits.

Note: The mantissa 0.9115 is not in Table IV. In this case we use the closest entry, namely 0.9117, corresponding to $x = 8.16$. Such approximations are good enough for our purposes.

EXAMPLE 3 Use logarithms to estimate $Q = \dfrac{0.00439}{0.705}$.

Solution We find $\log Q = \log 0.00439 - \log 0.705$ (by Law 2). Now use the table.

(This form is used to avoid a negative mantissa when subtracting in the next step.)

$$\left.\begin{array}{l} \log 0.00439 = 7.6425 - 10 = 17.6425 - 20 \\ \log 0.705 \quad = 9.8482 - 10 = 9.8482 - 10 \end{array}\right\} \text{(subtract)}$$

$$\log Q = 7.7943 - 10$$

$$Q = 6.23(10^{-3})$$

$$= 0.00623$$

EXAMPLE 4 Use logarithms to estimate $R = \sqrt[3]{0.0918}$.

Solution

$$\begin{aligned} \log R &= \log(0.0918)^{1/3} \\ &= \tfrac{1}{3}\log 0.0918 \qquad \text{(Law 3)} \\ &= \tfrac{1}{3}(8.9628 - 10) \\ &= \tfrac{1}{3}(28.9628 - 30) \\ &= 9.6543 - 10 \\ R &= 4.51(10^{-1}) = 0.451 \end{aligned}$$

(We avoid the fractional characteristic $-\frac{10}{3}$ by changing to $28.9628 - 30$.)

EXAMPLE 5 To determine how much a paint dealer should charge for a gallon of paint, he needs to find out how much the paint cost him per gallon in the first place. The paint is stored in a cylindrical drum $2\frac{1}{2}$ feet in diameter and $3\frac{3}{4}$ feet high. If he paid $400 for this quantity of paint, what did it cost him per gallon? (Use 1 cubic foot = 7.48 gallons.)

Solution The volume of the drum is the area of the base times the height. Thus there are

$$\pi(1.25)^2(3.75)$$

Volume of a cylinder: $V = \pi r^2 h$

cubic feet of paint in the drum. Then the number of gallons is

$$\pi(1.25)^2(3.75)(7.48)$$

Since the total cost was $400, the cost per gallon is given by

$$C = \frac{400}{\pi(1.25)^2(3.75)(7.48)}$$

We use $\pi = 3.14$ to do the computation, using logarithms:

$$\log C = \log 400 - (\log 3.14 + 2\log 1.25 + \log 3.75 + \log 7.48)$$

$$\begin{array}{rcll}
 & & & \log 400 = 2.6021 \\
 & & \log 3.14 = 0.4969 & \\
\log 1.25 = 0.0969 & \longrightarrow & 2\log 1.25 = 0.1938 \quad \text{(add)} & \quad \text{(subtract)} \\
 & & \log 3.75 = 0.5740 & \\
 & & \log 7.48 = 0.8739 & \\
\hline
 & & 2.1386 \quad \longrightarrow & 2.1386 \\
\hline
 & & & \log C = 0.4635 \\
 & & & C = 2.91 \times 10^0 \\
 & & & = 2.91
\end{array}$$

The paint cost the dealer approximately \$2.91 per gallon.

EXERCISES 7.8

In Exercises 1–12, estimate by using common logarithms.

1 $(512)(84{,}000)$

2 $(906)(2330)(780)$

3 $\dfrac{(927)(818)}{274}$

4 $\dfrac{274}{(927)(818)}$

5 $\dfrac{(0.421)(81.7)}{(368)(750)}$

6 $\dfrac{(579)(28.3)}{\sqrt{621}}$

7 $\dfrac{(28.3)\sqrt{621}}{579}$

8 $\left[\dfrac{28.3}{(579)(621)}\right]^2$

9 $\sqrt{\dfrac{28.3}{(579)(621)}}$

10 $\dfrac{(0.0941)^3(0.83)}{(7.73)^2}$

11 $\dfrac{\sqrt[3]{(186)^2}}{(600)^{1/4}}$

12 $\dfrac{\sqrt[4]{600}}{(186)^{2/3}}$

13 After running out of gasoline, a motorist had her gas tank filled at a cost of \$16.93. What was the cost per gallon if the gas tank's capacity is 14 gallons?

14 Suppose that a spaceship takes 3 days, 8 hours, and 20 minutes to travel from the earth to the moon. If the distance traveled was one-quarter of a million miles, what was the average speed of the spaceship in miles per hour?

15 A spaceship, launched from the earth, will travel 432,000,000 miles on its trip to the planet Jupiter. If its average velocity is 21,700 miles per hour, how long will the trip take? Give the answer in years.

16 When P dollars is invested in a bank that pays compound interest at the rate of r percent (expressed as a decimal) per year, the amount A after n years is given by the formula

$$A = P(1 + r)^n$$

(This formula will be derived in Chapter 10.)

(a) Find A for $P = 2500$, $r = 0.09$ (9%), and for $n = 3$.

(b) An investment of \$3750 earns compound interest at the rate of 11.2% per year. Find the amount A after 5 years.

17 The formula $P = \dfrac{A}{(1 + r)^n}$ gives the initial investment P in terms of the current amount of money A, the annual compound interest rate r, and the number of years n. How much money was invested at 12.8% if after 6 years there is now \$8440 in the bank?

18 If P dollars is invested at an interest rate r and the interest is compounded k times per year, the amount A after n years in given by

$$A = P\left(1 + \frac{r}{k}\right)^{kn}$$

(a) Use this formula to compute A for $P = \$5000$ and $r = 0.08$ if the interest is compounded semiannually for 3 years.

(b) Find A in part (a) with interest compounded quarterly.

(c) Find A in part (a) with $k = 8$.

***19** Explain why the mantissas in Table IV are between 0 and 1. (*Hint:* Take $1 \leq x < 10$ and now consider the common logarithms of 1, x, and 10.)

***20** Here is a computation for finding log 6.477. Study this procedure carefully and then find the logarithms below in the same manner.

	N		log N	
0.010 { 0.007 {	6.470	→	0.8109	} d } 0.0007
	6.477	→	?	
	6.480	→	0.8116	

$$\frac{0.007}{0.010} = \frac{d}{0.0007}$$

$$0.7 = \frac{d}{0.0007}$$

$$d = (0.7)(0.0007) = 0.00049$$

$$\log N = 0.8109 + 0.00049$$

$= 0.8114$ (rounded off to four decimal places)

(a) log 3.042 (b) log 7.849
(c) log 1.345 (d) log 5.444
(e) log 6.803 (f) log 2.711
(g) log 4.986 (h) log 9.008

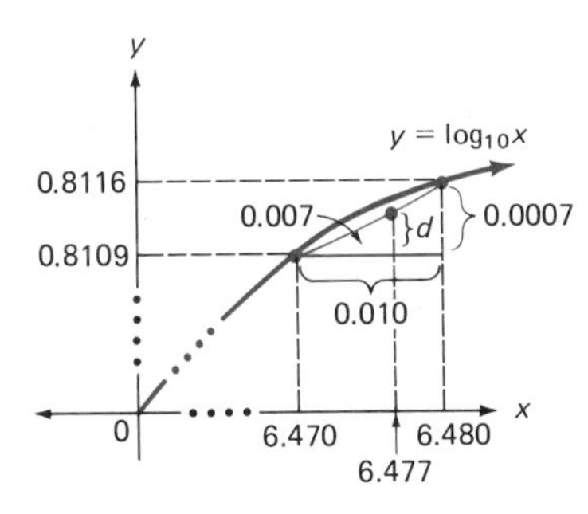

Note that the line segment approximates the curve of the log.

The method used here is called **linear interpolation.** The rationale behind this method is suggested by the accompanying figure.

*21 The method in Exercise 20 can be adapted for finding the number when the given logarithm is not an exact table entry. Study the following procedure for finding N in $\log N = 0.7534$, and then find the numbers N below in the same manner.

	log N		N	
0.0008 { 0.0006 {	0.7528	→	5.660	} d } 0.010
	0.7534	→	?	
	0.7536	→	5.670	

$$\frac{0.0006}{0.0008} = \frac{d}{0.01}$$

$$0.75 = \frac{d}{0.01}$$

$$d = (0.01)(0.75) = 0.0075$$

$$N = 5.660 + 0.0075$$

$= 5.668$ (rounded off to three decimal places)

(a) $\log N = 0.4510$ (b) $\log N = 0.9672$
(c) $\log N = 0.1391$ (d) $\log N = 0.7395$
(e) $\log N = 0.6527$ (f) $\log N = 0.8749$
(g) $\log N = 0.0092$ (h) $\log N = 0.9781$
(i) $\log N = 0.3547$

REVIEW EXERCISES FOR CHAPTER 7

The solutions to the following exercises can be found within the text of Chapter 7. Try to answer each question without referring to the text.

Section 7.1

1 State the definition of a one-to-one function.

2 Describe the horizontal line test for one-to-one functions.

3 Use the horizontal line test to determine which of the following are one-to-one functions.

(a) $y = \sqrt{x}$ (b) $y = \dfrac{1}{x}$
(c) $y = |x|$ (d) $y = x^2 - 2x + 1$
(e) $y = \sqrt[3]{x}$

4 Find the inverse g of $y = f(x) = 2x + 3$ and show that $(f \circ g)(x) = x$ and $(g \circ f)(x) = x$.

5 Find the inverse g of $f(x) = \sqrt{x}$ and show that $(f \circ g)(x) = x$ and $(g \circ f)(x) = x$.

Section 7.2

6 List the important properties of the exponential function $f(x) = b^x$ for $b > 1$, and for $0 < b < 1$.

7 Use a table of values to sketch $y = 8^x$ on the interval $[-1, 1]$.

8 Sketch $y = 2^x$ and $y = (\frac{1}{2})^x$ on the same axes.

9 Explain how to obtain the graphs of $y = 2^{x-3}$ and $y = 2^x - 1$ from $y = 2^x$.

10 If f is a one-to-one function and $f(x_1) = f(x_2)$, then what can you say about x_1 and x_2?

11 Use the one-to-one property of the function $f(x) = 5^x$ to solve the equation $5^{t^2} = 625$ for t.

Section 7.3

12 Which of the following statements are true?

(a) If $0 < b < 1$, the function $f(x) = b^x$ decreases.

(b) The point (0, 1) is on the curve $y = \log_b x$.

(c) $y = \log_b x$, for $b > 1$, increases and the curve is concave down.

(d) The domain of $y = b^x$ is the same as the range of $y = \log_b x$.

(e) The x-axis is an asymptote to $y = \log_b x$ and the y-axis is an asymptote to $y = b^x$.

13 Write the equation of the inverse of $y = 2^x$ and graph both on the same axes.

14 Find the domain for $y = \log_2(x - 3)$.

15 Explain how the graph of $g(x) = \log_5(x + 2)$ can be obtained from $f(x) = \log_5 x$.

16 (a) Change to logarithmic form: $27^{2/3} = 9$.

(b) Change to exponential form: $\log_6 \frac{1}{36} = -2$.

17 Solve for y: $\log_9 27 = y$.

18 Solve for b: $\log_b 8 = \frac{3}{4}$.

Section 7.4

19 Write the three laws of logarithms.

20 Express $\log_b \frac{AB^2}{C}$ in terms of $\log_b A$, $\log_b B$, and $\log_b C$.

21 Given that $\log_b 2 = 0.6931$ and $\log_b 3 = 1.0986$, find $\log_b \sqrt{12}$.

22 Express $\frac{1}{2} \log_b x - 3 \log_b(x - 1)$ as the logarithm of a single expression in x.

23 Solve for x: $\log_8(x - 6) + \log_8(x + 6) = 2$.

24 Solve for x:

$$\log_{10}(x^3 - 1) - \log_{10}(x^2 + x + 1) = 1.$$

25 Solve for x: $\log_3 2x - \log_3(x + 5) = 0$.

Section 7.5

26 Match the columns.

Curve	*Slope of tangent to curve through* (0, 1) *is*
(i) $y = 3^x$	(a) less than 1
(ii) $y = 2^x$	(b) equal to 1
(iii) $y = e^x$	(c) more than 1

27 Graph $y = e^x$ and $y = \ln x$ on the same axes.

28 Explain how the curve $y = \ln x^2$ can be obtained from $y = \ln x$.

29 Solve for t: $e^{2t-1} = 5$.

30 Solve for x: $\ln(x + 1) = 1 + \ln x$.

31 Determine the signs of $f(x) = x^2e^x + 2xe^x$.

32 Match the columns.

(i) $\ln x < 0$	(a) $x < 0$
(ii) $\ln x = 0$	(b) $x = 0$
(iii) $\ln x > 0$	(c) $x > 0$
(iv) $0 < e^x < 1$	(d) $0 < x < 1$
(v) $e^x = 1$	(e) $x = 1$
(vi) $e^x > 1$	(f) $x > 1$

Section 7.6

33 Get an approximate solution for x in $2^x = 35$.

34 A radioactive material is decreasing according to the formula $y = Ae^{-0.2x}$, where y is the amount of material remaining after x years. If the initial amount $A = 80$ grams, how much is left after 3 years?

35 Find the half-life of the radioactive substance in Exercise 34.

36 Solve for k in $\frac{A}{2} = Ae^{5750k}$. Leave your answer in terms of natural logs.

37 Use the formula $y = Ae^{(\ln 0.5/5750)x}$ to estimate the age of a skeleton that is found to contain one-fourth of its original amount of carbon-14.

Section 7.7

38 Convert to scientific notation:

(a) 2,070,000 (b) 0.00000084

39 Convert to standard notation:

(a) 1.21×10^4 (b) 1.21×10^{-2}

40 Evaluate $\frac{1}{800{,}000}$ using scientific notation.

41 Evaluate $\frac{(2{,}310{,}000)^2}{(11{,}200{,}000)(0.000825)}$ using scientific notation.

Section 7.8

42 Find log 0.0419.

43 Find N if $\log N = 6.1239$.

44 Use common logarithms to estimate $P = (963)(0.00847)$.

45 Use common logarithms to estimate $Q = \frac{0.00439}{0.705}$.

46 Use common logarithms to estimate $R = \sqrt[3]{0.0918}$.

47 To determine how much a paint dealer should charge for a gallon of paint, he needs to find out how much the paint cost him per gallon in the first place. The paint is stored in a cylindrical drum $2\frac{1}{2}$ feet in diameter and $3\frac{3}{4}$ feet high. If he paid \$400 for this quantity of paint, what did it cost him per gallon? (Use 1 cubic foot = 7.48 gallons.)

SAMPLE TEST QUESTIONS FOR CHAPTER 7

Use these questions to test your knowledge of the basic skills and concepts of Chapter 7. Then test your answers with those given at the end of the book.

1 Match each curve with one of the given equations listed at the top of the next column.

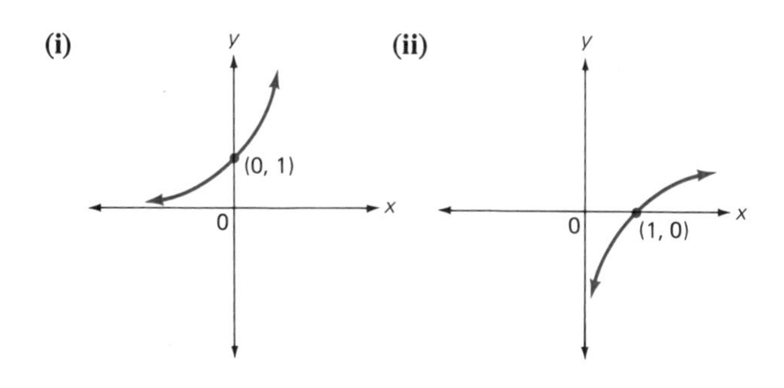

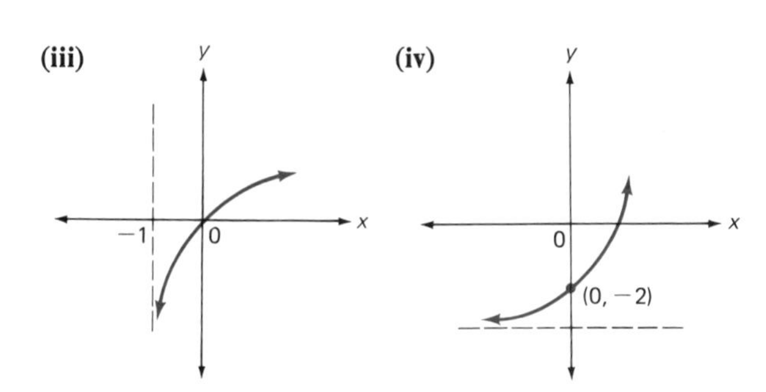

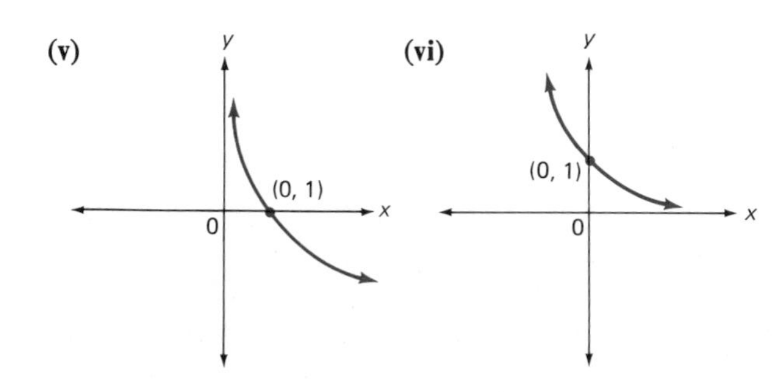

(a) $y = b^x;\ b > 1$
(b) $y = b^x;\ 0 < b < 1$
(c) $y = \log_b x;\ b > 1$
(d) $y = \log_b x;\ 0 < b < 1$
(e) $y = \log_b(x + 1);\ b > 1$
(f) $y = \log_b(x - 1);\ b > 1$
(g) $y = b^{x+2};\ b > 1$
(h) $y = b^x - 2;\ b > 1$

2 What does it mean to say that a function is one-to-one?

3 Find the inverse g of $y = f(x) = \sqrt[3]{x} - 1$ and show that $(f \circ g)(x) = x$.

4 Solve for x:
(a) $81^x = 9$
(b) $e^{\ln x} = 9$.

5 **(a)** Solve for b: $\log_b \frac{27}{8} = -3$.
(b) Evaluate $\log_{10} 0.01$.

Find the domain of each function and give the equation of the vertical or horizontal asymptote.

6 $y = f(x) = 2^x - 4$

7 $y = f(x) = \log_3(x + 4)$

8 Solve for x:

$$\log_{25} x^2 - \log_{25}(2x - 5) = \tfrac{1}{2}.$$

9 Sketch the graphs of $y = e^{-x}$ and its inverse on the same axes. Write an equation of the inverse in the form $y = g(x)$.

10 Use the laws of logarithms (as much as possible) to write $\ln \dfrac{x^3}{(x + 1)\sqrt{x^2 + 2}}$ as an expression involving sums and differences.

11 Solve for x in $4^{2x} = 5$ and express the answer in terms of natural logs.

12 A radioactive substance decays according to the formula $y = Ae^{-0.04t}$, where t is the time in years. If the initial amount $A = 50$ grams, find the half-life. Leave the answer in terms of natural logs.

13 Use scientific notation to evaluate $\dfrac{\sqrt{144{,}000{,}000}}{(2000)^2(0.0005)}$.

14 Use common logarithms to estimate

$$N = \frac{(2430)^2}{(0.842)\sqrt{27.9}}$$

15 Solve for x:

$$\ln 2 - \ln(1 - x) = 1 - \ln(x + 1).$$

ANSWERS TO THE TEST YOUR UNDERSTANDING EXERCISES

Page 240

The functions given in 2, 3, 5, 6, and 7 are one-to-one. The others are not.

Page 250

1

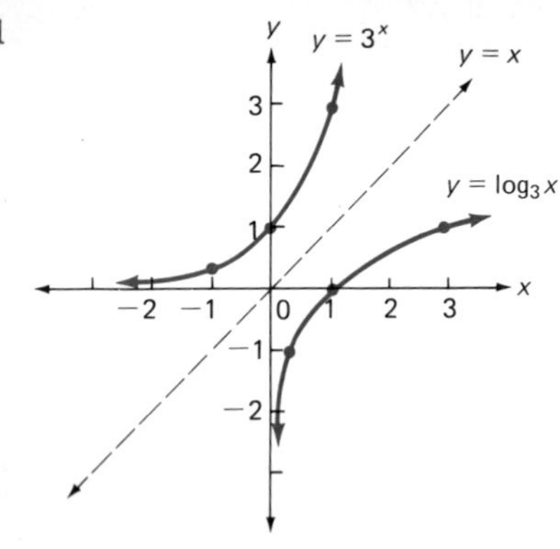

2

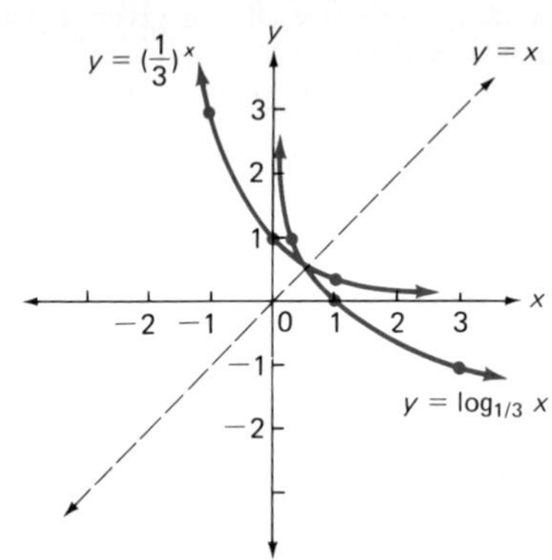

3 Shift 2 units left.

4 Shift 2 units up.

5 Reflect through x-axis.

6 Double the size of each ordinate.

Page 255

1 $\log_b A + \log_b B + \log_b C$

2 $\log_b A - \log_b B - \log_b C$

3 $2 \log_b A + 2 \log_b B - \log_b C$

4 $\log_b A + 2 \log_b B + 3 \log_b C$

5 $\log_b A + \frac{1}{2} \log_b B - \log_b C$

6 $\frac{1}{3} \log_b A - 3 \log_b B - 3 \log_b C$

7 $\log_b 3x^2$

8 $\log_b[\sqrt{x}(x-1)^2]$

9 $\log_b \frac{2x-1}{(x^2+1)^3}$

10 $\log_b \frac{x}{(x-1)(x-2)^2}$

11 2.8903

12 −0.5234

Page 264

(Correct to two decimal places.)

1 $\frac{\ln 5}{\ln 4} = \frac{1.609}{1.386} = 1.16$

2 $-\frac{\ln 5}{\ln 4} = -\frac{1.609}{1.386} = -1.16$

3 $\frac{\ln 12}{\ln 0.5} = \frac{2.485}{-0.693} = -3.59$

4 $\frac{\ln 10}{3 \ln 2} = \frac{2.303}{3(0.693)} = 1.11$

5 $\frac{\ln 15}{\ln 4} = \frac{2.708}{1.386} = 1.95$

6 $\frac{\ln 4}{\ln 67} = \frac{1.386}{4.205} = 0.33$

Page 272

1 2.4265 **2** 1.4265 **3** 0.4265 **4** 9.4265 − 10 **5** 8.4265 − 10

6 4.6232 **7** 6.9101 − 10 **8** 3.9025 **9** 7.0453 − 10 **10** 668

11 6.68 **12** 0.668 **13** 0.00668 **14** 51,600,000 **15** 0.0997

MATRICES, DETERMINANTS, AND LINEAR SYSTEMS

8

8.1

Solving linear systems using matrices

Methods of solving linear systems were studied in Section 3.6. In this section we develop another, more efficient method. We begin by forming the rectangular array of numbers, called a *matrix*, consisting of the coefficients and constants of the system. For example, the linear system

$$\begin{aligned} &\text{Equation 1:} \quad 2x + 5y + 8z = 11 \\ &\text{Equation 2:} \quad x + 4y + 7z = 10 \\ &\text{Equation 3:} \quad 3x + 6y + 12z = 15 \end{aligned}$$

is replaced by this corresponding matrix:

$$\begin{array}{l} \\ \\ \text{Row 1} \\ \text{Row 2} \\ \text{Row 3} \end{array} \quad \begin{array}{c} \text{Coefficients} \\ \begin{array}{ccc} x & y & z \\ \downarrow & \downarrow & \downarrow \end{array} \\ \left[\begin{array}{ccc:c} 2 & 5 & 8 & 11 \\ 1 & 4 & 7 & 10 \\ 3 & 6 & 12 & 15 \end{array}\right] \end{array}$$

The dashed vertical line serves as a reminder that the coefficients of the variables are to the left. The numbers to the right are the constants on the right-hand side of the equal signs in the given system.

Now compare the steps in the two columns that follow. As you read these steps keep in mind that the objective is to transform the given linear system into the form reached in step 5.

Working with the equations	*Working with the matrices*
Step 1 Write the system. $$\begin{aligned} 2x + 5y + 8z &= 11 \\ x + 4y + 7z &= 10 \\ 3x + 6y + 12z &= 15 \end{aligned}$$	Step 1 Write the corresponding matrix. $$\begin{array}{ccc:c} 2 & 5 & 8 & 11 \\ 1 & 4 & 7 & 10 \\ 3 & 6 & 12 & 15 \end{array}$$
Step 2 Interchange the first two equations. $$\begin{aligned} x + 4y + 7z &= 10 \\ 2x + 5y + 8z &= 11 \\ 3x + 6y + 12z &= 15 \end{aligned}$$	Step 2 Interchange the first two rows. $$\begin{array}{ccc:c} 1 & 4 & 7 & 10 \\ 2 & 5 & 8 & 11 \\ 3 & 6 & 12 & 15 \end{array}$$
Step 3 Add −2 times the first equation to the second, and −3 times the first to the third. $$\begin{aligned} x + 4y + 7z &= 10 \\ -3y - 6z &= -9 \\ -6y - 9z &= -15 \end{aligned}$$	Step 3 Add −2 times the first row to the second, and −3 times the first to the third. $$\begin{array}{ccc:c} 1 & 4 & 7 & 10 \\ 0 & -3 & -6 & -9 \\ 0 & -6 & -9 & -15 \end{array}$$
Step 4 Multiply the second equation by $-\frac{1}{3}$. $$\begin{aligned} x + 4y + 7z &= 10 \\ y + 2z &= 3 \\ -6y - 9z &= -15 \end{aligned}$$	Step 4 Multiply row 2 by $-\frac{1}{3}$. $$\begin{array}{ccc:c} 1 & 4 & 7 & 10 \\ 0 & 1 & 2 & 3 \\ 0 & -6 & -9 & -15 \end{array}$$
Step 5 Add 6 times the second equation to the third. $$\begin{aligned} x + 4y + 7z &= 10 \\ y + 2z &= 3 \\ 3z &= 3 \end{aligned}$$	Step 5 Add 6 times row 2 to row 3. $$\begin{array}{ccc:c} 1 & 4 & 7 & 10 \\ 0 & 1 & 2 & 3 \\ 0 & 0 & 3 & 3 \end{array}$$

The operations on the equations in the left column produced **equivalent systems** of equations. (Equivalent systems have the same solutions.) Since each of these systems has its corresponding matrix produced by comparable operations on the rows of the matrices, we say that the matrices are **row-equivalent**.

In step 5, we reached a row-equivalent matrix whose corresponding linear system

$$\begin{aligned} x + 4y + 7z &= 10 \\ y + 2z &= 3 \\ 3z &= 3 \end{aligned}$$

is said to be in **triangular form**. From this form we solve for the variables using **back-substitution**. That is, z is found from the last equation, then substitute back into the second to find y, and finally substitute back into the first to find x. Thus

$$3z = 3 \longrightarrow z = 1$$
$$y + 2(1) = 3 \longrightarrow y = 1$$
$$x + 4(1) + 7(1) = 10 \longrightarrow x = -1$$

The solution is $(-1, 1, 1)$, which can be checked in the given system.

To summarize, our new matrix method for solving a linear system has two major parts:

Part A can be completed by using a variety of row operations. The steps of a solution are therefore not unique.

Part A: Use the following **fundamental row operations** to transform the initial matrix corresponding to the linear system into a row-equivalent matrix as in step 5.
(1) Interchange two rows.
(2) Multiply a row by a nonzero constant.
(3) Add a multiple of a row to another row.

Part B: Convert the matrix obtained in Part A back into a linear system, which will be equivalent to the original system, and solve for the variables by back-substitution.

EXAMPLE 1 Solve the linear system using row-equivalent matrices.

$$\begin{aligned} 2x + 14y - 4z &= -2 \\ -4x - 3y + z &= 8 \\ 3x - 5y + 6z &= 7 \end{aligned}$$

Solution Begin by writing the matrix corresponding to the linear system and apply the fundamental row operations.

$$\left[\begin{array}{rrr|r} 2 & 14 & -4 & -2 \\ -4 & -3 & 1 & 8 \\ 3 & -5 & 6 & 7 \end{array}\right]$$ This is the matrix for the given system

Getting a 1 in the circled position will make it easier to get zeros *below* this 1 in the next step.

$$\left[\begin{array}{rrr|r} \textcircled{1} & 7 & -2 & -1 \\ -4 & -3 & 1 & 8 \\ 3 & -5 & 6 & 7 \end{array}\right] \longleftarrow \tfrac{1}{2} \times \text{(row 1)}$$

In the next step row operation (3) will be used to obtain 0's below the circled 1.

The explanations pointing to the rows state what row operations were applied to the rows of the *preceding matrix* to obtain the designated row.

$$\left[\begin{array}{rrr|r} 1 & 7 & -2 & -1 \\ 0 & 25 & -7 & 4 \\ 0 & -26 & 12 & 10 \end{array}\right] \begin{array}{l} \\ \longleftarrow \quad 4 \times (\text{row } 1) + \text{row } 2 \\ \longleftarrow -3 \times (\text{row } 1) + \text{row } 3 \end{array}$$

$$\left[\begin{array}{rrr|r} 1 & 7 & -2 & -1 \\ 0 & 25 & -7 & 4 \\ 0 & -1 & 5 & 14 \end{array}\right] \begin{array}{l} \\ \\ \longleftarrow \text{Row } 2 + \text{row } 3 \end{array}$$

Getting the -1 in the circled position will make it easier to get zero *below* this -1 in the next step.

$$\left[\begin{array}{rrr|r} 1 & 7 & -2 & -1 \\ 0 & \textcircled{-1} & 5 & 14 \\ 0 & 25 & -7 & 4 \end{array}\right] \begin{array}{l} \\ \longleftarrow \text{Interchange rows 2} \\ \longleftarrow \text{and 3} \end{array}$$

$$\left[\begin{array}{rrr|r} 1 & 7 & -2 & 1 \\ 0 & -1 & 5 & 14 \\ 0 & 0 & 118 & 354 \end{array}\right] \begin{array}{l} \\ \\ \longleftarrow 25 \times (\text{row } 2) + \text{row } 3 \end{array}$$

Now convert to the corresponding linear system and solve for the variables using back-substitution.

$$\begin{aligned} x + 7y - 2z &= -1 \\ -y + 5z &= 14 \\ 118z &= 354 \end{aligned}$$

$118z = 354 \longrightarrow z = 3$

$-y + 5(3) = 14 \longrightarrow y = 1$

$x + 7(1) - 2(3) = -1 \longrightarrow x = -2$

Thus the solution is $(-2, 1, 3)$.

This matrix procedure also reveals when a linear system has no solutions, that is, when it is an inconsistent system. In such a case we will obtain a row in a matrix of the form

$$0 \; 0 \; \cdots \; 0 \mid p$$

where $p \neq 0$. But when this row is converted to an equation, we get the false statement $0 = p$. For example, solving the system

$$\begin{aligned} 3x - 6y &= 9 \\ -2x + 4y &= -8 \end{aligned}$$

we have

$$\left[\begin{array}{rr|r} 3 & -6 & 9 \\ -2 & 4 & -8 \end{array}\right]$$

$$\left[\begin{array}{rr|r} 1 & -2 & 3 \\ -2 & 4 & -8 \end{array}\right] \begin{array}{l} \longleftarrow \frac{1}{3} \times (\text{row } 1) \\ \\ \end{array}$$

$$\left[\begin{array}{cc:c} 1 & -2 & 3 \\ 0 & 0 & -2 \end{array}\right] \longleftarrow 2 \times (\text{row } 1) + \text{row } 2$$

The last row gives the false equation $0 = -2$. Therefore, the system is inconsistent, there are no solutions.

The next example demonstrates how the matrix method can be used to solve a system that has infinitely many solutions, that is, a dependent system.

EXAMPLE 2 Solve the system

$$\begin{aligned} x + 2y - z &= 1 \\ 2x - y + 3z &= 4 \\ 5x + 5z &= 9 \end{aligned}$$

Solution

$$\left[\begin{array}{ccc:c} 1 & 2 & -1 & 1 \\ 2 & -1 & 3 & 4 \\ 5 & 0 & 5 & 9 \end{array}\right]$$

$$\left[\begin{array}{ccc:c} 1 & 2 & -1 & 1 \\ 0 & -5 & 5 & 2 \\ 0 & -10 & 10 & 4 \end{array}\right] \begin{array}{l} \\ \longleftarrow -2 \times (\text{row } 1) + \text{row } 2 \\ \longleftarrow -5 \times (\text{row } 1) + \text{row } 3 \end{array}$$

$$\left[\begin{array}{ccc:c} 1 & 2 & -1 & 1 \\ 0 & -5 & 5 & 2 \\ 0 & 0 & 0 & 0 \end{array}\right] \begin{array}{l} \\ \\ \longleftarrow -2 \times (\text{row } 2) + \text{row } 3 \end{array}$$

Since the last row contains all zeros, we have an equivalent linear system of two equations in three variables.

$$\begin{aligned} x + 2y - z &= 1 \\ -5y + 5z &= 2 \end{aligned}$$

This system has an *incomplete* triangular form.

Again we use back-substitution. First use the last equation to solve for y in terms of z to get $y = z - \frac{2}{5}$. Now let $z = c$ represent any number, giving $y = c - \frac{2}{5}$ and substitute back into the first equation.

$$x + 2y - z = x + 2(c - \tfrac{2}{5}) - c = 1 \longrightarrow x = \tfrac{9}{5} - c$$

The solutions are $(\frac{9}{5} - c, c - \frac{2}{5}, c)$ where c is any number. This result can be checked in the original system as follows.

Find the specific solutions of the system for the values $c = 0$, $c = 1$, $c = \frac{2}{5}$, and $c = -2$.

$$\begin{aligned} x + 2y - z &= \tfrac{9}{5} - c + 2(c - \tfrac{2}{5}) - c = 1 \\ 2x - y + 3z &= 2(\tfrac{9}{5} - c) - (c - \tfrac{2}{5}) + 3c = 4 \\ 5x + 5z &= 5(\tfrac{9}{5} - c) + 5c = 9 \end{aligned}$$

When using the matrix method of this section keep the following observations in mind.

1 If you reach a linear system in triangular form, as in Example 1, the system has a unique solution that can be found by back-substitution.

2 If you reach a linear system that has an incomplete triangular form, and there are no false equations, as in Example 2, then the system has many solutions that can be found by back-substitution.

3 If you reach a linear system having a false equation, then the system has no solutions.

EXERCISES 8.1

Use matrices to solve each system.

1 $$\begin{aligned} x + 5y &= -9 \\ 4x - 3y &= -13 \end{aligned}$$

2 $$\begin{aligned} 4x - y &= 6 \\ 2x + 3y &= 10 \end{aligned}$$

3 $$\begin{aligned} 3x + 2y &= 18 \\ 6x + 5y &= 45 \end{aligned}$$

4 $$\begin{aligned} 2x + 4y &= 24 \\ -3x + 5y &= -25 \end{aligned}$$

5 $$\begin{aligned} 4x - 5y &= -2 \\ 16x + 2y &= 3 \end{aligned}$$

6 $$\begin{aligned} x &= y - 7 \\ 3y &= 2x + 16 \end{aligned}$$

7 $$\begin{aligned} 2x &= -8y + 2 \\ 4y &= x - 1 \end{aligned}$$

8 $$\begin{aligned} 2x - 5y &= 4 \\ -10x + 25y &= -20 \end{aligned}$$

9 $$\begin{aligned} 30x + 45y &= 60 \\ 4x + 6y &= 8 \end{aligned}$$

10 $$\begin{aligned} x - y &= 3 \\ -\tfrac{1}{3}x + \tfrac{1}{3}y &= 1 \end{aligned}$$

11 $$\begin{aligned} 2x &= 8 - 3y \\ 6x + 9y &= 14 \end{aligned}$$

12 $$\begin{aligned} -10x + 5y &= 8 \\ 15x - 10y &= -4 \end{aligned}$$

13 $$\begin{aligned} x - 2y + 3z &= -2 \\ -4x + 10y + 2z &= -2 \\ 3x + y + 10z &= 7 \end{aligned}$$

14 $$\begin{aligned} 2x + 4y + 8z &= 14 \\ 4x - 2y + 2z &= 6 \\ -5x + 3y - z &= -4 \end{aligned}$$

15 $$\begin{aligned} -x + 2y + 3z &= 11 \\ 2x - 3y &= -6 \\ 3x - 3y + 3z &= 3 \end{aligned}$$

16 $$\begin{aligned} -2x + y + 2z &= 14 \\ 5x + z &= -10 \\ x - 2y - 3z &= -14 \end{aligned}$$

17 $$\begin{aligned} x - 2z &= 5 \\ 3y + 4z &= -2 \\ -2x + 3y + 8z &= 4 \end{aligned}$$

18 $$\begin{aligned} 2x + y &= 3 \\ 4x + 5z &= 6 \\ -2y + 5z &= -4 \end{aligned}$$

19 $$\begin{aligned} x - 2y + z &= 1 \\ -6x + y + 2z &= -2 \\ -4x - 3y + 4z &= 0 \end{aligned}$$

20 $$\begin{aligned} 4x - 3y + z &= 0 \\ -3x + y + 2z &= 0 \\ -2x - y + 5z &= 0 \end{aligned}$$

21 $$\begin{aligned} w - x + 2y + 2z &= 0 \\ 2w - y - 3z &= 0 \\ 4x - 3y + z &= -2 \\ -3w + 2x + 4z &= 1 \end{aligned}$$

22 $$\begin{aligned} 4w - 5x + 2z &= 0 \\ -2w + 10x + y - 3z &= 8 \\ 5x - 2y + 4z &= -16 \\ 6w + 3z &= 0 \end{aligned}$$

23 $$\begin{aligned} v + 2w - x + 3z &= 4 \\ -w + 5x - y - z &= 3 \\ 3v + 6x + 2z &= 1 \\ 2v + 3w + 3x - y + 5z &= 10 \\ v + 9x - 2y + z &= 5 \end{aligned}$$

24 $$\begin{aligned} w + 3y &= 10 \\ -x + 2y + 6z &= 2 \\ -3w - 2x - 4z &= 2 \\ 4x - y &= 8 \end{aligned}$$

8.2 Matrix algebra

The matrix method for solving linear systems of Section 8.1 uses row operations to transform a matrix into a row-equivalent matrix. That method, however, does not combine two or more matrices in any way. In this section we learn how to add and multiply matrices, and also learn that with these operations matrices have many, but not all, of the basic properties of the real numbers stated in Section 1.1.

A matrix A with m rows and n columns is said to have dimensions m and n and can be written as

$$A = \left[\begin{array}{cccc} a_{11} & a_{12} & \cdots & a_{1n} \\ a_{21} & a_{22} & \cdots & a_{2n} \\ \cdot & \cdot & \cdots & \cdot \\ \cdot & \cdot & \cdots & \cdot \\ \cdot & \cdot & \cdots & \cdot \\ a_{m1} & a_{m2} & \cdots & a_{mn} \end{array}\right] \begin{array}{l} \longleftarrow \text{1st row} \\ \longleftarrow \text{2nd row} \\ \\ \\ \\ \longleftarrow m\text{th row} \end{array}$$

$$\begin{array}{ccc} \uparrow & \uparrow & \uparrow \\ \text{1st} & \text{2nd} & n\text{th} \\ \text{column} & \text{column} & \text{column} \end{array}$$

Instead of the symbol A we sometimes use $A_{m \times n}$ to emphasize the dimensions of A, in which $m \times n$ means m rows by n columns. The preceding rectangular display can be abbreviated by writing

$$A_{m \times n} = [a_{ij}] \longleftarrow a_{ij} \text{ is the element in the } i\text{th row and } j\text{th column}$$

in which a_{ij} represents *each* element in the matrix for $1 \leq i \leq m$ and $1 \leq j \leq n$.

Two m by n matrices are equal if and only if they have precisely the same elements in the same positions. In symbols using $A = [a_{ij}]$ and $B = [b_{ij}]$, we have:

$A = B$ if and only if $a_{ij} = b_{ij}$ for each i and j $\longleftarrow$ Equality of matrices

To add two m by n matrices, we add corresponding elements. For example, here is the computation for the sum of two 3 by 2 matrices.

$$\begin{bmatrix} 2 & -5 \\ 3 & 7 \\ -1 & 0 \end{bmatrix} + \begin{bmatrix} -4 & 1 \\ 0 & 9 \\ 2 & -2 \end{bmatrix} = \begin{bmatrix} 2+(-4) & -5+1 \\ 3+0 & 7+9 \\ -1+2 & 0+(-2) \end{bmatrix} = \begin{bmatrix} -2 & -4 \\ 3 & 16 \\ 1 & -2 \end{bmatrix}$$

In general, the sum of $A_{m \times n} = [a_{ij}]$ and $B_{m \times n} = [b_{ij}]$ can be expressed as:

$A + B = [a_{ij}] + [b_{ij}] = [a_{ij} + b_{ij}]$ $\longleftarrow$ Addition of matrices

It is important to note that matrices can be added only when they have the same dimensions. Thus

$$\begin{bmatrix} 1 & 2 & 3 \\ -2 & 0 & 6 \end{bmatrix} \quad \text{and} \quad \begin{bmatrix} 6 & -5 \\ 3 & 3 \end{bmatrix}$$

cannot be added.

Matrix addition is both commutative and associative. That is, for matrices having the same dimensions:

$$A + B = B + A \quad \longleftarrow \text{Commutative property for addition}$$

$$(A + B) + C = A + (B + C) \quad \longleftarrow \text{Associative property for addition}$$

We will not present formal proofs of these and other properties. However, you can see that they make sense by verifying the properties for specific examples.

EXAMPLE 1 Let $A = \begin{bmatrix} 3 & 1 & -2 \\ -1 & 3 & 0 \end{bmatrix}$, $B = \begin{bmatrix} -5 & 5 & 6 \\ -2 & -2 & -2 \end{bmatrix}$, and $C = \begin{bmatrix} 0 & 1 & 0 \\ 1 & 0 & 1 \end{bmatrix}$. Show that $(A + B) + C = A + (B + C)$.

Solution

$$(A + B) + C = \left(\begin{bmatrix} 3 & 1 & -2 \\ -1 & 3 & 0 \end{bmatrix} + \begin{bmatrix} -5 & 5 & 6 \\ -2 & -2 & -2 \end{bmatrix}\right) + \begin{bmatrix} 0 & 1 & 0 \\ 1 & 0 & 1 \end{bmatrix}$$

$$= \begin{bmatrix} -2 & 6 & 4 \\ -3 & 1 & -2 \end{bmatrix} + \begin{bmatrix} 0 & 1 & 0 \\ 1 & 0 & 1 \end{bmatrix} = \begin{bmatrix} -2 & 7 & 4 \\ -2 & 1 & -1 \end{bmatrix}$$

You should also verify that $A + B = B + A$.

$$A + (B + C) = \begin{bmatrix} 3 & 1 & -2 \\ -1 & 3 & 0 \end{bmatrix} + \left(\begin{bmatrix} -5 & 5 & 6 \\ -2 & -2 & -2 \end{bmatrix} + \begin{bmatrix} 0 & 1 & 0 \\ 1 & 0 & 1 \end{bmatrix}\right)$$

$$= \begin{bmatrix} 3 & 1 & -2 \\ -1 & 3 & 0 \end{bmatrix} + \begin{bmatrix} -5 & 6 & 6 \\ -1 & -2 & -1 \end{bmatrix} = \begin{bmatrix} -2 & 7 & 4 \\ -2 & 1 & -1 \end{bmatrix}$$

Thus

$$(A + B) + C = A + (B + C)$$

The m by n zero matrix, denoted by O, is the matrix all of whose entries are 0. For example, the 2 by 3 zero matrix is

$$O_{2\times 3} = \begin{bmatrix} 0 & 0 & 0 \\ 0 & 0 & 0 \end{bmatrix}$$

The next rule is easy to verify:

$$A_{m\times n} + O_{m\times n} = A_{m\times n} \quad \longleftarrow \text{Identity property for addition}$$

The negative of an m by n matrix $A = [a_{ij}]$ is the matrix, denoted by $-A$, and is defined in this way:

$$-A = [-a_{ij}] \quad \longleftarrow \text{Negative of a matrix}$$

In other words, the negative of a matrix is formed by replacing each number in A by its opposite. Consequently, since $a_{ij} + (-a_{ij}) = 0$,

$$A + (-A) = O \quad \longleftarrow \text{Additive inverse property}$$

For example, if $A = \begin{bmatrix} 8 & -5 \\ -4 & 9 \\ 0 & 2 \end{bmatrix}$, then $-A = \begin{bmatrix} -8 & 5 \\ 4 & -9 \\ 0 & -2 \end{bmatrix}$ and

$$A + (-A) = \begin{bmatrix} 8 + (-8) & -5 + 5 \\ -4 + 4 & 9 + (-9) \\ 0 + 0 & 2 + (-2) \end{bmatrix} = \begin{bmatrix} 0 & 0 \\ 0 & 0 \\ 0 & 0 \end{bmatrix} = O$$

Subtraction of m by n matrices can be defined using the negative of a matrix.

$$A - B = A + (-B) \quad \longleftarrow \text{Subtraction of matrices}$$

A matrix A can be multiplied by a real number c by multiplying each element in A by c. This is referred to as **scalar multiplication**.

$$cA = c[a_{ij}] = [ca_{ij}] \quad \longleftarrow \text{Scalar multiplication}$$

It is customary to write the scalar c to the left of the matrix, thus we do *not* use the notation Ac for scalar multiplication.

For example,

$$7\begin{bmatrix} 8 & -5 \\ -4 & 9 \\ 0 & 2 \end{bmatrix} = \begin{bmatrix} 56 & -35 \\ -28 & 63 \\ 0 & 14 \end{bmatrix}$$

Using this property the negative of a matrix A may be written as $-A = -1A$.

Here is a list of scalar multiplication properties:

$$(cd)A = c(dA)$$
$$(c + d)A = cA + dA$$
$$c(A + B) = cA + cB$$

A, B have the same dimension and c, d are real numbers.

EXAMPLE 2 Verify $(cd)A = c(dA)$ using

$$c = -2, \qquad d = \tfrac{1}{3}, \qquad \text{and} \qquad A = \begin{bmatrix} 6 & -9 \\ 1 & 5 \end{bmatrix}.$$

Solution

$$(cd)A = (-2 \cdot \tfrac{1}{3})\begin{bmatrix} 6 & -9 \\ 1 & 5 \end{bmatrix} = -\tfrac{2}{3}\begin{bmatrix} 6 & -9 \\ 1 & 5 \end{bmatrix} = \begin{bmatrix} -4 & 6 \\ -\frac{2}{3} & -\frac{10}{3} \end{bmatrix}$$

$$c(dA) = -2\left(\tfrac{1}{3}\begin{bmatrix} 6 & -9 \\ 1 & 5 \end{bmatrix}\right) = -2\begin{bmatrix} 2 & -3 \\ \frac{1}{3} & \frac{5}{3} \end{bmatrix} = \begin{bmatrix} -4 & 6 \\ -\frac{2}{3} & -\frac{10}{3} \end{bmatrix}$$

TEST YOUR UNDERSTANDING

Let $A = \begin{bmatrix} 1 & 4 & -3 \\ 2 & 5 & 0 \\ 0 & 1 & 1 \end{bmatrix}, \quad B = \begin{bmatrix} -2 & 2 \\ 3 & -3 \\ 1 & 4 \end{bmatrix}, \quad C = \begin{bmatrix} 5 & 1 \\ -1 & 6 \\ 8 & 2 \end{bmatrix}.$

Evaluate each of the following, if possible.

1 $A + B$ **2** $B + C$ **3** $C + (B + C)$

4 $B - C$ **5** $C - B$ **6** $2A + 0A$

7 $2A + 3C$ **8** $(-5)(B + C)$ **9** $4((-\frac{1}{2})C)$

Multiplication of matrices is a more complicated operation than addition. First, let us consider the product of

$$A_{2\times3} = \begin{bmatrix} -4 & -1 & 2 \\ 3 & 8 & 1 \end{bmatrix} \quad \text{and} \quad B_{3\times2} = \begin{bmatrix} -2 & 1 \\ 7 & 0 \\ 4 & -6 \end{bmatrix}$$

Here is the completed product followed by the explanation as to how it was done.

$$\overset{A}{\begin{bmatrix} -4 & -1 & 2 \\ 3 & 8 & 1 \end{bmatrix}} \overset{B}{\begin{bmatrix} -2 & 1 \\ 7 & 0 \\ 4 & -6 \end{bmatrix}} = \overset{C}{\begin{bmatrix} 9 & -16 \\ 54 & -3 \end{bmatrix}} = \overset{C}{\begin{bmatrix} c_{11} & c_{12} \\ c_{21} & c_{22} \end{bmatrix}}$$

Row 1 of A and *column 1 of B* produce 9, the element in row 1, column 1 of C:

$$(-4)(-2) + (-1)(7) + (2)(4) = 9 = c_{11}$$

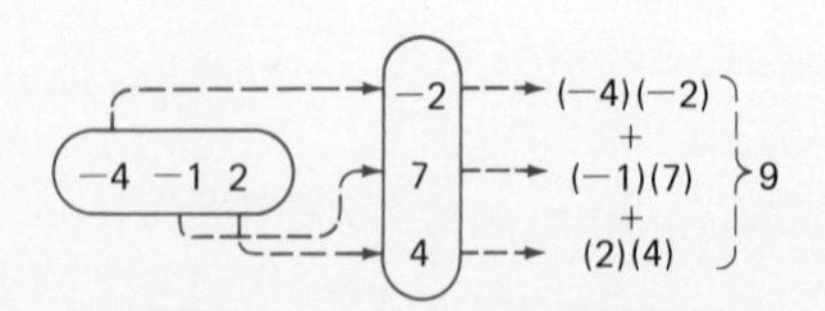

Row 1 of A and *column 2 of B* produce -16, the element in row 1, column 2 of C:

$$(-4)(1) + (-1)(0) + (2)(-6) = -16 = c_{12}$$

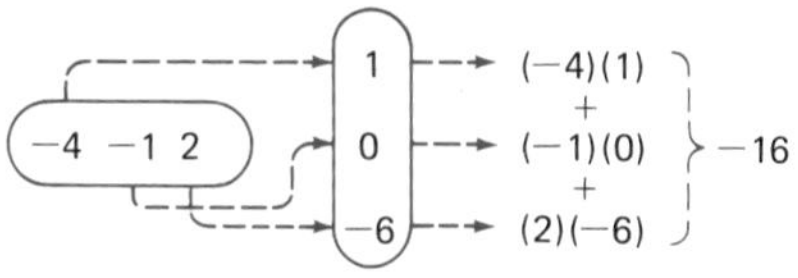

Row 2 of A and *column 1 of B* produce 54, the element in row 2, column 1 of C:

$$(3)(-2) + (8)(7) + (1)(4) = 54 = c_{21}$$

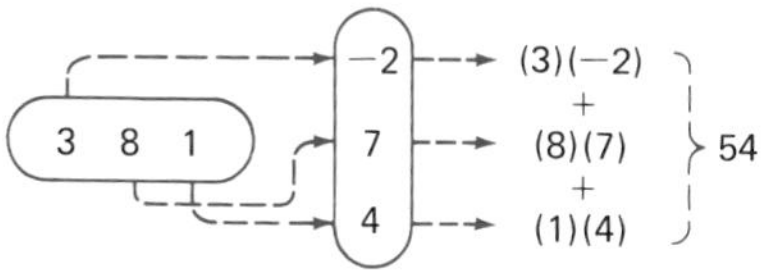

Row 2 of A and *column 2 of B* produce -3, the element in row 2, column 2 of C:

$$(3)(1) + (8)(0) + (1)(-6) = -3 = c_{22}$$

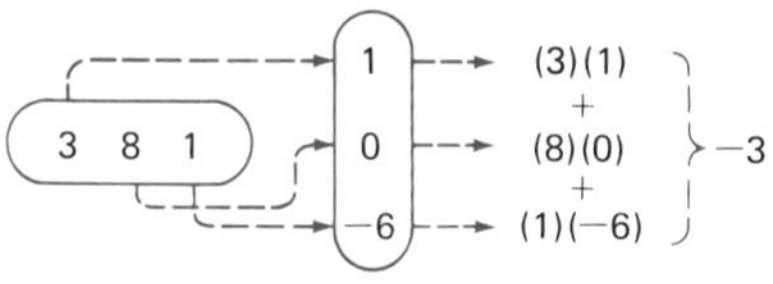

As you can observe in the preceding computations, an element in $AB = C$ is the sum of the products of numbers in a row of A by the numbers in a column of B. Furthermore, the ith row of A together with the jth column of B produces the element in row i, column j of C.

Here is a diagram that displays how the row 2, column 3 element in a product is completed.

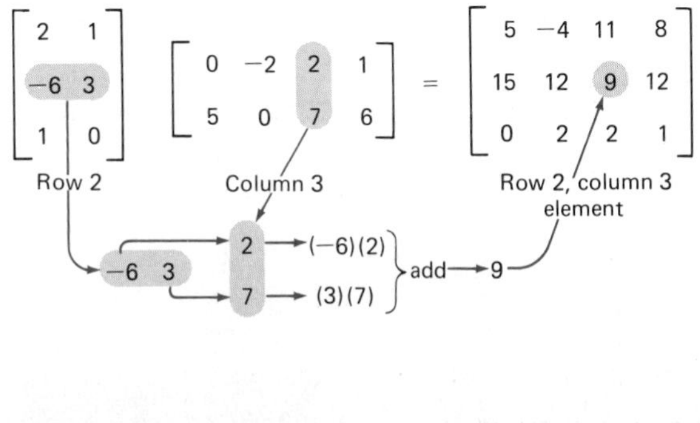

$$\begin{bmatrix} 2 & 1 \\ -6 & 3 \\ 1 & 0 \end{bmatrix} \begin{bmatrix} 0 & -2 & 2 & 1 \\ 5 & 0 & 7 & 6 \end{bmatrix} = \begin{bmatrix} 5 & -4 & 11 & 8 \\ 15 & 12 & 9 & 12 \\ 0 & 2 & 2 & 1 \end{bmatrix}$$

In each of the matrix products that have been found, you can see that it is possible to compute product AB only when *the number of columns of A equals the number of rows of B.* Furthermore, if A is m by n and B is n by p, then AB is n by p. For example,

$$(A_{3\times 5})(B_{5\times 7}) = C_{3\times 7}.$$

these must be the same — number of rows in A — number of columns in B

TEST YOUR UNDERSTANDING

Let $A = \begin{bmatrix} 1 & 2 \\ 3 & -5 \end{bmatrix}$ $\quad B = \begin{bmatrix} 6 & -4 & 9 \\ 1 & 2 & 0 \end{bmatrix}$

$C = \begin{bmatrix} 3 & 7 \\ -4 & 6 \\ 2 & -9 \end{bmatrix}$ $\quad D = \begin{bmatrix} 8 & -8 \\ -7 & -7 \end{bmatrix}$

Find each product if possible.

1 AD **2** AB **3** AC **4** CA

5 BD **6** BC **7** CB **8** $(AB)C$

Using A and D from the preceding set of exercises, we have

$$AD = \begin{bmatrix} 1 & 2 \\ 3 & -5 \end{bmatrix}\begin{bmatrix} 8 & -8 \\ -7 & -7 \end{bmatrix} = \begin{bmatrix} -6 & -22 \\ 59 & 11 \end{bmatrix}$$

Recall that a single counter-example is sufficient to prove that a property does *not* hold in general.

and

$$DA = \begin{bmatrix} 8 & -8 \\ -7 & -7 \end{bmatrix}\begin{bmatrix} 1 & 2 \\ 3 & -5 \end{bmatrix} = \begin{bmatrix} -16 & 56 \\ -28 & 21 \end{bmatrix}$$

Therefore, $AD \neq DA$, which proves that matrix multiplication is not commutative. We do have the following properties, though, assuming that the operations are possible.

In the exercises you will be asked to verify these properties for specific cases.

$(AB)C = A(BC)$ ⟵ Associative property for multiplication

$A(B + C) = AB + AC$ ⟵ Left distributive property

$(B + C)A = BA + CA$ ⟵ Right distributive property

Scalar multiplication can also be combined with the product of matrices according to this result in which c represents any real number.

$$c(AB) = (cA)B = A(cB)$$

There is also a multiplicative identity for **square matrices**, matrices that are n by n. This is the matrix whose main diagonal consists of 1's and all other entries are 0. We use I_n for this matrix.

In particular, when $n = 3$,

$$I_3 = \begin{bmatrix} 1 & 0 & 0 \\ 0 & 1 & 0 \\ 0 & 0 & 1 \end{bmatrix} \leftarrow \text{main diagonal}$$

Using $A = \begin{bmatrix} a & b & c \\ d & e & f \\ g & h & i \end{bmatrix}$, we have

$$I_3A = \begin{bmatrix} 1 & 0 & 0 \\ 0 & 1 & 0 \\ 0 & 0 & 1 \end{bmatrix}\begin{bmatrix} a & b & c \\ d & e & f \\ g & h & i \end{bmatrix} = \begin{bmatrix} a & b & c \\ d & e & f \\ g & h & i \end{bmatrix} = A$$

Similarly, $AI_3 = A$. In general, if A is n by n, then

$$AI_n = I_nA = A \quad \longleftarrow \text{Multiplication identity property}$$

I_n is the multiplicative identity for the n by n matrices. It is also called the ***n* th-order identity matrix.**

EXERCISES 8.2

In Exercises 1–24, evaluate the given matrix expression, if possible, using these matrices.

$A = \begin{bmatrix} 1 \\ -2 \end{bmatrix}$ $\quad B = \begin{bmatrix} 3 & 5 \end{bmatrix}$

$C = \begin{bmatrix} 5 & -1 \\ -3 & 4 \end{bmatrix}$ $\quad D = \begin{bmatrix} 1 & 2 \\ 3 & 4 \end{bmatrix}$

$E = \begin{bmatrix} 1 & -3 & 2 \\ -5 & 2 & 0 \\ 0 & -1 & 3 \end{bmatrix}$ $\quad F = \begin{bmatrix} 1 & 1 & 1 \\ -1 & 1 & 1 \\ -1 & -1 & 1 \end{bmatrix}$

$G = \begin{bmatrix} 1 & -4 & 0 \\ 2 & -3 & 5 \end{bmatrix}$ $\quad H = \begin{bmatrix} 5 & 2 & -3 & 0 \\ 1 & 0 & 3 & -1 \\ 4 & -2 & 0 & 1 \end{bmatrix}$

$J = \begin{bmatrix} 1 & 0 \\ 0 & -1 \\ 1 & 1 \\ 0 & 0 \end{bmatrix}$

1 $C + D$

2 $D - C$

3 $A + B$

4 $C + G$

5 $E + 2F$

6 $-2E + 5E$

7 BA

8 AB

9 CD

10 DC

11 CA

12 BC

13 GE

14 EG

15 GJ

16 EF

17 HJ

18 $(DA)C$

19 GH

20 $(\frac{1}{2}D)J$

21 $F\begin{bmatrix} 0 \\ 0 \\ 0 \end{bmatrix}$

22 $\begin{bmatrix} 0 \\ 0 \\ 0 \end{bmatrix}F$

23 I_3H

24 EI_3

In Exercises 25–30, verify the matrix equation using these matrices.

$$A = \begin{bmatrix} 2 & 1 \\ -4 & 1 \end{bmatrix} \quad B = \begin{bmatrix} 0 & 6 \\ 5 & -4 \end{bmatrix} \quad C = \begin{bmatrix} 3 & 7 \\ -1 & -2 \end{bmatrix}$$

25 $B + C = C + B$

26 $3(AB) = (3A)B = A(3B)$

27 $A + (B + C) = (A + B) + C$

28 $A(B + C) = AB + AC$

29 $(A + B)C = AC + BC$

30 $(AB)C = A(BC)$

31 Verify that $(AB)C = A(BC)$ using

$$A = \begin{bmatrix} 1 & -2 \\ 3 & 3 \\ 0 & 4 \end{bmatrix} \quad B = \begin{bmatrix} -1 & 6 & 8 \\ 2 & 1 & 3 \end{bmatrix} \quad C = \begin{bmatrix} 5 \\ -2 \\ 1 \end{bmatrix}$$

32 Arrange the four matrices with the indicated dimensions so that a product of all four can be formed. What will be the dimensions of this product?

$$A_{4\times 1}, \quad B_{3\times 5}, \quad C_{1\times 3}, \quad D_{2\times 4}$$

33 Let $E = \begin{bmatrix} 0 & 1 & 0 \\ 1 & 0 & 0 \\ 0 & 0 & 1 \end{bmatrix}$ and $A = \begin{bmatrix} a & b & c \\ d & e & f \\ g & h & i \end{bmatrix}$

(a) Evaluate EA.

(b) Compare the answer in part (a) to matrix A and find a matrix F so that FA is the same as A except that the first and third rows are interchanged.

34 (a) Use E and A in Exercise 33 and evaluate AE.

(b) Find a matrix F so that AF is the same as A except that the second and third columns are interchanged.

35 (a) Find x, y so that

$$\begin{bmatrix} 1 & 1 \\ 0 & 0 \end{bmatrix}\begin{bmatrix} 2 & 3 \\ x & y \end{bmatrix} = \begin{bmatrix} 0 & 0 \\ 0 & 0 \end{bmatrix}$$

(b) Explain why it is not possible to find x, y so that

$$\begin{bmatrix} 2 & 3 \\ x & y \end{bmatrix}\begin{bmatrix} 1 & 1 \\ 0 & 0 \end{bmatrix} = \begin{bmatrix} 0 & 0 \\ 0 & 0 \end{bmatrix}$$

36 Find $a, b, c,$ and d so that $\begin{bmatrix} 1 & 1 \\ 0 & 1 \end{bmatrix}\begin{bmatrix} a & b \\ c & d \end{bmatrix} = I_2$ and verify that $\begin{bmatrix} a & b \\ c & d \end{bmatrix}\begin{bmatrix} 1 & 1 \\ 0 & 1 \end{bmatrix} = I_2$.

37 (a) Let $A = \begin{bmatrix} x & 0 & 0 \\ 0 & y & 0 \\ 0 & 0 & z \end{bmatrix}$ and evaluate $A^2 = AA$.

(b) Use the result in part (a) to evaluate $A^3 = AAA$.

(c) If n is a positive integer, find A^n.

38 (a) Use $A = \begin{bmatrix} 3 & -1 \\ 2 & 1 \end{bmatrix}$, $B = \begin{bmatrix} 0 & -2 \\ 1 & 4 \end{bmatrix}$ to show that $(A + B)^2 \neq A^2 + 2AB + B^2$.

(b) Give a reason for each numbered step.

$$\begin{aligned} (A + B)^2 &= (A + B)(A + B) \\ &= (A + B)A + (A + B)B \qquad \text{(i)} \\ &= A^2 + BA + AB + B^2 \qquad \text{(ii)} \end{aligned}$$

(c) Verify the result of part (b) using matrices A and B in part (a).

8.3

Inverses

You may have wondered why nothing was said in the preceding section about division of matrices. The reason is simple. Division of matrices is undefined. There is, however, a process possible for some matrices that resembles the division of real numbers when the division is converted to multiplication using reciprocals as in

$$2 \div 5 = 2 \cdot \frac{1}{5} = 2 \cdot 5^{-1}$$

Similarly, some square matrices have inverses, denoted by A^{-1}, which makes it possible to compute products of the form BA^{-1}.

A square matrix B is said to be the inverse of a square matrix A if and only if we have the following:

$$AB = BA = I \quad \longleftarrow \text{Inverse matrices}$$

In such a case matrix A is also the inverse of B; A and B are inverses of one another.

The notation A^{-1} is frequently used for the inverse of a matrix A, so that

$$AA^{-1} = A^{-1}A = I_n$$

This compares to $x \cdot \frac{1}{x} = \frac{1}{x} \cdot x = 1$ for real numbers $x \neq 0$.

When a matrix has an inverse, it has only one inverse. That is, the inverse of a matrix is unique (see Exercise 37).

Does $A = \begin{bmatrix} 1 & -1 \\ 0 & 2 \end{bmatrix}$ have an inverse? If so, then there are numbers v, w, x, and y such that

$$\begin{bmatrix} 1 & -1 \\ 0 & 2 \end{bmatrix}\begin{bmatrix} v & w \\ x & y \end{bmatrix} = \begin{bmatrix} 1 & 0 \\ 0 & 1 \end{bmatrix}$$

Multiply to get

$$\begin{bmatrix} v - x & w - y \\ 2x & 2y \end{bmatrix} = \begin{bmatrix} 1 & 0 \\ 0 & 1 \end{bmatrix}$$

Since two matrices are equal if and only if their elements in the same positions are equal, we get

$$v - x = 1$$
$$w - y = 0$$
$$2x = 0 \longrightarrow x = 0$$
$$2y = 1 \longrightarrow y = \tfrac{1}{2}$$
$$v - x = v - 0 = 1 \longrightarrow v = 1$$
$$w - y = w - \tfrac{1}{2} = 0 \longrightarrow w = \tfrac{1}{2}$$

Therefore,

$$\begin{bmatrix} v & w \\ x & y \end{bmatrix} = \begin{bmatrix} 1 & \frac{1}{2} \\ 0 & \frac{1}{2} \end{bmatrix}$$

Now check by multiplying:

$$\begin{bmatrix} 1 & -1 \\ 0 & 2 \end{bmatrix}\begin{bmatrix} 1 & \frac{1}{2} \\ 0 & \frac{1}{2} \end{bmatrix} = \begin{bmatrix} 1 + 0 & \frac{1}{2} - \frac{1}{2} \\ 0 + 0 & 0 + 1 \end{bmatrix} = \begin{bmatrix} 1 & 0 \\ 0 & 1 \end{bmatrix} = I_2$$

This shows that $AA^{-1} = I_2$.

Also,

$$\begin{bmatrix} 1 & \frac{1}{2} \\ 0 & \frac{1}{2} \end{bmatrix}\begin{bmatrix} 1 & -1 \\ 0 & 2 \end{bmatrix} = \begin{bmatrix} 1 & 0 \\ 0 & 1 \end{bmatrix} = I_2$$

This shows that $A^{-1}A = I_2$.

Thus

$$A^{-1} = \begin{bmatrix} 1 & -1 \\ 0 & 2 \end{bmatrix}^{-1} = \begin{bmatrix} 1 & \frac{1}{2} \\ 0 & \frac{1}{2} \end{bmatrix}$$

Not all square matrices have inverses. For example, the matrix $\begin{bmatrix} 0 & 1 \\ 0 & 1 \end{bmatrix}$ has no inverse because if it did there would have to be v, w, x, and y so that

$$\begin{bmatrix} 0 & 1 \\ 0 & 1 \end{bmatrix}\begin{bmatrix} v & w \\ x & y \end{bmatrix} = \begin{bmatrix} 1 & 0 \\ 0 & 1 \end{bmatrix}$$

$$\begin{bmatrix} x & y \\ x & y \end{bmatrix} = \begin{bmatrix} 1 & 0 \\ 0 & 1 \end{bmatrix}$$

Equating the elements gives

$$x = 1 \quad \text{and} \quad x = 0$$

which is not possible. Consequently, $\begin{bmatrix} 0 & 1 \\ 0 & 1 \end{bmatrix}$ has no inverse.

The method of finding an inverse used in the preceding discussion is not practical for larger matrices. As you have seen, when $n = 2$ we have to solve 4 equations. With $n = 3$, there would be 9 equations, then 16 equations for $n = 4$, and so on.

Fortunately, there are other more efficient methods for finding inverses. We will now present such a procedure and first demonstrate how it works using the matrix $A = \begin{bmatrix} 1 & -1 \\ 0 & 2 \end{bmatrix}$, for which we already know the inverse.

Begin by constructing a 2 by 4 matrix that contains both A and the identity matrix I_2.

$$\begin{array}{c} \quad A \qquad\quad I_2 \\ \quad\downarrow \qquad\quad \downarrow \\ \left[\begin{array}{cc:cc} 1 & -1 & 1 & 0 \\ 0 & 2 & 0 & 1 \end{array}\right] \end{array}$$

Put A to the left and I_2 to the right of the dashed line.

Now apply fundamental row operations to the rows of this 2 by 4 matrix until the left half has been transformed into I_2. At this step the right half will be A^{-1}. Here are the details.

$$\begin{array}{rc} & \begin{array}{c} A \qquad\quad I_2 \\ \downarrow \qquad\quad \downarrow \end{array} \\ & \left[\begin{array}{cc:cc} 1 & -1 & 1 & 0 \\ 0 & 2 & 0 & 1 \end{array}\right] \\ \frac{1}{2} \times (\text{row } 2) \longrightarrow & \left[\begin{array}{cc:cc} 1 & -1 & 1 & 0 \\ 0 & 1 & 0 & \frac{1}{2} \end{array}\right] \\ \text{row } 2 + \text{row } 1 \longrightarrow & \left[\begin{array}{cc:cc} 1 & 0 & 1 & \frac{1}{2} \\ 0 & 1 & 0 & \frac{1}{2} \end{array}\right] \\ & \begin{array}{c} \uparrow \qquad\quad \uparrow \\ I_2 \qquad\quad A^{-1} \end{array} \end{array}$$

From our earlier work we know that $\begin{bmatrix} 1 & \frac{1}{2} \\ 0 & \frac{1}{2} \end{bmatrix}$ is the inverse of A.

This method can also be used to find out that a matrix does not have an inverse. The fundamental row operations are applied until we discover that it is not possible to reach I on the left. When this happens there is no inverse.

EXAMPLE 1 Show that $A = \begin{bmatrix} 2 & -4 \\ -1 & 2 \end{bmatrix}$ has no inverse.

Solution

$$\begin{array}{c} \\ \\ \end{array}\left[\begin{array}{rr|rr} 2 & -4 & 1 & 0 \\ -1 & 2 & 0 & 1 \end{array}\right] \quad (A \mid I_2)$$

$$\tfrac{1}{2} \times (\text{row } 1) \longrightarrow \left[\begin{array}{rr|rr} 1 & -2 & \tfrac{1}{2} & 0 \\ -1 & 2 & 0 & 1 \end{array}\right]$$

$$\text{row } 1 + \text{row } 2 \longrightarrow \left[\begin{array}{rr|rr} 1 & -2 & \tfrac{1}{2} & 0 \\ 0 & 0 & \tfrac{1}{2} & 1 \end{array}\right]$$

Because of the row of zeros to the left it is not possible to use row operations to obtain 1 0 in the first row on the left. Then I_2 cannot be reached on the left, and therefore there is no inverse of A.

TEST YOUR UNDERSTANDING

Find A^{-1}, if it exists, for the given A. When A^{-1} exists verify that $AA^{-1} = A^{-1}A = I_2$.

1 $A = \begin{bmatrix} 3 & -6 \\ -2 & 5 \end{bmatrix}$ **2** $A = \begin{bmatrix} -4 & 8 \\ -3 & 6 \end{bmatrix}$ **3** $A = \begin{bmatrix} -3 & -2 \\ -1 & 0 \end{bmatrix}$

4 $A = \begin{bmatrix} \frac{1}{2} & \frac{1}{4} \\ -\frac{1}{3} & -\frac{1}{3} \end{bmatrix}$ **5** $A = \begin{bmatrix} 7 & 7 \\ 7 & 7 \end{bmatrix}$ **6** $A = \begin{bmatrix} 10 & 100 \\ 100 & -100 \end{bmatrix}$

For larger matrices this method will require considerably more work; however, it is usually much more efficient than the procedure that would call for solving a large system of equations.

EXAMPLE 2 Let $A = \begin{bmatrix} 2 & 0 & -1 \\ -1 & 2 & 1 \\ 3 & -2 & -4 \end{bmatrix}$. Find A^{-1}.

Solution

$$\left[\begin{array}{rrr|rrr} 2 & 0 & -1 & 1 & 0 & 0 \\ -1 & 2 & 1 & 0 & 1 & 0 \\ 3 & -2 & -4 & 0 & 0 & 1 \end{array}\right] \quad (A \mid I_3)$$

Our first step will be to interchange the first two rows to get the -1 in the second row to the top left position.

The explanations of the row operations at the left use the abbreviation R_2 in place of row 2, etc. Recall that these notes describe what was done in the preceding matrix to get the indicated rows.

$$\begin{array}{l} \text{Interchange} \longrightarrow \\ R_1 \text{ and } R_2 \longrightarrow \\ \end{array} \left[\begin{array}{rrr|rrr} -1 & 2 & 1 & 0 & 1 & 0 \\ 2 & 0 & -1 & 1 & 0 & 0 \\ 3 & -2 & -4 & 0 & 0 & 1 \end{array}\right]$$

$$\begin{array}{l} \\ 2 \times R_1 + R_2 \longrightarrow \\ 3 \times R_1 + R_3 \longrightarrow \end{array} \left[\begin{array}{rrr|rrr} -1 & 2 & 1 & 0 & 1 & 0 \\ 0 & 4 & 1 & 1 & 2 & 0 \\ 0 & 4 & -1 & 0 & 3 & 1 \end{array}\right]$$

$$\begin{array}{l} \\ \\ R_3 + (-R_2) \longrightarrow \end{array} \left[\begin{array}{rrr|rrr} -1 & 2 & 1 & 0 & 1 & 0 \\ 0 & 4 & 1 & 1 & 2 & 0 \\ 0 & 0 & -2 & -1 & 1 & 1 \end{array}\right]$$

$$\begin{array}{l} \\ \\ -\frac{1}{2} \times R_3 \longrightarrow \end{array} \left[\begin{array}{rrr|rrr} -1 & 2 & 1 & 0 & 1 & 0 \\ 0 & 4 & 1 & 1 & 2 & 0 \\ 0 & 0 & 1 & \frac{1}{2} & -\frac{1}{2} & -\frac{1}{2} \end{array}\right]$$

$$\begin{array}{l} -1 \times R_3 + R_1 \longrightarrow \\ -1 \times R_3 + R_2 \longrightarrow \\ \end{array} \left[\begin{array}{rrr|rrr} -1 & 2 & 0 & -\frac{1}{2} & \frac{3}{2} & \frac{1}{2} \\ 0 & 4 & 0 & \frac{1}{2} & \frac{5}{2} & \frac{1}{2} \\ 0 & 0 & 1 & \frac{1}{2} & -\frac{1}{2} & -\frac{1}{2} \end{array}\right]$$

$$\begin{array}{l} -1 \times R_1 \longrightarrow \\ \frac{1}{4} \times R_2 \longrightarrow \\ \end{array} \left[\begin{array}{rrr|rrr} 1 & -2 & 0 & \frac{1}{2} & -\frac{3}{2} & -\frac{1}{2} \\ 0 & 1 & 0 & \frac{1}{8} & \frac{5}{8} & \frac{1}{8} \\ 0 & 0 & 1 & \frac{1}{2} & -\frac{1}{2} & -\frac{1}{2} \end{array}\right]$$

$$\begin{array}{l} 2 \times R_2 + R_1 \longrightarrow \\ \\ \end{array} \left[\begin{array}{rrr|rrr} 1 & 0 & 0 & \frac{3}{4} & -\frac{1}{4} & -\frac{1}{4} \\ 0 & 1 & 0 & \frac{1}{8} & \frac{5}{8} & \frac{1}{8} \\ 0 & 0 & 1 & \frac{1}{2} & -\frac{1}{2} & -\frac{1}{2} \end{array}\right]$$

$$\uparrow I_3 \qquad\qquad \uparrow A^{-1}$$

You should now verify that the indicated matrix A^{-1} satisfies

$$AA^{-1} = A^{-1}A = I$$

One application of inverses is to the solution of linear systems in which the number of variables is the same as the number of equations. For such a system the coefficients of the variables form a square matrix A. Then, as you will see, if A^{-1} exists the system can be solved using this inverse.

The first step is to convert the linear system into a matrix equation. For example, consider this system:

$$\begin{aligned} 2x \phantom{{}+2y} - z &= 2 \\ -x + 2y + z &= 0 \\ 3x - 2y - 4z &= 10 \end{aligned}$$

In terms of matrices it can be written as

$$\begin{bmatrix} 2 & 0 & -1 \\ -1 & 2 & 1 \\ 3 & -2 & 4 \end{bmatrix} \begin{bmatrix} x \\ y \\ z \end{bmatrix} = \begin{bmatrix} 2 \\ 0 \\ 10 \end{bmatrix}$$

↑ matrix of coefficients ↑ matrix of variables ↑ matrix of constants

You should multiply the matrices on the left to get a 3 by 1 product. Then equate elements to get the given system.

Using A for the matrix of coefficients, X for the matrix of variables, and C for the matrix of constants, we have

$$AX = C$$

Now assume that A^{-1} exists, and multiply the preceding equation by A^{-1}. Then

$$A^{-1}(AX) = A^{-1}C$$

$$(A^{-1}A)X = A^{-1}C \qquad \text{(Matrix multiplication is associative)}$$

$$IX = A^{-1}C \qquad (A^{-1}A = I)$$

$$X = A^{-1}C \qquad \text{(}I \text{ is the matrix identity for multiplication)}$$

The solution can now be found by equating the entries in $X = \begin{bmatrix} x \\ y \\ z \end{bmatrix}$ with the three numbers in $A^{-1}C$.

$A^{-1}C$ has dimensions 3 by 1.

EXAMPLE 3 Solve the linear system using the inverse of the matrix of coefficients.

$$3x + 4y = 5$$
$$x + 2y = 3$$

Solution The matrix of coefficients is $A = \begin{bmatrix} 3 & 4 \\ 1 & 2 \end{bmatrix}$. Its inverse is found to be $A^{-1} = \begin{bmatrix} 1 & -2 \\ -\frac{1}{2} & \frac{3}{2} \end{bmatrix}$.

Then, using $X = A^{-1}C$, we have

$$\begin{bmatrix} x \\ y \end{bmatrix} = \begin{bmatrix} 1 & -2 \\ -\frac{1}{2} & \frac{3}{2} \end{bmatrix} \begin{bmatrix} 5 \\ 3 \end{bmatrix} = \begin{bmatrix} -1 \\ 2 \end{bmatrix}$$

Therefore, $x = -1$, $y = 2$. Check this in the original system.

EXAMPLE 4 Solve the linear system using A^{-1}, where A is the matrix of coefficients.

$$2x \qquad - z = 2$$
$$-x + 2y + z = 0$$
$$3x - 2y - 4z = 10$$

Our first step will be to write this system in matrix form.

Solution The matrix of coefficient A is the same as the matrix in Example 2. Then, using A^{-1} from Example 2 and the result $X = A^{-1}C$, we get

$$\begin{bmatrix} x \\ y \\ z \end{bmatrix} = \begin{bmatrix} \frac{3}{4} & -\frac{1}{4} & -\frac{1}{4} \\ \frac{1}{8} & \frac{5}{8} & \frac{1}{8} \\ \frac{1}{2} & -\frac{1}{2} & -\frac{1}{2} \end{bmatrix} \begin{bmatrix} 2 \\ 0 \\ 10 \end{bmatrix} = \begin{bmatrix} -1 \\ \frac{3}{2} \\ -4 \end{bmatrix}$$

Therefore, $x = -1$, $y = \frac{3}{2}$, $z = -4$.

The system in Example 4 was solved with little effort because A^{-1} was already available. In fact, any system having the same A as its matrix of coefficients can be solved just as quickly, regardless of the constants. Therefore, this inverse method is particularly useful in solving more than one linear system having the same matrix of coefficients. If, as in Example 4, we want to solve only one system and A^{-1} were not available, then the method of Section 8.1 is usually less work.

EXERCISES 8.3

In Exercises 1–20, find A^{-1}, if it exists.

1 $A = \begin{bmatrix} 4 & -1 \\ 2 & 0 \end{bmatrix}$

2 $A = \begin{bmatrix} -\frac{1}{5} & -\frac{2}{5} \\ 0 & \frac{1}{4} \end{bmatrix}$

3 $A = \begin{bmatrix} \frac{1}{3} & -\frac{4}{3} \\ -2 & 8 \end{bmatrix}$

4 $A = \begin{bmatrix} -\frac{3}{2} & \frac{5}{3} \\ \frac{9}{4} & -\frac{5}{2} \end{bmatrix}$

5 $A = \begin{bmatrix} 2 & -5 \\ -3 & 4 \end{bmatrix}$

6 $A = \begin{bmatrix} 10 & 15 \\ -5 & -1 \end{bmatrix}$

7 $A = \begin{bmatrix} \frac{1}{3} & \frac{2}{3} \\ \frac{2}{3} & \frac{1}{3} \end{bmatrix}$

8 $A = \begin{bmatrix} -3 & 6 \\ 6 & -3 \end{bmatrix}$

9 $A = \begin{bmatrix} 0 & a \\ b & 0 \end{bmatrix}$, $ab \neq 0$

10 $A = \begin{bmatrix} 2 & 1 & 0 \\ 0 & 0 & -2 \\ 4 & 4 & 0 \end{bmatrix}$

11 $A = \begin{bmatrix} 0 & 1 & 0 \\ 1 & 0 & 0 \\ 0 & 0 & 1 \end{bmatrix}$

12 $A = \begin{bmatrix} 1 & 2 & -1 \\ -1 & 3 & 4 \\ 1 & 7 & 2 \end{bmatrix}$

13 $A = \begin{bmatrix} 1 & 0 & 2 \\ 2 & -1 & 0 \\ 0 & 3 & 4 \end{bmatrix}$

14 $A = \begin{bmatrix} 1 & 1 & -1 \\ 1 & -1 & -1 \\ -1 & -1 & -1 \end{bmatrix}$

15 $A = \begin{bmatrix} 4 & -3 & 1 \\ 0 & -1 & 9 \\ -2 & 1 & 4 \end{bmatrix}$

16 $A = \begin{bmatrix} 8 & -13 & 2 \\ -4 & 7 & -1 \\ 3 & -5 & 1 \end{bmatrix}$

17 $A = \begin{bmatrix} -11 & 2 & 2 \\ -4 & 0 & 1 \\ 6 & -1 & -1 \end{bmatrix}$

18 $A = \begin{bmatrix} 1 & 2 & 0 \\ -1 & 1 & 4 \\ 2 & 3 & -1 \end{bmatrix}$

19 $A = \begin{bmatrix} 1 & 1 & 0 & 2 \\ -1 & 0 & 2 & -1 \\ 0 & 2 & 0 & -2 \\ 2 & 0 & 0 & 5 \end{bmatrix}$

20 $A = \begin{bmatrix} 2 & -3 & 0 & 4 \\ -4 & 1 & -1 & 0 \\ -2 & 7 & -2 & 12 \\ 10 & 0 & 3 & -4 \end{bmatrix}$

In Exercises 21 and 22, write the system of linear equations obtained from the matrix equation $AX = C$ for the given matrices.

21 $A = \begin{bmatrix} 3 & 1 \\ 2 & -2 \end{bmatrix}$, $X = \begin{bmatrix} x \\ y \end{bmatrix}$, $C = \begin{bmatrix} 9 \\ 14 \end{bmatrix}$

22 $A = \begin{bmatrix} 5 & -2 & 3 \\ 0 & 4 & 1 \\ 2 & -1 & 6 \end{bmatrix}$, $X = \begin{bmatrix} x \\ y \\ z \end{bmatrix}$, $C = \begin{bmatrix} -2 \\ 7 \\ 0 \end{bmatrix}$

In Exercises 23–34, solve the linear system using the inverse of the matrix of coefficients. Observe that each matrix of coefficients is one of the matrices given in Exercises 1–20.

23 $\begin{aligned} 2x - 5y &= 7 \\ -3x + 4y &= -14 \end{aligned}$

24 $\begin{aligned} 2x - 5y &= -21 \\ -3x + 4y &= -7 \end{aligned}$

25 $\begin{aligned} \tfrac{1}{3}x + \tfrac{2}{3}y &= -8 \\ \tfrac{2}{3}x + \tfrac{1}{3}y &= 5 \end{aligned}$

26 $\begin{aligned} \tfrac{1}{3}x + \tfrac{2}{3}y &= 0 \\ \tfrac{2}{3}x + \tfrac{1}{3}y &= 1 \end{aligned}$

27 $\begin{aligned} x \qquad + 2z &= 4 \\ 2x - y \qquad &= -8 \\ 3y + 4z &= 0 \end{aligned}$

28 $\begin{aligned} x \qquad + 2z &= -2 \\ 2x - y \qquad &= 2 \\ 3y + 4z &= 1 \end{aligned}$

29 $\begin{aligned} x + y - z &= 1 \\ x - y - z &= 2 \\ -x - y - z &= 3 \end{aligned}$

30 $\begin{aligned} x + y - z &= -4 \\ x - y - z &= 6 \\ -x - y - z &= 10 \end{aligned}$

31 $\begin{aligned} 8x - 13y + 2z &= 1 \\ -4x + 7y - z &= 3 \\ 3x - 5y + z &= -2 \end{aligned}$

32 $\begin{aligned} -11x + 2y + 2z &= 0 \\ -4x \qquad + z &= 5 \\ 6x - y - z &= -1 \end{aligned}$

33 $\begin{aligned} w + x \qquad + 2z &= 2 \\ -w \qquad + 2y - z &= -6 \\ 2x \qquad - 2z &= 0 \\ 2w \qquad + 5z &= 8 \end{aligned}$

34 $\begin{aligned} w + x \qquad + 2z &= 1 \\ -w \qquad + 2y - z &= 3 \\ 2x \qquad - 2z &= 2 \\ 2w \qquad + 5z &= -1 \end{aligned}$

35 For $A = \begin{bmatrix} 2 & 1 \\ 0 & -4 \end{bmatrix}$ and $B = \begin{bmatrix} 0 & 2 \\ 1 & -2 \end{bmatrix}$ show that $(AB)^{-1} = B^{-1}A^{-1}$.

36 For matrix A in Exercise 35, verify that

$$(A^{-1})^2 = (A^2)^{-1}$$

37 **(a)** Assume that the matrix A has an inverse and let B and C be any matrices such that

$$AB = BA = I \quad \text{and} \quad AC = CA = I$$

Begin with the equation $AB = I$ and prove that $B = C$.

(b) What has been proven about the inverse of a matrix in part (a)?

38 **(a)** Let A and B be n by n matrices having inverses. Simplify the products $(AB)(B^{-1}A^{-1})$ and $(B^{-1}A^{-1})(AB)$.

(b) As a result of part (a) what can you say about the matrix $B^{-1}A^{-1}$?

8.4 Introduction to determinants and Cramer's rule

The general linear system S of two equations in two variables may be written in this standard form:

$$S: \quad \begin{aligned} a_1x + b_1y &= c_1 \\ a_2x + b_2y &= c_2 \end{aligned}$$

in which a_1, a_2, b_1, b_2, c_1, and c_2 are constants.

To solve this system we may begin as follows:

$$a_1b_2x + b_1b_2y = c_1b_2 \longleftarrow \text{Multiply first equation by } b_2$$

$$-a_2b_1x - b_1b_2y = -c_2b_1 \longleftarrow \text{Multiply second equation by } -b_1$$

Now add to eliminate y:

$$a_1b_2x - a_2b_1x = c_1b_2 - c_2b_1$$

$$(a_1b_2 - a_2b_1)x = c_1b_2 - c_2b_1$$

See the multiplication-addition method discussed in Section 3.6.

If $a_1b_2 - a_2b_1 \neq 0$, we may divide to get

$$x = \frac{c_1b_2 - c_2b_1}{a_1b_2 - a_2b_1}$$

Similarly, multiplying the first and second equations by a_2 and $-a_1$, respectively, will produce

$$y = \frac{a_1c_2 - a_2c_1}{a_1b_2 - a_2b_1}$$

Summarizing the preceding, we have the following result.

If the system S is given by

$$a_1x + b_1y = c_1$$
$$a_2x + b_2y = c_2$$

then its solution is

$$\left(\frac{c_1b_2 - c_2b_1}{a_1b_2 - a_2b_1}, \frac{a_1c_2 - a_2c_1}{a_1b_2 - a_2b_1}\right)$$

provided that $a_1b_2 - a_2b_1 \neq 0$.

It is important to realize that this solution applies to *any* consistent system and that the values for x and y are given in terms of the coefficients and constants of the system. For example, the system

$$8x - 20y = 3$$
$$4x + 10y = \tfrac{3}{2}$$

has $a_1 = 8$, $b_1 = -20$, $c_1 = 3$, $a_2 = 4$, $b_2 = 10$, $c_2 = \frac{3}{2}$. Then

$$x = \frac{(3)(10) - (\frac{3}{2})(-20)}{(8)(10) - (4)(-20)} = \frac{30 + 30}{80 + 80} = \frac{60}{160} = \frac{3}{8}$$

$$y = \frac{(8)(\frac{3}{2}) - (4)(3)}{(8)(10) - (4)(-20)} = \frac{12 - 12}{160} = \frac{0}{160} = 0$$

As you can see, keeping track of the position of each constant in the general form is tedious. We can simplify this process with the introduction of some special symbolism.

First notice that the denominator for both x and y is $a_1b_2 - a_2b_1$. This difference uses the matrix of coefficients of system S.

$$a_1x + b_1y = c_1$$
$$a_2x + b_2y = c_2$$

The arrow from top to bottom gives the product a_1b_2. This product comes *first* in $a_1b_2 - a_2b_1$. The arrow from bottom to top gives the *second* product a_2b_1. The number $a_1b_2 - a_2b_1$ is called the **determinant** of the matrix $A = \begin{bmatrix} a_1 & b_1 \\ a_2 & b_2 \end{bmatrix}$ and is symbolized by putting vertical bars instead of brackets around the entries in A, as follows.

For $A = \begin{bmatrix} a_1 & b_1 \\ a_2 & b_2 \end{bmatrix}$ the determinant of A is given by

$$|A| = \begin{vmatrix} a_1 & b_1 \\ a_2 & b_2 \end{vmatrix} = a_1b_2 - a_2b_1$$

Here are some illustrations:

$$\begin{vmatrix} 8 & -20 \\ 4 & 10 \end{vmatrix} = (8)(10) - (4)(-20) = 160$$

$$\begin{vmatrix} -20 & 8 \\ 10 & 4 \end{vmatrix} = (-20)(4) - (10)(8) = -160$$

$$\begin{vmatrix} 10 & -20 \\ 8 & 4 \end{vmatrix} = (10)(4) - (8)(-20) = 200$$

TEST YOUR UNDERSTANDING

Evaluate each of the determinants.

1 $\begin{vmatrix} 1 & 2 \\ 3 & 4 \end{vmatrix}$ **2** $\begin{vmatrix} 1 & 3 \\ 2 & 4 \end{vmatrix}$ **3** $\begin{vmatrix} 2 & 1 \\ 4 & 3 \end{vmatrix}$

4 $\begin{vmatrix} 1 & 3 \\ 4 & 2 \end{vmatrix}$ **5** $\begin{vmatrix} 2 & 3 \\ 1 & 4 \end{vmatrix}$ **6** $\begin{vmatrix} 4 & 3 \\ 1 & 2 \end{vmatrix}$

7 $\begin{vmatrix} 10 & -5 \\ 2 & 1 \end{vmatrix}$ **8** $\begin{vmatrix} \frac{1}{2} & 6 \\ 0 & 4 \end{vmatrix}$ **9** $\begin{vmatrix} -8 & -4 \\ -7 & -3 \end{vmatrix}$

10 $\begin{vmatrix} -1 & 0 \\ 0 & -1 \end{vmatrix}$ **11** $\begin{vmatrix} 0 & 2 \\ 2 & 0 \end{vmatrix}$ **12** $\begin{vmatrix} \frac{1}{3} & -\frac{1}{4} \\ 8 & -6 \end{vmatrix}$

The determinant $\begin{vmatrix} a_1 & b_1 \\ a_2 & b_2 \end{vmatrix}$ is said to be of order 2 because it involves two rows and two columns. We also describe it as being a **second-order determinant.**

In the general solution for S, the numerator for x is $c_1b_2 - c_2b_1$. Using our new symbolism, this number is the second-order determinant:

$$\begin{vmatrix} c_1 & b_1 \\ c_2 & b_2 \end{vmatrix} = c_1b_2 - c_2b_1$$

Similarly, the numerator for y is

$$\begin{vmatrix} a_1 & c_1 \\ a_2 & c_2 \end{vmatrix} = a_1c_2 - a_2c_1$$

In summary, we have the following, known as *Cramer's rule*, for solving such a *system* of linear equations.

CRAMER'S RULE

Given the consistent system S,

In a consistent system, $a_1b_2 - a_2b_1 \neq 0$.

$$a_1x + b_1y = c_1$$
$$a_2x + b_2y = c_2$$

the solutions for x and y are

$$x = \frac{\begin{vmatrix} c_1 & b_1 \\ c_2 & b_2 \end{vmatrix}}{\begin{vmatrix} a_1 & b_1 \\ a_2 & b_2 \end{vmatrix}} \quad \text{and} \quad y = \frac{\begin{vmatrix} a_1 & c_1 \\ a_2 & c_2 \end{vmatrix}}{\begin{vmatrix} a_1 & b_1 \\ a_2 & b_2 \end{vmatrix}}$$

EXAMPLE 1 Use Cramer's rule to solve the system

$$8x - 20y = 3$$
$$4x + 10y = \tfrac{3}{2}$$

Solution

$$a_1 = 8 \qquad b_1 = -20 \qquad c_1 = 3$$
$$a_2 = 4 \qquad b_2 = 10 \qquad c_2 = \tfrac{3}{2}$$

Now use determinants as follows:

$$x = \frac{\begin{vmatrix} 3 & -20 \\ \frac{3}{2} & 10 \end{vmatrix}}{\begin{vmatrix} 8 & -20 \\ 4 & 10 \end{vmatrix}} = \frac{(3)(10) - (\frac{3}{2})(-20)}{(8)(10) - (4)(-20)} = \frac{60}{160} = \frac{3}{8}$$

$$y = \frac{\begin{vmatrix} 8 & 3 \\ 4 & \frac{3}{2} \end{vmatrix}}{160} = \frac{(8)(\frac{3}{2}) - (4)(3)}{160} = \frac{0}{160} = 0$$

This style of solution becomes easier to set up by making a few simple observations. First, the determinant of the matrix of coefficients

$$\begin{vmatrix} a_1 & b_1 \\ a_2 & b_2 \end{vmatrix}$$

is the denominator for *each* fraction. To write the fraction for x, first record the denominator. Then, to get the determinant in the numerator, simply remove the first column (the coefficients of x) and replace it with constants c_1 and c_2, as shown in the following figure. Similarly, the replacement of the second column by c_1, c_2 gives the numerator in the fraction for y.

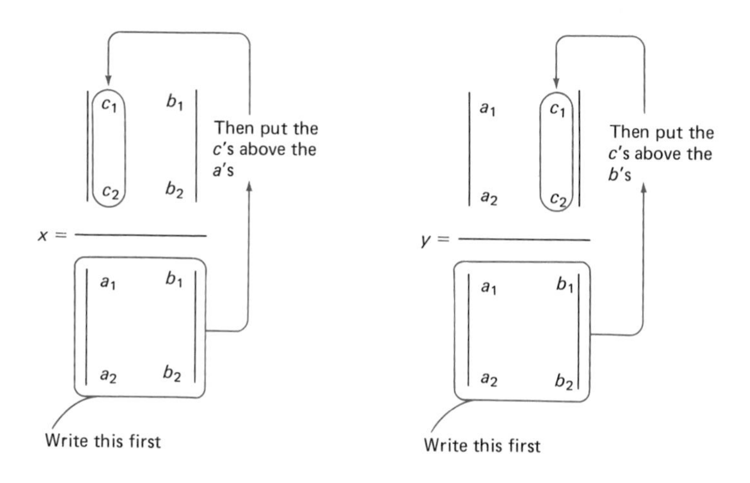

EXAMPLE 2 Solve the following system using determinants.

$$\begin{aligned} 5x - 9y &= 7 \\ -8x + 10y &= 2 \end{aligned}$$

Solution

$$x = \frac{\begin{vmatrix} 7 & -9 \\ 2 & 10 \end{vmatrix}}{\begin{vmatrix} 5 & -9 \\ -8 & 10 \end{vmatrix}} = \frac{70 - (-18)}{50 - 72} = \frac{88}{-22} = -4$$

$$y = \frac{\begin{vmatrix} 5 & 7 \\ -8 & 2 \end{vmatrix}}{\begin{vmatrix} 5 & -9 \\ -8 & 10 \end{vmatrix}} = \frac{10 - (-56)}{-22} = -3$$

EXAMPLE 3 Use Cramer's rule to solve the given system.

$$\begin{aligned} 3x &= 2y + 22 \\ 2(x + y) &= x - 2y - 2 \end{aligned}$$

Solution First write the system in standard form.

$$\begin{aligned} 3x - 2y &= 22 \\ x + 4y &= -2 \end{aligned}$$

$$x = \frac{\begin{vmatrix} 22 & -2 \\ -2 & 4 \end{vmatrix}}{\begin{vmatrix} 3 & -2 \\ 1 & 4 \end{vmatrix}} = \frac{84}{14} = 6 \qquad y = \frac{\begin{vmatrix} 3 & 22 \\ 1 & -2 \end{vmatrix}}{14} = \frac{-28}{14} = -2$$

The general system

$$\begin{aligned} a_1x + b_1y &= c_1 \\ a_2x + b_2y &= c_2 \end{aligned}$$

Recall that a system of two linear equations in two variables is dependent when the two equations are equivalent. Inconsistent means that there are no (common) solutions.

is either dependent or inconsistent when the determinant of the matrix of coefficients is zero, that is, when

$$\begin{vmatrix} a_1 & b_1 \\ a_2 & b_2 \end{vmatrix} = 0$$

For example, consider these two systems:

(a) $\begin{aligned} 2x - 3y &= 5 \\ -10x + 15y &= 8 \end{aligned}$ (b) $\begin{aligned} 2x - 3y &= 8 \\ -10x + 15y &= -40 \end{aligned}$

Use the methods in Section 8.1 or in Section 3.7 to verify these results for systems (a) and (b).

In each case we have the following:

$$\begin{vmatrix} a_1 & b_1 \\ a_2 & b_2 \end{vmatrix} = \begin{vmatrix} 2 & -3 \\ -10 & 15 \end{vmatrix} = 0$$

System (a) turns out to be inconsistent and system (b) is dependent.

EXERCISES 8.4

Evaluate each determinant.

1 $\begin{vmatrix} 5 & -1 \\ -3 & 4 \end{vmatrix}$ **2** $\begin{vmatrix} 1 & 7 \\ 3 & 4 \end{vmatrix}$

3 $\begin{vmatrix} 17 & -3 \\ 20 & 2 \end{vmatrix}$ **4** $\begin{vmatrix} -7 & 9 \\ -5 & 5 \end{vmatrix}$

5 $\begin{vmatrix} 10 & 5 \\ 6 & -3 \end{vmatrix}$ **6** $\begin{vmatrix} 6 & 11 \\ 0 & -9 \end{vmatrix}$

7 $\begin{vmatrix} 16 & 0 \\ -9 & 0 \end{vmatrix}$ **8** $\begin{vmatrix} a & b \\ 3a & 3b \end{vmatrix}$

Solve each system using determinants.

9 $\begin{aligned} 3x + 9y &= 15 \\ 6x + 12y &= 18 \end{aligned}$ **10** $\begin{aligned} x - y &= 7 \\ -2x + 5y &= -8 \end{aligned}$

11 $\begin{aligned} -4x + 10y &= 8 \\ 11x - 9y &= 15 \end{aligned}$ **12** $\begin{aligned} 7x + 4y &= 5 \\ -x + 2y &= -2 \end{aligned}$

13 $\begin{aligned} 5x + 2y &= 3 \\ 2x + 3y &= -1 \end{aligned}$ **14** $\begin{aligned} 3x + 3y &= 6 \\ 4x - 2y &= -1 \end{aligned}$

15 $\begin{aligned} \tfrac{1}{3}x + \tfrac{3}{8}y &= 13 \\ x - \tfrac{9}{4}y &= -42 \end{aligned}$ **16** $\begin{aligned} x - 3y &= 7 \\ -\tfrac{1}{2}x + \tfrac{1}{4}y &= 1 \end{aligned}$

17 $\begin{aligned} 3x + y &= 20 \\ y &= x \end{aligned}$ **18** $\begin{aligned} 3x + 2y &= 5 \\ 2x + 3y &= 0 \end{aligned}$

19 $\begin{aligned} \tfrac{1}{2}x - \tfrac{2}{7}y &= -\tfrac{1}{2} \\ -\tfrac{1}{3}x - \tfrac{1}{2}y &= \tfrac{31}{6} \end{aligned}$ **20** $\begin{aligned} -4x + 3y &= -20 \\ 2x + 6y &= -15 \end{aligned}$

21 $\begin{aligned} 9x - 12 &= 4y \\ 3x + 2y &= 3 \end{aligned}$

22 $\begin{aligned} \frac{x-y}{3} - \frac{y}{6} &= \frac{2}{3} \\ 2x + 9(y - 2x) &= 8 \end{aligned}$

Verify that the determinant of the coefficients is zero for each of the following. Then decide whether the system is dependent or inconsistent.

23 $\begin{aligned} 5x - 2y &= 3 \\ -15x + 6y &= -4 \end{aligned}$ **24** $\begin{aligned} 2x - 3y &= 5 \\ 10 - 4x &= -6y \end{aligned}$

25 $\begin{aligned} 16x - 4y &= 20 \\ 12x - 3y &= 15 \end{aligned}$ **26** $\begin{aligned} 2x - 6y &= -12 \\ -3x + 9y &= 18 \end{aligned}$

27 $\begin{aligned} 3x &= 5y - 10 \\ 6x - 10y &= -25 \end{aligned}$ **28** $\begin{aligned} 10y &= 2x - 4 \\ x - 5y &= 2 \end{aligned}$

29 When variables are used for some of the entries in the symbolism of a determinant, the determinant itself can be used to state equations. Solve for x in the following.

$$\begin{vmatrix} x & 2 \\ 5 & 3 \end{vmatrix} = 8$$

30 Solve the given system.

$$\begin{vmatrix} x & y \\ 2 & 4 \end{vmatrix} = 5$$

$$\begin{vmatrix} 1 & y \\ -1 & x \end{vmatrix} = -\tfrac{1}{2}$$

Exercises 31–40 deal with some general properties of determinants.

***31** Show that if the rows and columns of a second-order determinant are interchanged, the value of the determinant remains the same.

*32 Show that if one of the rows of $\begin{vmatrix} a_1 & b_1 \\ a_2 & b_2 \end{vmatrix}$ is a nonzero multiple of the other, then the determinant is zero.

*33 Use Exercises 31 and 32 to demonstrate that the determinant is zero if one column is a nonzero multiple of the other.

*34 Show that if each element of a row (or column) of a second-order determinant is multiplied by the same number k, the value of the determinant is multiplied by k.

*35 Make repeated use of the result in Exercise 34 to show the following:

$$\begin{vmatrix} 27 & 3 \\ 105 & -75 \end{vmatrix} = (45)\begin{vmatrix} 9 & 1 \\ 7 & -5 \end{vmatrix}$$

or

$$\begin{vmatrix} 27 & 3 \\ 105 & -75 \end{vmatrix} = (45)\begin{vmatrix} 3 & 1 \\ 7 & -15 \end{vmatrix}$$

Then evaluate each side to check.

*36 Prove:

$$\begin{vmatrix} a_1 + t_1 & b_1 \\ a_2 + t_2 & b_2 \end{vmatrix} = \begin{vmatrix} a_1 & b_1 \\ a_2 & b_2 \end{vmatrix} + \begin{vmatrix} t_1 & b_1 \\ t_2 & b_2 \end{vmatrix}$$

*37 Prove that if to each element of a row (or column) of a second-order determinant we add k times the corresponding element of the other row (or column), then the value of the resulting determinant is the same as the original one.

38 (a) Evaluate $\begin{vmatrix} 3 & 5 \\ -6 & -1 \end{vmatrix}$ by definition.

(b) Evaluate the same determinant using the result of Exercise 37 by adding 2 times row 1 to row 2.

(c) Evaluate the same determinant using the result in Exercise 37 by adding -6 times column 2 to column 1.

Evaluate using the results of Exercises 34 and 37.

39 $\begin{vmatrix} 12 & -42 \\ -6 & 27 \end{vmatrix}$

40 $\begin{vmatrix} 45 & 75 \\ 40 & -25 \end{vmatrix}$

8.5 Third-order determinants

Just as a system of two linear equations in two variables can be solved using second-order determinants, so can a system of three linear equations in three variables be solved by using *third-order* determinants.

A third-order determinant may be defined in terms of second-order determinants as follows:

$$\begin{vmatrix} a_1 & b_1 & c_1 \\ a_2 & b_2 & c_2 \\ a_3 & b_3 & c_3 \end{vmatrix} = a_1\begin{vmatrix} b_2 & c_2 \\ b_3 & c_3 \end{vmatrix} - a_2\begin{vmatrix} b_1 & c_1 \\ b_3 & c_3 \end{vmatrix} + a_3\begin{vmatrix} b_1 & c_1 \\ b_2 & c_2 \end{vmatrix}$$

This is an expansion along the elements in column 1.

Note that the first term on the right is the product of a_1 times a second-order determinant that can be found by drawing vertical and horizontal lines through a_1.

$$\begin{vmatrix} a_1 & b_1 & c_1 \\ a_2 & b_2 & c_2 \\ a_3 & b_3 & c_3 \end{vmatrix}$$

The four entries left over give the appropriate determinant. Similar schemes can be used to find the second-order determinants for a_2 and a_3.

EXAMPLE 1 Evaluate:

$$\begin{vmatrix} 2 & -2 & 2 \\ 3 & 1 & 0 \\ 2 & -1 & 1 \end{vmatrix}$$

We will begin the solution by expanding along the first column.

Solution

$$\begin{vmatrix} 2 & -2 & 2 \\ 3 & 1 & 0 \\ 2 & -1 & 1 \end{vmatrix} = 2\begin{vmatrix} 1 & 0 \\ -1 & 1 \end{vmatrix} - 3\begin{vmatrix} -2 & 2 \\ -1 & 1 \end{vmatrix} + 2\begin{vmatrix} -2 & 2 \\ 1 & 0 \end{vmatrix}$$
$$= 2(1 - 0) - 3(-2 + 2) + 2(0 - 2)$$
$$= 2 - 0 - 4$$
$$= -2$$

A third-order determinant can be evaluated and simplified as follows:

$$\begin{vmatrix} a_1 & b_1 & c_1 \\ a_2 & b_2 & c_2 \\ a_3 & b_3 & c_3 \end{vmatrix} = a_1(b_2c_3 - b_3c_2) - a_2(b_1c_3 - b_3c_1) + a_3(b_1c_2 - b_2c_1)$$
$$= a_1b_2c_3 + a_2b_3c_1 + a_3b_1c_2 - a_1b_3c_2 - a_2b_1c_3 - a_3b_2c_1$$

Show that the following alternative expansion produces the same result:

This is an expansion along the elements in row 1.

$$a_1\begin{vmatrix} b_2 & c_2 \\ b_3 & c_3 \end{vmatrix} - b_1\begin{vmatrix} a_2 & c_2 \\ a_3 & c_3 \end{vmatrix} + c_1\begin{vmatrix} a_2 & b_2 \\ a_3 & b_3 \end{vmatrix}$$

There are six such expansions, one for each row and each column, all giving the same result. The preceding expansion was along the first row, and the expansion done at the beginning of this section was along the first column. Here is the expansion and simplification done along the second column.

$$-b_1\begin{vmatrix} a_2 & c_2 \\ a_3 & c_3 \end{vmatrix} + b_2\begin{vmatrix} a_1 & c_1 \\ a_3 & c_3 \end{vmatrix} - b_3\begin{vmatrix} a_1 & c_1 \\ a_2 & c_2 \end{vmatrix}$$
$$= -b_1(a_2c_3 - a_3c_2) + b_2(a_1c_3 - a_3c_1) - b_3(a_1c_2 - a_2c_1)$$
$$= a_1b_2c_3 + a_2b_3c_1 + a_3b_1c_2 - a_1b_3c_2 - a_2b_1c_3 - a_3b_2c_1$$

When you expand along a row or column keep the following display of signs in mind. It gives the signs preceding the row or column elements in the expansion.

Observe how the signs in the second column have been used in the preceding expansion along the second column.

$$\begin{vmatrix} + & - & + \\ - & + & - \\ + & - & + \end{vmatrix}$$

Here is another procedure that can be used to evaluate a third-order determinant. Rewrite the first two columns at the right as shown. Follow the arrows pointing downward to get the three products having a plus sign, and the arrows pointing upward give the three products having a negative sign.

$$\begin{vmatrix} a_1 & b_1 & c_1 \\ a_2 & b_2 & c_2 \\ a_3 & b_3 & c_3 \end{vmatrix} \begin{matrix} a_1 & b_1 \\ a_2 & b_2 \\ a_3 & b_3 \end{matrix} = a_1b_2c_3 + b_1c_2a_3 + c_1a_2b_3 - a_3b_2c_1 - b_3c_2a_1 - c_3a_2b_1$$

EXAMPLE 2 Evaluate:

$$\begin{vmatrix} -1 & 3 & 4 \\ 2 & 1 & 2 \\ 5 & 1 & 3 \end{vmatrix}$$

Solution

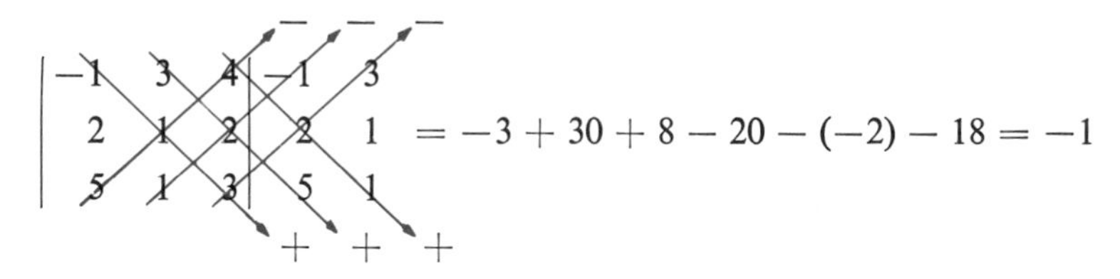

$$= -3 + 30 + 8 - 20 - (-2) - 18 = -1$$

TEST YOUR UNDERSTANDING

Evaluate each determinant.

1 $\begin{vmatrix} 2 & 1 & 0 \\ -1 & 0 & 3 \\ 0 & 4 & -5 \end{vmatrix}$ **2** $\begin{vmatrix} 3 & 6 & 9 \\ 0 & 0 & 0 \\ -2 & -4 & -6 \end{vmatrix}$ **3** $\begin{vmatrix} 3 & -2 & 7 \\ 0 & 3 & -5 \\ 0 & 0 & 3 \end{vmatrix}$

4 $\begin{vmatrix} -1 & 4 & -2 \\ 6 & -6 & 1 \\ 3 & 3 & 2 \end{vmatrix}$ **5** $\begin{vmatrix} 1 & -2 & 3 \\ -4 & 5 & -4 \\ 3 & -2 & 1 \end{vmatrix}$ **6** $\begin{vmatrix} 2 & -1 & 9 \\ -7 & 3 & -4 \\ 2 & -1 & 9 \end{vmatrix}$

Third-order determinants can be used to extend Cramer's rule for the solution of three linear equations in three unknowns. Consider this system:

$$a_1x + b_1y + c_1z = d_1$$
$$a_2x + b_2y + c_2z = d_2$$
$$a_3x + b_3y + c_3z = d_3$$

By completing the tedious computations involved, it can be shown that the solution for this system is the following:

$$x = \frac{\begin{vmatrix} d_1 & b_1 & c_1 \\ d_2 & b_2 & c_2 \\ d_3 & b_3 & c_3 \end{vmatrix}}{D} \qquad y = \frac{\begin{vmatrix} a_1 & d_1 & c_1 \\ a_2 & d_2 & c_2 \\ a_3 & d_3 & c_3 \end{vmatrix}}{D} \qquad z = \frac{\begin{vmatrix} a_1 & b_1 & d_1 \\ a_2 & b_2 & d_2 \\ a_3 & b_3 & d_3 \end{vmatrix}}{D}$$

where D is the following determinant.

$$D = \begin{vmatrix} a_1 & b_1 & c_1 \\ a_2 & b_2 & c_2 \\ a_3 & b_3 & c_3 \end{vmatrix} \quad \text{and} \quad D \neq 0$$

Note that D is the determinant of the matrix of coefficients of the linear system. Then in the numerator for x, the column of c's has replaced the column of a's in D. A similar column replacement is made in the numerators for y and z.

EXAMPLE 3 Use Cramer's rule to solve this system.

$$\begin{aligned} x + 2y + z &= 3 \\ 2x - y - z &= 4 \\ -x - y + 2z &= -5 \end{aligned}$$

Solution First we find D and note that $D \neq 0$.

$$D = \begin{vmatrix} 1 & 2 & 1 \\ 2 & -1 & -1 \\ -1 & -1 & 2 \end{vmatrix} = 1\begin{vmatrix} -1 & -1 \\ -1 & 2 \end{vmatrix} - 2\begin{vmatrix} 2 & 1 \\ -1 & 2 \end{vmatrix} + (-1)\begin{vmatrix} 2 & 1 \\ -1 & -1 \end{vmatrix}$$

$$= -12$$

Verify each of the following computations.

$$x = \frac{\begin{vmatrix} 3 & 2 & 1 \\ 4 & -1 & -1 \\ -5 & -1 & 2 \end{vmatrix}}{D} = \frac{-24}{-12} = 2$$

$$y = \frac{\begin{vmatrix} 1 & 3 & 1 \\ 2 & 4 & -1 \\ -1 & -5 & 2 \end{vmatrix}}{D} = \frac{-12}{-12} = 1$$

$$z = \frac{\begin{vmatrix} 1 & 2 & 3 \\ 2 & -1 & 4 \\ -1 & -1 & -5 \end{vmatrix}}{D} = \frac{12}{-12} = -1$$

Check this solution in the original system.

Thus $x = 2$, $y = 1$, and $z = -1$.

EXERCISES 8.5

Evaluate each determinant.

1 $\begin{vmatrix} 2 & 2 & -1 \\ -1 & 3 & -3 \\ 1 & 2 & 3 \end{vmatrix}$

2 $\begin{vmatrix} 2 & 0 & -1 \\ 3 & -2 & 1 \\ -3 & 0 & 4 \end{vmatrix}$

3 $\begin{vmatrix} 1 & -3 & 2 \\ -5 & 2 & 0 \\ 4 & -1 & 3 \end{vmatrix}$

4 $\begin{vmatrix} 1 & 2 & 3 \\ 4 & 5 & 6 \\ 7 & 8 & 9 \end{vmatrix}$

5 $\begin{vmatrix} 1 & 1 & 1 \\ -1 & 1 & 1 \\ -1 & -1 & 1 \end{vmatrix}$

6 $\begin{vmatrix} 1 & 1 & 4 \\ 2 & 2 & -5 \\ 3 & 3 & 6 \end{vmatrix}$

***7** Show that if the rows and columns of a third-order determinant are interchanged, the value of the determinant remains the same.

***8** **(a)** Show that if one of the rows of a third-order determinant is a nonzero multiple of the other, then the determinant is zero.

(b) Use part (a) and Exercise 7 to demonstrate that the determinant is 0 if one column is a nonzero multiple of the other.

***9** Show that if each element of a row (or column) of a third-order determinant is multiplied by the same number k, the value of the determinant is multiplied by k.

10 Use the result of Exercise 8(a) to evaluate

$$\begin{vmatrix} 3 & 5 & 5 \\ 1 & 2 & 3 \\ -4 & -8 & -12 \end{vmatrix}$$

11 Use the result of Exercise 8(b) to evaluate

$$\begin{vmatrix} 7 & 2 & -14 \\ 0 & 6 & 0 \\ -3 & 1 & 6 \end{vmatrix}$$

12 Use the results of Exercise 9 to explain each step.

(i) $$\begin{vmatrix} 8 & -10 & 2 \\ 4 & 25 & -1 \\ 2 & 10 & 0 \end{vmatrix} = 2\begin{vmatrix} 4 & -10 & 2 \\ 2 & 25 & -1 \\ 1 & 10 & 0 \end{vmatrix}$$

(ii) $$= 10\begin{vmatrix} 4 & -2 & 2 \\ 2 & 5 & -1 \\ 1 & 2 & 0 \end{vmatrix}$$

(iii) $$= 20\begin{vmatrix} 2 & -1 & 1 \\ 2 & 5 & -1 \\ 1 & 2 & 0 \end{vmatrix}$$

13 Evaluate the determinant given on the left side of (i) in Exercise 12, and also evaluate the result given in (iii).

Use Cramer's rule to solve each system.

14 $$\begin{aligned} x + y + z &= 2 \\ x - y + 3z &= 12 \\ 2x + 5y + 2z &= -2 \end{aligned}$$

15 $$\begin{aligned} x + 2y + 3z &= 5 \\ 3x - y \quad &= -3 \\ -4x \quad + z &= 6 \end{aligned}$$

16 $$\begin{aligned} x - 8y - 2z &= 12 \\ -3x + 3y + z &= -10 \\ 4x + y + 5z &= 2 \end{aligned}$$

17 $$\begin{aligned} 2x + y \quad &= 5 \\ 3x \quad - 2z &= -7 \\ -3y + 8z &= -5 \end{aligned}$$

18 $$\begin{aligned} 4x - 2y - z &= 1 \\ 2x + y + 2z &= 9 \\ x - 3y - z &= \tfrac{3}{2} \end{aligned}$$

19 $$\begin{aligned} 6x + 3y - 4z &= 5 \\ \tfrac{3}{2}x + y - 4z &= 0 \\ 3x - y + 8z &= 5 \end{aligned}$$

20 A fourth-order determinant can be found by expanding along a row or column as was done for a third-order determinant, and using a similar alternating pattern for the signs. Write this expansion along the first row for

$$\begin{vmatrix} 2 & -1 & 3 & 0 \\ 1 & 0 & 5 & -3 \\ 0 & 2 & -4 & 6 \\ -5 & 3 & 0 & 1 \end{vmatrix}$$

21 Simplify the expansion found in Exercise 20 to evaluate the determinant.

22 Let

$$A = \begin{bmatrix} -3 & 4 & 0 & 1 \\ 9 & -1 & 2 & 0 \\ 0 & 0 & -6 & 0 \\ 0 & 3 & 0 & -2 \end{bmatrix}$$

Along which row or column would you expand to evaluate $|A|$ in the most efficient way? Use this expansion to find $|A|$ and check by using a different expansion.

8.6 Linear programming

Systems of linear inequalities were introduced at the end of Section 3.7. You should review that earlier work before proceeding.

All the systems of linear inequalities that will be used here will be in the variables x and y, as in the system at the top of the next page.

$$x \geq 0$$
$$y \geq 0$$
$$\tfrac{3}{4}x + \tfrac{1}{2}y \leq 6$$
$$\tfrac{1}{2}x + y \leq 6$$

The graph of this system is a region R consisting of all points (x, y) that satisfy *each* of the inequalities in the system. First note that the conditions $x \geq 0$ and $y \geq 0$ require the points to be in quadrant I, or on the nonnegative parts of each axis. Now graph the lines $\tfrac{3}{4}x + \tfrac{1}{2}y = 6$, $\tfrac{1}{2}x + y = 6$ and determine their point of intersection (6, 3). Then, by the procedure used on page 112, the (x, y) that satisy both inequalities $\tfrac{3}{4}x + \tfrac{1}{2}y \leq 6$ and $\tfrac{1}{2}x + y \leq 6$ are found to be below or on these lines. The completed region R looks like this.

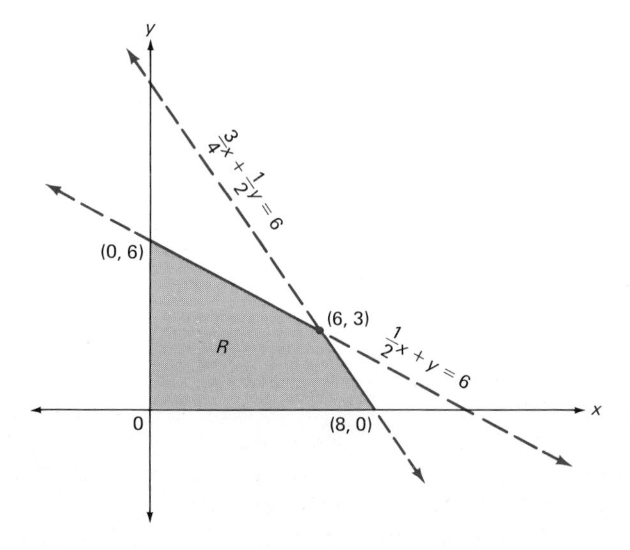

The region R is said to be a **convex set of points**, because any two points inside R can be joined by a line segment that is totally inside R. Roughly speaking, the boundary of R "bends outward." Here is a figure of a region that is *not* convex; note, in this figure, that points A and B cannot be connected by a line segment that is totally inside the region.

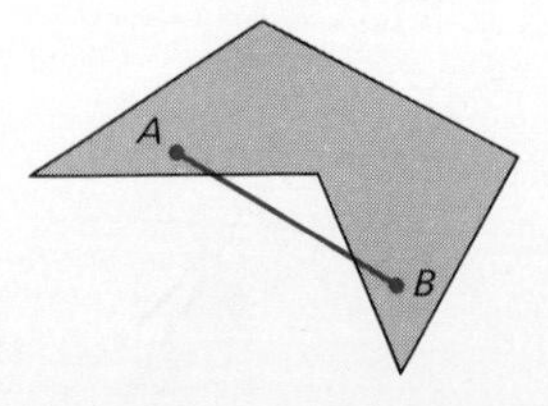

Draw the line $\ell_1 : x + y = 4$ in the same coordinate system as region, R. (See the next figure.) All points (x, y) in R that are on this line have coordinates whose sum is 4. Now draw lines ℓ_2 and ℓ_3 parallel to ℓ_1, and also having equations $x + y = 2$ and $x + y = 8$, respectively. Obviously, all points in R on ℓ_2 or on ℓ_3 have coordinates whose sum is *not* 4.

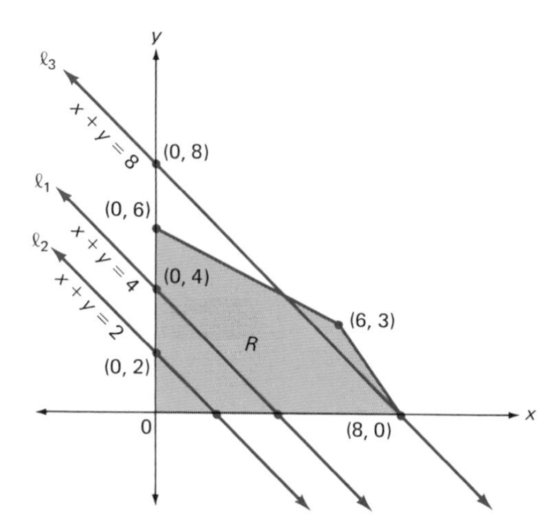

These three lines all have the form $x + y = k$, where k is some constant. There are an infinite number of such parallel lines, depending on the constant k. They all have slope -1. Some of these lines intersect R and others do not; we will be interested only in those that do.

Can you guess which line of this form will intersect R and have the largest possible value k?

It is not difficult to see the answer, because all these lines are parallel; and the "higher" the line, the larger will be the value k. As a matter of fact, the line $x + y = k$ has k as its y-intercept. So the line we are looking for will be the line that intersects the y-axis as high up as possible, has slope -1, and still meets region R. Because of the shape of R (it is convex), this will be the line through (6, 3) with slope -1, as shown in the following figure.

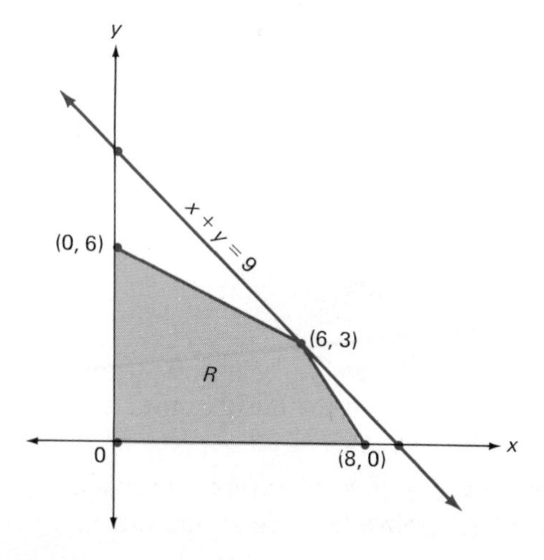

The equation of the line through (6, 3) with slope -1 is $x + y = 9$. Any other parallel line with a larger k-value will be higher and cannot intersect R; and those others that do will be lower and have a smaller k-value.

To sum up, we can say that of all the lines with form $x + y = k$ that intersect region R, the one that has the largest k-value is the line $x + y = 9$. Putting it another way, we can say that of all the points in R, the point (6, 3) is the point whose coordinates give the maximum value for the quantity $k = x + y$.

Suppose that we now look for the point in R that produces the maximum value for k, where $k = 2x + 3y$. First we draw a few parallel lines with equation of the form $k = 2x + 3y$. Below are such lines for k taking on the values 6, 10, and 18.

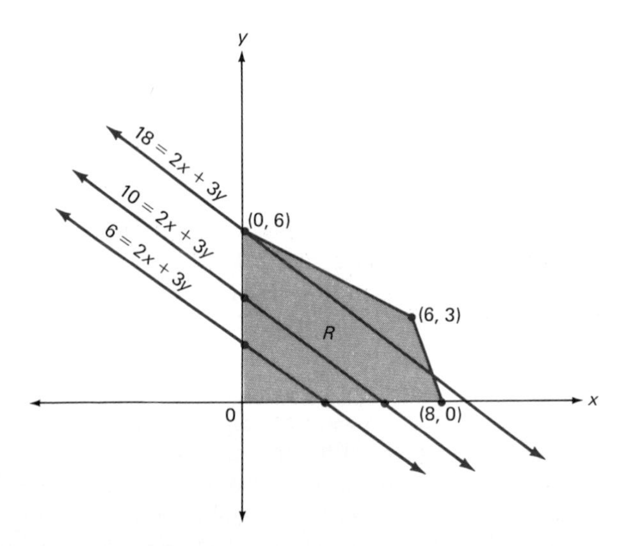

We see that the higher the line, the larger will be the value of k. Because of the convex shape of R, the highest such parallel line (intersecting R) must pass through the vertex (6, 3). For this line we get $2(6) + 3(3) = 21 = k$. This is the largest value of $k = 2x + 3y$ for the (x, y) in R.

The preceding observations should be convincing evidence for the following result:

> Whenever we have a convex-shaped region R, then the point in R that produces the maximum value of a quantity of the form $k = ax + by$, where a and b are positive, will be at a vertex of R.

Because of this result it is no longer necessary to draw lines through R. All that needs to be done to find the maximum of $k = ax + by$ is to graph the region R, find the coordinates of all vertices on the boundary, and see which one produces the largest value for k.

EXAMPLE 1 Maximize the quantity $k = 4x + 5y$ for (x, y) in the region S given by this system:

$$\begin{aligned} -x + 2y &\leq 2 \\ 3x + 2y &\leq 10 \\ x + 6y &\geq 6 \\ 3x + 2y &\geq 6 \end{aligned}$$

Solution First graph the corresponding four lines and shade in the required region. Next find the four vertices of the region S by solving the appropriate pairs of equations. Since $k = 4x + 5y$ will be a maximum only at a vertex, the listing below shows that 18 is the maximum value.

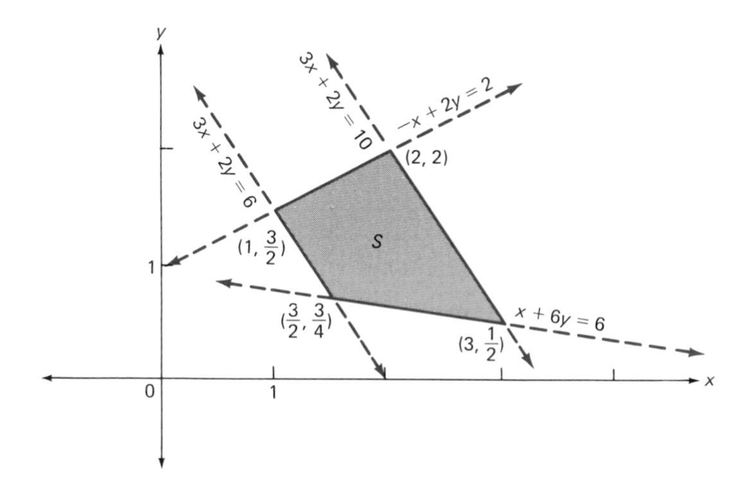

Vertex	$k = 4x + 5y$
$(1, \frac{3}{2})$	$4(1) + 5(\frac{3}{2}) = 11\frac{1}{2}$
$(\frac{3}{2}, \frac{3}{4})$	$4(\frac{3}{2}) + 5(\frac{3}{4}) = 9\frac{3}{4}$
$(2, 2)$	$4(2) + 5(2) = 18$
$(3, \frac{1}{2})$	$4(3) + 5(\frac{1}{2}) = 14\frac{1}{2}$

Instead of asking for the maximum of $k = 4x + 5y$, as in Example 1, it is also possible to find the minimum value of $k = 4x + 5y$ for the same region S. The earlier discussions explaining why a vertex of a convex region gives the maximum can be easily adjusted to show that a vertex will also give the minimum. Essentially, it becomes a matter of finding the line with correct slope, that has the lowest possible y-intercept, and that still intersects S. Because of this, the table in the solution to the preceding example shows that $9\frac{3}{4}$ is the minimum value of $k = 4x + 5y$. It occurs at the vertex $(\frac{3}{2}, \frac{3}{4})$.

TEST YOUR UNDERSTANDING

Find the required values for the region R, where R is the graph of the system.

$$5x + 4y \geq 40$$
$$x + 4y \leq 40$$
$$7x + 4y \leq 112$$

1 Maximum of $k = x + y$	**2** Minimum of $k = x + y$
3 Maximum of $k = x + 2y$	**4** Minimum of $k = x + 2y$
5 Maximum of $k = x + 5y$	**6** Minimum of $k = x + 5y$
7 Maximum of $k = 2x + y$	**8** Minimum of $k = 2x + y$
9 Maximum of $k = 3x + 20y$	**10** Minimum of $k = 3x + 20y$

The preceding method of finding the maximum or minimum of a quantity $k = ax + by$, relative to a convex region, can be applied to a variety of applied situations. Here is a typical problem.

Read this problem several times before proceeding.

A store sells two kind of bicycles, model A and model B. The store buys them unassembled from a wholesaler. Two employees are hired just to assemble the bicycles. Working together to assemble model A, employee I works $\frac{3}{4}$ hour and employee II works $\frac{1}{2}$ hour. Model B requires $\frac{1}{2}$ hour's work by employee I as well as 1 hour's work by employee II. There is a \$25 profit on each model A sold and \$22 on each model B. Because of the popularity of the sport, the store is able to sell as many bicycles as they decide to assemble. If each employee works no more than 6 hours per day, how many bicycles of each model should they assemble in order to get the maximum profit?

Let x be the number of model A bicycles assembled per day, and y the number of model B's per day. Then $25x$ is the daily profit earned for model A, and $22y$ is the daily profit for model B. The total daily profit p is given by $p = 25x + 22y$. It is this quantity we need to maximize.

Next we find the time each employee works. For employee I, $\frac{3}{4}x + \frac{1}{2}y$ will be the number of hours worked per day, because each model A (there are x of these) requires $\frac{3}{4}$ hour, and each model B (there are y of these) requires $\frac{1}{2}$ hour. But each employee works *no more than* 6 hours per day. Thus the total time for this worker satisfies

$$\tfrac{3}{4}x + \tfrac{1}{2}y \leq 6$$

By very similar reasoning, the total time for employee II is $\frac{1}{2}x + y$, which satisfies

$$\tfrac{1}{2}x + y \leq 6$$

It is also known that $x \geq 0$ and $y \geq 0$ because there cannot be a negative number of either model.

Collecting the preceding conditions, we have that x and y must satisfy this system:

$$x \geq 0$$
$$y \geq 0$$
$$\tfrac{3}{4}x + \tfrac{1}{2}y \leq 6$$
$$\tfrac{1}{2}x + y \leq 6$$

We want to find the (x, y) for this system so that $p = 25x + 22y$ is a maximum.

The graph of this system is the same region R given at the beginning of this section. The points (x, y) in R represent all the possibilities, because for all such points the numbers x and y satisfy *all* the conditions of the stated problem; and any (x, y) not in R cannot be a possibility because the x- and y-values would not satisfy *all* the stated conditions.

The table below shows that the vertex (6, 3) gives the answer.

Vertex	$p = 25x + 22y$
(0, 0)	0
(0, 6)	132
(6, 3)	216
(8, 0)	200

Therefore, a maximum daily profit is realized by assembling 6 bicycles of model A and 3 of model B.

Problems such as this are called **linear programming** problems. The inequalities that give the region R are sometimes called the **constraints**. The region R is called the set of **feasible points**, and (6, 3) is referred to as the **optimal point**.

EXERCISES 8.6

1 **(a)** Graph the region given by the system $x \geq 0$, $y \geq 0$, $x + 3y \leq 6$.

(b) Find all the vertices.

(c) Find the maximum and minimum of $p = x + y$.

(d) Find the maximum and minimum of $q = 6x + 10y$.

(e) Find the maximum and minimum of $r = 2x + 9y$.

2 Follow the instructions given in Exercise 1 for the following system: $x \geq 0$, $y \geq 0$, $x + 4y \leq 72$, $5x + 4y \leq 120$.

3 Follow the instructions given in Exercise 1 for the following system: $x \geq 0$, $y \geq 0$, $x + 6y \leq 96$, $4x + 5y \leq 118$, $3x + y \leq 72$.

4 Follow the instructions given in Exercise 1 for the following system: $x \geq 0$, $y \geq 0$, $x - 2y \geq -10$, $2x + 7y \leq 57$, $5x + 6y \leq 85$, $5x + 2y \leq 75$.

5 Follow the instructions given in Exercise 1 for the following system: $x \geq 0$, $y - x \leq 3$, $x + 4y \leq 22$, $2x + y \leq 16$, $x - 3y \leq 1$, $x + 4y \geq 8$.

The regions for the systems in Exercises 6 and 7 are "open-ended"; that is, their borders do not form a closed polygon and the graph extends endlessly in the direction of the open end. These regions are still convex in the sense discussed in the text.

6 Graph the region given by the system $y \geq 0$, $4x - y \geq 20$, $2x - 3y \geq 0$, $x - 4y \geq -20$, and evaluate the maximum and minimum (when they exist) of the following:

(a) $p = x + y$ (b) $q = 6x + 10y$

(c) $r = 2x + y$

7 Follow the instructions of Exercise 6 for $x \geq 0$, $5x + 2y \geq 22$, $6x + 7y \geq 54$, $2x + 11y \geq 44$, $4x - 5y \leq 34$.

8 A publisher prints and sells both hardcover and paperback copies of the same book. Two machines, M_1 and M_2, are needed jointly to manufacture these books. To produce one hardcover copy, machine M_1 works $\frac{1}{6}$ hour and machine M_2 works $\frac{1}{12}$ hour. For a paperback copy, machines M_1 and M_2 work $\frac{1}{15}$ hour and $\frac{1}{10}$ hour, respectively. Each machine may be operated no more than 12 hours per day. If the profit is \$2 on a hardcover copy and \$1 on a paperback copy, how many of each type should be made per day to earn the maximum profit?

9 A manufacturer produces two models of a certain product: model I and model II. There is a \$5 profit on model I and an \$8 profit on model II. Three machines, M_1, M_2, and M_3, are used jointly to manufacture these models. The number of hours that each machine operates to produce 1 unit of each model is given in the table:

	Model I	*Model II*
Machine M_1	$1\frac{1}{2}$	1
Machine M_2	$\frac{3}{4}$	$1\frac{1}{2}$
Machine M_3	$1\frac{1}{3}$	$1\frac{1}{3}$

No machine is in operation more than 12 hours per day.

(a) If x is the number of model I made per day, and y the number of model II per day, show that x and y satisfy the following constraints.

$$x \geq 0$$
$$y \geq 0$$
$$\tfrac{3}{2}x + y \leq 12$$
$$\tfrac{3}{4}x + \tfrac{3}{2}y \leq 12$$
$$\tfrac{4}{3}x + \tfrac{4}{3}y \leq 12$$

(b) Express the daily profit p in terms of x and y.

(c) Graph the feasible region given by the constraints in part (a) and find the coordinates of the vertices.

(d) What is the maximum profit, and how many of each model are produced daily to realize it?

(e) Find the maximum profit possible for the constraints stated in part (a) if the unit profits are \$8 and \$5 for models I and II, respectively.

10 A farmer buys two varieties of animal feed. Type A contains 8 ounces of corn and 4 ounces of oats per pound; type B contains 6 ounces of corn and 8 ounces of oats per pound. The farmer wants to combine the two feeds so that the resulting mixture has at least 60 pounds of corn and at least 50 pounds of oats. Feed A costs him 5¢ per pound and feed B costs 6¢ per pound. How many pounds of each type should the farmer buy to minimize the cost?

REVIEW EXERCISES FOR CHAPTER 8

The solutions of the following exercises can be found within the text of Chapter 8. Try to answer each question without referring to the text.

Section 8.1

Solve each system by using the row-equivalent matrix procedure.

1
$$\begin{aligned} 2x + 5y + 8z &= 11 \\ x + 4y - 7z &= 10 \\ 3x + 6y + 12z &= 15 \end{aligned}$$

2
$$\begin{aligned} 2x + 14y - 4z &= -2 \\ -4x - 3y + z &= 8 \\ 3x - 5y + 6z &= 7 \end{aligned}$$

3
$$\begin{aligned} 3x - 6y &= 9 \\ -2x + 4y &= -8 \end{aligned}$$

4
$$\begin{aligned} x + 2y - z &= 1 \\ 2x - y + 3z &= 4 \\ 5x \quad\quad + 5z &= 9 \end{aligned}$$

Section 8.2

5 Add: $\begin{bmatrix} 2 & -5 \\ 3 & 7 \\ -1 & 0 \end{bmatrix} + \begin{bmatrix} -4 & 1 \\ 0 & 9 \\ 2 & -2 \end{bmatrix}$.

6 Verify $(A + B) + C = A + (B + C)$ using

$A = \begin{bmatrix} 3 & 1 & -2 \\ -1 & 3 & 0 \end{bmatrix}$, $B = \begin{bmatrix} -5 & 5 & 6 \\ -2 & -2 & -2 \end{bmatrix}$, and

$$C = \begin{bmatrix} 0 & 1 & 0 \\ 1 & 0 & 1 \end{bmatrix}.$$

7 Verify $A + (-A) = 0$ using $A = \begin{bmatrix} 8 & -5 \\ -4 & 9 \\ 0 & 2 \end{bmatrix}$.

8 Find $7A$ using A in Exercise 7.

9 Verify $(cd)A = c(dA)$ using $c = -2$, $d = \frac{1}{3}$, and $A = \begin{bmatrix} 6 & -9 \\ 1 & 5 \end{bmatrix}$.

10 Find AB for $A = \begin{bmatrix} -4 & -1 & 2 \\ 3 & 8 & 1 \end{bmatrix}$ and $B = \begin{bmatrix} -2 & 1 \\ 7 & 0 \\ 4 & -6 \end{bmatrix}$.

11 Multiply: $\begin{bmatrix} 2 & 1 \\ -6 & 3 \\ 1 & 0 \end{bmatrix} \begin{bmatrix} 0 & -2 & 2 & 1 \\ 5 & 0 & 7 & 6 \end{bmatrix}$.

12 Find I_3A for $A = \begin{bmatrix} a & b & c \\ d & e & f \\ g & h & i \end{bmatrix}$.

Section 8.3

Find A^{-1}, if it exists.

13 $A = \begin{vmatrix} 1 & -1 \\ 0 & 2 \end{vmatrix}$

14 $A = \begin{vmatrix} 2 & -4 \\ -1 & 2 \end{vmatrix}$

15 $A = \begin{bmatrix} 2 & 0 & -1 \\ -1 & 2 & 1 \\ 3 & -2 & -4 \end{bmatrix}$

16 Write the linear system as a matrix equation.

$$\begin{aligned} 2x \qquad - z &= 2 \\ -x + 2y + z &= 0 \\ 3x - 2y - 4z &= 10 \end{aligned}$$

17 Solve the system using the inverse of the matrix of coefficients.

$$\begin{aligned} 3x + 4y &= 5 \\ x + 2y &= 3 \end{aligned}$$

18 Solve the system in Exercise 16 using the inverse of the matrix of coefficients.

19 If A^{-1} exists, prove that the solution to the matrix equation $AX = C$ is given by $X = A^{-1}C$.

Section 8.4

Evaluate the determinants.

20 $\begin{vmatrix} 8 & -20 \\ 4 & 10 \end{vmatrix}$

21 $\begin{vmatrix} -20 & 8 \\ 10 & 4 \end{vmatrix}$

22 $\begin{vmatrix} 10 & -20 \\ 8 & 4 \end{vmatrix}$

Solve the given system using Cramer's rule.

23 $\begin{aligned} 8x - 20y &= 3 \\ 4x + 10y &= \tfrac{3}{2} \end{aligned}$

24 $\begin{aligned} 5x - 9y &= 7 \\ -8x + 10y &= 2 \end{aligned}$

25 $\begin{aligned} 3x &= 2y + 22 \\ 2(x + y) &= x - 2y - 2 \end{aligned}$

Section 8.5

26 Evaluate the determinant $\begin{vmatrix} 2 & -2 & 2 \\ 3 & 1 & 0 \\ 2 & -1 & 1 \end{vmatrix}$.

27 Evaluate the determinant $\begin{vmatrix} a_1 & b_1 & c_1 \\ a_2 & b_2 & c_2 \\ a_3 & b_3 & c_3 \end{vmatrix}$ by expanding along the second column and simplify.

28 Solve the system using Cramer's rule.

$$\begin{aligned} x + 2y + z &= 3 \\ 2x - y - z &= 4 \\ -x - y + 2z &= -5 \end{aligned}$$

Section 8.6

29 Maximize the quantity $k = 4x + 5y$ for (x, y) in the region S given by this system:

$$\begin{aligned} -x + 2y &\leq 2 \\ 3x + 2y &\leq 10 \\ x + 6y &\geq 6 \\ 3x + 2y &\geq 6 \end{aligned}$$

30 Find the minimum of $k = 4x + 5y$ for the system in Exercise 29.

31 A store sells two kind of bicycles, model A and model B. The store buys them unassembled from a wholesaler. Two employees are hired just to assemble the bicycles. Working together to assemble model A, employee I works $\frac{3}{4}$ hour and employee II works $\frac{1}{2}$ hour. Model B requires $\frac{1}{2}$ hour's work by employee I as well as 1 hour's work by employee II. There is a \$25 profit on each model A sold and \$22 on each model B. Because of the popularity of the sport, the store is able to sell as many bicycles as they decide to assemble. If each employee works no more than 6 hours per day, how many bicycles of each model should they assemble in order to get the maximum profit?

SAMPLE TEST QUESTIONS FOR CHAPTER 8

Use the questions to test your knowledge of the basic skills and concepts of Chapter 8. Then check your answers with those given at the end of the book.

In Questions 1–4, solve the linear system using row-equivalent matrices.

1 $\begin{aligned} 2x + 5y &= 4 \\ 4x - 3y &= -18 \end{aligned}$

2 $\begin{aligned} -6x + 3y &= 9 \\ 10x - 5y &= -12 \end{aligned}$

3 $\begin{aligned} 2x - y + 2z &= -3 \\ x + 4y - 3z &= 18 \\ -4x + 2y - 3z &= 0 \end{aligned}$

4 $\begin{aligned} x + y + 2z &= 3 \\ 3x + 2z &= 2 \\ -x + 2y + 2z &= 4 \end{aligned}$

5 Write the system in Question 1 as a matrix equation.

6 Write the system in Question 3 as a matrix equation.

In Questions 7–19, evaluate whenever possible, using these matrices.

$$A = \begin{bmatrix} 2 & 0 & 3 \\ -1 & 4 & 9 \end{bmatrix} \qquad B = \begin{bmatrix} 1 & 0 & -1 \\ 0 & -1 & 1 \\ -1 & 0 & 11 \end{bmatrix}$$

$$C = \begin{bmatrix} 1 & 2 & 3 \\ 4 & 5 & 6 \\ 7 & 8 & 9 \\ 1 & 0 & 1 \end{bmatrix} \qquad D = \begin{bmatrix} -1 & -1 \\ 2 & 2 \\ -3 & -3 \end{bmatrix}$$

$$E = \begin{bmatrix} 3 & 0 \\ 1 & 1 \end{bmatrix} \qquad F = \begin{bmatrix} -5 & 7 \\ 4 & -9 \end{bmatrix}$$

$$G = \begin{bmatrix} 0 & -1 & 1 \\ 6 & 2 & 3 \\ -5 & -3 & -8 \end{bmatrix} \qquad H = \begin{bmatrix} -1 \\ 1 \\ 2 \end{bmatrix}$$

$$J = [2 \quad -2 \quad 4]$$

7 $B + G$

8 $A - B$

9 $5E - 2F$

10 AB

11 m_{32} in the product $CB = M$.

12 AD

13 $A(EF)$

14 E^3 (means EEE)

15 HJ

16 The dimension of $CBDA$

17 The dimension of $DBAE$

18 $(|E|)(|F|)$

19 $|G|$

20 Evaluate: $(A + \frac{1}{2}B)(C - 2D)$ for $A = [11, 7, 2]$, $B = [-2, 6, -2]$, $C = \begin{bmatrix} -5 \\ -2 \\ 4 \end{bmatrix}$, $D = \begin{bmatrix} -8 \\ 5 \\ 1 \end{bmatrix}$.

21 Find the inverse of $A = \begin{bmatrix} 3 & 1 \\ 2 & -2 \end{bmatrix}$.

22 Use the result of Question 21 to solve this system.

$$\begin{aligned} 3x + y &= 9 \\ 2x - 2y &= 14 \end{aligned}$$

23 Find the inverse of $A = \begin{bmatrix} 1 & 0 & -2 \\ 0 & 1 & 0 \\ 3 & 2 & 0 \end{bmatrix}$.

24 The matrix $A = \begin{bmatrix} -11 & 2 & 2 \\ -4 & 0 & 1 \\ 6 & -1 & -1 \end{bmatrix}$ has inverse $A^{-1} = \begin{bmatrix} 1 & 0 & 2 \\ 2 & -1 & 3 \\ 4 & 1 & 8 \end{bmatrix}$. Use this information to solve this linear system.

$$\begin{aligned} -11x + 2y + 2z &= 3 \\ -4x + z &= -1 \\ 6x - y - z &= -2 \end{aligned}$$

In Questions 25 and 26, use Cramer's rule to solve the system.

25 $\begin{aligned} 4x - 2y &= 15 \\ 3x + 2y &= -8 \end{aligned}$

26 $\begin{aligned} 2x + y - z &= -3 \\ x - 2y + z &= 8 \\ 3x - y - 2z &= -1 \end{aligned}$

27 Graph the system

$x \geq 0, \quad y \geq 0, \quad x + 2y \leq 8, \quad x + y \leq 5,$
$2x + y \leq 8$

28 Find the maximum value of $k = 3x + 2y$ for the region in Question 27.

ANSWERS TO TEST YOUR UNDERSTANDING EXERCISES

Page 288

1 Not possible.

2 $\begin{bmatrix} 3 & 3 \\ 2 & 3 \\ 9 & 6 \end{bmatrix}$

3 $\begin{bmatrix} 8 & 4 \\ 1 & 9 \\ 17 & 8 \end{bmatrix}$

4 $\begin{bmatrix} -7 & 1 \\ 4 & -9 \\ -7 & 2 \end{bmatrix}$ 5 $\begin{bmatrix} 7 & -1 \\ -4 & 9 \\ 7 & -2 \end{bmatrix}$ 6 $\begin{bmatrix} 2 & 8 & -6 \\ 4 & 10 & 0 \\ 0 & 2 & 2 \end{bmatrix}$

7 Not possible. 8 $\begin{bmatrix} -15 & -15 \\ -10 & -15 \\ -45 & -30 \end{bmatrix}$ 9 $\begin{bmatrix} -10 & -2 \\ 2 & -12 \\ -16 & -4 \end{bmatrix}$

Page 290

1 $\begin{bmatrix} -6 & -22 \\ 59 & 11 \end{bmatrix}$ 2 $\begin{bmatrix} 8 & 0 & 9 \\ 13 & -22 & 27 \end{bmatrix}$ 3 Not possible. 4 $\begin{bmatrix} 24 & -29 \\ 14 & -38 \\ -25 & 49 \end{bmatrix}$

5 Not possible. 6 $\begin{bmatrix} 52 & -63 \\ -5 & 19 \end{bmatrix}$ 7 $\begin{bmatrix} 25 & 2 & 27 \\ -18 & 28 & -36 \\ 3 & -26 & 18 \end{bmatrix}$ 8 $\begin{bmatrix} 42 & -25 \\ 181 & -284 \end{bmatrix}$

Page 295

1 $\begin{bmatrix} \frac{5}{3} & 2 \\ \frac{2}{3} & 1 \end{bmatrix}$ 2 Does not exist. 3 $\begin{bmatrix} 0 & -1 \\ -\frac{1}{2} & \frac{3}{2} \end{bmatrix}$

4 $\begin{bmatrix} 4 & 3 \\ -4 & -6 \end{bmatrix}$ 5 Does not exist. 6 $\begin{bmatrix} \frac{1}{110} & \frac{1}{110} \\ \frac{1}{110} & -\frac{1}{1100} \end{bmatrix}$

Page 301

1 −2 2 −2 3 2 4 −10 5 5 6 5
7 20 8 2 9 −4 10 1 11 −4 12 0

Page 307

1 −29 2 0 3 27 4 −93 5 −8 6 0

Page 314

1 19 2 1 3 26 4 −34 5 50 6 −139 7 37 8 10 9 200 10 −592

SEQUENCES AND SERIES

9

9.1 Sequences

The same equation can be used to define a variety of functions by changing the domain. For example, below are the graphs of three functions all of whose range values are given by the equation $y = x^2$ for the indicated domains.

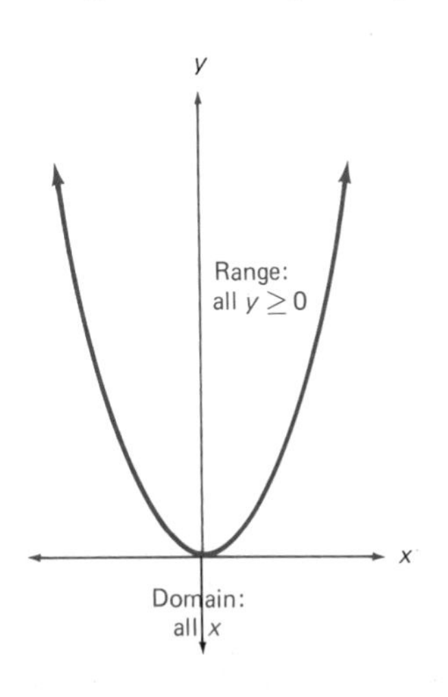

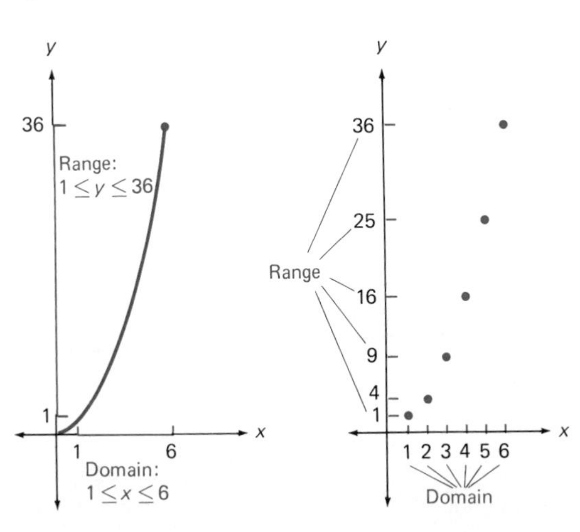

The type of function that is studied in this chapter is illustrated by the preceding graph at the right, where the domain consists of the consecutive integers 1, 2, 3, 4, 5, 6. This kind of function is called a **sequence.**

> A sequence is a function whose domain is a set of consecutive integers, usually a set of positive consecutive integers beginning with 1.

s_n is read "s-sub-n" and has the same meaning as the functional notation $s(n)$, that is, "s of n."

Instead of using the variable x, letters, such as n, k, i are normally used for the domain variable of a sequence. Frequently, sequences (functions) will be denoted by the lowercase letter s and the range values by s_n, which are also called the **terms** of the sequence.

Sequences are often given by stating their **general** or ***n*th terms.** Thus the general term of the sequence, previously given by $y = x^2$, becomes $s_n = n^2$.

EXAMPLE 1 Find the range values of the sequence given by $s_n = \dfrac{1}{n}$ for the domain $\{1, 2, 3, 4, 5\}$ and graph.

Solution The range values and graph are as follows.

This is an example of a *finite* sequence since the domain is finite. That is, the domain is a set of integers having a last element.

$$s_1 = 1$$
$$s_2 = \tfrac{1}{2}$$
$$s_3 = \tfrac{1}{3}$$
$$s_4 = \tfrac{1}{4}$$
$$s_5 = \tfrac{1}{5}$$

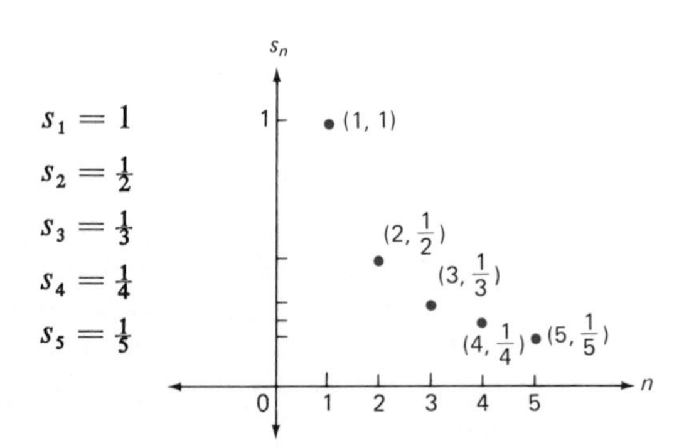

EXAMPLE 2 List the first seven terms of the sequence given by $s_k = \dfrac{(-1)^k}{k^2}$.

Solution

$$s_1 = \frac{(-1)^1}{1^2} = -1$$

$$s_2 = \frac{(-1)^2}{2^2} = \frac{1}{4}$$

$$s_3 = \frac{(-1)^3}{3^2} = -\frac{1}{9}$$

$$s_4 = \frac{(-1)^4}{4^2} = \frac{1}{16}$$

$$s_5 = \frac{(-1)^5}{5^2} = -\frac{1}{25}$$

$$s_6 = \frac{(-1)^6}{6^2} = \frac{1}{36}$$

$$s_7 = \frac{(-1)^7}{7^2} = -\frac{1}{49}$$

Sometimes a sequence is given by a verbal description. If, for example, we ask for the increasing sequence of odd integers beginning with -3, then this implies the infinite sequence whose first few terms are

$$-3, -1, 1, \ldots$$

A sequence can also be given by presenting a listing of its first few terms, possibly including the general term. Thus the preceding sequence is

$$-3, -1, 1, \ldots, 2n - 5, \ldots$$

This is an example of an *infinite* sequence since the domain is infinite. That is, the domain consists of an endless collection of integers.

EXAMPLE 3 Find the tenth term of the sequence

$$-3, 4, \frac{5}{3}, \ldots, \frac{n+2}{2n-3}, \ldots$$

Solution Since the first term -3 is obtained by letting $n = 1$ in the general term $\frac{n+2}{2n-3}$, the tenth term is

$$s_{10} = \frac{10+2}{2(10)-3} = \frac{12}{17}$$

EXAMPLE 4 Write the first four terms of the sequence given by $s_n = \left(1 + \frac{1}{n}\right)^n$. Round off to two decimal places when appropriate.

Solution

$$s_1 = (1 + \tfrac{1}{1})^1 = 2$$
$$s_2 = (1 + \tfrac{1}{2})^2 = (\tfrac{3}{2})^2 = \tfrac{9}{4} = 2.25$$
$$s_3 = (1 + \tfrac{1}{3})^3 = (\tfrac{4}{3})^3 = \tfrac{64}{27} = 2.37$$
$$s_4 = (1 + \tfrac{1}{4})^4 = (\tfrac{5}{4})^4 = \tfrac{625}{256} = 2.44$$

The terms of the sequence in Example 4 are getting successively larger. But the increase from term to term is getting smaller. That is, the differences between successive terms are decreasing:

$$s_2 - s_1 = 0.25$$
$$s_3 - s_2 = 0.12$$
$$s_4 - s_3 = 0.07$$

Use a calculator to verify these table entries to four decimal places.

n	$s_n = \left(1 + \frac{1}{n}\right)^n$
10	2.5937
50	2.6916
100	2.7048
500	2.7156
1000	2.7169
5000	2.7180
10,000	2.7181

If more terms of $s_n = \left(1 + \frac{1}{n}\right)^n$ were computed, you would see that while the terms keep on increasing, the amount by which each new term increases keeps getting smaller.

It turns out that no matter how large n is, the value of $\left(1 + \frac{1}{n}\right)^n$ is never more than 2.72. In fact, the larger the n that is taken, the closer $\left(1 + \frac{1}{n}\right)^n$ gets to the irrational value $e = 2.71828\ldots$. This is the number that was introduced in Chapter 7 in reference to natural logarithms and exponential functions.

EXERCISES 9.1

The domain of each sequence in Exercises 1–6 consists of the integers 1, 2, 3, 4, 5. Write the corresponding range values and graph the sequence.

1 $s_n = 2n - 1$

2 $s_n = 10 - n^2$

3 $s_n = (-1)^n$

4 $s_k = -\frac{6}{k}$

5 $s_i = 8(-\frac{1}{2})^i$

6 $s_i = (\frac{1}{2})^{i-3}$

Write the first four terms of the sequence given by the formula in each of Exercises 7–30.

7 $s_k = (-1)^k k^2$

8 $s_j = 3(\frac{1}{10})^{j-1}$

9 $s_j = 3(\frac{1}{10})^j$

10 $s_j = 3(\frac{1}{10})^{j+1}$

11 $s_j = 3(\frac{1}{10})^{2j}$

12 $s_n = \frac{(-1)^{n+1}}{n+3}$

13 $s_n = \frac{1}{n} - \frac{1}{n+1}$

14 $s_n = \frac{n^2 - 4}{n+2}$

15 $s_k = (2k - 10)^2$

16 $s_k = 1 + (-1)^k$

17 $s_n = -2 + (n-1)(3)$

18 $s_n = a + (n-1)(d)$

19 $s_i = \frac{i-1}{i+1}$

20 $s_i = 64^{1/i}$

21 $s_n = \left(1 + \frac{1}{n}\right)^{n-1}$

22 $s_n = \frac{1}{2^n}$

23 $s_n = -2(\frac{3}{4})^{n-1}$

24 $s_k = ar^{k-1}$

25 $s_k = \frac{k}{2^k}$

26 $s_n = \frac{(-1)^n}{n} + n$

27 $s_k = \frac{k}{k+1} - \frac{k+1}{k}$

28 $s_n = \left(1 + \frac{1}{n+1}\right)^n$

29 $s_n = 4$

30 $s_n = \frac{n}{(n+1)(n+2)}$

31 Find the sixth term of $1, 2, 5, \ldots, \frac{1}{2}(1 + 3^{n-1}), \ldots$.

32 Find the ninth and tenth terms of $0, 4, 0, \ldots, \frac{2^n + (-2)^n}{n}, \ldots$.

33 Find the seventh term of $s_k = 3(0.1)^{k-1}$.

34 Find the twentieth term of $s_n = (-1)^{n-1}$.

35 Find the twelfth term of $s_i = i$.

36 Find the twelfth term of $s_i = (i-1)^2$.

37 Find the twelfth term of $s_i = (1-i)^3$.

38 Find the hundredth term of $s_n = \frac{n+1}{n^2 + 5n + 4}$.

39 Write the first four terms of the sequence of even increasing integers beginning with 4.

40 Write the first four terms of the sequence of decreasing odd integers beginning with 3.

41 Write the first five positive multiples of 5 and find the formula for the nth term.

42 Write the first five powers of 5 and find the formula for the nth term.

43 Write the first five powers of -5 and find the formula for the nth term.

44 Write the first five terms of the sequence of reciprocals of the negative integers and find the formula for the nth term.

45 The numbers 1, 3, 6, and 10 are called **triangular numbers** because they correspond to the number of dots in the triangular arrays.

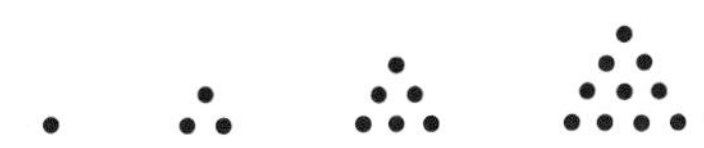

Find the next three triangular numbers.

46 Write the first eight terms of $s_n = \sin \frac{n\pi}{2}$.

47 Write the first five terms of $s_n = \frac{3^n}{2^n + 1}$.

48 Write the first seven terms of $s_n = n!$ ($n!$ is read as "n factorial" and is defined by

$$n! = n(n-1) \cdot (n-2) \cdots 3 \cdot 2 \cdot 1)$$

49 Write the first four terms of

$$s_n = \frac{(2n+1)(2n-1) \cdots 5 \cdot 3}{(2n)(2n-2) \cdots 4 \cdot 2}.$$

***50** When an investment earns **simple interest** it means the interest is earned only on the original investment. For example, if P dollars are invested in a bank that pays simple interest at the annual rate of r percent, then the interest for the first year is Pr, and the amount in the bank at the end of the year is $P + Pr$. For the second year, the interest is again Pr; the amount now would be $(P + Pr) + Pr = P + 2Pr$.

(a) What is the amount after n years?

(b) What is the amount in the bank if an investment of \$750 has been earning simple interest for 5 years at the annual rate of 12%?

(c) If the amount in the bank is \$5395 after 12 years, what was the original investment P if it has been earning simple interest at the annual rate of $12\frac{1}{2}$%?

9.2 Sums of finite sequences

How long would it take you to add up the integers from 1 to 1000? Here is a quick way. List the sequence displaying the first few and last few terms.

$$1, 2, 3, \ldots, 998, 999, 1000$$

Add them in pairs, the first and last, the second and second from last, and so on.

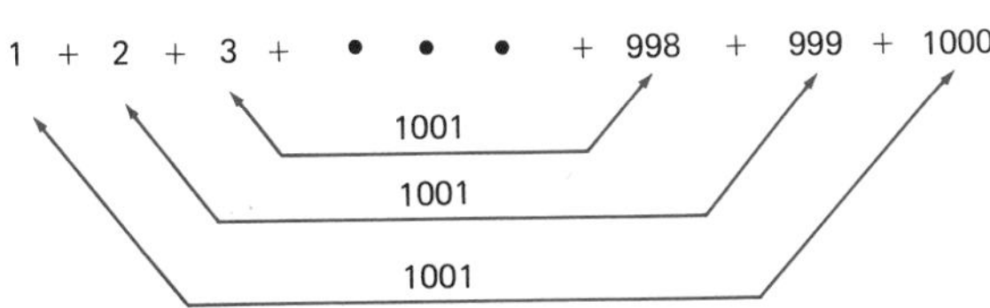

It is told that Carl Friedrich Gauss (1777–1855) discovered how to compute such sums when he was 10 years old. He became one of the greatest mathematicians of all time.

Since there are 500 such pairs to be added, the total is

$$500(1001) = 500{,}500$$

For any finite sequence we can add up all its terms and say that we have found the *sum of the sequence*. The sum of a sequence is called a **series**. For example, the sequence

$$1, 3, 5, 7, 9, 11$$

can be associated with the series

$$1 + 3 + 5 + 7 + 9 + 11$$

The sum of the terms in this series can easily be found, by adding, to be 36. As another example, the sequence $s_n = \frac{1}{n}$ for $n = 1, 2, 3, 4, 5$ has the sum

$$1 + \frac{1}{2} + \frac{1}{3} + \frac{1}{4} + \frac{1}{5} = \frac{60 + 30 + 20 + 15 + 12}{60} = \frac{137}{60}$$

EXAMPLE 1 Find the sum of the first seven terms of $s_k = 2k$.

Solution

$$s_1 + s_2 + s_3 + s_4 + s_5 + s_6 + s_7 = 2 + 4 + 6 + 8 + 10 + 12 + 14 = 56$$

There is a very handy notational device available for expressing the sum of a sequence. The Greek letter $\sum$ (capital sigma) is used for this purpose. Referring to Example 1, the sum of the seven terms is expressed by the symbol $\sum_{k=1}^{7} s_k$; that is,

Just think of the sigma as the command to add.

$$\sum_{k=1}^{7} s_k = s_1 + s_2 + s_3 + s_4 + s_5 + s_6 + s_7$$

Add the terms s_k for consecutive values of k, starting with $k = 1$ up to and including $k = 7$. With this symbolism, the question in Example 1 can now be stated by asking for the value of $\sum_{k=1}^{7} s_k$, where $s_k = 2k$, or by asking for the value of $\sum_{k=1}^{7} 2k$.

EXAMPLE 2 Find $\sum_{n=1}^{5} s_n$, where $s_n = \frac{2}{n}$.

Solution

$$\begin{aligned}
\sum_{n=1}^{5} s_n &= s_1 + s_2 + s_3 + s_4 + s_5 \\
&= \frac{2}{1} + \frac{2}{2} + \frac{2}{3} + \frac{2}{4} + \frac{2}{5} \\
&= \frac{120 + 60 + 40 + 30 + 24}{60} \\
&= \frac{274}{60} = \frac{137}{30}
\end{aligned}$$

EXAMPLE 3 Find $\sum_{k=1}^{4} (2k + 1)$.

Solution It is understood here that we are to find the sum of the first four terms of the sequence whose general term is $s_k = 2k + 1$.

$$\begin{aligned}
\sum_{k=1}^{4} (2k + 1) &= (2 \cdot 1 + 1) + (2 \cdot 2 + 1) + (2 \cdot 3 + 1) + (2 \cdot 4 + 1) \\
&= 3 + 5 + 7 + 9 \\
&= 24
\end{aligned}$$

EXAMPLE 4 Find $\sum_{k=1}^{5} s_i$, where $s_i = (-1)^i(i+1)$.

Solution

$$\begin{aligned}\sum_{i=1}^{5} s_i &= s_1 + s_2 + s_3 + s_4 + s_5 \\ &= (-1)^1(1+1) + (-1)^2(2+1) + (-1)^3(3+1) \\ &\qquad + (-1)^4(4+1) + (-1)^5(5+1) \\ &= -2 + 3 - 4 + 5 - 6 \\ &= -4\end{aligned}$$

EXERCISES 9.2

Find the sum of the first five terms of the sequence given by the formula in each of Exercises 1–6.

1 $s_n = 3n$

2 $s_k = (-1)^k \frac{1}{k}$

3 $s_i = i^2$

4 $s_i = i^3$

5 $s_k = \frac{3}{10^k}$

6 $s_n = -6 + 2(n-1)$

7 Find $\sum_{n=1}^{8} s_n$ where $s_n = 2^n$.

8 Find $\sum_{n=0}^{8} s_n$ where $s_n = \frac{1}{2^n}$.

9 Find $\sum_{k=1}^{20} s_k$ where $s_k = 3$.

Find each of the following sums for $n = 7$.

10 $2 + 4 + \ldots + 2n$

11 $2 + 4 + \ldots + 2^n$

12 $-7 + 2 + \ldots + (9n - 16)$

13 $3 + \frac{3}{2} + \ldots + 3(\frac{1}{2})^{n-1}$

Compute each of the following.

14 $\sum_{k=1}^{6} (5k)$

15 $5\left(\sum_{k=1}^{6} k\right)$

16 $\sum_{n=1}^{4} (n^2 + n)$

17 $\sum_{n=1}^{4} n^2 + \sum_{n=1}^{4} n$

18 $\sum_{i=1}^{8} (i - 2i^2)$

19 $\sum_{k=1}^{4} \frac{k}{2^k}$

20 $\sum_{k=1}^{7} (-1)^k$

21 $\sum_{k=1}^{8} (-1)^k$

22 $\sum_{k=3}^{7} (2k - 5)$

23 $\sum_{j=1}^{6} [-3 + (j-1)5]$

24 $\sum_{k=-3}^{3} 10^k$

25 $\sum_{k=-3}^{3} \frac{1}{10^k}$

26 $\sum_{k=1}^{5} 4(-\frac{1}{2})^{k-1}$

27 $\sum_{i=1}^{4} (-1)^i 3^i$

28 $\sum_{n=1}^{3} \left(\frac{n+1}{n} - \frac{n}{n+1}\right)$

29 $\sum_{n=1}^{3} \frac{n+1}{n} - \sum_{n=1}^{3} \frac{n}{n+1}$

30 $\sum_{k=1}^{8} \frac{1 + (-1)^k}{2}$

31 $\sum_{k=1}^{3} (0.1)^{2k}$

32 Read the discussion at the beginning of this section, where we found the sum of the first 1000 positive integers, and find a formula for the sum of the first n positive integers for n even.

33 **(a)** Find $\sum_{k=1}^{n} (2k - 1)$ for each of the following values of n: 2, 3, 4, 5, 6.
(b) On the basis of the results in part (a), find a formula for the sum of the first n odd numbers.

34 The sequence 1, 1, 2, 3, 5, 8, 13, . . . is called the *Fibonacci sequence*. Its first two terms are ones, and each term thereafter is computed by adding the preceding two terms.
(a) Write the next seven terms of this sequence.
(b) Let $u_1, u_2, u_3, \ldots, u_n, \ldots$ be the Fibonacci sequence. Evaluate $S_n = \sum_{k=1}^{n} u_k$ for these values of n: 1, 2, 3, 4, 5, 6, 7, 8.
***(c)** Note that $u_1 = u_3 - u_2$, $u_2 = u_4 - u_3$, $u_3 = u_5 - u_4$, and so on. Use this form for the first n numbers to derive a formula for the sum of the first n Fibonacci numbers.

*35 Let s_n be a sequence with $s_1 = 2$ and $s_m + s_n = s_{m+n}$, where m and n are any positive integers. Show that $s_n = 2n$ for any n.

*36 Show that $\sum_{k=1}^{9} \log \frac{k+1}{k} = 1$.

$\left(\textit{Hint: } \log \frac{a}{b} = \log a - \log b.\right)$

*37 Prove: $\sum_{k=1}^{n} s_k + \sum_{k=1}^{n} t_k = \sum_{k=1}^{n} (s_k + t_k)$.

*38 Prove: $\sum_{k=1}^{n} cs_k = c \sum_{k=1}^{n} s_k$, c a constant.

*39 Prove: $\sum_{k=1}^{n} (s_k + c) = \left(\sum_{k=1}^{n} s_k\right) + nc$, c a constant.

*40 Evaluate $\sum_{k=1}^{10} \frac{1}{k(k+1)}$ using the result

$$\frac{1}{k(k+1)} = \frac{1}{k} - \frac{1}{k+1}$$

9.3

Arithmetic sequences and series

Here are the first five terms of the sequence whose general term is $s_k = 7k - 2$:

$$5, 12, 19, 26, 33$$

Do you notice any special pattern? It does not take long to observe that each term, after the first, is 7 more than the preceding term. This sequence is an example of an *arithmetic sequence.*

An arithmetic sequence is also referred to as an *arithmetic progression.*

ARITHMETIC SEQUENCE

A sequence is said to be *arithmetic* if each term, after the first, is obtained from the preceding term by adding a common value.

Let us consider the first four terms of three different arithmetic sequences:

$$2, 4, 6, 8, \ldots$$
$$-\tfrac{1}{2}, -1, -\tfrac{3}{2}, -2, \ldots$$
$$11, 2, -7, -16, \ldots$$

For the first sequence, the common value (or difference) that is added to each term to get the next is 2. Thus it is easy to see that 10, 12, and 14 are the next three terms. You might guess that the nth term is $s_n = 2n$.

The second sequence has the common difference $-\frac{1}{2}$. This can be found by subtracting the first term from the second, or the second from the third, and so forth. The nth term is $s_n = -\frac{1}{2}n$.

The third sequence has -9 as its common difference, but it is not so easy to see what the general term is. Rather than employ a hit-or-miss process in trying to find the general term of this sequence, we will instead consider arithmetic sequences in general, thus making it possible to write the general term of any such sequence.

Let s_n be the nth term of an arithmetic sequence. Denote its first term by the letter a; that is, $s_1 = a$. Also, let d be the common difference. Then the first four terms are

$$s_1 = a$$
$$s_2 = s_1 + d = a + d$$
$$s_3 = s_2 + d = (a + d) + d = a + 2d$$
$$s_4 = s_3 + d = (a + 2d) + d = a + 3d$$

The pattern is clear. Without further computation we see that

$$s_5 = a + 4d$$
$$s_6 = a + 5d$$

Since the coefficient of d is always 1 less than the number of the term, the nth term is given as follows.

The nth term of an arithmetic sequence is

$$s_n = a + (n - 1)d$$

where a is the first term and d is common difference.

This formula says that the nth term of an arithmetic sequence is completely identified by its first term a and its common difference d.

By substituting the values $n = 1, 2, 3, 4, 5, 6$, you can check that this formula gives the preceding terms s_1 through s_6.

Let us return to the sequence given earlier: $11, 2, -7, -16, \ldots$. It is now easy to find its nth term, with $a = 11$ and $d = -9$:

$$s_n = 11 + (n - 1)(-9)$$
$$= -9n + 20$$

TEST YOUR UNDERSTANDING

Each of the following gives the first few terms of an arithmetic sequence. Find the nth term in each case.

1 $5, 10, 15, \ldots$

2 $6, 2, -2, \ldots$

3 $\frac{1}{10}, \frac{1}{5}, \frac{3}{10}, \ldots$

4 $-5, -13, -21, \ldots$

5 $1, 2, 3, \ldots$

6 $-3, -2, -1, \ldots$

Find the nth term s_n of the arithmetic sequence with the given values for the first term and the common difference.

7 $a = \frac{2}{3}; d = \frac{2}{3}$

8 $a = 53; d = -12$

9 $a = 0; d = \frac{1}{5}$

10 $a = 2; d = 1$

Adding the terms of a finite sequence may not be much work when the number of terms to be added is small. When many terms are to be added, however, the amount of time and effort needed can be overwhelming. For example, to add the first 10,000 terms of the arithmetic sequence beginning with

$$246, 261, 276, \ldots$$

The sum of an arithmetic sequence is called an *arithmetic series.*

would call for an enormous effort, unless some shortcut could be found. Fortunately, there is an easy way available to find such sums. This method (in disguise) was already used in the question at the start of Section 10.2. Let us look at the general situation. Let S_n denote the sum of the first n terms of the arithmetic sequence given by $s_k = a + (k - 1)d$:

$$S_n = \sum_{k=1}^{n} [a + (k - 1)d]$$

$$= a + [a + d] + [a + 2d] + \cdots + [a + (n - 1)d]$$

Put this sum in reverse order and write the two equalities together as follows:

$$\begin{array}{l} S_n = a + [a+d] + \cdots + [a+(n-2)d] + [a+(n-1)d] \\ \quad\updownarrow \qquad\quad \updownarrow \qquad\qquad\quad \updownarrow \qquad\qquad\quad \updownarrow \\ S_n = [a+(n-1)d] + [a+(n-2)d] + \cdots + [a+d] + a \end{array}$$

Now add to get

$$2S_n = [2a+(n-1)d] + [2a+(n-1)d] + \cdots + [2a+(n-1)d] + [2a+(n-1)d]$$

On the right-hand side of this equation there are n terms of the form $2a + (n - 1)d$. Therefore,

$$2S_n = n[2a + (n - 1)d]$$

Divide by 2 to solve for S_n:

$$\mathbf{S_n = \frac{n}{2}[2a + (n - 1)d]}$$

Returning to the sigma notation, we can summarize our results this way:

ARITHMETIC SERIES

$$\sum_{k=1}^{n} [a + (k - 1)d] = \frac{n}{2}[2a + (n - 1)d]$$

EXAMPLE 1 Find S_{20} for the arithmetic sequence whose first term is $a = 3$ and whose common difference is $d = 5$.

Solution Substituting $a = 3$, $d = 5$, and $n = 20$ into the formula for S_n, we have

$$S_{20} = \frac{20}{2}[2(3) + (20 - 1)5]$$

$$= 10(6 + 95)$$

$$= 1010$$

EXAMPLE 2 Find the sum of the first 10,000 terms of the arithmetic sequence beginning with 246, 261, 276, ...

Solution Since $a = 246$ and $d = 15$,

$$S_{10,000} = \frac{10,000}{2}[2(246) + (10,000 - 1)15]$$
$$= 5000(150,477)$$
$$= 752,385,000$$

EXAMPLE 3 Find the sum of the first n positive integers.

Solution First observe that the problem calls for the sum of the sequence $s_k = k$ for $k = 1, 2, \ldots, n$. This is an arithmetic sequence with $a = 1$ and $d = 1$. Therefore,

$$\sum_{k=1}^{n} k = \frac{n}{2}[2 + (n - 1)1] = \frac{n(n + 1)}{2}$$

With the result of Example 3 we are able to check the answer for the sum of the first 1000 positive integers, found at the beginning of Section 10.2, as follows:

$$\sum_{k=1}^{1000} k = \frac{1000(1001)}{2} = 500,500$$

The form $s_k = a + (k - 1)d$ for the general term of an arithmetic sequence easily converts to

$$s_k = dk + (a - d)$$

It is this latter form that is ordinarily used when the general term of a *specific* arithmetic sequence is given. For example, we would usually begin with the form

$$s_k = 3k + 5$$

instead of

$$s_k = 8 + (k - 1)3$$

The important thing to notice in the form $s_k = dk + (a - d)$ is that the common difference is the coefficient of k.

EXAMPLE 4 Find: $\sum_{k=1}^{50} (-6k + 10)$.

Solution First note that $s_k = -6k + 10$ is an arithmetic sequence with $d = -6$ and with $a = s_1 = 4$.

$$\sum_{k=1}^{50} (-6k + 10) = \frac{50}{2}[2(4) + (50 - 1)(-6)]$$
$$= -7150$$

EXERCISES 9.3

Each of the following gives the first two terms of an arithmetic sequence. Write the next three terms; find the nth term; and find the sum of the first 20 terms.

1 $1, 3, \ldots$ **2** $2, 4, \ldots$

3 $2, -4, \ldots$ **4** $1, -3, \ldots$

5 $\frac{15}{2}, 8, \ldots$ **6** $-\frac{4}{3}, -\frac{11}{3}, \ldots$

7 $\frac{2}{5}, -\frac{1}{5}, \ldots$ **8** $-\frac{1}{2}, \frac{1}{4}, \ldots$

9 $50, 100, \ldots$ **10** $-27, -2, \ldots$

11 $-10, 10, \ldots$ **12** $225, 163, \ldots$

Find the indicated sum by using ordinary addition; also find the sum by using the formula for the sum of an arithmetic sequence.

13 $5 + 10 + 15 + 20 + 25 + 30 + 35 + 40 + 45 + 50 + 55 + 60 + 65$

14 $-33 - 25 - 17 - 9 - 1 + 7 + 15 + 23 + 31 + 39$

15 $\frac{3}{4} + 1 + \frac{5}{4} + \frac{3}{2} + \frac{7}{4} + 2 + \frac{9}{4} + \frac{5}{2} + \frac{11}{4}$

16 $128 + 71 + 14 - 43 - 100 - 157$

In Exercises 17–24, find S_{100} for the arithmetic sequence with the given values for a and d.

17 $a = 3; d = 3$ **18** $a = 1; d = 8$

19 $a = -91; d = 21$ **20** $a = -7; d = -10$

21 $a = \frac{1}{7}; d = 5$ **22** $a = \frac{2}{5}; d = -4$

23 $a = 725; d = 100$ **24** $a = 0.1; d = 10$

25 Find S_{28} for the sequence $-8, 8, \ldots, 16n - 24, \ldots$.

26 Find S_{25} for the sequence $96, 100, \ldots, 4n + 92, \ldots$.

27 Find the sum of the first 50 positive multiples of 12.

28 (a) Find the sum of the first 100 positive even numbers.

(b) Find the sum of the first n positive even numbers.

29 (a) Find the sum of the first 100 positive odd numbers.

(b) Find the sum of the first n positive odd numbers.

Evaluate the series in each of Exercises 30–37.

30 $\sum_{k=1}^{12} [3 + (k - 1)9]$ **31** $\sum_{k=1}^{9} [-6 + (k - 1)\frac{1}{2}]$

32 $\sum_{k=1}^{20} (4k - 15)$ **33** $\sum_{k=1}^{30} (10k - 1)$

34 $\sum_{k=1}^{40} (-\frac{1}{3}k + 2)$ **35** $\sum_{k=1}^{49} (\frac{3}{4}k - \frac{1}{2})$

36 $\sum_{k=1}^{20} 5k$ **37** $\sum_{k=1}^{n} 5k$

38 Find u such that $7, u, 19$ is an arithmetic sequence.

39 Find u such that $-7, u, \frac{5}{2}$ is an arithmetic sequence.

40 Find the twenty-third term of the arithmetic sequence $6, -4, \ldots$.

41 Find the thirty-fifth term of the arithmetic sequence $-\frac{2}{3}, -\frac{1}{5}, \ldots$.

42 An object is dropped from an airplane and falls 32 feet during the first second. During each successive second it falls 48 feet more than in the preceding second. How many feet does it travel during the first 10 seconds? How far does it fall during the tenth second?

43 Suppose you save \$10 one week and that each week thereafter you save 50¢ more than the preceding week. How much will you have saved by the end of 1 year?

44 A pyramid of blocks has 26 blocks in the bottom row and 2 fewer blocks in each successive row thereafter. How many blocks are there in the pyramid?

***45** Find $\sum_{n=6}^{20} (5n - 3)$.

***46** Find the sum of all the even numbers between 33 and 427.

***47** If $\sum_{k=1}^{30} [a + (k - 1)d] = -5865$ and $\sum_{k=1}^{20} [a + (k - 1)d] = -2610$, find a and d.

48 Listing the first few terms of a sequence like $2, 4, 6, \ldots$, without stating its general term or describing just what kind of sequence it is makes it impossible to predict the next term. Show that both $t_n = 2n$ and $s_n = 2n + (n - 1)(n - 2)(n - 3)$ produce these first three terms but that their fourth terms are different.

***49** Find u and v such that $3, u, v, 10$ is an arithmetic sequence.

***50** Show that the sum of an arithmetic sequence of n terms is n times the average of the first and last terms.

Use the result in Exercise 50 to find S_{80} for the given arithmetic sequences.

51 $s_k = 3k - 8$

52 $s_k = \frac{1}{2}k + 10$

53 $s_k = -5k$

54 What is the connection between arithmetic sequences and linear functions?

***55** The function f defined by $f(x) = 3x + 7$ is a linear function. Evaluate the series $\sum_{k=1}^{16} f_k$, where $f_k = f(k)$ is the arithmetic sequence associated with f.

9.4 Geometric sequences and series

Suppose that a ball is dropped from a height of 4 feet and bounces straight up and down, always bouncing up exactly one-half the distance it just came down. How far will the ball have traveled if you catch it after it reaches the top of the fifth bounce? The following figure will help you to answer this question. For the sake of clarity, the bounces have been separated in the figure.

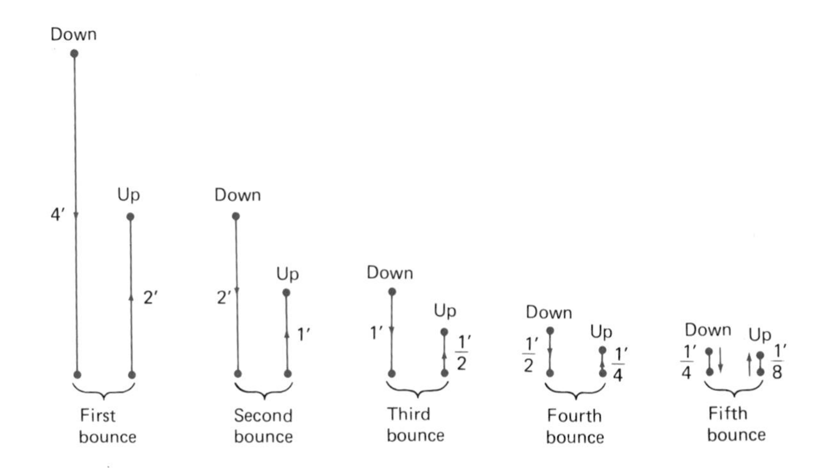

From this diagram we can determine how far the ball has traveled on each bounce. On the first bounce it goes 4 feet down and 2 feet up, for a total of 6 feet; on the second bounce the total distance is $2 + 1 = 3$ feet; and so on. These distances form the following sequence of five terms (one for each bounce):

$$6, \ 3, \ \tfrac{3}{2}, \ \tfrac{3}{4}, \ \tfrac{3}{8}$$

This sequence has the special property that, after the first term, each successive term can be obtained by multiplying the preceding term by $\frac{1}{2}$; that is, the second term, 3, is half the first, 6, and so on. This is an example of a *geometric sequence*. Later we will develop a formula for finding the sum of such a sequence; in the meantime, we can find the total distance the ball has traveled during the five bounces by adding the five terms as follows:

$$6 + 3 + \frac{3}{2} + \frac{3}{4} + \frac{3}{8} = \frac{48 + 24 + 12 + 6 + 3}{8} = 11\frac{5}{8}$$

A geometric sequence is also referred to as a *geometric progression.*

GEOMETRIC SEQUENCE

A sequence is said to be *geometric* if each term, after the first, is obtained by multiplying the preceding term by a common value.

Here are the first four terms of three different geometric sequences.

$$2, -4, 8, -16, \ldots$$
$$1, \tfrac{1}{3}, \tfrac{1}{9}, \tfrac{1}{27}, \ldots$$
$$5, -5, 5, -5, \ldots$$

By inspection you can determine that the common multipliers for these sequences are -2, $\frac{1}{3}$, and -1, respectively. We will find their nth terms by deriving the formula for the nth term of any geometric sequence.

Let s_n be the nth term of a geometric sequence, and let a be its first term. The common multiplier, which is also called the **common ratio**, is denoted by r. Here are the first five terms:

$$s_1 = a$$
$$s_2 = ar$$
$$s_3 = ar^2$$
$$s_4 = ar^3$$
$$s_5 = ar^4$$

Notice that the exponent of r is 1 less than the number of the term. This observation allows us to write the nth term as follows:

This formula says that the nth term of a geometric sequence is completely determined by its first term a and common ratio r.

The nth term of a geometric sequence is

$$s_n = ar^{n-1}$$

where a is the first term and r is the common ratio.

With this result, the first four terms and the nth terms of the three given geometric sequences are as follows:

$$2, -4, 8, -16, \ldots, 2(-2)^{n-1}$$
$$1, \tfrac{1}{3}, \tfrac{1}{9}, \tfrac{1}{27}, \ldots, 1(\tfrac{1}{3})^{n-1}$$
$$5, -5, 5, -5, \ldots, 5(-1)^{n-1}$$

You can substitute the values $n = 1, 2, 3$, and 4 into the forms for the nth terms and see that the given first four terms are obtained in each case.

EXAMPLE 1 Find the hundredth term of the geometric sequence having $r = \frac{1}{2}$ and $a = \frac{1}{2}$.

Solution The nth term of this sequence is given by

$$s_n = \frac{1}{2}\left(\frac{1}{2}\right)^{n-1}$$
$$= \frac{1}{2}\left(\frac{1}{2^{n-1}}\right)$$
$$= \frac{1}{2^n}$$

Thus

$$s_{100} = \frac{1}{2^{100}}$$

The reason r is called the common ratio of a geometric sequence $s_n = ar^{n-1}$ is that for each n the ratio of the $(n+1)$st term to the nth term equals r. Thus

$$\frac{s_{n+1}}{s_n} = \frac{ar^n}{ar^{n-1}} = r$$

EXAMPLE 2 Find the nth term of the geometric sequence beginning with

$$6, 9, \tfrac{27}{2}, \ldots$$

and find the seventh term.

Solution First find r.

$$r = \frac{s_2}{s_1} = \frac{9}{6} = \frac{3}{2}$$

r can also be found using s_2 and s_3.

$$\frac{s_3}{s_2} = \frac{\frac{27}{2}}{9} = \frac{27}{2 \cdot 9} = \frac{3}{2}$$

Then the nth term is

$$s_n = 6\left(\frac{3}{2}\right)^{n-1}$$

Let $n = 7$ to get

$$s_7 = 6\left(\frac{3}{2}\right)^{7-1} = 6\left(\frac{3}{2}\right)^6$$
$$= (3 \cdot 2) \cdot \frac{3^6}{2^6} = \frac{3^7}{2^5} = \frac{2187}{32}$$

EXAMPLE 3 Write the kth term of the geometric sequence $s_k = (\frac{1}{2})^{2k}$ in the form ar^{k-1} and find the value of a and r.

Solution

$$s_k = (\tfrac{1}{2})^{2k} = [(\tfrac{1}{2})^2]^k = (\tfrac{1}{4})^k$$
$$= \tfrac{1}{4}(\tfrac{1}{4})^{k-1} \quad \longleftarrow \quad \text{this is now in the form } ar^{n-1}$$

Then $a = \frac{1}{4}$ and $r = \frac{1}{4}$.

Note: The first term a can *also* be found by simply computing s_1 in the given formula for s_k; the value of r, however, is not so obvious in this original form.

TEST YOUR UNDERSTANDING

Write the first five terms of the geometric sequences with the given general term. Also write the nth term in the form ar^{n-1} and find r.

1 $s_n = (\frac{1}{2})^{n-1}$ **2** $s_n = (\frac{1}{2})^{n+1}$

3 $s_n = (-\frac{1}{2})^n$ **4** $s_n = (-\frac{1}{3})^{3n}$

Find r and the nth term of the geometric sequence with the given first two terms.

5 $\frac{1}{5}, 2$ **6** $27, -12$

Let us return to the original problem of this section. We found that the total distance the ball traveled was $11\frac{5}{8}$ feet. This is the sum of the first five terms of the geometric sequence whose nth term is $6(\frac{1}{2})^{n-1}$. Adding these five terms was easy. But what about adding the first 100 terms? There is a formula for the sum of a geometric sequence that will enable us to find such answers efficiently.

The sum of a geometric sequence is called a **geometric series**. Just as with arithmetic series, there is a formula for finding such sums. To discover this formula, let $s_k = ar^{k-1}$ be a geometric sequence and denote the sum of the first n terms by $S_n = \sum_{k=1}^{n} ar^{k-1}$. Then

$$S_n = a + ar + ar^2 + \cdots + ar^{n-2} + ar^{n-1}$$

Multiplying this equation by r gives

$$rS_n = ar + ar^2 + \cdots + ar^{n-1} + ar^n$$

Now consider these two equations:

$$\begin{aligned} S_n &= a + ar + ar^2 + \cdots + ar^{n-2} + ar^{n-1} \\ &\quad\;\;\;\; \updownarrow \quad\; \updownarrow \qquad\qquad\; \updownarrow \qquad\;\; \updownarrow \\ rS_n &= \qquad ar + ar^2 + \cdots + ar^{n-2} + ar^{n-1} + ar^n \end{aligned}$$

Subtract and factor:

$$S_n - rS_n = a - ar^n$$

$$(1 - r)S_n = a(1 - r^n)$$

Divide by $1 - r$ to solve for S_n:

Here $r \neq 1$. However, when $r = 1$, $s_k = ar^{k-1} = a$, which is an arithmetic sequence having $d = 0$.

$$S_n = \frac{a(1 - r^n)}{1 - r}$$

Returning to sigma notation, we can summarize our results this way:

GEOMETRIC SERIES

$$\sum_{k=1}^{n} ar^{k-1} = \frac{a(1 - r^n)}{1 - r}$$

This formula can be used to verify the earlier result for the bouncing ball:

$$\sum_{k=1}^{5} 6\left(\frac{1}{2}\right)^{k-1} = \frac{6[1-(\frac{1}{2})^5]}{1-\frac{1}{2}}$$

$$= \frac{6(1-\frac{1}{32})}{\frac{1}{2}}$$

$$= \frac{93}{8}$$

$$= 11\frac{5}{8}$$

EXAMPLE 4 Find the sum of the first 100 terms of the geometric sequence given by $s_k = 6(\frac{1}{2})^{k-1}$ and show that the answer is very close to 12.

Solution

$$S_{100} = \frac{6\left(1-\frac{1}{2^{100}}\right)}{1-\frac{1}{2}}$$

$$= 12\left(1-\frac{1}{2^{100}}\right)$$

Next observe that the fraction $\frac{1}{2^{100}}$ is so small that $1-\frac{1}{2^{100}}$ is very nearly equal to 1, and therefore S_{100} is very close to 12.

EXAMPLE 5 Evaluate $\sum_{k=1}^{8} 3(\frac{1}{10})^{k+1}$.

Solution We get $3(\frac{1}{10})^{k+1} = \frac{3}{100}(\frac{1}{10})^{k-1}$. Then $a = 0.03$, $r = 0.1$, and

$$S_8 = \frac{0.03[1-(0.1)^8]}{1-0.1}$$

$$= \frac{0.03(1-0.00000001)}{0.9}$$

$$= 0.033333333$$

Geometric sequences have many applications, as illustrated in the next example. You will find others in the exercises at the end of this section.

EXAMPLE 6 Suppose that you save \$128 in January and that each month thereafter you only manage to save half of what you saved the previous month. How much do you save in the tenth month, and what are your total savings after 10 months?

Solution The amounts saved each month form a geometric sequence with $a = 128$ and $r = \frac{1}{2}$. Then $s_n = 128(\frac{1}{2})^{n-1}$ and

$$s_{10} = 128\left(\frac{1}{2}\right)^9 = \frac{2^7}{2^9} = \frac{1}{4} = 0.25$$

This means that you saved 25¢ in the tenth month. Your total savings is given by

$$\begin{aligned} S_{10} &= \frac{128\left(1 - \frac{1}{2^{10}}\right)}{1 - \frac{1}{2}} \\ &= 256\left(1 - \frac{1}{2^{10}}\right) \\ &= 256 - \frac{256}{2^{10}} \\ &= 256 - \frac{2^8}{2^{10}} \\ &= 255.75 \end{aligned}$$

EXERCISES 9.4

The first three terms of a geometric sequence are given in Exercises 1–12. Write the next three terms and also find the formula for the nth term.

1 $2, 4, 8, \ldots$

2 $2, -4, 8, \ldots$

3 $1, 3, 9, \ldots$

4 $2, -2, 2, \ldots$

5 $-3, 1, -\frac{1}{3}, \ldots$

6 $100, 10, 1, \ldots$

7 $-1, -5, -25, \ldots$

8 $12, -6, 3, \ldots$

9 $-6, -4, -\frac{8}{3}, \ldots$

10 $-64, 16, -4, \ldots$

11 $\frac{1}{1000}, \frac{1}{10}, 10, \ldots$

12 $\frac{27}{8}, \frac{3}{2}, \frac{2}{3}, \ldots$

In Exercises 13–15, find the sum of the first six terms of the indicated sequence by using ordinary addition and also by using the formula for a geometric series.

13 The sequence in Exercise 1.

14 The sequence in Exercise 5.

15 The sequence in Exercise 9.

16 Find the tenth term of the geometric sequence $2, 4, 8, \ldots$.

17 Find the fourteenth term of the geometric sequence $\frac{1}{8}, \frac{1}{4}, \frac{1}{2}, \ldots$.

18 Find the fifteenth term of the geometric sequence $\frac{1}{100{,}000}, \frac{1}{10{,}000}, \frac{1}{1000}, \ldots$.

19 What is the one-hundred-first term of the geometric sequence having $a = 3$ and $r = -1$?

20 For the geometric sequence with $a = 100$ and $r = \frac{1}{10}$, use the formula $s_n = ar^{n-1}$ to find which term is equal to $\frac{1}{10^{10}}$.

Evaluate the series in Exercises 21–29.

21 $\sum_{k=1}^{10} 2^{k-1}$

22 $\sum_{j=1}^{10} 2^{j+2}$

23 $\sum_{k=1}^{n} 2^{k-1}$

24 $\sum_{k=1}^{8} 3(\frac{1}{10})^{k-1}$

25 $\sum_{k=1}^{5} 3^{k-4}$

26 $\sum_{k=1}^{6} (-3)^{k-2}$

27 $\sum_{j=1}^{5} (\frac{2}{3})^{j-2}$

28 $\sum_{k=1}^{8} 16(\frac{1}{2})^{k+2}$

29 $\sum_{k=1}^{8} 16(-\frac{1}{2})^{k+2}$

30 Find $u > 0$ such that $2, u, 98$ forms a geometric sequence.

31 Find $u < 0$ such that $\frac{1}{7}, u, \frac{25}{63}$ forms a geometric sequence.

32 Suppose that the amount you save in any given month is twice the amount you saved in the previous months. How much will you have saved at the end of 1 year if you save \$1 in January? How much if you saved 25¢ in January?

33 A certain bacterial culture doubles in number

every day. If there were 1000 bacteria at the end of the first day, how many will there be after 10 days? How many after n days?

***34** A radioactive substance is decaying so that at the end of each month there is only one-third as much as there was at the beginning of the month. If there were 75 grams of the substance at the beginning of the year, how much is left at midyear?

***35** Suppose that an automobile depreciates 10% in value each year for the first 5 years. What is it worth after 5 years if its original cost was \$5280? (*Hint:* Use $a = 5280$ and $n = 6$.)

***36** Compound-interest problems can be explained in terms of geometric sequences. The basic idea is that an investment P will earn i percent interest for the first year. Then the new total, consisting of the original investment P plus the interest Pi, will earn i percent interest for the second year, and so on. In such a situation we say that P earns i percent interest **compounded annually**.

(a) If \$1000 is invested in a bank paying 6% interest compounded annually, the amount at the end of the first year is

$$1000 + 1000(0.06) = 1000(1.06) = 1060$$

After 2 years the amount becomes

$$1060 + 1060(0.06) = 1060(1.06) = 1123.60$$

What is the amount after 3 years?

(b) Let A_1 be the amount after 1 year on an investment of P dollars at i percent interest compounded annually. Then

$$A_1 = P + Pi = P(1 + i)$$

After the second year:

$$\begin{aligned} A_2 &= P(1 + i) + P(1 + i)i \\ &= P(1 + i)(1 + i) \\ &= P(1 + i)^2 \end{aligned}$$

Find A_3.

(c) Referring to part (b), find the formula for A_n, the amount after n years, and show that this is the nth term of a geometric sequence having ratio $r = 1 + i$.

***37** A sum of \$800 is invested at 11% interest compounded annually.

(a) What is the amount after n years?

(b) What is the amount after 5 years? (Use common logarithms to get an approximation.)

***38** How much money must be invested at the interest rate of 12%, compounded annually, so that after 3 years the amount is \$1000?

***39** Use a calculator to find the amount of money that an investment of \$1500 earns at the interest rate of 8% compounded annually for 5 years.

9.5 Infinite geometric series

In decimal form the fraction $\frac{3}{4}$ becomes 0.75, which means $\frac{75}{100}$. This can also be written as $\frac{7}{10} + \frac{5}{100}$. What about $\frac{1}{3}$? As a decimal we can write

$$\tfrac{1}{3} = 0.333\ldots$$

where the dots mean that the 3 repeats endlessly. We can express this decimal as the sum of fractions whose denominators are powers of 10:

$$\tfrac{1}{3} = \tfrac{3}{10} + \tfrac{3}{100} + \tfrac{3}{1000} + \cdots$$

The sum of an infinite sequence is an infinite series.

The numbers being added here are the terms of the *infinite geometric sequence* with first term $a = \frac{3}{10}$ and common ratio $r = \frac{1}{10}$. Thus the nth term is

$$\begin{aligned} ar^{n-1} &= \frac{3}{10}\left(\frac{1}{10}\right)^{n-1} = 3\left(\frac{1}{10}\right)\left(\frac{1}{10}\right)^{n-1} \\ &= 3\left(\frac{1}{10}\right)^n \\ &= \frac{3}{10^n} \end{aligned}$$

The sum of the first n terms is found by using the formula

$$S_n = \sum_{k=1}^{n} ar^{k-1} = \frac{a(1-r^n)}{1-r}$$

Here are some cases:

$$S_1 = \frac{\frac{3}{10}\left(1-\frac{1}{10}\right)}{1-\frac{1}{10}} = \frac{1}{3}\left(1-\frac{1}{10}\right) = 0.3$$

$$S_2 = \frac{\frac{3}{10}\left(1-\frac{1}{10^2}\right)}{1-\frac{1}{10}} = \frac{1}{3}\left(1-\frac{1}{10^2}\right) = 0.33$$

$$S_3 = \frac{\frac{3}{10}\left(1-\frac{1}{10^3}\right)}{1-\frac{1}{10}} = \frac{1}{3}\left(1-\frac{1}{10^3}\right) = 0.333$$

$$S_{10} = \frac{\frac{3}{10}\left(1-\frac{1}{10^{10}}\right)}{1-\frac{1}{10}} = \frac{1}{3}\left(1-\frac{1}{10^{10}}\right) = 0.3333333333$$

$$S_n = \frac{\frac{3}{10}\left(1-\frac{1}{10^n}\right)}{1-\frac{1}{10}} = \frac{1}{3}\left(1-\frac{1}{10^n}\right) = 0.\underbrace{333\ldots3}_{n \text{ places}}$$

You can see that as more and more terms are added, the closer and closer the answer gets to $\frac{1}{3}$. This can be seen by studying the form for the sum of the first n terms:

$$S_n = \frac{1}{3}\left(1-\frac{1}{10^n}\right)$$

It is clear that the bigger n is, the closer $\frac{1}{10^n}$ is to zero, the closer $1-\frac{1}{10^n}$ is to 1 and, finally, the closer S_n is to $\frac{1}{3}$. Although it is true that S_n is never exactly equal to $\frac{1}{3}$, for very large n the difference between S_n and $\frac{1}{3}$ is very small. Saying this another way:

By taking n large enough, S_n can be made as close to $\frac{1}{3}$ as we like.

This is what we mean when we say that the sum of all the terms is $\frac{1}{3}$. In symbols:

$$\frac{3}{10}+\frac{3}{10^2}+\frac{3}{10^3}+\cdots+\frac{3}{10^n}+\cdots = \frac{1}{3}$$

The summation symbol, $\sum$, can also be used here after an adjustment in notation is made. Traditionally, the symbol ∞ has been used to suggest an infinite number of objects. So we use this and make the transition from the sum of a finite number of terms

$$S_n = \sum_{k=1}^{n} \frac{3}{10^k} = \frac{3}{10} + \frac{3}{10^2} + \cdots + \frac{3}{10^n} = \frac{1}{3}\left(1 - \frac{3}{10^n}\right)$$

to the sum of an infinite number of terms:

$$S_\infty = \sum_{k=1}^{\infty} \frac{3}{10^k} = \frac{3}{10} + \frac{3}{10^2} + \cdots + \frac{3}{10^n} + \cdots = \frac{1}{3}$$

In calculus the symbol S_∞ is replaced by $\lim_{n\to\infty} S_n = \frac{1}{3}$, which is read as "the limit of S_n as n gets arbitrarily large is $\frac{1}{3}$."

Not every geometric sequence produces an infinite geometric series that has a finite sum. For instance, the sequence

$$2, 4, 8, \ldots, 2^n, \ldots$$

is geometric, but the corresponding geometric series

$$2 + 4 + 8 + \cdots + 2^n + \cdots$$

cannot have a finite sum.

By now you might suspect that the common ratio r determines whether or not an infinite geometric sequence can be added. This turns out to be true. To see how this works, the general case will be considered next.

Let

$$a, ar, ar^2, \ldots, ar^{n-1}, \ldots$$

be an infinite geometric sequence. The sum of the first n terms is

$$S_n = \frac{a(1 - r^n)}{1 - r}$$

Rewrite in this form:

$$S_n = \frac{a}{1 - r}(1 - r^n)$$

At this point the importance of r^n becomes clear. If, as n gets larger, r^n gets very large, then the infinite geometric series will not have a finite sum. But when r^n gets arbitrarily close to zero as n gets larger, then S_n gets closer and closer to $\frac{a}{1 - r}$.

Use a calculator to verify the powers of $r = 0.9$ and $r = 1.1$ to the indicated decimal places.

$(0.9)^1 = 0.9$
$(0.9)^{10} = 0.35$
$(0.9)^{20} = 0.12$
$(0.9)^{40} = 0.015$
$(0.9)^{80} = 0.0002$
$(0.9)^{100} = 0.00003$
↓
getting close to 0

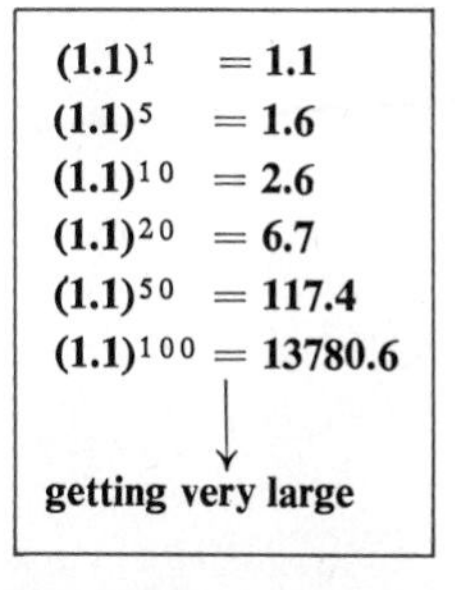
$(1.1)^1 = 1.1$
$(1.1)^5 = 1.6$
$(1.1)^{10} = 2.6$
$(1.1)^{20} = 6.7$
$(1.1)^{50} = 117.4$
$(1.1)^{100} = 13780.6$
↓
getting very large

The values of r for which r^n gets arbitrarily close to zero are precisely those values between -1 and 1; that is, $|r| < 1$. For instance, $\frac{3}{5}$, $-\frac{1}{10}$, and 0.09 are values of r for which r^n gets close to zero; and 1.01, -2, and $\frac{3}{2}$ are values for which r^n does not get close to zero.

To sum up, we have the following useful result:

SUM OF AN INFINITE GEOMETRIC SERIES

If $|r| < 1$, then $\sum_{k=1}^{\infty} ar^{k-1} = \frac{a}{1 - r}$. For other values of r the series has no finite sum.

EXAMPLE 1 Find the sum of the infinite geometric series

$$27 + 3 + \frac{1}{3} + \cdots$$

Solution Since $r = \frac{3}{27} = \frac{1}{9}$ and $a = 27$, the preceding result gives

$$27 + 3 + \frac{1}{3} + \cdots = \frac{27}{1 - \frac{1}{9}} = \frac{27}{\frac{8}{9}}$$

$$= \frac{243}{8}$$

EXAMPLE 2 Why does the infinite geometric series $\sum_{k=1}^{\infty} 5(\frac{4}{3})^{k-1}$ have no finite sum?

Solution The series has no finite sum because the common ratio $r = \frac{4}{3}$ is not between -1 and 1.

EXAMPLE 3 Find: $\sum_{k=1}^{\infty} \frac{7}{10^{k+1}}$.

Solution Since $\frac{7}{10^{k+1}} = 7\left(\frac{1}{10^{k+1}}\right) = 7\left(\frac{1}{10^2}\right)\left(\frac{1}{10^{k-1}}\right) = \frac{7}{100}\left(\frac{1}{10}\right)^{k-1}$, it follows that $a = \frac{7}{100}$ and $r = \frac{1}{10}$. Therefore, by the formula for the sum of an infinite geometric series we have

$$S_\infty = \sum_{k=1}^{\infty} \frac{7}{10^{k+1}} = \frac{\frac{7}{100}}{1 - \frac{1}{10}} = \frac{7}{100 - 10}$$

$$= \frac{7}{90}$$

TEST YOUR UNDERSTANDING

Find the common ratio r, and then find the sum if the given infinite geometric series has one.

1 $10 + 1 + \frac{1}{10} + \cdots$

2 $\frac{1}{64} + \frac{1}{16} + \frac{1}{4} + \cdots$

3 $36 - 6 + \frac{1}{6} - \cdots$

4 $-16 - 4 - 1 - \cdots$

5 $\sum_{k=1}^{\infty} (\frac{4}{3})^{k-1}$

6 $\sum_{k=1}^{\infty} 3(0.01)^k$

7 $\sum_{i=1}^{\infty} (-1)^i 3^i$

8 $\sum_{n=1}^{\infty} 100(-\frac{9}{10})^{n+1}$

9 $101 - 102.01 + 103.0301 - \cdots$

The introduction to this section indicated how the endless repeating decimal 0.333. . . can be regarded as an infinite geometric series. The next example illustrates how such decimal fractions can be written in the rational form $\frac{a}{b}$ (the ratio of two integers) by using the formula for the sum of an infinite geometric series.

EXAMPLE 4 Express the repeating decimal 0.242424. . . in rational form.

Compare the method shown in Example 4 with that developed in Exercise 45 of Section 1.2.

Solution First write

$$0.242424\ldots = \frac{24}{100} + \frac{24}{10{,}000} + \frac{24}{1{,}000{,}000} + \cdots$$
$$= \frac{24}{10^2} + \frac{24}{10^4} + \frac{24}{10^6} + \cdots + \frac{24}{10^{2k}} + \cdots$$
$$= \sum_{k=1}^{\infty} \frac{24}{10^{2k}}$$

Next observe that

$$\frac{24}{10^{2k}} = 24\left(\frac{1}{10^{2k}}\right) = 24\left(\frac{1}{10^2}\right)^k$$
$$= 24\left(\frac{1}{100}\right)^k = \frac{24}{100}\left(\frac{1}{100}\right)^{k-1}$$

Then $a = \frac{24}{100}$, $r = \frac{1}{100}$, and

$$0.242424\ldots = \sum_{k=1}^{\infty} \frac{24}{10^{2k}}$$
$$= \frac{\frac{24}{100}}{1 - \frac{1}{100}}$$
$$= \frac{24}{99} = \frac{8}{33}$$

You can check this answer by dividing 33 into 8.

EXAMPLE 5 A racehorse running at the constant rate of 30 miles per hour will finish a 1-mile race in 2 minutes. Now consider the race broken down into the following parts: before the racehorse can finish the 1-mile race it must first reach the halfway mark; having done that, the horse must next reach the quarter pole; then it must reach the eighth pole; and so on. That is, it must always cover half the distance remaining before it can cover the whole distance. Show that the sum of the infinite number of time intervals is also 2 minutes.

It seems as if the horse cannot finish the race this way. But read on to see that there is really no contradiction with this interpretation.

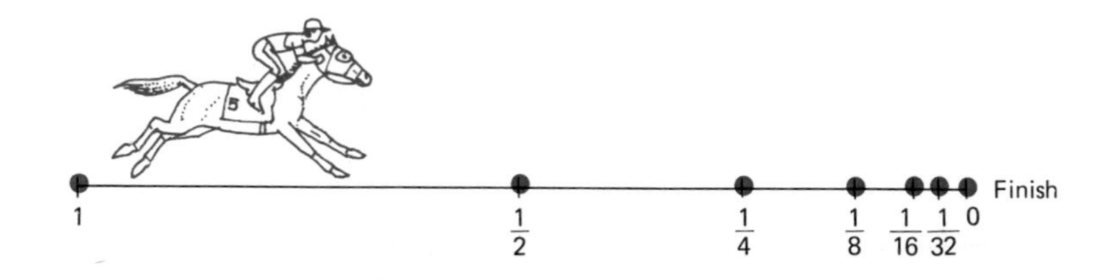

Solution For the first $\frac{1}{2}$ mile the time will be $\frac{\frac{1}{2}}{\frac{1}{2}} = 1$ minute; for the next $\frac{1}{4}$ mile the time will be $\frac{\frac{1}{4}}{\frac{1}{2}} = \frac{1}{2}$ minute; for the next $\frac{1}{8}$ mile the time will be

$T = \frac{D}{R}\left(\text{time} = \frac{\text{distance}}{\text{rate}}\right)$

$\frac{\frac{1}{8}}{\frac{1}{2}} = \frac{1}{4}$ minute; and for the nth distance, which is $\frac{1}{2^n}$ miles, the time will be $\frac{\frac{1}{2^n}}{\frac{1}{2}} = \frac{1}{2^{n-1}}$.

Thus the total time is given by this series:

$$\sum_{k=1}^{\infty} \frac{1}{2^{k-1}} = 1 + \frac{1}{2} + \frac{1}{4} + \cdots + \frac{1}{2^{n-1}} + \cdots$$

This is an infinite geometric series having $a = 1$ and $r = \frac{1}{2}$. Thus

$$\sum_{k=1}^{\infty} 1\left(\frac{1}{2}\right)^{k-1} = \frac{1}{1 - \frac{1}{2}} = 2$$

which is the same result as before.

CAUTION: LEARN TO AVOID MISTAKES LIKE THESE

WRONG	RIGHT
$\sum_{k=1}^{\infty} \left(\frac{1}{2}\right)^{n+1} = \frac{1}{1 - \frac{1}{2}}$	$\sum_{n=1}^{\infty} \left(\frac{1}{2}\right)^{n+1} = \sum_{n=1}^{\infty} \frac{1}{4}\left(\frac{1}{2}\right)^{n-1} = \frac{\frac{1}{4}}{1 - \frac{1}{2}}$
$\sum_{n=1}^{\infty} \left(-\frac{1}{3}\right)^{n-1} = \frac{1}{1 - \frac{1}{3}}$	$\sum_{n=1}^{\infty} \left(-\frac{1}{3}\right)^{n-1} = \frac{1}{1 - \left(-\frac{1}{3}\right)}$
$\sum_{n=1}^{\infty} 3(1.02)^{n-1} = \frac{3}{1 - 1.02}$	$\sum_{n=1}^{\infty} 3(1.02)^{n-1}$ is not a finite sum since $r = 1.02 > 1$

EXERCISES 9.5

Find the sum, if it exists, of each infinite geometric series.

1 $2 + 1 + \frac{1}{2} + \cdots$

2 $8 + 4 + 2 + \cdots$

3 $25 + 5 + 1 + \cdots$

4 $1 + \frac{4}{3} + \frac{16}{9} + \cdots$

5 $1 - \frac{1}{2} + \frac{1}{4} - \cdots$

6 $100 - 1 + \frac{1}{100} - \cdots$

7 $1 + 0.1 + 0.01 + \cdots$

8 $52 + 0.52 + 0.0052 + \cdots$

9 $-2 - \frac{1}{4} - \frac{1}{32} - \cdots$

10 $-729 + 81 - 9 + \cdots$

Decide whether or not the given infinite geometric series has a sum. If it does, find it using $S_\infty = \frac{a}{1 - r}$.

11 $\sum_{k=1}^{\infty} (\frac{1}{3})^{k-1}$

12 $\sum_{k=1}^{\infty} (\frac{1}{3})^{k}$

13 $\sum_{k=1}^{\infty} (\frac{1}{3})^{k+1}$

14 $\sum_{n=1}^{\infty} \frac{1}{2^{n+1}}$

15 $\sum_{n=1}^{\infty} \frac{1}{2^{n-2}}$

16 $\sum_{k=1}^{\infty} (\frac{1}{10})^{k-1}$

17 $\sum_{k=1}^{\infty} 2(0.1)^{k-1}$

18 $\sum_{k=1}^{\infty} (-\frac{1}{2})^{k-1}$

19 $\sum_{n=1}^{\infty} (\frac{3}{2})^{n-1}$

20 $\sum_{n=1}^{\infty} (-\frac{1}{3})^{n+2}$

21 $\sum_{k=1}^{\infty} (0.7)^{k-1}$

22 $\sum_{k=1}^{\infty} 5(0.7)^{k}$

23 $\sum_{k=1}^{\infty} 5(1.01)^{k}$

24 $\sum_{k=1}^{\infty} (\frac{1}{10})^{k-4}$

25 $\sum_{k=1}^{\infty} 10(\frac{2}{3})^{k-1}$

26 $\sum_{k=1}^{\infty} (-1)^{k}$

27 $\sum_{k=1}^{\infty} (0.45)^{k-1}$

28 $\sum_{k=1}^{\infty} (-0.9)^{k+1}$

29 $\sum_{n=1}^{\infty} 7(-\frac{3}{4})^{n-1}$

30 $\sum_{k=1}^{\infty} (0.1)^{2k}$

31 $\sum_{k=1}^{\infty} (-\frac{2}{5})^{2k}$

Find a rational form for each of the following repeating decimals in a manner similar to that in Example 4. Check your answers.

32 0.444 . . .

33 0.777 . . .

34 7.777 . . .

35 0.131313 . . .

36 13.131313 . . .

37 0.0131313 . . .

38 0.050505 . . .

39 0.999 . . .

40 0.125125125 . . .

***41** Suppose that a 1-mile distance a racehorse must run is divided into an infinite number of parts, obtained by always considering $\frac{2}{3}$ of the remaining distance to be covered. Then the lengths of these parts form the sequence $\frac{2}{3}, \frac{2}{9}, \frac{2}{27}, \ldots, \frac{2}{3^n}, \ldots$.

(a) Find the sequence of times corresponding to these distances. (Assume that the horse is moving at a rate of $\frac{1}{2}$ mile per minute.)

(b) Show that the sum of the times in part (a) is 2 minutes.

***42** A certain ball always rebounds $\frac{1}{3}$ the distance it falls. If the ball is dropped from a height of 9 feet, how far does it travel before coming to rest? (See the similar situation at the beginning of Section 10.4)

***43** A substance initially weighing 64 grams is decaying at a rate such that after 4 hours there are only 32 grams left. In another 2 hours only 16 grams remain; in another 1 hour after that only 8 grams remain; and so on. How long does it take altogether until nothing of the substance is left?

***44** After it is set in motion, each swing in either direction of a particular pendulum is 40% as long as the preceding swing. What is the total distance that the end of the pendulum travels before coming to rest if the first swing is 30 inches long?

***45** Assume that a racehorse takes 1 minute to go the first $\frac{1}{2}$ mile of a 1-mile race. After that, the horse's speed is no longer constant: for the next $\frac{1}{4}$ mile it takes $\frac{2}{3}$ minute; for the next $\frac{1}{8}$ mile it takes $\frac{4}{9}$ minute; for the next $\frac{1}{16}$ mile it takes $\frac{40}{81}$ minute; and so on, so that the time intervals form a geometric sequence. Why can't the horse finish the race?

9.6 Mathematical induction

Study these statements:

$$1 = 1^2$$
$$1 + 3 = 2^2$$
$$1 + 3 + 5 = 3^2$$
$$1 + 3 + 5 + 7 = 4^2$$
$$1 + 3 + 5 + 7 + 9 = 5^2$$

Do you see the pattern? The last statement shows that the sum of the first five positive odd integers is 5^2. What about the sum of the first six positive odd integers? The pattern is the same:

$$1 + 3 + 5 + 7 + 9 + 11 = 6^2$$

It would be reasonable to guess that the sum of the first n positive odd integers is n^2. That is,

$$1 + 3 + 5 + \cdots + (2n - 1) = n^2$$

But a guess is not a proof. It is our objective in this section to learn how to prove a statement that involves an infinite number of cases.

Let us refer to the nth statement above by $\mathcal{S}_n$. Thus $\mathcal{S}_1$, $\mathcal{S}_2$, $\mathcal{S}_3$, $\mathcal{S}_4$, $\mathcal{S}_5$, and $\mathcal{S}_6$ are the first six cases of $\mathcal{S}_n$ that we know are true.

Does the truth of the first six cases allow us to conclude that $\mathcal{S}_n$ is true for all positive integer n? No! We cannot assume that a few special cases guarantee an infinite number of cases. If we allowed "proving by a finite number of cases," then the following example is such a "proof" that all positive even integers are less than 100.

> The first positive even integer is 2, and we know that $2 < 100$. The second is 4, and we know that $4 < 100$. The third is 6, and $6 < 100$. Therefore, since $2n < 100$ for a finite number of cases, we might conclude that $2n < 100$ for all n.

This false result should convince you that in trying to prove a collection of statements $\mathcal{S}_n$ for all positive integers $n = 1, 2, 3, \ldots$, we need to do more than just check it out for a finite number of cases. We need to call on a type of proof known as **mathematical induction**.

Suppose that we had a long (endless) row of dominoes each 2 inches long all standing up in a straight row so that the distance between any two of them is $1\frac{1}{2}$ inches. How can you make them all fall down with the least effort?

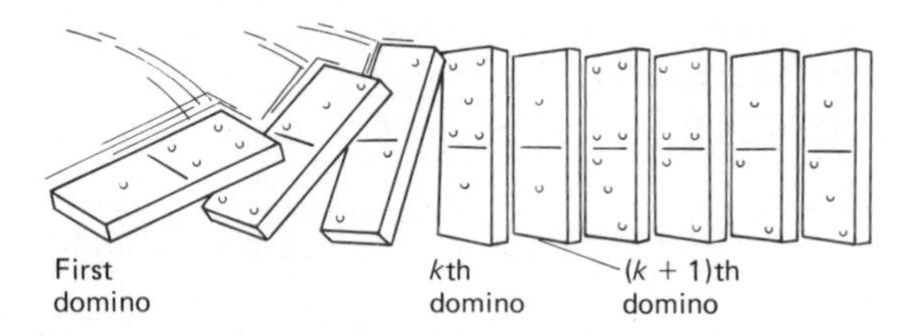

The answer is obvious. Push the first domino down toward the second. Since the first one must fall, and because the space between each pair is less than the length of a domino, they will all (eventually) fall down. The first knocks down the second, the second knocks down the third, and, in general, the kth domino knocks down the $(k + 1)$st. Two things guaranteed this "chain reaction":

1 The first domino will fall.

2 If any domino falls, then so will the next.

These two conditions are guidelines in forming the **principle of mathematical induction**.

THE PRINCIPLE OF MATHEMATICAL INDUCTION

Let $\mathcal{S}_n$ be a statement for each positive integer n. Suppose that the following two conditions hold:

1 $\mathcal{S}_1$ is true.

2 If $\mathcal{S}_k$ is true, then $\mathcal{S}_{k+1}$ is true, where k is any positive integer.

Then $\mathcal{S}_n$ is a true statement for all n.

Condition 1 starts the "chain reaction" and condition 2 keeps it going.

Note that we are not proving this principle; rather, it is a basic principle that we accept and use to construct proofs. It is very important to realize that in condition 2 we are not proving $\mathcal{S}_k$ to be true; rather, we must prove this proposition:

$$\text{If } \mathcal{S}_k \text{ is true, then } \mathcal{S}_{k+1} \text{ is true}$$

Consequently, a proof by mathematical induction *includes* a proof of the proposition that $\mathcal{S}_k$ implies $\mathcal{S}_{k+1}$, a proof within a proof. Within that inner proof, we are allowed to *assume* and use $\mathcal{S}_k$.

EXAMPLE 1 Prove by mathematical induction that $\mathcal{S}_n$ is true for all positive integers n, where $\mathcal{S}_n$ is the statement

$$1 + 3 + 5 + \cdots + (2n - 1) = n^2$$

At the beginning of this section, we guessed at the formula for the sum of the squares of the first n odd integers. Now, using mathematical induction, this formula is proved in Example 1.

Proof Both conditions 1 and 2 of the principle of mathematical induction must be satisfied. We begin with the first.

(1) $\mathcal{S}_1$ is true because $1 = 1^2$.

(2) Suppose that $\mathcal{S}_k$ is true, where k is a positive integer. That is, we assume

$$1 + 3 + 5 + \cdots + (2k - 1) = k^2$$

Proving S_1 starts the chain. The first domino has fallen.

We assume $\mathcal{S}_k$ to be true to see what effect it has on the next case, $\mathcal{S}_{k+1}$. This is comparable to considering what happens when the kth domino falls.

The next odd number after $2k - 1$ is $2(k + 1) - 1 = 2k + 1$, which is added to the preceding equation:

$$\begin{array}{rl} 1 + 3 + 5 + \cdots + (2k - 1) &= k^2 \\ 2k + 1 &= 2k + 1 \\ \hline 1 + 3 + 5 + \cdots + (2k - 1) + (2k + 1) &= k^2 + 2k + 1 \end{array}$$

Factor the right side.

$$1 + 3 + 5 + \cdots + (2k + 1) = (k + 1)^2$$

This is the statement $\mathcal{S}_{k+1}$. Therefore, we have shown that if $\mathcal{S}_k$ is given, then $\mathcal{S}_{k+1}$ must follow. This, together with the fact that $\mathcal{S}_1$ is true, allows us to say that $\mathcal{S}_n$ is true for all n by the principle of mathematical induction.

Now that $\mathcal{S}_k$ implies $\mathcal{S}_{k+1}$; the chain reaction keeps going.

EXAMPLE 2 Prove that the sum of the squares of the first n consecutive positive integers is given by $\dfrac{n(n+1)(2n+1)}{6}$.

Proof Let $\mathcal{S}_n$ be the statement

$$\underbrace{1^2 + 2^2 + 3^2 + \cdots + n^2}_{\text{(sum of first } n \text{ squares)}} = \frac{n(n+1)(2n+1)}{6}.$$

for any positive integer n.

(1) $\mathcal{S}_k$ is true because $1^2 = \dfrac{1(1+1)(2 \cdot 1 + 1)}{6}$.

(2) Suppose that $\mathcal{S}_k$ is true for any k. That is, we assume

$$1^2 + 2^2 + 3^2 + \cdots + k^2 = \frac{k(k+1)(2k+1)}{6}$$

Add the next square $(k+1)^2$ to both sides.

$$\text{(A)}\quad 1^2 + 2^2 + 3^2 + \cdots + k^2 + (k+1)^2 = \frac{k(k+1)(2k+1)}{6} + (k+1)^2$$

Combine the right side.

$$\frac{k(k+1)(2k+1)}{6} + (k+1)^2 = \frac{k(k+1)(2k+1) + 6(k+1)^2}{6}$$

$$= \frac{2k^3 + 9k^2 + 13k + 6}{6}$$

$$= \frac{(k+1)(k+2)(2k+3)}{6}$$

Substitute back into Equation (A).

$$1^2 + 2^2 + 3^2 + \cdots + (k+1)^2 = \frac{(k+1)(k+2)(2k+3)}{6}$$

To see that this is $\mathcal{S}_{k+1}$, rewrite the right side:

$$1^2 + 2^2 + 3^2 + \cdots + (k+1)^2 = \frac{(k+1)[(k+1)+1][2(k+1)+1]}{6}$$

Now both conditions of the principle of mathematical induction have been satisfied, and it follows that

$$1^2 + 2^2 + 3^2 + \cdots + n^2 = \frac{n(n+1)(2n+1)}{6}$$

is true for all integers $n \geq 1$.

Examples 1 and 2 involved equations that were established by mathematical induction. However, this principle is also used for other types of mathematical situations. The next example demonstrates the application of this principle when an inequality is involved.

EXAMPLE 3 Let $t > 0$ and use mathematical induction to prove that $(1 + t)^n > 1 + nt$ for all positive integers $n \geq 2$.

Try a few specific cases. Use a calculator to verify these:

$$(1.02)^2 > 1 + 2(0.02)$$
$$(1.001)^2 > 1 + 2(0.001)$$
$$(1.00054)^2 > 1 + 2(0.00054)$$

Proof Let $\mathcal{S}_n$ be the statement $(1 + t)^n > 1 + nt$, where $t > 0$ and n is any integer where $n \geq 2$.

(1) For $n = 2$, $(1 + t)^2 = 1 + 2t + t^2$. Since $t^2 > 0$, we get

$$1 + 2t + t^2 > 1 + 2t$$

because

$$(1 + 2t + t^2) - (1 + 2t) = t^2 > 0$$

Recall that $a > b$ if and only if $a - b > 0$.

(2) Suppose that for $k \geq 2$ we have $(1 + t)^k > 1 + kt$. Multiply both sides by $(1 + t)$.

$$(1 + t)^k(1 + t) > (1 + kt)(1 + t)$$

Then

$$(1 + t)^{k+1} > 1 + (k + 1)t + kt^2$$

But $1 + (k + 1)t + kt^2 > 1 + (k + 1)t$. Therefore, by transitivity of $>$,

$$(1 + t)^{k+1} > 1 + (k + 1)t$$

By (1) and (2) above, the principle of mathematical induction implies that $(1 + t)^n > 1 + nt$ for all integers $n \geq 2$.

EXERCISES 9.6

Use mathematical induction to prove the following statements for all positive integers n.

1 $1 + 2 + 3 + \cdots + n = \dfrac{n(n + 1)}{2}$

2 $2 + 4 + 6 + \cdots + 2n = n(n + 1)$

3 $\sum_{i=1}^{n} 3i = \dfrac{3n(n + 1)}{2}$

4 $1 + 4 + 7 + \cdots + (3n - 2) = \dfrac{n(3n - 1)}{2}$

5 $\dfrac{5}{3} + \dfrac{4}{3} + 1 + \cdots + \left(-\dfrac{1}{3}n + 2\right) = \dfrac{n(11 - n)}{6}$

6 $1 \cdot 2 + 2 \cdot 3 + 3 \cdot 4 + \cdots + n(n + 1) = \dfrac{n(n + 1)(n + 2)}{3}$

7 $\dfrac{1}{1 \cdot 2} + \dfrac{1}{2 \cdot 3} + \dfrac{1}{3 \cdot 4} + \cdots + \dfrac{1}{n(n + 1)} = \dfrac{n}{n + 1}$

8 $3 + 3^2 + 3^3 + \cdots + 3^n = \dfrac{3^{n+1} - 3}{2}$

9 $-2 - 4 - 6 - \cdots - 2n = -n - n^2$

10 $1 + \dfrac{1}{2} + \dfrac{1}{2^2} + \cdots + \dfrac{1}{2^{n-1}} = 2\left(1 - \dfrac{1}{2^n}\right)$

11 $1 + \frac{2}{5} + \frac{4}{25} + \cdots + (\frac{2}{5})^{n-1} = \frac{5}{3}[1 - (\frac{2}{5})^n]$

12 $1 - \frac{1}{3} + \frac{1}{9} - \cdots + (-\frac{1}{3})^{n-1} = \frac{3}{4}[1 - (-\frac{1}{3})^n]$

13 $1^3 + 2^3 + 3^3 + \cdots + n^3 = \dfrac{n^2(n + 1)^2}{4}$

14 Which of Exercises 1–13 can also be proved using the formula for an arithmetic series? Which ones can be proved using the formula for a geometric series?

Use mathematical induction to prove the following for all positive integers $n \geq 1$.

15 $\sum_{i=1}^{n} ar^{i-1} = \dfrac{a(1 - r^n)}{1 - r}, r \neq 1$

16 $\sum_{i=1}^{n} [a + (i - 1)d] = \dfrac{n}{2}[2a + (n - 1)d]$

17 $a^n < 1$, where $0 < a < 1$.

18 Let $0 < a < 1$. Use mathematical induction to prove that $a^n < a$ for all integers $n \geq 2$.

19 Let a and b be real numbers. Then use mathematical induction to prove that $(ab)^n = a^n b^n$ for all positive integers n.

20 Use mathematical induction to prove the generalized distributive property

$$a(b_0 + b_1 + \cdots + b_n) = ab_0 + ab_1 + \cdots + ab_n$$

for all positive integers n, where a and b_i are real numbers. (Assume that parentheses may be inserted into or extracted from an indicated sum of real numbers.)

*21 Give an inductive proof of

$$|a_0 + a_1 + \cdots + a_n| \leq |a_0| + |a_1| + \cdots + |a_n|$$

for all positive integers n, where the a_i are real numbers.

*22 Use induction to prove that if $a_0 a_1 \cdots a_n = 0$, then at least one of the factors is zero for all positive integers n. (Assume that parentheses may be inserted into or extracted from an indicated product of real numbers.)

*23 (a) Prove by induction that

$$\frac{a^n - b^n}{a - b} = a^{n-1} + a^{n-2}b + \cdots + ab^{n-2} + b^{n-1}$$

for all integers $n \geq 2$.

$$\left(\textit{Hint: } \text{Consider } \frac{a^{n+1} - b^{n+1}}{a - b} = \frac{a^n a - b^n b}{a - b} = \frac{a^n a - b^n a + b^n a - b^n b}{a - b}\right)$$

(b) How does the result in part (a) give the factorization of the difference of two nth powers?

*24 Let u_n be the sequence such that $u_1 = 1$, $u_2 = 1$, and $u_{n+1} = u_{n-1} + u_n$ for $n \geq 2$. (This is the Fibonacci sequence; see Exercise 34, page 327.) Use mathematical induction to prove that for any positive integer n, $\sum_{i=1}^{n} u_i = u_{n+2} - 1$.

In Exercises 25 and 26, you are first asked to investigate a pattern leading to a formula and then asked to prove the formula by mathematical induction.

25 (a) Complete the statements.

$$1 =$$

$$1 + 2 + 1 =$$

$$1 + 2 + 3 + 2 + 1 =$$

$$1 + 2 + 3 + 4 + 3 + 2 + 1 =$$

$$1 + 2 + 3 + 4 + 5 + 4 + 3 + 2 + 1 =$$

(b) Use the results in part (a) to *guess* the sum in the general case.

$$1 + 2 + 3 + \cdots + (n-1) + n + (n-1) + \cdots + 3 + 2 + 1 =$$

(c) Prove part (b) by mathematical induction.

26 (a) The left-hand column contains a number of points in a plane, no three of which are collinear. The right-hand column contains the number of distinct lines that the points determine. Complete this information for the last three cases.

Number of Points	Number of Lines
2	1
3	3
4	
5	
6	

(b) Observe that when there are 4 points, there are $6 = \frac{4(3)}{2}$ lines. Write similar statements when there are 2, 3, 5, or 6 points.

(c) Conjecture (guess) what the number of distinct lines is for n points, no three of which are collinear. Prove the conjecture by mathematical induction.

REVIEW EXERCISES FOR CHAPTER 9

The solutions to the following exercises can be found within the text of Chapter 9. Try to answer each question without referring to the text.

Section 9.1

1 Find the range values of the sequence $s_n = \frac{1}{n}$ for the domain {1, 2, 3, 4, 5,} and graph.

2 List the first seven terms of the sequence $s_k = \frac{(-1)^k}{k^2}$.

3 Find the tenth term of the sequence $s_n = \frac{n+2}{2n-3}$.

4 Write the first four terms of the sequence $s_n = \left(1 + \frac{1}{n}\right)^n$ and round off the terms to two decimal places.

Section 9.2

5 Find the sum of the first seven terms of $s_k = 2k$.

6 Find $\sum_{n=1}^{5} s_n$, where $s_n = \frac{2}{n}$.

7 Find $\sum_{k=1}^{4} (2k + 1)$.

8 Find $\sum_{i=1}^{5} (-1)^i(i + 1)$.

Section 9.3

9 What is the nth term of an arithmetic sequence whose first term is a and whose common difference is d?

10 Find the nth term of the arithmetic sequence 11, 2, $-7, \ldots$.

11 Write the formula for the sum S_n of the arithmetic sequence $s_k = a + (k - 1)d$.

12 Find S_{20} for the arithmetic sequence with $a = 3$ and $d = 5$.

13 Find the sum of the first 10,000 terms of the arithmetic sequence 246, 261, 276, ...

14 Find the sum of the first n positive integers.

15 Find $\sum_{k=1}^{50} (-6k + 10)$.

Section 9.4

16 What is a geometric sequence?

In Exercises 17–19, write the nth term of the geometric sequence.

17 2, -4, 8, . . .

18 5, -5, 5, . . .

19 6, 9, $\frac{27}{2}$, . . .

20 Find the hundredth term of the geometric sequence having $r = \frac{1}{2}$ and $a = \frac{1}{2}$.

21 Write the kth term of the geometric sequence $s_k = (\frac{1}{2})^{2k}$ in the form ar^{k-1} and find the values of a and r.

22 Write the formula for the sum S_n of a geometric sequence $s_k = ar^{k-1}$.

23 Find the sum of the first 100 terms of the geometric sequence $s_k = 6(\frac{1}{2})^{k-1}$.

24 Evaluate $\sum_{k=1}^{8} 3\left(\frac{1}{10}\right)^{k+1}$.

25 Suppose that you save \$128 in January and that each month thereafter you only manage to save half of what you saved the previous month. How much do you save in the tenth month? What are your total savings after 10 months?

Section 9.5

26 For which values of r does $\sum_{k=1}^{\infty} ar^{k-1} = \frac{a}{1-r}$?

27 Find the sum of the infinite geometric series $27 + 3 + \frac{1}{3} + \cdots$.

28 Why does the infinite series $\sum_{k=1}^{\infty} 5(\frac{4}{3})^{k-1}$ have no finite sum?

29 Evaluate $\sum_{k=1}^{\infty} \frac{7}{10^{k+1}}$.

30 Express the repeating decimal 0.242424 . . . in rational form (the ratio of two integers).

31 A racehorse running at the constant rate of 30 miles per hour will finish a 1-mile race in 2 minutes. Now consider the race broken down into the following parts. Before the racehorse can finish the 1-mile race it must first reach the halfway mark; having done that, the horse must next reach the quarter pole; then it must reach the eighth pole; and so on. That is, it must always cover half the distance before it can cover the whole distance. Show that the sum of the infinite number of time intervals is also 2 minutes.

Section 9.6

32 To prove that a statement $\mathcal{S}_n$ is true for each positive integer n, what two conditions must be established according to the principle of mathematical induction?

33 Prove by mathematical induction:

$$1 + 3 + 5 + \cdots + (2n - 1) = n^2$$

34 Use mathematical induction to prove:

$$1^2 + 2^2 + 3^2 + \cdots + n^2 = \frac{n(n + 1)(2n + 1)}{6}$$

35 Let $t > 0$ and use mathematical induction to prove that $(1 + t)^n > 1 + nt$ for all positive integers $n \geq 2$.

SAMPLE TEST QUESTIONS FOR CHAPTER 9

Use these questions to test your knowledge of the basic skills and concepts of Chapter 9. Then check your answers with those given at the end of the book.

1 Find the first four terms of the sequence given by $s_n = \frac{n^2}{6 - 5n}$.

2 Find the tenth term of the sequence in Question 1.

In questions 3 and 4, an arithmetic sequence has $a = -3$ and $d = \frac{1}{2}$.

3 Find the forty-ninth term.

4 What is the sum of the first 20 terms?

5 Write the next three terms of the geometric sequence beginning $-768, 192, -48, \ldots$.

6 Write the nth term of the sequence in Question 5.

7 Use the formula for the sum of a finite geometric sequence to show that

$$\sum_{k=1}^{4} 8(\tfrac{1}{2})^k = \tfrac{15}{2}$$

8 Find $\sum_{j=1}^{101} (4j - 50)$.

Decide whether each of the given infinite geometric series in Exercises 9–11 has a sum. Find the sum if it exists; otherwise, give a reason why there is no sum.

9 $\sum_{k=1}^{\infty} 8(\frac{3}{4})^{k+1}$

10 $1 + \frac{3}{2} + \frac{9}{4} + \cdots$

11 $0.06 - 0.009 + 0.00135 - \cdots$

12 Change the repeating decimal $0.363636\ldots$ into rational form.

13 Suppose you save \$10 one week and that each week thereafter you save 10¢ more than the week before. How much will you have saved after 1 year?

14 An object is moving along a straight line such that each minute it travels one-third as far as it did during the preceding minute. How far will the object have moved before coming to rest if it moves 24 feet during the first minute?

15 Prove by mathematical induction:

$$5 + 10 + 15 + \cdots + 5n = \frac{5n(n+1)}{2}$$

16 Prove by mathematical induction:

$$\frac{1}{1 \cdot 3} + \frac{1}{3 \cdot 5} + \frac{1}{5 \cdot 7} + \cdots + \frac{1}{(2n-1)(2n+1)} = \frac{n}{2n+1}$$

ANSWERS TO THE TEST YOUR UNDERSTANDING EXERCISES

Page 329

1 $5n$ **2** $-4n + 10$ **3** $\frac{1}{10}n$ **4** $-8n + 3$ **5** n

6 $n - 4$ **7** $\frac{2}{3}n$ **8** $-12n + 65$ **9** $\frac{1}{5}n - \frac{1}{5}$ **10** $n + 1$

Page 336

1 $1, \frac{1}{2}, \frac{1}{4}, \frac{1}{8}, \frac{1}{16}$; $1(\frac{1}{2})^{n-1}$; $r = \frac{1}{2}$

2 $\frac{1}{4}, \frac{1}{8}, \frac{1}{16}, \frac{1}{32}, \frac{1}{64}$; $\frac{1}{4}(\frac{1}{2})^{n-1}$; $r = \frac{1}{2}$

3 $-\frac{1}{2}, \frac{1}{4}, -\frac{1}{8}, \frac{1}{16}, -\frac{1}{32}$; $-\frac{1}{2}(-\frac{1}{2})^{n-1}$; $r = -\frac{1}{2}$

4 $-\frac{1}{27}, \frac{1}{27^2}, -\frac{1}{27^3}, \frac{1}{27^4}, -\frac{1}{27^5}$; $-\frac{1}{27}\left(-\frac{1}{27}\right)^{n-1}$; $r = -\frac{1}{27}$

5 $r = 10$; $\frac{1}{5}(10)^{n-1}$

6 $r = -\frac{4}{9}$; $27(-\frac{4}{9})^{n-1}$

Page 342

1 $r = \frac{1}{10}$; $S_\infty = 11\frac{1}{9}$ **2** $r = 4$; no finite sum. **3** $r = -\frac{1}{6}$; $S_\infty = 30\frac{6}{7}$

4 $r = \frac{1}{4}$; $S_\infty = -21\frac{1}{3}$ **5** $r = \frac{4}{3}$; no finite sum. **6** $r = 0.01$; $S_\infty = \frac{1}{33}$

7 $r = -3$; no finite sum. **8** $r = -\frac{9}{10}$; $S_\infty = \frac{810}{19}$ **9** $r = -1.01$; no finite sum.

PERMUTATIONS, COMBINATIONS, PROBABILITY

10.1 Permutations

Suppose that there are 30 students in your class, and you all decide to become acquainted by shaking hands. Each person shakes hands with every other person. How many handshakes take place? Although this problem may not be an important or realistic one, it does suggest types of problems we may solve with counting procedures. As a start, we consider the following important principle of counting.

FUNDAMENTAL PRINCIPLE OF COUNTING

Suppose that a first task can be completed in m_1 ways, a second task in m_2 ways, and so on, until we reach the rth task that can be done in m_r ways; then total number of ways in which these tasks can be completed together is the product

$$m_1 m_2 \cdots m_r$$

To illustrate this fundamental principle, let us assume that you are planning a trip that will consist of visiting three cities, A, B, and C. You have your choice as to the order in which you are to visit the cities. How

Sometimes the use of the fundamental principle can be avoided by simply listing all the possible cases, as is done for the six trips shown here. However, this principle is very useful when the numbers involved are very large and such listings become impractical. This is illustrated in Examples 1 and 2.

many different trips are possible? A trip begins with a stop at any one of the three cities, the second stop will be at any one of the remaining two cities, and the trip is completed by stopping at the remaining city. Using the fundamental principle of counting, there are $3 \cdot 2 \cdot 1 = 6$ trips possible.

Another way to obtain the solution is to draw a *tree diagram* that illustrates all possible routes. From the diagram we can read the six possible trips. The arrangement ABC means that the trip begins with city A, goes to B, and ends at C. The other arrangements have similar interpretations.

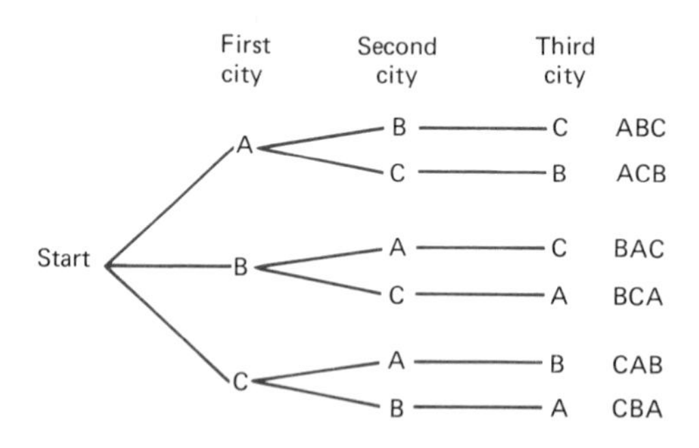

EXAMPLE 1 A club consists of 15 boys and 20 girls. They wish to elect officers consisting of a girl as president and a boy as vice-president. They also wish to elect a treasurer and a secretary who may be of either sex. How many sets of officers are possible?

We must assume here that no one person may hold two offices at the same time.

Solution There are 20 choices for president and 15 choices for vice-president. Thereafter, since two club members have been chosen, and the remaining positions can be filled by either a boy or a girl, 33 members are left for the post of treasurer, and then 32 choices for secretary. Then by the fundamental principle of counting, the total number of choices is

$$20 \cdot 15 \cdot 33 \cdot 32 = 316,800$$

EXAMPLE 2 How many three-digit whole numbers can be formed if zero is not an acceptable digit in the hundreds place and repetitions of digits are allowed?

Solution Imagine that you must place a digit in each of three positions, as in this display.

If repetition of digits were not allowed in this problem, then the solution would be $9 \cdot 9 \cdot 8 = 648$.

_____ _____ _____ Digits to use:

0 cannot be used here ↗ 0, 1, 2, 3, 4, 5, 6, 7, 8, 9

There are 9 choices for the first position, and 10 for each of the others. By the fundamental principle of counting, the solution is $9 \cdot 10 \cdot 10 = 900$.

In our first illustration where the six trips to the three cities were listed, the order of the elements A, B, C was crucial; trip ABC is different from trip

BAC. We say that each of the six arrangements is a *permutation* of three objects taken three at a time. In general, for n elements (n a positive integer) and r an integer satisfying $0 \leq r \leq n$, we have this definition.

PERMUTATION

A *permutation* of n elements taken r at a time is an arrangement, without repetitions, of r of the n elements. The number of permutations of n elements taken r at a time is denoted by ${}_nP_r$.

When considering permutations, the order of the elements is important. Thus 213 and 312 are two different permutations of the digits 1, 2, 3.

EXAMPLE 3 How many three-letter "words" from the 26 letters of the alphabet are possible? No duplication of letters is permitted.

Solution The first letter may be any one of the 26 letters of the alphabet, the second may be any one of the remaining 25, and the third is chosen from the remaining 24. Thus the total number of different "words" is $26 \cdot 25 \cdot 24 = 15{,}600$. Since we have taken 3 elements out of 26, without repetitions and in all possible orders, we may say that there are 15,600 permutations; that is, ${}_{26}P_3 = 26 \cdot 25 \cdot 24 = 15{,}600$.

Note that a "word" here is to be interpreted as any collection of three letters. That is, "nlm" is considered to be a word in this example.

Note that ${}_{26}P_3 = 26 \cdot 25 \cdot 24$ has three factors beginning with 26 and with each successive factor decreasing by 1. In general, for ${}_nP_r$ there will be r factors beginning with n, as follows.

$$
\begin{aligned}
{}_nP_r &= n(n-1)(n-2)(n-3)\cdots[n-(r-1)] \\
&= n(n-1)(n-2)(n-3)\cdots(n-r+1)
\end{aligned}
$$

This is a formula that gives the number of permutations of n elements taken r at a time. Begin with n, and use r factors. For example: $${}_{10}P_4 = \underbrace{10 \cdot 9 \cdot 8 \cdot 7}_{\text{four factors}}$$

A specific application of this formula occurs when $n = r$. In this case we have the permutation of n elements taken n at a time, and the product has n factors.

$${}_nP_n = n(n-1)(n-2)(n-3)\cdots 3 \cdot 2 \cdot 1$$

We may abbreviate this formula by using **factorial notation.**

$${}_nP_n = n!$$

For example:

$$
\begin{aligned}
{}_3P_3 &= 3! = 3 \cdot 2 \cdot 1 = 6 \\
{}_4P_4 &= 4! = 4 \cdot 3 \cdot 2 \cdot 1 = 24 \\
{}_5P_5 &= 5! = 5 \cdot 4 \cdot 3 \cdot 2 \cdot 1 = 120
\end{aligned}
$$

For future consistency we find it convenient to define $0!$ as equal to 1.

EXAMPLE 4 How many different ways can the four letters of the word MATH be arranged using each letter only once in each arrangement?

Solution Here we have the permutations of four elements taken four at a time. Thus ${}_4P_4 = 4! = 4 \cdot 3 \cdot 2 \cdot 1 = 24$. This includes such arrangements as MATH, AMTH, TMAH, and HMAT. Can you list all 24 possibilities? Draw a tree diagram to help you find all possible words.

TEST YOUR UNDERSTANDING

Evaluate.

1 ${}_{10}P_4$ **2** ${}_8P_3$ **3** ${}_6P_6$ **4** $\frac{10!}{8!}$ **5** $\frac{12!}{9!\,3!}$

6 How many four-letter "words" from the 26 letters of the alphabet are possible? No duplication of letters is permitted.

7 Answer Exercise 6 if duplications are permitted.

8 How many different ways can the letters of the word EAT be arranged using each letter once in each arrangement? List all the possibilities.

9 Draw a tree diagram that shows all the whole numbers that can be formed using the digits 2, 5, 8 so that each of the digits is used once in each number.

10 How many four-digit whole numbers greater than 5000 can be formed using each of the digits 3, 5, 6, 8 once in each number?

You have seen that $n!$ consists of n factors, beginning with n and successively decreasing to 1. At times it will be useful to display only some of the specific factors in $n!$ For example:

$$n! = n(n-1)! = n(n-1)(n-2)! = n(n-1)(n-2)(n-3)!$$

In particular,

$$5! = 5 \cdot 4! = 5 \cdot 4 \cdot 3! = 5 \cdot 4 \cdot 3 \cdot 2!$$

We can use the formula for ${}_nP_n$ to obtain a different form for ${}_nP_r$. To do so we write the formula for ${}_nP_r$ and then multiply numerator and denominator by $(n-r)!$ as follows.

$${}_nP_r = \frac{n(n-1)(n-2)\cdots[n-(r-1)]}{1} \cdot \frac{(n-r)!}{(n-r)!}$$

Now observe that the numerator can be simplified as $n!$ to produce the following useful formula for the permutation of n elements taken r at a time.

It is important to keep in mind that a permutation implies that there are to be no repetitions.

PERMUTATION OF n ELEMENTS TAKEN r AT A TIME

$${}_nP_r = \frac{n!}{(n-r)!}$$

EXAMPLE 5 Evaluate ${}_7P_4$ by using each of the formulas for ${}_nP_r$.

Solution

(a) Using ${}_nP_r = n(n-1)(n-2)\cdots(n-r+1)$ we have

$${}_7P_4 = 7 \cdot 6 \cdot 5 \cdot 4 = 840$$

(b) Using $_nP_r = \frac{n!}{(n-r)!}$ we have

$$_7P_4 = \frac{7!}{3!} = \frac{7 \cdot 6 \cdot 5 \cdot 4 \cdot \cancel{3!}}{\cancel{3!}} = 840$$

EXAMPLE 6 A class contains 10 members. They wish to elect officers consisting of a president, vice-president, and secretary-treasurer. How many sets of officers are possible?

Solution We may think of the three offices to be filled in terms of a first office (president), a second office (vice-president), and a third office (secretary-treasurer). Thus we need to select 3 out of 10 members and arrange them in all possible orders; we need to find the permutations of 10 elements taken 3 at a time.

$$_{10}P_3 = \frac{10!}{7!} = \frac{10 \cdot 9 \cdot 8 \cdot \cancel{7!}}{\cancel{7!}} = 720$$

EXERCISES 10-1

Evaluate.

1 $\frac{7!}{6!}$

2 $\frac{12!}{10!}$

3 $\frac{12!}{2!\,10!}$

4 $\frac{15!}{10!\,5!}$

5 $_5P_4$

6 $_5P_5$

7 $_4P_1$

8 $_8P_5$

9 How many different batteries consisting of a pitcher and a catcher can be formed by a baseball club that includes five pitchers and three catchers?

10 How many different outfits can Laura wear if she is able to match any one of five blouses, four skirts, and three pairs of shoes?

For Exercises 11–18, consider three-letter "words" to be formed by using the vowels a, e, i, o, and u.

11 How many different three letter words can be formed if repetitions are not allowed?

12 How many different words can be formed if repetitions are allowed?

13 How many different words without repetitions can be formed whose middle letter is *o*?

14 How many different words without repetitions can be formed whose first letter is *e*?

15 How many different words without repetitions can be formed whose letters at the ends are *u* and *i*?

16 If repetitions are allowed, how many different words can be formed whose middle letter is *a*?

***17** How many different words can be formed containing the letter *a* and two other letters?

***18** How many different words can be formed containing the letters *a* and *e* so that these letters are not next to each other?

For Exercises 19–24, consider three-digit numbers to be formed using the digits 1, 2, 3, 4, 5, 6, 7, 8, and 9. Also assume that repetition of digits is not allowed unless specified otherwise.

19 How many three-digit whole numbers can be formed?

20 How many three-digit whole numbers can be formed if repetition of digits is allowed?

21 How many three-digit whole numbers can be formed that are even? (*Hint:* An even number has a units digit of 2, 4, 6, or 8.)

22 How many three-digit whole numbers can be formed that are divisible by 5?

23 How many three-digit whole numbers can be formed that are greater than 600?

24 How many three-digit whole numbers can be formed that are less than 400 and are divisible by 5?

25 In how many different ways can the letters of

STUDY be arranged using each letter only once in each arrangement?

26 A class consists of 20 members. In how many different ways can the class select a set of officers consisting of a president, a vice-president, a secretary, and treasurer?

27 A baseball team consists of nine players. How many different batting orders are possible? How many are possible if the pitcher bats last? (Use a calculator for your computations.)

28 (a) Each question in a multiple-choice exam has the four choices indicated by the letters a, b, c, and d. If there are eight questions, how many ways can the test be answered? (*Hint:* Use the fundamental counting principle.)

(b) How many ways are there if no two consecutive answers can have the same answer?

29 When people are seated at a circular table, we consider only their positions relative to each other and are not concerned with the particular seat that a person occupies. How many arrangements are there for seven people to seat themselves around a circular table? (*Hint:* Consider one person's position as fixed.)

30 Review Exercise 29 and conjecture a formula for the number of different permutations of n distinct objects placed around a circle.

31 A license plate is formed by listing two letters of the alphabet followed by three digits. How many different license plates are possible (a) if repetitions of letters and digits are not allowed; (b) if repetitions are allowed?

32 Write an expression that gives the number of arrangements of n objects taken r at a time if repetitions are allowed.

33 Solve for n: (a) ${}_nP_1 = 10$; (b) ${}_nP_2 = 12$.

34 Show that $2({}_nP_{n-2}) = {}_nP_{n-1}$.

35 A pair of dice is rolled. Each die has six faces on each of which is one of the numbers 1 through 6. One possible outcome is (3, 5), where the first digit shows the outcome of one die and the second digit shows the outcome of the other die. How many different outcomes are possible? List all possible outcomes as pairs of numbers (x, y).

10.2

Combinations

A permutation may be regarded as an *ordered* collection of elements. For example, a visit to cities A, B, and C in the order ABC is different from a visit in the order ACB. On the other hand, there are times that we need to consider situations where the order of the elements is not essential, In general, letting n be a positive integer and r be any integer satisfying $0 \leq r \leq n$, we define a *combination* as follows.

COMBINATION

A *combination* is a collection of r distinct elements selected out of n elements without regard to order. The number of combinations of n elements taken r at a time is denoted by ${}_nC_r$.

As an example, suppose that a class of 10 members wishes to elect a committee consisting of three of its members. These three members are not to be designated as holding any special office. Then a committee consisting of members David, Ellen, and Robert is the same regardless of the order in which they are selected. In other words, using D, E, and R as abbreviations for their names, each of the following permutations is the same combination.

DER DRE EDR ERD RDE RED

Each combination of three members in this illustration actually gives rise to $3! = 6$ permutations. To find the number of possible committees we need to find the combinations of 10 elements taken 3 at a time. The *number* of such combinations is expressed as ${}_{10}C_3$ and is read as "the number of combinations of 10 elements taken 3 at a time." Since each of these combinations produces $3! = 6$ permutations, it follows that

Note that the one combination DER gives rise to these six permutations: DER, DRE, EDR, ERD, RDE, and RED.

$$3!({}_{10}C_3) = {}_{10}P_3 \qquad \text{or} \qquad {}_{10}C_3 = \frac{{}_{10}P_3}{3!} = \frac{10 \cdot 9 \cdot 8}{3 \cdot 2 \cdot 1} = 120$$

Using ${}_nP_r = \dfrac{n!}{(n-r)!}$, we obtain these general forms for ${}_nC_r$.

COMBINATION OF n ELEMENTS TAKEN r AT A TIME

$${}_nC_r = \frac{{}_nP_r}{r!} = \frac{n!}{r!(n-r)!}$$

Frequently, the symbol $\binom{n}{r}$ is used in place of ${}_nC_r$. Note that $\binom{n}{0} = \dfrac{n!}{0!\,n!} = 1$; also $\binom{n}{n} = \dfrac{n!}{n!\,0!} = 1$. Can you prove that $\binom{n}{r} = \binom{n}{n-r}$?

Recall that we have defined $0! = 1$. Do you see here why this definition was made?

EXAMPLE 1 Evaluate $\binom{10}{2}$.

Solution

$$\binom{10}{2} = \frac{10!}{2!\,8!} = \frac{10 \cdot 9 \cdot \cancel{8!}}{2 \cdot \cancel{8!}} = 45$$

EXAMPLE 2 Using the digits 1 through 9, how many different four-digit whole numbers can be formed if repetition of digits is not allowed?

Solution Order is important here; thus 4923 is a different number from 9432. Therefore we need to find the *permutations* of nine elements taken four at a time.

$${}_9P_4 = \frac{9!}{5!} = 9 \cdot 8 \cdot 7 \cdot 6 = 3024$$

The question of order is the essential ingredient that determines whether a problem involves permutations or combinations. Examples 2 and 3 illustrate this distinction.

EXAMPLE 3 A student has a penny, a nickel, a dime, a quarter, and a half-dollar and wishes to leave a tip consisting of exactly two coins. How many different amounts as tips are possible?

Solution Order is not important here; a tip of 10¢ + 25¢ is the same as one of 25¢ + 10¢. Therefore, we need to find the *combinations* of five things taken two at a time.

$$\binom{5}{2} = {}_5C_2 = \frac{5!}{2!\,3!} = \frac{5 \cdot 4}{2} = 10$$

List all 10 possibilities.

The next example illustrates how the fundamental principle of counting is used in a problem involving combinations.

EXAMPLE 4 A class consists of 10 boys and 15 girls. How many committees of five can be selected if each committee is to consist of two boys and three girls?

Solution The order is not essential here since the committee members do not hold any special offices.

To select two boys: $_{10}C_2 = \frac{10!}{2!\,8!} = 45$

To select three girls: $_{15}C_3 = \frac{15!}{3!\,12!} = 455$

Since there are 45 pairs of boys possible, and since each of these pairs can be matched with any of the possible 455 triples of girls, the fundamental principle of counting gives

$$45 \cdot 455 = 20{,}475$$

as the total number of committees that can be formed.

TEST YOUR UNDERSTANDING

Evaluate.

1 $_{10}C_4$ 2 $_5C_5$ 3 $_8C_0$ 4 $\binom{12}{3}$ 5 $\binom{8}{5}$

6 Show that $_{10}C_3 = {}_{10}C_7$.

7 How many different ways can a committee of 4 be selected from a group of 12 students?

8 A supermarket carries 6 brands of canned peas and 8 brands of canned corn. A shopper wants to try 2 different brands of peas and 3 different brands of corn. How many ways can the shopper select the 5 items?

9 How many lines are determined by eight points in a plane if no three points are on the same line?

10 How many triangles can be formed using five points in a plane no three of which are on the same line?

Many problems in probability will be solved through the use of combinations. For example, let us consider an ordinary deck of playing cards consisting of 52 different cards. These are divided into four suits: spades, hearts, diamonds, and clubs. There are 13 cards in each suit: from 1 (ace) through 10, jack, queen, and king.

EXAMPLE 5 A "poker hand" consists of 5 cards. How many different hands can be dealt from a deck of 52 cards?

Solution The order of the 5 cards dealt is not important, so that this becomes a problem involving combinations rather than permutations. We wish to find the number of combination of 52 elements taken 5 at a time.

$$_{52}C_5 = \frac{52!}{5!\,47!} = \frac{52 \cdot 51 \cdot 50 \cdot 49 \cdot 48 \cdot \cancel{47!}}{5 \cdot 4 \cdot 3 \cdot 2 \cdot 1 \cdot \cancel{47!}} = 2{,}598{,}960$$

If available, you should use a calculator to solve problems with extensive computations such as shown here. First try to simplify the fraction.

Consider Example 5 and let us determine the *probability* of being dealt a hand that consists of four aces and the jack of spades. Since there is only one such hand possible, the chances of obtaining this hand is given by the ratio $\frac{1}{2{,}598{,}960}$. Obviously, being dealt such a hand is an extremely rare occurrence! (We explore the topic of probability in greater detail in Section 10.4.)

EXAMPLE 6 An ice cream parlor advertises that you may have your choice of five different toppings, and you may choose none, one, two, three, four, or all five toppings. How many choices are there in all?

Solution There are several ways to approach this problem. From one point of view you may consider yourself on a cafeteria line with five stations. At each one you have two choices, to accept the topping or not to accept it. Thus the total number of choices is $2 \cdot 2 \cdot 2 \cdot 2 \cdot 2 = 32$. From another point of view, the solution is the number of different ways that we can select none, one, two, three, four, or five elements from a total of five possibilities; that is,

$$\binom{5}{0} + \binom{5}{1} + \binom{5}{2} + \binom{5}{3} + \binom{5}{4} + \binom{5}{5}$$

Show that this sum is also equal to 32.

Since, for example, choosing one topping or choosing two toppings are not done together, we add the number of possibilities rather than multiply.

EXERCISES 10.2

Evaluate.

1 $_5C_2$

2 $_{10}C_1$

3 $_{10}C_0$

4 $_4C_3$

5 $\binom{15}{15}$

6 $\binom{30}{3}$

7 $\binom{30}{27}$

8 $\binom{n}{3}$

9 A class consists of 20 members. In how many different ways can the class select **(a)** A committee of 4? **(b)** A set of 4 officers?

10 On a test a student must select 8 questions out of a total of 10. In how many different ways can this be done?

11 How many different ways can six people be split up into two teams of three each?

12 How many straight lines are determined by five points, no three of which are collinear?

13 There are 15 women on a basketball team. In how many different ways can a coach field a team of 5 players?

14 Answer Exercise 13 if two of the players can only play center and the others can play any of the remaining positions. (Assume that exactly one center is in a game at one time.)

15 Answer the question stated at the beginning of Section 10.1: How many handshakes take place when each person in a group of 30 shakes hands with every other person?

16 Box A contains 8 balls and box B contains 10 balls. In how many different ways can 5 balls be selected from these boxes if 2 are to be taken from box A and 3 from box B?

17 A class consists of 12 women and 10 men. A committee is to be selected consisting of 3 women and 2 men. How many different committees are possible?

18 Solve for n: **(a)** ${}_nC_1 = 6$; **(b)** ${}_nC_2 = 6$.

19 Convert to fraction form and simplify: **(a)** ${}_nC_{n-1}$; **(b)** ${}_nC_{n-2}$.

20 Prove: $\binom{n}{r} = \binom{n}{n-r}$.

21 Evaluate ${}_nC_4$ given that ${}_nP_4 = 1680$.

22 Solve for n: $35\binom{n-3}{3} = 4\binom{n}{3}$.

23 Consider this expression: $\binom{n}{0} + \binom{n}{1} + \binom{n}{2} + \cdots + \binom{n}{n-1} + \binom{n}{n}$. Evaluate for **(a)** $n = 2$; **(b)** $n = 3$; **(c)** $n = 4$; and **(d)** $n = 5$.

24 Use the results of Exercise 23 and conjecture the value of the expression for any positive integer n.

25 How many ways can 5-card hands be selected out of a deck of 52 cards so that all 5 cards are in the same suit?

26 How many ways can 5-card hands be selected out of a deck of 52 cards such that 4 of the cards have the same face value? (Four 10's and some fifth card is one such hand.)

27 How many ways can 5-card hands be selected out of a deck of 52 cards so that the 5 cards consist of a pair and three of a kind? (Two kings and three 7's is one such hand.)

28 A student wants to form a schedule consisting of 2 mathematics courses, 2 history courses, and 1 art course. The student can make these selections from 6 mathematics courses, 10 history courses, and 5 art courses. Assuming that there are no time conflicts, how many ways can the student select the 5 courses?

29 Suppose that in the U.S. Senate 25 Republicans and 19 Democrats are eligible for membership on a new committee. If this new committee is to consist of 9 senators, how many committees would be possible if the committee contained:

(a) Five Republicans and four Democrats?

(b) Five Democrats and four Republicans?

30 In Exercise 29, how many committees are possible if the chairperson of the committee is an eligible Republican senator appointed by the Vice-President, and the rest of the committee is evenly divided between Democrats and Republicans?

***31** Suppose that a 9-member committee is to vote on an amendment. How many different ways can the votes be cast so that the amendment passes by a simple majority? (A simple majority means 5 or more yes votes.)

***32** In Exercise 31, how many different favorable ways can the votes be cast if the amendment needs at least a $\frac{2}{3}$ majority?

***33** Prove: $\binom{n}{r-1} + \binom{n}{r} = \binom{n+1}{r}$.

10.3

The binomial expansion

The factored form of the trinomial square $a^2 + 2ab + b^2$ is $(a + b)^2$. Turning this around, we say that the *expanded form* of $(a + b)^2$ is $a^2 + 2ab + b^2$. And if $(a + b)^2$ is multiplied by $a + b$, we get the expansion of $(a + b)^3$. Here is a list of the expansions of the first five powers of the binomial $a + b$.

You can verify these results by multiplying the expansion in each row by a + b to get the expansion in the next row.

$$(a + b)^1 = a + b$$

$$(a + b)^2 = a^2 + 2ab + b^2$$

$$(a + b)^3 = a^3 + 3a^2b + 3ab^2 + b^2$$

$$(a + b)^4 = a^4 + 4a^3b + 6a^2b^2 + 4ab^3 + b^4$$

$$(a + b)^5 = a^5 + 5a^4b + 10a^3b^2 + 10a^2b^3 + 5ab^4 + b^5$$

Our objective here is to learn how to find such expansions directly without having to multiply. That is, we want to be able to expand $(a + b)^n$, especially for larger values of n, without having to multiply $a + b$ by itself repeatedly.

Let n represent a positive integer. As seen in the display, each expansion begins with a^n and ends with b^n. Moreover, each expansion has $n + 1$ terms that are all preceded by plus signs. Now look at the case for $n = 5$. Replace the first term a^5 by a^5b^0 and use a^0b^5 in place of b^5. Then

$$(a + b)^5 = a^5b^0 + 5a^4b + 10a^3b^2 + 10a^2b^3 + 5ab^4 + a^0b^5$$

In this form it becomes clear that (from left to right) the exponents of a successively decrease by 1, beginning with 5 and ending with zero. At the same time, the exponents of b increase from zero to 5. Also note that the sum of the exponents for each term is 5. Verify that similar patterns also hold for the other cases shown.

Using the preceding observations we would *expect* the expansion of $(a + b)^6$ to have seven terms that, except for the unknown coefficients, look like this:

$$a^6 + \underline{\quad}a^5b + \underline{\quad}a^4b^2 + \underline{\quad}a^3b^3 + \underline{\quad}a^2b^4 + \underline{\quad}ab^5 + b^6$$

Our list of expansions reveals that the second coefficient, as well as the coefficient of the next to the last term, is the number n. Filling in these coefficients for the case $n = 6$ gives

$$a^6 + 6a^5b + \underline{\quad}a^4b^2 + \underline{\quad}a^3b^3 + \underline{\quad}a^2b^4 + 6ab^5 + b^6$$

To get the remaining coefficients we return to the case $n = 5$ and learn how such coefficients can be generated. Look at the second and third terms.

$$\underbrace{\textcircled{5}\,a^{\textcircled{4}}b}_{\text{2nd term}} \qquad\qquad \underbrace{10a^3b^{\textcircled{2}}}_{\text{3rd term}}$$

If the exponent 4 of a in the *second* term is multiplied by the coefficient 5 of the *second* term and then divided by the exponent 2 of b in the *third* term, the result is 10, the coefficient of the third term.

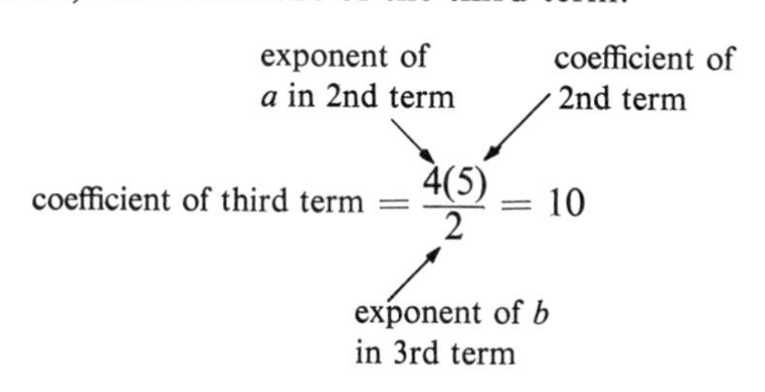

Verify that this procedure works for the next coefficient.

On the basis of the evidence we expect the missing coefficients for the case $n = 6$ to be obtainable in the same way. Here are the computations:

Use $\textcircled{6}\,a^{\textcircled{5}}b + \underline{\quad}a^4b^{\textcircled{2}}$: 3rd coefficient $= \dfrac{5(6)}{2} = 15$

Use $\textcircled{15}\,a^{\textcircled{4}}b^2 + \underline{\quad}a^3b^{\textcircled{3}}$: 4th coefficient $= \dfrac{4(15)}{3} = 20$

Use ⑳ a③b^3 + ____a^2b④: 5th coefficient $= \frac{3(20)}{4} = 15$

Finally, we may write the following expansion:

$$(a + b)^6 = a^6 + 6a^5b + 15a^4b^2 + 20a^3b^3 + 15a^2b^4 + 6ab^5 + b^6$$

You can verify that this equality is correct by multiplying the expansion for $(a + b)^5$ by $a + b$.

More labor can be saved by observing the symmetry in the expansions of $(a + b)^n$. For instance, when $n = 6$ the coefficients around the middle term are symmetric. Similarly, when $n = 5$ the coefficients around the two middle terms are symmetric.

EXAMPLE 1 Write the expansion of $(x + 2)^7$.

Solution Let x and 2 play the role of a and b in $(a + b)^7$, respectively.

$$(x + 2)^7 = x^7 + 7x^6 2 + \_\_\_\_x^5 2^2 + \_\_\_\_x^4 2^3 + \_\_\_\_x^3 2^4 + \_\_\_\_x^2 2^5 + 7x2^6 + 2^7$$

Now find the missing coefficients as follows:

$$\text{3rd coefficient} = \frac{6(7)}{2} = 21 = \text{6th coefficient}$$

$$\text{4th coefficient} = \frac{5(21)}{3} = 35 = \text{5th coefficient}$$

The completed expansion may now be given as follows:

$$\begin{aligned}(x + 2)^7 &= x^7 + 7x^6 2 + 21x^5 2^2 + 35x^4 2^3 + 35x^3 2^4 + 21x^2 2^5 + 7x2^6 + 2^7 \\ &= x^7 + 14x^6 + 84x^5 + 280x^4 + 560x^3 + 672x^2 + 448x + 128\end{aligned}$$

After some experience with these computations you should be able to write the expansion for the general case $(a + b)^n$, the **binomial formula**:

The *general* or rth term can be expressed as $\frac{n(n-1)(n-2)\cdots(n-r+1)}{r!}a^{n-r}b^r$

$$(a + b)^n = a^n + \frac{n}{1}a^{n-1}b + \frac{n(n-1)}{1 \cdot 2}a^{n-2}b^2 + \frac{n(n-1)(n-2)}{1 \cdot 2 \cdot 3}a^{n-3}b^3 + \cdots + \frac{n}{1}ab^{n-1} + b^n$$

EXAMPLE 2 Use the binomial formula to write the expansion of $(x + 2y)^4$.

Solution Use the formula with $a = x$, $b = 2y$ and $n = 4$. Then simplify.

$$\begin{aligned}(x + 2y)^4 &= x^4 + \frac{4}{1}x^3(2y) + \frac{4 \cdot 3}{1 \cdot 2}x^2(2y)^2 + \frac{4 \cdot 3 \cdot 2}{1 \cdot 2 \cdot 3}x(2y)^3 + (2y)^4 \\ &= x^4 + 8x^3y + 24x^2y^2 + 32xy^3 + 16y^4\end{aligned}$$

To get an expansion of the binomial $a - b$, write $a - b = a + (-b)$ and substitute into the previous form. For example, with $n = 6$,

$$(a - b)^6 = [a + (-b)]^6 = a^6 + 6a^5(-b) + 15a^4(-b)^2 + 20a^3(-b)^3 + 15a^2(-b)^4 + 6a(-b)^5 + (-b)^6$$

$$= a^6 - 6a^5b + 15a^4b^2 - 20a^3b^3 + 15a^2b^4 - 6ab^5 + b^6$$

This result indicates that the expansion of $(a - b)^n$ is the same as the expansion of $(a + b)^n$ except that the signs alternate, beginning with plus.

Be certain that you recognize the difference between the expansion of $(a - b)^n$ and the factorization of $a^n - b^n$. (See page 60.)

EXAMPLE 3 Expand: $(x - 1)^7$.

Solution Use the coefficients found in Example 1, with alternating signs.

$$(x - 1)^7 = x^7 - 7x^6 1^1 + 21x^5 1^2 - 35x^4 1^3 + 35x^3 1^4 - 21x^2 1^5 + 7x^1 1^6 - 1^7$$
$$= x^7 - 7x^6 + 21x^5 - 35x^4 + 35x^3 - 21x^2 + 7x - 1$$

We have developed the binomial formula by generalizing from specific cases to a general form. However, we can also make use of our knowledge of combinations to find this formula in a different way. Let us consider the expansion of $(a + b)^5$ again from a different point of view.

$$(a + b)^5 = \underbrace{(a + b)(a + b)(a + b)(a + b)(a + b)}_{\text{5 factors}}$$

To expand $(a + b)^5$, consider each term as follows.

First term: Multiply all the a's together to obtain a^5.

Second term: We need to combine all terms of the form a^4b. How are these terms formed in the multiplication process that gives the expansion? One of these terms is formed by multiplying the a's in the first four factors times the b in the last factor.

$$(\textcircled{a} + b)(\textcircled{a} + b)(\textcircled{a} + b)(\textcircled{a} + b)(a + \boxed{b})$$

Product $= a^4b$

Another of these terms is formed like this:

$$(\textcircled{a} + b)(\textcircled{a} + b)(\textcircled{a} + b)(a + \boxed{b})(\textcircled{a} + b)$$

Product $= a^4b$

Now you can see that the number of such terms is the same as the number of ways we can select just one of the b's from the five factors. This can be done in five ways, which can be expressed as ${}_5C_1$ or as $\binom{5}{1}$, the coefficient of a^4b.

Third term: Search for all terms of the form a^3b^2. The number of ways of selecting two b's from the five terms is ${}_5C_2$ or $\binom{5}{2}$.

Fourth term: The number of terms of the form a^2b^3 is the number of ways of selecting three b's from five terms, that is, ${}_5C_3$ or $\binom{5}{3}$.

Fifth term: The number of ways of selecting four b's from the five terms is ${}_5C_4$ or $\binom{5}{4}$, the coefficient of ab^4.

Sixth term: Multiply all the b's together to obtain b^5.

Thus we may write the expansion of $(a + b)^5$ in this form:

$$a^5 + \binom{5}{1}a^4b + \binom{5}{2}a^3b^2 + \binom{5}{3}a^2b^3 + \binom{5}{4}ab^4 + b^5$$

For consistency of form, we may write the coefficient of a^5 as $\binom{5}{0}$, and that of b^5 as $\binom{5}{5}$. In each case note that $\binom{5}{0} = \binom{5}{5} = 1$.

A similar argument can be made for each of the terms of the expansion of $(a + b)^n$. For example, to find the coefficient of the term that has the factors $a^{n-r}b^r$, we need to find the number of different ways of selecting r b's from n terms. This can be expressed as ${}_nC_r$ or $\binom{n}{r}$. Now we are ready to generalize and write this second form of the *binomial formula*:

This expansion can be written in sigma notation as

$$(a + b)^n = \sum_{r=0}^{n} \binom{n}{r} a^{n-r}b^r$$

$$(a + b)^n = \binom{n}{0}a^nb^0 + \binom{n}{1}a^{n-1}b^1 + \binom{n}{2}a^{n-2}b^2 + \cdots + \binom{n}{r}a^{n-r}b^r + \cdots + \binom{n}{n-1}a^1b^{n-1} + \binom{n}{n}a^0b^n$$

In the binomial formula note that in each term the sum of the exponents is equal to n. Also, the general term shown is actually the $(r + 1)$st term of the expansion. These observations are used in the examples that follow.

The numbers $\binom{n}{r}$ are referred to as *binomial coefficients.*

EXAMPLE 4 Find the sixth term of $(a + b)^8$.

Solution Note that the exponent of b in the general expansion is always one less than the number of the term. Furthermore, the sum of the exponents for each term will be equal to 8. The sixth term is of the form a^3b^5, the coefficient of this term is $\binom{8}{5}$.

$$\binom{8}{5} = {}_8C_5 = \frac{8!}{5!\,3!} = 56$$

The sixth term is $56a^3b^5$.

We could also have solved Example 4 by looking at the general form of the expansion and the $(r + 1)$st term shown there. Thus to find the sixth term, let $r + 1 = 6$; then $r = 5$ and $\binom{n}{r}a^{n-r}b^r = \binom{8}{5}a^3b^5$.

EXAMPLE 5 Find the fourth term of $(x - 2y)^{10}$.

Solution Use the general term $\binom{n}{r}a^{n-r}b^r$, which is the $(r+1)$st term. Then $r+1=4$ and $r=3$, $n=10$, and $n-r=7$. Thus the fourth term is

$$\binom{10}{3}x^7(-2y)^3 = 120x^7(-8y^3) = -960x^7y^3$$

Note that in Example 5 we may think of $(x-2y)^{10}$ as $[x+(-2y)]^{10}$ so that we may apply the binomial formula for $(a+b)^n$.

EXERCISES 10.3

Expand and simplify.

1 $(x+1)^5$

2 $(x-1)^6$

3 $(x+1)^7$

4 $(x-1)^8$

5 $(a-b)^4$

6 $(3x-2)^4$

7 $(3x-y)^5$

8 $(x+y)^8$

9 $(a^2+1)^5$

10 $(2+h)^9$

11 $(1-h)^{10}$

12 $(x-2)^7$

Simplify.

13 $\dfrac{(1+h)^3-1}{h}$

14 $\dfrac{(3+h)^4-81}{h}$

15 $\dfrac{(c+h)^3-c^3}{h}$

16 $\dfrac{(x+h)^6-x^6}{h}$

17 $\dfrac{2(x+h)^5-2x^5}{h}$

18 $\dfrac{\dfrac{1}{(2+h)^2}-\dfrac{1}{4}}{h}$

19 Evaluate 2^{10} by expanding $(1+1)^{10}$.

20 Write the first five terms in the expansion of $(x+1)^{15}$. What are the last five terms?

21 Write the first five terms and the last five terms in the expansion of $(c+h)^{20}$.

22 Write the first four terms and the last four terms in the expansion of $(a-1)^{30}$.

23 Study this triangular array of numbers and discover the connection with the expansions of $(a+b)^n$, where $n = 1, 2, 3, 4, 5, 6$.

```
            1   1
          1   2   1
        1   3   3   1
      1   4   6   4   1
    1   5  10  10   5   1
  1   6  15  20  15   6   1
```

This triangular array of numbers is called **Pascal's triangle**, named after French mathematician Blaise Pascal (1623–1662). However, the triangle appeared in Chinese writings as early as 1303. How many properties of the triangle can you find? For example, find the sum of the entries in each row.

24 Discover how the 6th row of the triangle in Exercise 23 can be obtained from the 5th row by studying the connection between the 4th and 5th rows indicated by this scheme.

```
   1   4   6   4   1
     +   +   +   +
 1   5  10  10   5   1
```

25 Using the result of Exercise 24, write the 7th, 8th, 9th, and 10th rows of the triangle.

26 Use the 9th row found in Exercise 25 to expand $(x+h)^9$.

27 Use the 10th row found in Exercise 25 to expand $(x-h)^{10}$.

28 Why does the sum of all the numbers in one line of Pascal's triangle equal twice the sum of the numbers in the preceding line?

29 Find the sixth term in the expansion of $(a+2b)^{10}$.

30 Find the fifth term in the expansion of $(2x-y)^8$.

31 Find the fourth term in the expansion of $\left(\frac{1}{x}+\sqrt{x}\right)^7$.

32 Find the eighth term in the expansion of $\left(\frac{a}{2}+\frac{b^2}{3}\right)^{10}$.

33 Find the term that contains x^5 in the expansion of $(2x+3y)^8$.

34 Find the term that contains y^{10} in the expansion of $(x-2y^2)^8$.

35 Write the middle term of the expansion of $\left(3a-\frac{b}{2}\right)^{10}$.

36 Write the last three terms of the expansion of $(a^2-2b^3)^7$.

10.4

Probability

Concepts of probability are encountered frequently in daily life, such as a weather forecaster's statement that there is "a 20% chance of rain." Actually, the formal study of probability started in the seventeenth century when two famous mathematicians, Pascal and Fermat, considered the following problem that was posed to them by a gambler. Two people are involved in a game of chance and are forced to quit before either one has won. The number of points needed to win the game is known, and the number of points that each player has at the time is known. The problem was to determine how the stakes should be divided.

From this beginning mathematicians developed the theory of probability that has far-reaching effects in many fields of endeavor. In this section we explore a few elementary aspects of probability based on the principles that have been developed in this chapter.

First coin Second coin
H: H HH, T HT
T: H TH, T TT

Let us begin by considering the situation of tossing two coins. What is the probability that both coins will be heads? We can approach this problem by forming a list of all possible outcomes. (A tree diagram is helpful in identifying all the possibilities.)

First coin	*Second coin*
Heads	Heads
Heads	Tails
Tails	Heads
Tails	Tails

Event	*Probability*
2 heads	$\frac{1}{4}$
1 heads	$\frac{2}{4}$
0 heads	$\frac{1}{4}$

Notice that there are four possible outcomes {HH, HT, TH, TT} and that only one of these gives the required two heads. Thus we say that the probability that both coins will be heads is $\frac{1}{4}$. In symbols we may write $P(\text{HH}) = \frac{1}{4}$. This reflects what happens *in the long run*. If we continue to toss two coins repeatedly, we would expect that on the average one out of four tosses will show two heads. Of course, there may be times when we toss double heads several times in a row; but if the experiment were to be repeated 1000 times, we can expect to have *about* 250 cases of double heads.

Let us assume that an experiment can have n different outcomes, each equally likely, and that m of these outcomes produce the event E. Then the **probability** that the event E will occur, $P(E)$, is given as

Note: This definition assumes that all events are equally likely to occur; that is, each event has an equally likely chance to happen.

$$P(E) = = \frac{m}{n} = \frac{\text{number of successful outcomes}}{\text{total number of outcomes}}$$

EXAMPLE 1 A die is tossed. What is the probability of tossing a 5?

Solution There are six possible outcomes {1, 2, 3, 4, 5, 6}. Each one has the same chance of occurring as the others. There is only one way to succeed, namely by tossing a 5. Thus $P(5) = \frac{1}{6}$.

EXAMPLE 2 A die is tossed. What is the probability of tossing a 7?

Solution None of the outcomes is 7. Thus $P(7) = \frac{0}{6} = 0$.

We can phrase the result of Example 2 in another way: $P(\text{a 7 cannot occur}) = 1$.

From Example 2 we see that the probability that an event *cannot* occur is 0. Furthermore, the probability for an event that will *always* occur is 1. For example, the probability of tossing a number less than 7 on a single toss of a die is $\frac{6}{6} = 1$ since all numbers are less than 7. This leads to the following observation for the probability that an event E will occur:

$$0 \leq P(E) \leq 1$$

As an extension of this idea, we note that every event will either occur or will fail to occur. That is, $P(E) + P(\text{not } E) = 1$. Therefore,

$$P(\text{not } E) = 1 - P(E)$$

EXAMPLE 3 A single card is drawn from a deck of cards. What is the probability that the card is *not* a spade?

Solution There are 52 cards in a deck, and 13 of these are spades. The probability of drawing a spade is $\frac{13}{52}$. Thus the probability of not drawing a spade is

$$1 - \frac{13}{52} = \frac{39}{52} = \frac{3}{4}$$

We need to be careful when adding probabilities since we may only do this when events are *disjoint*, that is, when they cannot both happen at the same time. For example, consider the probability of drawing a king or a queen when a single card is drawn from a deck of cards.

$$P(\text{king}) = \frac{4}{52} \qquad P(\text{queen}) = \frac{4}{52}$$

$$P(\text{king or a queen}) = \frac{4}{52} + \frac{4}{52} = \frac{8}{52} = \frac{2}{13}$$

When two event E and F are disjoint, then $P(E \text{ or } F) = P(E) + P(F)$.

This seems to agree with our intuition since there are 8 cards in a deck of 52 cards that meet the necessary conditions. These conditions are disjoint because a card cannot be a king and a queen at the same time. Note, however, the difference in the conditions of Example 4.

EXAMPLE 4 A single card is drawn from a deck of cards. What is the probability that the card is either a queen or a spade?

Solution The probability of drawing a queen is $\frac{4}{52}$, and the probability of drawing a spade is $\frac{13}{52}$. However there is one card that is being counted twice in these probabilities, namely the queen of spades. Thus we account for this as follows:

$$P(\text{queen or spade}) = \frac{13}{52} + \frac{4}{52} - \frac{1}{52} = \frac{16}{52} = \frac{4}{13}$$

In Example 4 the events are *not* disjoint because it is possible for a card to be both a queen and a spade.

In general, for events E and F,

$$P(E \text{ or } F) = P(E) + P(F) - P(E \text{ and } F).$$

TEST YOUR UNDERSTANDING

A pair of coins is tossed. Find the probability of each outcome.

1 Both are tails.
2 One is heads and one is tails.
3 At least one is heads.
4 Both are heads or both are tails.

A single die is tossed. Find the probability of each outcome.

5 An even number comes up.
6 A number less than 5 comes up.
7 An even number or a number greater than 3 comes up.

A single card is drawn from a deck of cards. Find the probability of each outcome.

We consider the picture cards to be the jacks, queens, and kings.

8 A red card is drawn.
9 A heart or a spade is drawn.
10 An ace or a heart is drawn.
11 A picture card is drawn.
12 A picture card or an ace is drawn.

Sometimes a probability example can be solved in a variety of ways. For example, consider the probability of drawing two aces when two cards are selected from a deck of cards. In all such cases, unless stated otherwise, we shall assume that a card is drawn and *not* replaced in the deck before the second card is drawn.

Probability that the first card is an ace $= \frac{4}{52}$.

Assume that an ace is drawn. Then there are only 51 cards left in the deck, of which 3 are aces. Thus the probability that the second card is an ace $= \frac{3}{51}$.

We now make use of the following principle. Suppose that $P(E)$ represents the probability that an event E will occur, and $P(F \text{ given } E)$ is the probability that event F occurs after E has occurred. Then

$$P(E \text{ and } F) = P(E) \times P(F \text{ given } E)$$

Thus to find the probability that both cards are aces in our example, we must multiply:

$$\frac{4}{52} \cdot \frac{3}{51} = \frac{1}{221}$$

This example can also be expressed through the use of combinations. Thus the total number of ways to select two cards from the deck is ${}_{52}C_2$. Furthermore, the number of ways of selecting two aces from the four aces in a deck of cards is ${}_4C_2$. Therefore,

Show that $\frac{{}_4C_2}{{}_{52}C_2} = \frac{1}{221}$.

$$\text{probability of selecting two aces} = \frac{{}_4C_2}{{}_{52}C_2}$$

EXAMPLE 5 A single card is drawn from a deck of cards. It is then replaced and a second card is drawn. What is the probability that both cards are aces?

Solution These two events are said to be *independent*; neither outcome depends on the other. In each case the probability is $\frac{4}{52}$. The probability that both are aces is $\frac{4}{52} \cdot \frac{4}{52} = \frac{1}{169}$.

EXAMPLE 6 Five cards are dealt from a deck of 52 cards. What is the probability that exactly two of the cards are aces?

In Example 6 we assume that we are to draw *exactly* two aces. A more difficult problem is to find the probability of drawing *at least* two aces. Can you solve that problem?

Solution The number of ways of selecting two aces from the four aces in the deck is ${}_4C_2$. However, we also need to select three additional cards from the remaining 48 cards in the deck that are *not* aces; this can be done in ${}_{48}C_3$ ways. Thus, by the fundamental counting principle, the total number of ways of drawing two aces (and three other cards) is the product $({}_4C_2) \cdot ({}_{48}C_3)$.

$$\text{probability of drawing two aces in five cards} = \frac{({}_4C_2) \cdot ({}_{48}C_3)}{{}_{52}C_5}$$

Show that this ratio reduces to $\frac{2162}{54{,}145}$.

EXAMPLE 7 A class consists of 10 men and 8 women. Four members are to be selected *at random* to represent the class. What is the probability that the selection will consist of two men and two women?

If the members are selected *at random*, then each one has an equally likely chance of being selected. For example, a random selection could involve having all 18 names placed in a hat and 4 names drawn as in a lottery.

Solution Number of ways of selecting two men $= {}_{10}C_2$.
Number of ways of selecting two women $= {}_8C_2$.
Number of ways of selecting four members of the class $= {}_{18}C_4$.
Thus the probability that the committee will consist of two men and two women is given as

$$\frac{({}_{10}C_2) \cdot ({}_8C_2)}{{}_{18}C_4} = \frac{7}{17}$$

EXERCISES 10.4

Use the tree diagram showing the results of tossing three coins to find the probability of each outcome.

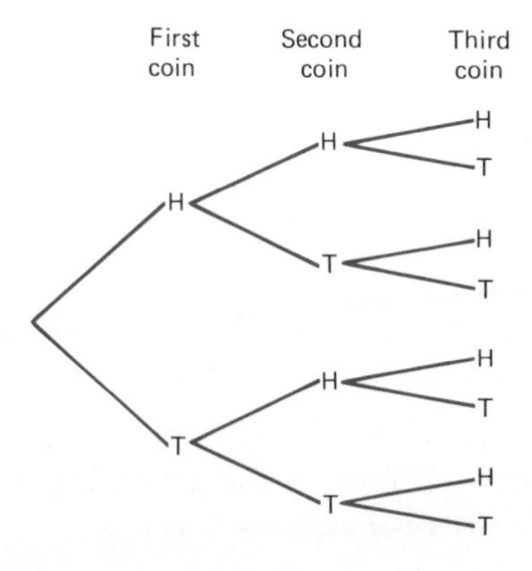

1 All three coins are heads.

2 Exactly one coin is heads.

3 At least one coin is heads.

4 None of the coins is heads.

5 At most one coin is heads.

6 At least two coins are heads.

There are 36 different ways that a pair of dice can land, as shown in the diagram on the next page. Each outcome is shown as a pair of numbers showing the outcome on each die. Use this diagram to find the probability of each outcome in Exercises 7–14 when a pair of dice are tossed.

(1, 1)	(1, 2)	(1, 3)	(1, 4)	(1, 5)	(1, 6)
(2, 1)	(2, 2)	(2, 3)	(2, 4)	(2, 5)	(2, 6)
(3, 1)	(3, 2)	(3, 3)	(3, 4)	(3, 5)	(3, 6)
(4, 1)	(4, 2)	(4, 3)	(4, 4)	(4, 5)	(4, 6)
(5, 1)	(5, 2)	(5, 3)	(5, 4)	(5, 5)	(5, 6)
(6, 1)	(6, 2)	(6, 3)	(6, 4)	(6, 5)	(6, 6)

The first number in each pair shows the outcome of tossing one die; the other number is the outcome of the second die.

7 Both die show the same number.

8 The sum is 11.

9 The sum is 7.

10 The sum is 7 or 11.

11 The sum is 2, 3, or 12.

12 The sum is 6 or 8.

13 The sum is an odd number.

14 The sum is not 7.

Two cards are drawn from a deck of 52 playing cards without replacement. Find the probability of each outcome.

15 Both cards are red.

16 Both cards are spades.

17 Both cards are the ace of hearts.

***18** Both cards are the same suit.
(*Hint:* A successful outcome is to have both cards spades *or* both hearts *or* both diamonds *or* both clubs. The sum of these probabilities gives the solution.)

Two cards are drawn from a deck of 52 playing cards with the first card replaced before the second card is drawn. Find the probability of each outcome.

19 Both cards are black or both are red.

20 Both cards are hearts.

21 Both cards are the ace of hearts.

22 Both cards are picture cards.

23 Neither card is an ace.

24 Neither card is a spade.

25 The first card is an ace and the second card is a king.

26 The first card is an ace and the second card is not an ace.

A bag of marbles contains 8 red marbles and 5 green marbles. Three marbles are drawn at random. Find the probability of each outcome.

27 All are red.

28 All are green.

29 Two are red and one is green.

30 One is red and two are green.

31 A student takes a true–false test consisting of 10 questions by just guessing at each answer. Find the probability that the student's score will be (**a**) 100%; (**b**) 0%; (**c**) 80% or better.

***32** A die is tossed three times in succession. Find the probability that (a) all three tosses show 5; (**b**) exactly one of the tosses shows 5; (**c**) at least one of the tosses shows 5.

***33** Five cards are dealt from a deck of 52 cards. Find the probability of obtaining (**a**) four aces; (**b**) four of a kind (that is, four aces or four twos or four threes, etc.); (**c**) a flush (that is, all five cards of the same suit).

***34** You are given 10 red marbles and 10 green marbles to distribute into two boxes. You will then be blindfolded and asked to select one of the boxes and then draw one marble from that box. You win $10,000 if the marble you select is red. Finally, you are allowed to distribute the 20 marbles into the two boxes in any way that you wish before you begin to make the selection. Try to determine what is the best strategy for distributing the marbles; that is, how many of each color should you place in each box?

Mathematical expectation is defined as the product of the probability that an event will occur and the amount to be received if it does occur. For example, if you are to receive $12 if you toss a 6 on one throw of a die, your expectation is $\frac{1}{6}(\$12) = \$2$. *What is your expectation for each of the events described in Exercises 35–38?*

35 You toss a coin twice and receive $20 if both tosses show heads.

36 You select a single card from a deck of cards and receive $26 if it is an ace or a king.

***37** You toss two coins. If both land heads, you receive $20; if only one coin lands heads, you receive $10; and you receive nothing if both coins are tails.

***38** You throw a pair of dice. If the sum of the outcomes is 7, you receive $30; if the sum is more than 7, you receive $12; and you pay $24 (you receive minus $24) if the sum is less than 7.

The odds in favor of an event occurring is the ratio of the probability that it will occur to the probability that it will not occur. For example, the odds in favor of tossing two heads in two tosses of a coin are $\frac{1/4}{3/4} = \frac{1}{3}$ *or 1 to 3. The odds against this event are 3 to 1.*

39 A card is drawn from a deck of cards. What are the odds in favor of obtaining an ace? What are the odds against this event occurring?

40 What are the odds in favor of tossing three successive heads with a coin?

41 What are the odds against tossing a 7 or an 11 in a single throw of a pair of dice?

42 Two cards are drawn simultaneously from a deck of cards. What are the odds in favor of both cards being hearts?

REVIEW EXERCISES FOR CHAPTER 10

The solutions to the following exercises can be found within the text of Chapter 10. Try to answer each question without referring to the text.

Section 10.1

1 You are planning a trip that will consist of visiting three cities, A, B, and C. You have your choice as to the order in which you are to visit the cities. List all possible different trips.

2 A club consists of 15 boys and 20 girls. They wish to elect officers consisting of a girl as president and a boy as vice-president. They also wish to elect a treasurer and a secretary who may be of either sex. How many sets of officers are possible?

3 How many three-digit whole numbers can be formed if zero is not an acceptable digit in the hundreds place and repetitions are allowed?

4 How many three-letter "words" from the 26 letters of the alphabet are possible? No duplication of letters is permitted.

5 How many different ways can the four letters of the word MATH be arranged using each letter only once in each arrangement?

6 Evaluate ${}_7P_4$ by using each of the formulas for ${}_nP_r$.

7 A class contains 10 members. They wish to elect officers consisting of a president, vice-president, and secretary-treasurer. How many sets of officers are possible?

Section 10.2

8 Evaluate: **(a)** $\binom{10}{2}$; **(b)** ${}_{52}C_5$.

9 Using the digits 1 through 9, how many different four-digit whole numbers can be formed if repetition of digits is not allowed?

10 A student has a penny, a nickel, a dime, a quarter, and a half-dollar and wishes to leave a tip consisting of exactly two coins. How many different amounts as tips are possible?

11 A class consists of 10 boys and 15 girls. How many committees of five can be selected if each committee is to consist of two boys and three girls?

12 An ice cream parlor advertises that you may have your choice of five different toppings, and you may choose none, one, two, three, four, or all five toppings. How many choices are there in all?

Section 10.3

13 Write the expansion of $(x + 2)^7$.

14 Write the expansion of $(x + 2y)^4$.

15 Expand: $(x - 1)^7$.

16 Write the general or rth term of $(a + b)^n$.

17 Find the sixth term of $(a + b)^8$.

18 Find the fourth term of $(x - 2y)^{10}$.

Section 10.4

19 A die is tossed. **(a)** What is the probability of tossing a 5? **(b)** What is the probability of tossing a 7?

20 A single card is drawn from a deck of cards. What is the probability that the card is *not* a spade?

21 A single card is drawn from a deck of cards. What is the probability that the card is either a queen or a spade?

22 A single card is drawn from a deck of cards. It is then replaced and a second card is drawn. What is the probability that both cards are aces?

23 Five cards are dealt from a deck of 52 cards. What is the probability that exactly two of the cards are aces?

24 A class consists of 10 men and 8 women. Four members are to be selected at random to represent the class. What is the probability that the selection will consist of two men and two women?

SAMPLE TEST QUESTIONS FOR CHAPTER 10

Use these questions to test your knowledge of the basic skills and concepts of Chapter 10. Then check your answers with those given at the end of the book.

1 Evaluate: **(a)** ${}_{10}P_3$; **(b)** ${}_{10}C_3$.

2 How many different ways can the letters of the

word TODAY be arranged so that each arrangement uses each of the letters once?

3 How many three-digit whole numbers can be formed if zero is not an acceptable hundreds digit and **(a)** repetitions are allowed; **(b)** repetitions are not allowed?

4 In how many different ways can a class of 15 students select a committee of three students?

5 A class consists of 12 boys and 10 girls. How many committees of four can be selected if each committee is to consist of two boys and two girls?

6 How many different four-letter "words" can be formed from the letters m, o, n, d, a, y if no letter may be used more than once?

7 A student takes a five-question true–false test just by guessing at each answer.
 (a) How many different sets of answers are possible?
 (b) What is the probability of obtaining a score of 100% on the test?
 (c) What is the probability that the score will be 0%?

8 A single die is tossed. What is the probability that it will show a even number or a number less than 3?

9 A pair of dice is tossed. What is the probability that the sum will be **(a)** 12; **(b)** not 12; **(c)** less than 5?

10 Two cards are drawn from a deck of 52 playing cards. What is the probability that they are both aces or both kings if **(a)** the first card is replaced before the second card is drawn; **(b)** no replacements are made?

11 Five cards are dealt from a deck of 52 playing cards. What is the probability that exactly four of the cards are picture cards?

12 A box contains 20 red chips and 15 green chips. Five chips are selected from the box at random. What is the probability that exactly four of the chips will be red?

13 Four coins are tossed. What is the probability that **(a)** all four coins will land showing heads; **(b)** none of the coins will show heads?

14 Two integers from 0 through 10 are selected simultaneously and at random. What is the probability that they will both be odd?

15 Box A contains 5 green and 3 red chips. Box B contains 4 green and 6 red chips. You are to select one chip from each box. What is the probability that both chips will be red?

16 In Exercise 15, suppose that you are blindfolded and told to select just one chip from one of the boxes. What is the probability that the chip you select will be red?

17 Expand: $(x - 2y)^5$.

18 Write the first four terms of the expansion of $\left(\frac{1}{2a} - b\right)^{10}$.

19 Write the seventh term of $(3a + b)^{11}$.

20 Write the middle term of $(2x - y)^{16}$.

ANSWERS TO THE TEST YOUR UNDERSTANDING EXERCISES

Page 356

1 5040 **2** 336 **3** 720 **4** 90 **5** 220

6 358,800 **7** 456,976 **8** 6; EAT, ETA, AET, ATE, TEA, TAE

9

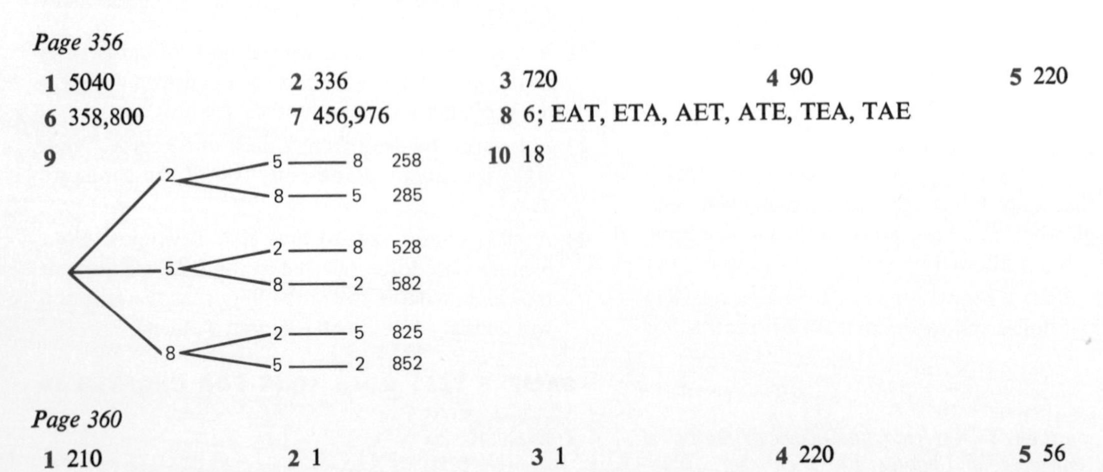

10 18

Page 360

1 210 **2** 1 **3** 1 **4** 220 **5** 56

6 ${}_{10}C_3 = \frac{10!}{3!\,7!} = \frac{10!}{7!\,3!} = {}_{10}C_7$ **7** 495 **8** 840 **9** 28 **10** 10

Page 370

1 $\frac{1}{4}$ **2** $\frac{1}{2}$ **3** $\frac{3}{4}$ **4** $\frac{1}{2}$ **5** $\frac{1}{2}$ **6** $\frac{2}{3}$
7 $\frac{2}{3}$ **8** $\frac{1}{2}$ **9** $\frac{1}{2}$ **10** $\frac{4}{13}$ **11** $\frac{3}{13}$ **12** $\frac{4}{13}$

TABLES

Table I: Square Roots and Cube Roots

N	$\sqrt{N}$	$\sqrt[3]{N}$	N	$\sqrt{N}$	$\sqrt[3]{N}$	N	$\sqrt{N}$	$\sqrt[3]{N}$	N	$\sqrt{N}$	$\sqrt[3]{N}$
1	1.000	1.000	51	7.141	3.708	101	10.050	4.657	151	12.288	5.325
2	1.414	1.260	52	7.211	3.733	102	10.100	4.672	152	12.329	5.337
3	1.732	1.442	53	7.280	3.756	103	10.149	4.688	153	12.369	5.348
4	2.000	1.587	54	7.348	3.780	104	10.198	4.703	154	12.410	5.360
5	2.236	1.710	55	7.416	3.803	105	10.247	4.718	155	12.450	5.372
6	2.449	1.817	56	7.483	3.826	106	10.296	4.733	156	12.490	5.383
7	2.646	1.913	57	7.550	3.849	107	10.344	4.747	157	12.530	5.395
8	2.828	2.000	58	7.616	3.871	108	10.392	4.762	158	12.570	5.406
9	3.000	2.080	59	7.681	3.893	109	10.440	4.777	159	12.610	5.418
10	3.162	2.154	60	7.746	3.915	110	10.488	4.791	160	12.649	5.429
11	3.317	2.224	61	7.810	3.936	111	10.536	4.806	161	12.689	5.440
12	3.464	2.289	62	7.874	3.958	112	10.583	4.820	162	12.728	5.451
13	3.606	2.351	63	7.937	3.979	113	10.630	4.835	163	12.767	5.463
14	3.742	2.410	64	8.000	4.000	114	10.677	4.849	164	12.806	5.474
15	3.873	2.466	65	8.062	4.021	115	10.724	4.863	165	12.845	5.485
16	4.000	2.520	66	8.124	4.041	116	10.770	4.877	166	12.884	5.496
17	4.123	2.571	67	8.185	4.062	117	10.817	4.891	167	12.923	5.507
18	4.243	2.621	68	8.246	4.082	118	10.863	4.905	168	12.961	5.518
19	4.359	2.668	69	8.307	4.102	119	10.909	4.919	169	13.000	5.529
20	4.472	2.714	70	8.367	4.121	120	10.954	4.932	170	13.038	5.540
21	4.583	2.759	71	8.426	4.141	121	11.000	4.946	171	13.077	5.550
22	4.690	2.802	72	8.485	4.160	122	11.045	4.960	172	13.115	5.561
23	4.796	2.844	73	8.544	4.179	123	11.091	4.973	173	13.153	5.572
24	4.899	2.884	74	8.602	4.198	124	11.136	4.987	174	13.191	5.583
25	5.000	2.924	75	8.660	4.217	125	11.180	5.000	175	13.229	5.593
26	5.099	2.962	76	8.718	4.236	126	11.225	5.013	176	13.267	5.604
27	5.196	3.000	77	8.775	4.254	127	11.269	5.027	177	13.304	5.615
28	5.292	3.037	78	8.832	4.273	128	11.314	5.040	178	13.342	5.625
29	5.385	3.072	79	8.888	4.291	129	11.358	5.053	179	13.379	5.636
30	5.477	3.107	80	8.944	4.309	130	11.402	5.066	180	13.416	5.646
31	5.568	3.141	81	9.000	4.327	131	11.446	5.079	181	13.454	5.657
32	5.657	3.175	82	9.055	4.344	132	11.489	5.092	182	13.491	5.667
33	5.745	3.208	83	9.110	4.362	133	11.533	5.104	183	13.528	5.677
34	5.831	3.240	84	9.165	4.380	134	11.576	5.117	184	13.565	5.688
35	5.916	3.271	85	9.220	4.397	135	11.619	5.130	185	13.601	5.698
36	6.000	3.302	86	9.274	4.414	136	11.662	5.143	186	13.638	5.708
37	6.083	3.332	87	9.327	4.431	137	11.705	5.155	187	13.675	5.718
38	6.164	3.362	88	9.381	4.448	138	11.747	5.168	188	13.711	5.729
39	6.245	3.391	89	9.434	4.465	139	11.790	5.180	189	13.748	5.739
40	6.325	3.420	90	9.487	4.481	140	11.832	5.192	190	13.784	5.749
41	6.403	3.448	91	9.539	4.498	141	11.874	5.205	191	13.820	5.759
42	6.481	3.476	92	9.592	4.514	142	11.916	5.217	192	13.856	5.769
43	6.557	3.503	93	9.644	4.531	143	11.958	5.229	193	13.892	5.779
44	6.633	3.530	94	9.695	4.547	144	12.000	5.241	194	13.928	5.789
45	6.708	3.557	95	9.747	4.563	145	12.042	5.254	195	13.964	5.799
46	6.782	3.583	96	9.798	4.579	146	12.083	5.266	196	14.000	5.809
47	6.856	3.609	97	9.849	4.595	147	12.124	5.278	197	14.036	5.819
48	6.928	3.634	98	9.899	4.610	148	12.166	5.290	198	14.071	5.828
49	7.000	3.659	99	9.950	4.626	149	12.207	5.301	199	14.107	5.838
50	7.071	3.684	100	10.000	4.642	150	12.247	5.313	200	14.142	5.848

Table II: Exponential Functions

x	e^x	e^{-x}	x	e^x	e^{-x}
0.0	1.00	1.000	3.1	22.2	0.045
0.1	1.11	0.905	3.2	24.5	0.041
0.2	1.22	0.819	3.3	27.1	0.037
0.3	1.35	0.741	3.4	30.0	0.033
0.4	1.49	0.670	3.5	33.1	0.030
0.5	1.65	0.607	3.6	36.6	0.027
0.6	1.82	0.549	3.7	40.4	0.025
0.7	2.01	0.497	3.8	44.7	0.022
0.8	2.23	0.449	3.9	49.4	0.020
0.9	2.46	0.407	4.0	54.6	0.018
1.0	2.72	0.368	4.1	60.3	0.017
1.1	3.00	0.333	4.2	66.7	0.015
1.2	3.32	0.301	4.3	73.7	0.014
1.3	3.67	0.273	4.4	81.5	0.012
1.4	4.06	0.247	4.5	90.0	0.011
1.5	4.48	0.223	4.6	99.5	0.010
1.6	4.95	0.202	4.7	110	0.0091
1.7	5.47	0.183	4.8	122	0.0082
1.8	6.05	0.165	4.9	134	0.0074
1.9	6.69	0.150	5.0	148	0.0067
2.0	7.39	0.135	5.5	245	0.0041
2.1	8.17	0.122	6.0	403	0.0025
2.2	9.02	0.111	6.5	665	0.0015
2.3	9.97	0.100	7.0	1097	0.00091
2.4	11.0	0.091	7.5	1808	0.00055
2.5	12.2	0.082	8.0	2981	0.00034
2.6	13.5	0.074	8.5	4915	0.00020
2.7	14.9	0.067	9.0	8103	0.00012
2.8	16.4	0.061	9.5	13360	0.00075
2.9	18.2	0.055	10.0	22026	0.000045
3.0	20.1	0.050			

Table III: Natural Logarithms (Base *e*)

x	$\ln x$	x	$\ln x$	x	$\ln x$
0.0		3.4	1.224	6.8	1.917
0.1	−2.303	3.5	1.253	6.9	1.932
0.2	−1.609	3.6	1.281	7.0	1.946
0.3	−1.204	3.7	1.308	7.1	1.960
0.4	−0.916	3.8	1.335	7.2	1.974
0.5	−0.693	3.9	1.361	7.3	1.988
0.6	−0.511	4.0	1.386	7.4	2.001
0.7	−0.357	4.1	1.411	7.5	2.015
0.8	−0.223	4.2	1.435	7.6	2.028
0.9	−0.105	4.3	1.459	7.7	2.041
1.0	0.000	4.4	1.482	7.8	2.054
1.1	0.095	4.5	1.504	7.9	2.067
1.2	0.182	4.6	1.526	8.0	2.079
1.3	0.262	4.7	1.548	8.1	2.092
1.4	0.336	4.8	1.569	8.2	2.104
1.5	0.405	4.9	1.589	8.3	2.116
1.6	0.470	5.0	1.609	8.4	2.128
1.7	0.531	5.1	1.629	8.5	2.140
1.8	0.588	5.2	1.649	8.6	2.152
1.9	0.642	5.3	1.668	8.7	2.163
2.0	0.693	5.4	1.686	8.8	2.175
2.1	0.742	5.5	1.705	8.9	2.186
2.2	0.788	5.6	1.723	9.0	2.197
2.3	0.833	5.7	1.740	9.1	2.208
2.4	0.875	5.8	1.758	9.2	2.219
2.5	0.916	5.9	1.775	9.3	2.230
2.6	0.956	6.0	1.792	9.4	2.241
2.7	0.993	6.1	1.808	9.5	2.251
2.8	1.030	6.2	1.825	9.6	2.262
2.9	1.065	6.3	1.841	9.7	2.272
3.0	1.099	6.4	1.856	9.8	2.282
3.1	1.131	6.5	1.872	9.9	2.293
3.2	1.163	6.6	1.887	10.0	2.303
3.3	1.194	6.7	1.902		

Table IV: Four-Place Common Logarithms (Base 10)

x	0	1	2	3	4	5	6	7	8	9
1.0	.0000	.0043	.0086	.0128	.0170	.0212	.0253	.0294	.0334	.0374
1.1	.0414	.0453	.0492	.0531	.0569	.0607	.0645	.0682	.0719	.0755
1.2	.0792	.0828	.0864	.0899	.0934	.0969	.1004	.1038	.1072	.1106
1.3	.1139	.1173	.1206	.1239	.1271	.1303	.1335	.1367	.1399	.1430
1.4	.1461	.1492	.1523	.1553	.1584	.1614	.1644	.1673	.1703	.1732
1.5	.1761	.1790	.1818	.1847	.1875	.1903	.1931	.1959	.1987	.2014
1.6	.2041	.2068	.2095	.2122	.2148	.2175	.2201	.2227	.2253	.2279
1.7	.2304	.2330	.2355	.2380	.2405	.2430	.2455	.2480	.2504	.2529
1.8	.2553	.2577	.2601	.2625	.2648	.2672	.2695	.2718	.2742	.2765
1.9	.2788	.2810	.2833	.2856	.2878	.2900	.2923	.2945	.2967	.2989
2.0	.3010	.3032	.3054	.3075	.3096	.3118	.3139	.3160	.3181	.3201
2.1	.3222	.3243	.3263	.3284	.3304	.3324	.3345	.3365	.3385	.3404
2.2	.3424	.3444	.3464	.3483	.3502	.3522	.3541	.3560	.3579	.3598
2.3	.3617	.3636	.3655	.3674	.3692	.3711	.3729	.3747	.3766	.3784
2.4	.3802	.3820	.3838	.3856	.3874	.3892	.3909	.3927	.3945	.3962
2.5	.3979	.3997	.4014	.4031	.4048	.4065	.4082	.4099	.4116	.4133
2.6	.4150	.4166	.4183	.4200	.4216	.4232	.4249	.4265	.4281	.4298
2.7	.4314	.4330	.4346	.4362	.4378	.4393	.4409	.4425	.4440	.4456
2.8	.4472	.4487	.4502	.4518	.4533	.4548	.4564	.4579	.4594	.4609
2.9	.4624	.4639	.4654	.4669	.4683	.4698	.4713	.4728	.4742	.4757
3.0	.4771	.4786	.4800	.4814	.4829	.4843	.4857	.4871	.4886	.4900
3.1	.4914	.4928	.4942	.4955	.4969	.4983	.4997	.5011	.5024	.5038
3.2	.5051	.5065	.5079	.5092	.5105	.5119	.5132	.5145	.5159	.5172
3.3	.5185	.5198	.5211	.5224	.5237	.5250	.5263	.5276	.5289	.5302
3.4	.5315	.5328	.5340	.5353	.5366	.5378	.5391	.5403	.5416	.5428
3.5	.5441	.5453	.5465	.5478	.5490	.5502	.5514	.5527	.5539	.5551
3.6	.5563	.5575	.5587	.5599	.5611	.5623	.5635	.5647	.5658	.5670
3.7	.5682	.5694	.5705	.5717	.5729	.5740	.5752	.5763	.5775	.5786
3.8	.5798	.5809	.5821	.5832	.5843	.5855	.5866	.5877	.5888	.5899
3.9	.5911	.5922	.5933	.5944	.5955	.5966	.5977	.5988	.5999	.6010
4.0	.6021	.6031	.6042	.6053	.6064	.6075	.6085	.6096	.6107	.6117
4.1	.6128	.6138	.6149	.6160	.6170	.6180	.6191	.6201	.6212	.6222
4.2	.6232	.6243	.6253	.6263	.6274	.6284	.6294	.6304	.6314	.6325
4.3	.6335	.6345	.6355	.6365	.6375	.6385	.6395	.6405	.6415	.6425
4.4	.6435	.6444	.6454	.6464	.6474	.6484	.6493	.6503	.6513	.6522
4.5	.6532	.6542	.6551	.6561	.6571	.6580	.6590	.6599	.6609	.6618
4.6	.6628	.6637	.6646	.6656	.6665	.6675	.6684	.6693	.6702	.6712
4.7	.6721	.6730	.6739	.6749	.6758	.6767	.6776	.6785	.6794	.6803
4.8	.6812	.6821	.6830	.6839	.6848	.6857	.6866	.6875	.6884	.6893
4.9	.6902	.6911	.6920	.6928	.6937	.6946	.6955	.6964	.6972	.6981
5.0	.6990	.6998	.7007	.7016	.7024	.7033	.7042	.7050	.7059	.7067
5.1	.7076	.7084	.7093	.7101	.7110	.7118	.7126	.7135	.7143	.7152
5.2	.7160	.7168	.7177	.7185	.7193	.7202	.7210	.7218	.7226	.7235
5.3	.7243	.7251	.7259	.7267	.7275	.7284	.7292	.7300	.7308	.7316
5.4	.7324	.7332	.7340	.7348	.7356	.7364	.7372	.7380	.7388	.7396
N	**0**	**1**	**2**	**3**	**4**	**5**	**6**	**7**	**8**	**9**

Table IV: Continued

x	0	1	2	3	4	5	6	7	8	9
5.5	.7404	.7412	.7419	.7427	.7435	.7443	.7451	.7459	.7466	.7474
5.6	.7482	.7490	.7497	.7505	.7513	.7520	.7528	.7536	.7543	.7551
5.7	.7559	.7566	.7574	.7582	.7589	.7597	.7604	.7612	.7619	.7627
5.8	.7634	.7642	.7649	.7657	.7664	.7672	.7679	.7686	.7694	.7701
5.9	.7709	.7716	.7723	.7731	.7738	.7745	.7752	.7760	.7767	.7774
6.0	.7782	.7789	.7796	.7803	.7810	.7818	.7825	.7832	.7839	.7846
6.1	.7853	.7860	.7868	.7875	.7882	.7889	.7896	.7903	.7910	.7917
6.2	.7924	.7931	.7938	.7945	.7952	.7959	.7966	.7973	.7980	.7987
6.3	.7993	.8000	.8007	.8014	.8021	.8028	.8035	.8041	.8048	.8055
6.4	.8062	.8069	.8075	.8082	.8089	.8096	.8102	.8109	.8116	.8122
6.5	.8129	.8136	.8142	.8149	.8156	.8162	.8169	.8176	.8182	.8189
6.6	.8195	.8202	.8209	.8215	.8222	.8228	.8235	.8241	.8248	.8254
6.7	.8261	.8267	.8274	.8280	.8287	.8293	.8299	.8306	.8312	.8319
6.8	.8325	.8331	.8338	.8344	.8351	.8357	.8363	.8370	.8376	.8382
6.9	.8388	.8395	.8401	.8407	.8414	.8420	.8426	.8432	.8439	.8445
7.0	.8451	.8457	.8463	.8470	.8476	.8482	.8488	.8494	.8500	.8506
7.1	.8513	.8519	.8525	.8531	.8537	.8543	.8549	.8555	.8561	.8567
7.2	.8573	.8579	.8585	.8591	.8597	.8603	.8609	.8615	.8621	.8627
7.3	.8633	.8639	.8645	.8651	.8657	.8663	.8669	.8675	.8681	.8686
7.4	.8692	.8698	.8704	.8710	.8716	.8722	.8727	.8733	.8739	.8745
7.5	.8751	.8756	.8762	.8768	.8774	.8779	.8785	.8791	.8797	.8802
7.6	.8808	.8814	.8820	.8825	.8831	.8837	.8842	.8848	.8854	.8859
7.7	.8865	.8871	.8876	.8882	.8887	.8893	.8899	.8904	.8910	.8915
7.8	.8921	.8927	.8932	.8938	.8943	.8949	.8954	.8960	.8965	.8971
7.9	.8976	.8982	.8987	.8993	.8998	.9004	.9009	.9015	.9020	.9025
8.0	.9031	.9036	.9042	.9047	.9053	.9058	.9063	.9069	.9074	.9079
8.1	.9085	.9090	.9096	.9101	.9106	.9112	.9117	.9122	.9128	.9133
8.2	.9138	.9143	.9149	.9154	.9159	.9165	.9170	.9175	.9180	.9186
8.3	.9191	.9196	.9201	.9206	.9212	.9217	.9222	.9227	.9232	.9238
8.4	.9243	.9248	.9253	.9258	.9263	.9269	.9274	.9279	.9284	.9289
8.5	.9294	.9299	.9304	.9309	.9315	.9320	.9325	.9330	.9335	.9340
8.6	.9345	.9350	.9355	.9360	.9365	.9370	.9375	.9380	.9385	.9390
8.7	.9395	.9400	.9405	.9410	.9415	.9420	.9425	.9430	.9435	.9440
8.8	.9445	.9450	.9455	.9460	.9465	.9469	.9474	.9479	.9484	.9489
8.9	.9494	.9499	.9504	.9509	.9513	.9518	.9523	.9528	.9533	.9538
9.0	.9542	.9547	.9552	.9557	.9562	.9566	.9571	.9576	.9581	.9586
9.1	.9590	.9595	.9600	.9605	.9609	.9614	.9619	.9624	.9628	.9633
9.2	.9638	.9643	.9647	.9652	.9657	.9661	.9666	.9671	.9675	.9680
9.3	.9685	.9689	.9694	.9699	.9703	.9708	.9713	.9717	.9722	.9727
9.4	.9731	.9736	.9741	.9745	.9750	.9754	.9759	.9763	.9768	.9773
9.5	.9777	.9782	.9786	.9791	.9795	.9800	.9805	.9809	.9814	.9818
9.6	.9823	.9827	.9832	.9836	.9841	.9845	.9850	.9854	.9859	.9863
9.7	.9868	.9872	.9877	.9881	.9886	.9890	.9894	.9899	.9903	.9908
9.8	.9912	.9917	.9921	.9926	.9930	.9934	.9939	.9943	.9948	.9952
9.9	.9956	.9961	.9965	.9969	.9974	.9978	.9983	.9987	.9991	.9996
N	0	1	2	3	4	5	6	7	8	9

ANSWERS TO ODD-NUMBERED EXERCISES AND SAMPLE TEST QUESTIONS

CHAPTER 1: REAL NUMBERS, EQUATIONS, INEQUALITIES

1.1 REAL NUMBERS AND THEIR PROPERTIES (*Page 6*)

1 Closure property for addition. **3** Inverse property for addition. **5** Commutative property for multiplication.
7 Commutative property for addition. **9** Identity property for addition. **11** Additive inverse.
13 Multiplication property of zero. **15** Distributive property. **17** 3 **19** 5 **21** 7
23 If $ab = 0$, then at least one of a or b is zero. Since $5 \neq 0$, $x - 2 = 0$ which implies $x = 2$.
25 (i) $a(-b) = -(ab)$; (ii) Definition of $\div$; (iii) associative of $\times$.
27 (i) Distributive; (ii) distributive; (iii) associative for $+$; (iv) commutative for $\times$; (v) commutative for $+$; (vi) distributive; (vii) commutative for $+$.
29 No; as a counterexample $2^3 \neq 3^2$.
31 If $10 - x = 7$, then $7 + x = 10$; if $7 + x = 10$, then $10 - x = 7$.
33 If n is an even integer, then n^2 is an even integer; if n^2 is an even integer, then n is an even integer.
35 If $\triangle ABC$ is congruent to $\triangle DEF$, then the three sides of one triangle are congruent to the three sides of the other triangle; if three sides of $\triangle ABC$ are congruent to the three sides of $\triangle DEF$, then $\triangle ABC$ is congruent to $\triangle DEF$.
37 n is a multiple of 3 if and only if n^2 is a multiple of 3.
39 Let $\frac{0}{0} = x$, where x is some number. Then the definition of division gives $0 \cdot x = 0$. Since any number x will work, the answer to $\frac{0}{0}$ is not unique; therefore, $\frac{0}{0}$ is undefined.

1.2 SETS OF NUMBERS (*Page 10*)

1 0, 1, 2 **3** 3, 4, 5, 6 **5** $-2, -1$ **7** $\ldots, -3, -2, -1, 0$ **9** There are none. **11** True. **13** True. **15** False. **17** False. **19** True. **21** b, c, e **23** d, e **25** a, b, c, e **27** a, b, c, e **29** d, e **31** c, e **33**

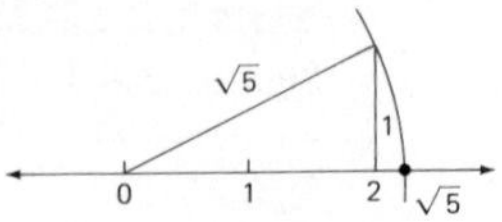

35 π; circumference $= 2\pi(\frac{1}{2}) = \pi$ **37** 0.8 **39** $0.\overline{285714}$ **41** $3.\overline{461538}$ **43** $0.01\overline{4}$ **45** **(a)** $\frac{5}{11}$; **(b)** $\frac{37}{99}$; **(c)** $\frac{26}{111}$; **(d)** 1

1.3 INTRODUCTION TO EQUATIONS AND PROBLEM SOLVING (*Page 16*)

1 $x = 4$ **3** $x = -4$ **5** $x = -4$ **7** $x = -8$ **9** $x = \frac{9}{2}$ **11** $x = -9$ **13** $x = \frac{20}{3}$ **15** $x = 15$ **17** $x = -10$ **19** $x = -\frac{22}{3}$ **21** $x = 5$ **23** $x = 3$ **25** $x = 90$ **27** $\dfrac{P - 2\ell}{2}$ **29** $\dfrac{N - u}{10}$ **31** $\dfrac{C}{2\pi}$ **33** $\dfrac{w - 7}{4}$ **35** 50°F **37** 0°C

39 $w = 7$, $l = 21$ **41** 10 at 10¢, 13 at 20¢, 5 at 18¢. **43** $3\frac{1}{2}$ hours.

45 Let $x + (x + 2) + (x + 4) = 180$. Then $3x = 174$ and $x = 58$. Therefore, the integers must be even.

47 $w = 5$, $l = 14$ **49** 77, 79, 81

1.4 PROPERTIES OF ORDER (*Page 20*)

1 $<$ **3** $<$ **5** $<$ **7** $=$ **9** **(a)** $x < 1$; **(b)** $x > -\frac{2}{3}$; **(c)** $x \geq 5$; **(d)** $-10 < x \leq 7$ **11** True.

13 False; $-10 < 5$ and $-4 < 6$, but $(-10)(-4) = 40 > 30$ **15** True. **17** True.

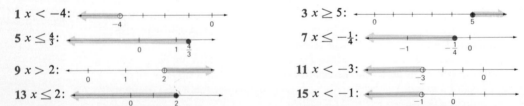

19 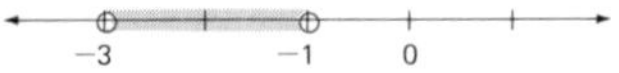

21 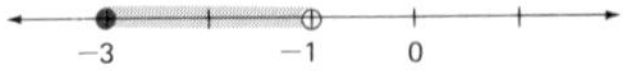

23 

25 

27 $[-5, 2]$ **29** $[-6, 0)$ **31** $(-10, 10)$ **33** $(-\infty, 5)$ **35** $[-2, \infty)$

37 $(-\infty, -1]$ **39** **(a)** $b - a$ is positive; **(b)** $(b - a)c$; **(c)** $bc - ac$; **(d)** $bc - ac$; **(e)** $ac < bc$

41 We are given that $1 < a$. Since $0 < 1$ the transitive property gives $0 < a$. Now use Rule 3 to get $a \cdot 1 < a \cdot a$, or $a < a^2$.

1.5 CONDITIONAL INEQUALITIES (*Page 25*)

1 $x < -4$: 

3 $x \geq 5$: 

5 $x \leq \frac{4}{3}$: 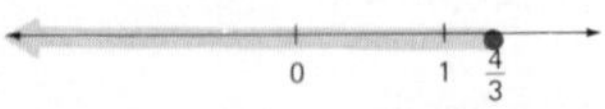

7 $x \leq -\frac{1}{4}$: 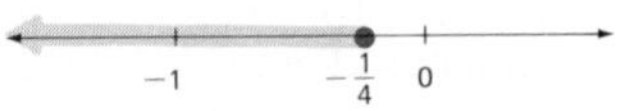

9 $x > 2$: 

11 $x < -3$: 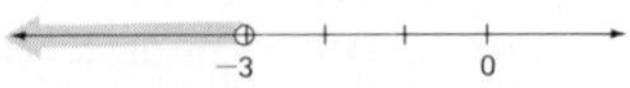

13 $x \leq 2$: 

15 $x < -1$: 

17 **(a)** Negative; **(b)** negative. **19** **(a)** Positive; **(b)** positive; **(c)** negative; **(d)** negative.

21 $x \leq 2$: 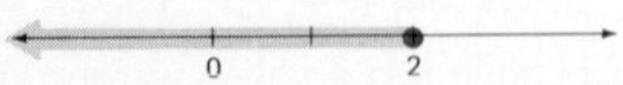

23 $x \neq 13$: 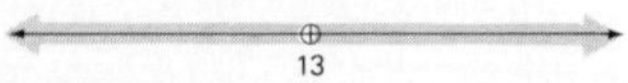

25 $x \leq 3$: 

27 $x \leq -8$: 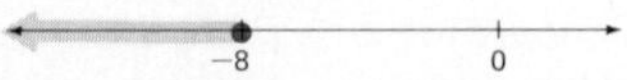

29

Interval	$x < -1$	$-1 < x < 1$	$1 < x$
Sign of $x + 1$	$-$	$+$	$+$
Sign of $x - 1$	$-$	$-$	$+$
Sign of $(x + 1)(x - 1)$	$+$	$-$	$+$

Solution: $-1 < x < 1$

31

Interval	$x < -5$	$-5 < x < \frac{1}{2}$	$\frac{1}{2} < x$
Sign of $x + 5$	$-$	$+$	$+$
Sign of $2x - 1$	$-$	$-$	$+$
Sign of $(x + 5)(2x - 1)$	$+$	$-$	$+$

Solution: $-5 \leq x \leq \frac{1}{2}$

33

Interval	$x < -1$	$-1 < x < 0$	$0 < x$
Sign of $x + 1$	$-$	$+$	$+$
Sign of x	$-$	$-$	$+$
Sign of $\frac{x}{x+1}$	$+$	$-$	$+$

Solution: $x < -1$ or $x > 0$

35

Interval	$x < -5$	$-5 < x < 2$	$2 < x$
Sign of $x + 5$	$-$	$+$	$+$
Sign of $2 - x$	$+$	$+$	$-$
Sign of $\frac{x+5}{2-x}$	$-$	$+$	$-$

Solution: $x < -5$ or $x > 2$

37

Interval	$x<-3$	$-3<x<1$	$1<x<2$	$2<x$
Sign of $x+3$	$-$	$+$	$+$	$+$
Sign of $x-1$	$-$	$-$	$+$	$+$
Sign of $x-2$	$-$	$-$	$-$	$+$
Sign of $(x+3)(x-1)(x-2)$	$-$	$+$	$-$	$+$

Solution: $-3 \leq x \leq 1$ or $x \geq 2$

39 $x<2$ or $x>5$

1.6 ABSOLUTE VALUE (*Page 30*)

1 False. **3** False. **5** True. **7** True. **9** False. **11** $x=\pm\frac{1}{2}$ **13** $x=-2, x=4$ **15** $x=-2, x=5$ **17** $x=1, x=7$ **19** $x=-\frac{20}{3}, x=4$ **21** $x>0$

23 $x=-4$ or $x=2$: −4 0 2

25 $x \leq -2$ or $x \geq 4$: −2 0 4

27 $-5 \leq x \leq 1$: −5 0 1

29 $x=-5$ or $x=5$: −5 0 5

31 $x \leq -5$ or $x \geq 5$: −5 0 5

33 $2 \leq x \leq 8$: 0 2 8

35 $2.9<x<3.1$: 2.9 3 3.1

37 $-3<x<4$: −3 0 4

39 $2<x<6$: −1 0 2 6

41 $3 \leq x \leq 5$: 0 3 5

43 All $x \neq 3$: 0 3

45 -5 **47** $\frac{11}{2}$ **49** 0

51 $|5-2| \geq ||5|-|2||;\ 3 \geq 3$
$|5-(-2)| \geq ||5|-|-2||;\ 7 \geq 3$
$|-5-2| \geq ||-5|-|2||;\ 7 \geq 3$
$|-5-(-2)| \geq ||-5|-|-2||;\ 3 \geq 3$

53
$$\left|\frac{x}{y}\right| = -\frac{x}{y} \quad \text{since } \frac{x}{y}<0$$
$$= \frac{-x}{y}$$
$$= \frac{|x|}{|y|} \quad \text{since } x<0 \text{ and } y>0$$

55 **(a)** $|x|<3$; **(b)** $|2x|<\frac{1}{2}$; **(c)** $|x-2|<3$; **(d)** $|x-\frac{1}{2}|<1$

57 **(a)** $\frac{9}{10}<x<\frac{11}{10}$; **(b)** $\frac{27}{10}<3x<\frac{33}{10}$; **(c)** $-\frac{3}{10}<3x-3<\frac{3}{10}$; **(d)** $-\frac{3}{10}<y-8<\frac{3}{10}$; **(e)** $|y-8|<\frac{3}{10}$

SAMPLE TEST QUESTIONS FOR CHAPTER 1 (*Page 32*)

1

	Whole numbers	*Integers*	*Rational numbers*	*Irrational numbers*	*Negative integers*
$\frac{-6}{3}$		✓	✓		✓
0.231			✓		
$\sqrt{12}$				✓	
$\sqrt{\frac{9}{4}}$			✓		
1983	✓	✓	✓		

2 Construct a perpendicular of unit length at the point with coordinate 3. The hypotenuse of the resulting right triangle has length $\sqrt{3^2 + 1^2} = \sqrt{10}$. Now use a compass with this hypotenuse as radius and the origin as center to locate $\sqrt{10}$ on the number line.
3 (a) False; (b) true; (c) false; (d) true; (e) false; (f) false; (g) true; (h) false.
4 (a) Distributive property; (b) identity property for addition; (c) identity property for multiplication; (d) definition of subtraction; (e) associative property for multiplication; (f) definition of $a < b$.
5 (a) If $a = b$, then $a + c = b + c$; (b) if $a = b$, then $ac = bc$. 6 (a) 1; (b) 3 7 $x = 16$ 8 $x = -11$
9 $x = -\frac{9}{2}$ 10 Width = 7 inches; length = 19 inches. 11 2:27 P.M.

12 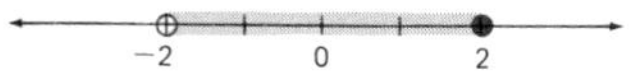

13 

14 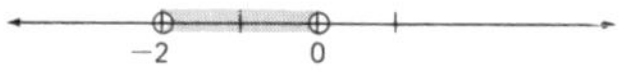

15 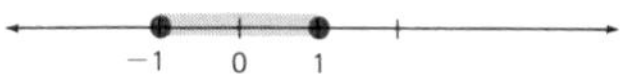

16 $x = \frac{5}{4}, x = \frac{1}{4}$ 17 $x < -2$ 18 $x > 2$

19 $x < \frac{2}{9}$: 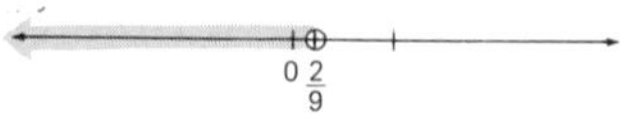

20 $x \leq -1$ or $x \geq 2$: 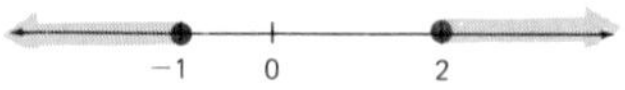

21 $x > 2$: 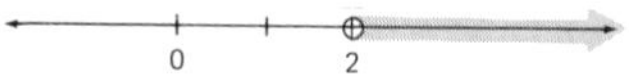

22 $-5 < x < \frac{1}{2}$: 

CHAPTER 2: FUNDAMENTALS OF ALGEBRA

2.1 INTEGRAL EXPONENTS (*Page 41*)

1 False; 3^6. 3 False; 2^7. 5 True. 7 True. 9 False; $2 \cdot 3^4$. 11 False; 1. 13 False; 2^3. 15 False; 2^{-8}.
17 7 19 $\frac{5}{3}$ 21 1 23 4 25 24 27 4 29 $\frac{15}{4}$ 31 $\frac{1}{x^6}$ 33 x^{12} 35 $72a^5$ 37 $16a^8$ 39 x^6y^2 41 $\frac{x^8}{y^6}$ 43 $\frac{y^7}{x^5}$
45 $5x$ 47 $\frac{b^2}{9}$ 49 $(a+b)^6$ 51 $-3x^3y^3$ 53 $\frac{1}{(a+3b)^{22}}$ 55 $\frac{1}{x^2} + \frac{1}{y^2}$ 57 x^2y^3 59 $\frac{10}{(4-5x)^3}$ 61 9 63 8
65 -3 67 $\frac{ab}{a+b}$ 69 $\frac{1}{a^{-n}} = \frac{1}{\frac{1}{a^n}} = 1 \cdot \frac{a^n}{1} = a^n$

2.2 RADICALS AND RATIONAL EXPONENTS (*Page 47*)

1 $11^{1/2}$ 3 $9^{1/4}$ 5 $6^{2/3}$ 7 $(-\frac{1}{3})^{3/5}$ 9 $\sqrt{3}$ 11 $\sqrt[3]{-19}$ 13 $\frac{1}{\sqrt{2}}$ 15 $\frac{1}{\sqrt[4]{\frac{3}{4}}}$ 17 True. 19 True.
21 False; -2. 23 False; 1.2. 25 True. 27 5 29 $\frac{1}{9}$ 31 $\frac{1}{16}$ 33 -3 35 $\frac{1}{2}$ 37 9 39 50 41 $\frac{1}{32}$ 43 5
45 13 47 $\frac{\sqrt[3]{35}}{6}$ 49 $\frac{148}{135}$ 51 $-\frac{11}{4}$ 53 $\frac{4a^2}{b^6}$ 55 $\frac{1}{a^3b^6}$ 57 1 59 $16a^4b^6$ 61 $\frac{a^2}{b}$ 63 $\frac{3x}{(3x^2+2)^{1/2}}$
65 $\frac{x+2}{(x^3+4x)^{1/2}}$ 67 x 69 $\frac{1}{xy^n}$
71 Let $x = \sqrt[n]{a}$, $y = \sqrt[n]{b}$. Then $x^n = a$ and $y^n = b$. Now we get $\frac{a}{b} = \frac{x^n}{y^n} = \left(\frac{x}{y}\right)^n$. Thus, by definition, $\sqrt[n]{\frac{a}{b}} = \frac{x}{y}$.
But $x = \sqrt[n]{a}$ and $y = \sqrt[n]{b}$. Therefore, $\sqrt[n]{\frac{a}{b}} = \frac{\sqrt[n]{a}}{\sqrt[n]{b}}$.

2.3 SIMPLIFYING RADICALS (*Page 51*)

1 $4\sqrt{2}$ 3 12 5 $-\sqrt{3}$ 7 $10\sqrt{2}$ 9 $17\sqrt{5}$ 11 $10\sqrt{2}$ 13 $6\sqrt[3]{2}$ 15 $7\sqrt{2}$ 17 $14\sqrt{2}$ 19 $3\sqrt[3]{7x}$
21 $|x|\sqrt{2}$ 23 $5\sqrt[4]{2}$ 25 $5\sqrt{10}$ 27 $\frac{33\sqrt{2}}{2}$ 29 $5\sqrt{6} - 3\sqrt{2}$ 31 $7\sqrt{2x}$ 33 $12|x|$

35 $|x|\sqrt{y} + 12|x|\sqrt{2y}$ 37 $7a\sqrt{5a}$ 39 $4\sqrt{5}$ 41 $4x\sqrt{2}$ 43 $2\sqrt{2} + 2\sqrt{3}$ 45 $\frac{\sqrt{2}}{6}$ 47 $\frac{8\sqrt{3}}{|x|}$ 49 $4\sqrt[3]{4}$
51 $|xy| = \sqrt{(xy)^2} = \sqrt{x^2y^2} = \sqrt{x^2}\sqrt{y^2} = |x||y|$

2.4 FUNDAMENTAL OPERATIONS WITH POLYNOMIALS (*Page 56*)

1 $8x^2 - 2x + 7$ 3 $4x^3 - 9x^2 + 3$ 5 $2x^3 + x^2 + 11x - 28$ 7 $x^3 - 7x^2 - 10x + 8$ 9 $4x^3 - x^2 - 5x - 22$
11 $3x + 1$ 13 $6y + 6$ 15 $x^3 + 2x^2 + x$ 17 $7y + 8$ 19 $-4x^2 - 5x + 3$ 21 $-20x^4 + 4x^3 + 2x^2$
23 $x^2 + 2x + 1$ 25 $4x^2 + 26x - 14$ 27 $20x^2 - 42x + 18$ 29 $x^2 + \frac{1}{4}x - \frac{1}{8}$ 31 $-6x^2 - 3x + 18$
33 $-6x^2 + 3x + 18$ 35 $\frac{4}{9}x^2 + 8x + 36$ 37 $-63 + 55x - 12x^2$ 39 $a^2x^2 - b^2$ 41 $x^2 - 0.01$
43 $\frac{1}{25}x^2 - \frac{1}{10}x + \frac{1}{16}$ 45 $x - 100$ 47 $x - 2$ 49 1 51 $x^4 - 2x^3 + 2x^2 - 31x - 36$
53 $8x^4 + 12x^3 + 6x^2 - 18x - 27$ 55 $x^3 - 8$ 57 $x^5 - 32$ 59 $x^{4n} - x^{2n} - 2$ 61 $3x - 6x^2 + 3x^3$
63 $-6x^3 + 19x^2 - x - 6$ 65 $5x^4 - 12x^2 + 2x + 4$ 67 $6x^5 + 15x^4 - 12x^3 - 27x^2 + 10x + 15$
69 $x^3 - 3x^2 + 3x - 1$ 71 $a^4 + 4a^3b + 6a^2b^2 + 4ab^3 + b^4$ 73 $8x^3 + 36x^2 + 54x + 27$
75 $\frac{1}{27}x^3 + x^2 + 9x + 27$ 77 $6(\sqrt{5} + \sqrt{3})$ 79 $-2(\sqrt{2} + 3)$ 81 $\frac{x + 2\sqrt{xy} + y}{x - y}$ 83 $\frac{-4}{5 - 3\sqrt{5}}$
85 $\frac{x - y}{x - 2\sqrt{xy} + y}$ 87 Multiply numerator and denominator of the first fraction by $\sqrt{4 + h} + 2$ and simplify.

2.5 FACTORING POLYNOMIALS (*Page 62*)

1 $5(x - 1)$ 3 $2x(8x^3 + 4x^2 + 2x + 1)$ 5 $2ab(2a - 3)$ 7 $2x(y + 2x + 4x^3)$ 9 $5(a - b)(2x + y)$
11 $(9 + x)(9 - x)$ 13 $(2x + 3)(2x - 3)$ 15 $(a + 11b)(a - 11b)$ 17 $(x + 4)(x^2 - 4x + 16)$
19 $(5x - 4)(25x^2 + 20x + 16)$ 21 $(2x + 7y)(4x^2 - 14xy + 49y^2)$ 23 $(\sqrt{3} + 2x)(\sqrt{3} - 2x)$
25 $(\sqrt{x} + 6)(\sqrt{x} - 6)$ 27 $(2\sqrt{2} + \sqrt{3x})(2\sqrt{2} - \sqrt{3x})$ 29 $(\sqrt[3]{7} + a)[(\sqrt[3]{7})^2 - a\sqrt[3]{7} + a^2]$
31 $(3\sqrt[3]{x} + 1)(9\sqrt[3]{x^2} - 3\sqrt[3]{x} + 1)$ 33 $(\sqrt[3]{3x} - \sqrt[3]{4})(\sqrt[3]{9x^2} + \sqrt[3]{12x} + \sqrt[3]{16})$ 35 $(x - 1)(x + y)$
37 $(x - 1)(y - 1)$ 39 $(2 - y^2)(1 + x)$ 41 $7(x + h)(x^2 - hx + h^2)$ 43 $xy(x + y)(x - y)$
45 $(x^2 + y^2)(x + y)(x - y)$ 47 $(a^4 + b^4)(a^2 + b^2)(a + b)(a - b)$
49 $(a - 2)(a^4 + 2a^3 + 4a^2 + 8a + 16)$ 51 $(1 - h)(1 + h + h^2 + h^3 + h^4 + h^5 + h^6)$
53 $3(a + 1)(x + 2)(x^2 - 2x + 4)$ 55 $(x - y)(a + b)(a^2 - ab + b^2)$ 57 $(x^2 + 4y^2)(x + 2y)(x - 2y)^2$
59 $(x + 2)^2(8x + 7)$ 61 $x(x^3 + 1)^2(11x^3 - 9x + 2)$ or $x(x + 1)^2(x^2 - x + 1)^2(11x^3 - 9x + 2)$
63 **(a)** $(x + 2)(x^4 - 2x^3 + 4x^2 - 8x + 16)$; **(b)** $x(2x + y)(64x^6 - 32x^5y + 16x^4y^2 - 8x^3y^3 + 4x^2y^4 - 2xy^5 + y^6)$.

2.6 FACTORING TRINOMIALS (*Page 66*)

1 $(x + 2)^2$ 3 $(a - 7)^2$ 5 $(1 + b)^2$ 7 $4(a + 1)^2$ 9 $9(x - y)^2$ 11 $(3x + 2y)^2$ 13 $(x + 2)(x + 3)$
15 $(x + 17)(x + 3)$ 17 $(5a - 1)(4a - 1)$ 19 $(x + 2)(x + 18)$ 21 $(3x + 1)^2$ 23 $(5a - 1)^2$
25 $(3x + 2)(x + 6)$ 27 $(7x + 1)(2x + 5)$ 29 $(8x - 1)(x - 1)$ 31 $(2a + 5)^2$ 33 $(b + 15)(b + 3)$
35 $2(2x - 3)(2x - 1)$ 37 $(4a - 3)(3a - 4)$ 39 $(5x + 1)(3x - 2)$ 41 $(3a + 7)(2a - 3)$ 43 $(2x - 1)(2x + 3)$
45 $(8a + 3b)(3a + 2b)$ 47 $5(x + 1)(x + 4)$ 49 $a(2x + 1)^2$ 51 Not factorable. 53 $2(3x - 5)(x + 2)$
55 Not factorable. 57 $2(x - 8)(x + 7)$ 59 $2(b + 2)(b + 4)$ 61 $ab(a - b)^2$ 63 $8(2x - 1)(x - 1)$
65 $25(a + b)^2$ 67 $(a - 1)^2(a^2 + a + 1)^2$ 69 $3xy(x^2 + 2y)(2x^2 - 5y)$ 71 $(a - b)^2(a^2 + ab + b^2)^2$

2.7 FUNDAMENTAL OPERATIONS WITH RATIONAL EXPRESSIONS (*Page 72*)

1 False; $\frac{1}{21}$. 3 False; $\frac{ax}{2} - \frac{5b}{6}$. 5 True. 7 $\frac{2x}{3z}$ 9 $-x$ 11 $\frac{n + 1}{n^2 + 1}$ 13 $\frac{3(x - 1)}{2(x + 1)}$ 15 $\frac{2x + 3}{2x - 3}$
17 $\frac{a - 4b}{a^2 - 4ab + 16b^2}$ 19 $\frac{2y}{x}$ 21 $\frac{2}{a^2}$ 23 $\frac{x}{y}$ 25 $\frac{-3a - 8b}{6}$ 27 $\frac{x^2 + 1}{3(x + 1)}$ 29 1 31 $\frac{2 - xy}{x}$ 33 $\frac{5y^2 - y}{y^2 - 1}$
35 $\frac{3x}{x + 1}$ 37 $\frac{8 - x}{x^2 - 4}$ 39 $\frac{3(x + 9)}{(x - 3)(x + 3)^2}$ 41 $x - 3$ 43 $\frac{2(n^2 + 4)}{n + 2}$ 45 $\frac{1}{2(x + 2)}$ 47 $-\frac{1}{4x}$ 49 $-\frac{1}{4(4 + h)}$

51 $-\dfrac{1}{3(x+3)}$ 53 $\dfrac{4-x}{16x^2}$ 55 $\dfrac{xy}{y+x}$ 57 $\dfrac{-4x}{(1+x^2)^2}$ 59 $-\dfrac{4}{x^2}$ 61 $-\dfrac{1}{ab}$ 63 $\dfrac{y^2-x^2}{x^3y^3}$

65 (a) $\dfrac{\dfrac{AD}{B}+C}{D} = \dfrac{AD+BC}{BD} = \dfrac{AD}{BD} + \dfrac{BC}{BD} = \dfrac{A}{B} + \dfrac{C}{D}$

(b) $\left[\dfrac{\left(\dfrac{AB}{D}+C\right)D}{F} + E\right] \cdot F = \left[\dfrac{AB+CD}{F} + E\right] \cdot F = AB + CD + EF$

67 $\dfrac{2xy^2 - 2x^2y\left(\dfrac{x^2}{y^2}\right)}{y^4} = \dfrac{2xy^2 - \dfrac{2x^4}{y}}{y^4} = \dfrac{2xy^3 - 2x^4}{y^5} = \dfrac{2x(x^3+8) - 2x^4}{y^5} = \dfrac{16x}{y^5}$

SAMPLE TEST QUESTIONS FOR CHAPTER 2 (*Page 75*)

1 False. 2 True. 3 False. 4 False. 5 True. 6 False. 7 (b) 8 (a) 9 (b) 10 $7cd - 6a^2$
11 $11x^2 - 14x - 21$ 12 $5x^4 + 12x^3 - 2x - 3$ 13 $(4-3b)(16+12b+9b^2)$ 14 $3x(x-2)(x+2)(x^2+4)$
15 $(2x-3)(3x+1)$ 16 $(x-2)(x^4+2x^3+4x^2+8x+16)$ 17 $(2x+y^2)(x-3y)$ 18 $(x+3)(ax-2)$
19 $\dfrac{2(x+3)}{(x+2)(x+5)}$ 20 x 21 $-\dfrac{7+x}{49x^2}$ 22 $\dfrac{1}{x+1}$ 23 $\dfrac{3x^2-18x+25}{(x-3)(x+3)(2x-5)}$

CHAPTER 3: LINEAR FUNCTIONS AND EQUATIONS

3.1 INTRODUCTION TO THE FUNCTION CONCEPT (*Page 77*)

1 Function: all reals. 3 Function: all $x > 0$. 5 Not a function. 7 Function: all $x \neq -1$. 9 Not a function.
11 Function: $x < -2$ or $x > 2$. 13 Not a function; for $x = 3$, $y = 4$ or $y = 1$. 15 True. 17 False; -3.
19 True. 21 False: $-x^2 + 16$. 23 True. 25 (a) -3; (b) -1; (c) 0. 27 (a) 1; (b) 0; (c) $\frac{1}{4}$.
29 (a) -2; (b) -1; (c) $-\frac{7}{8}$. 31 (a) 2; (b) 0; (c) $\frac{5}{16}$. 33 (a) $-\frac{1}{2}$; (b) -1; (c) -2.
35 (a) -1; (b) does not exist; (c) $\sqrt[3]{2}$. 37 (a) 81; (b) 5; (c) $\frac{25}{36}$; (d) $\frac{1}{36}$. 39 $3h(2) = 24 \neq 48 = h(6)$
41 $x + 3$ 43 $-\dfrac{1}{3x}$ 45 2 47 1 49 $-4-h$ 51 $-\dfrac{4+h}{4(2+h)^2}$

3.2 THE RECTANGULAR COORDINATE SYSTEM (*Page 83*)

1

x	−3	−2	−1	0	1	2
y	−5	−4	−3	−2	−1	0

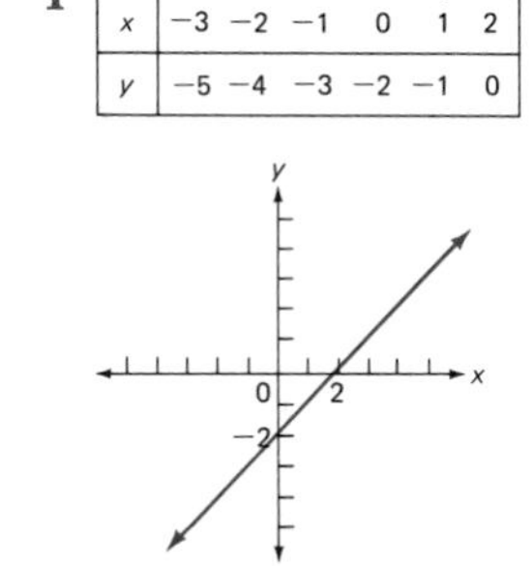

3

x	−2	−1	0	1	2
y	−8	−6	−4	−2	0

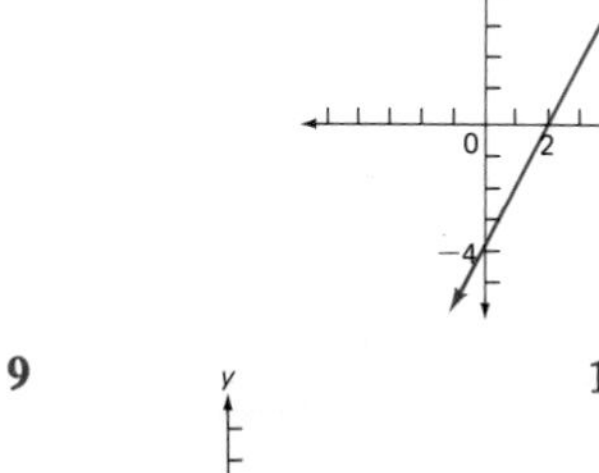

5

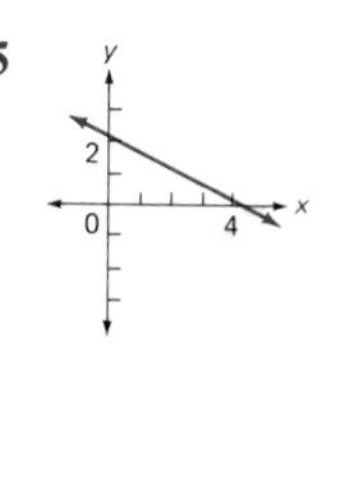

7

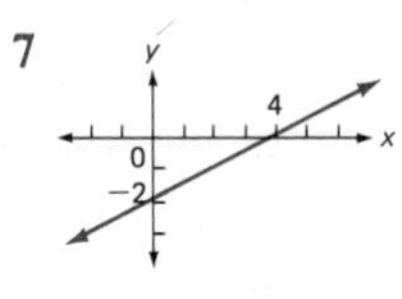

9

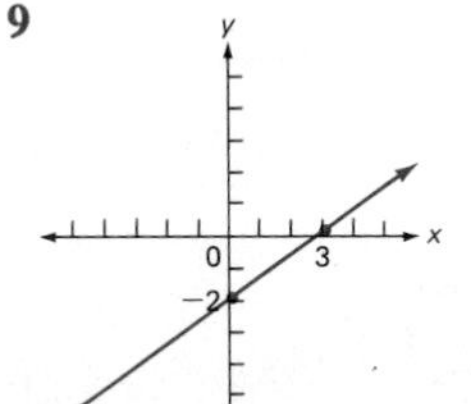

11

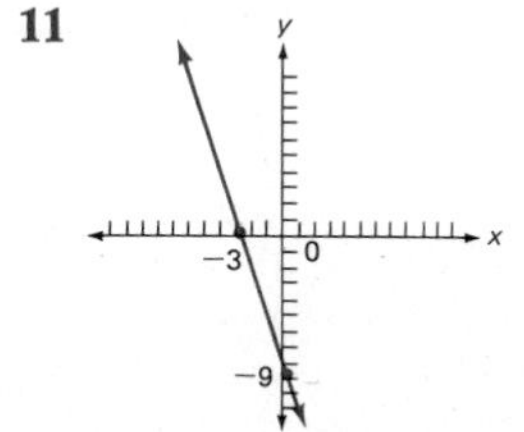

13

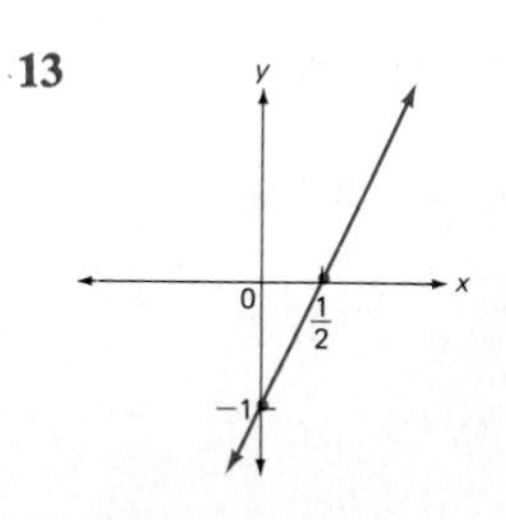

15

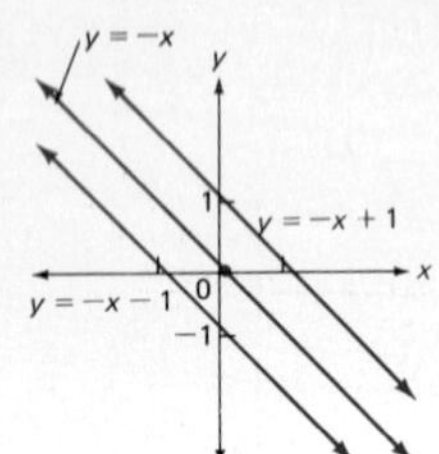

17

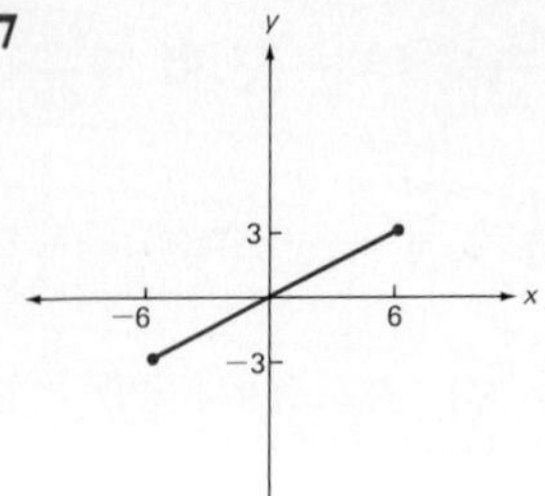

19

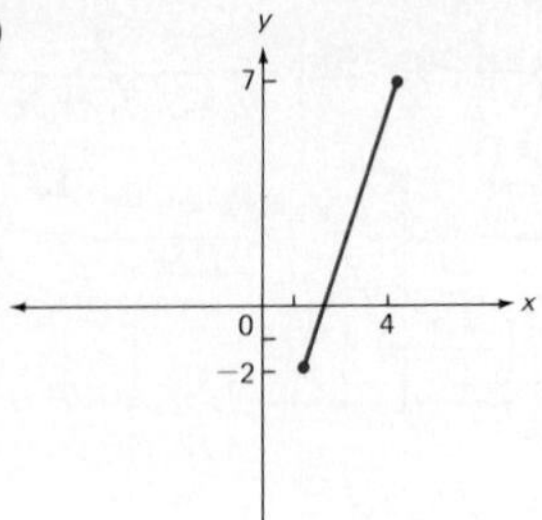

21 (a)

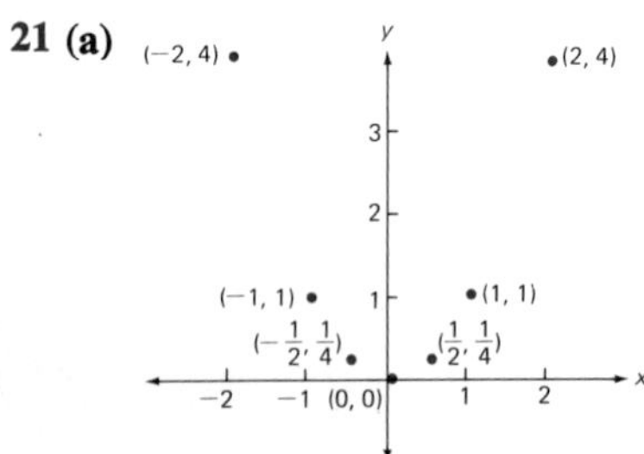

(b)

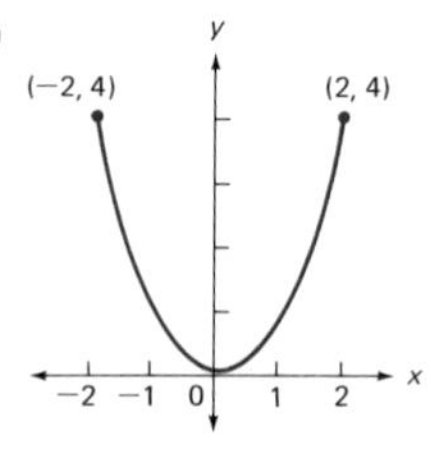

23 (a)

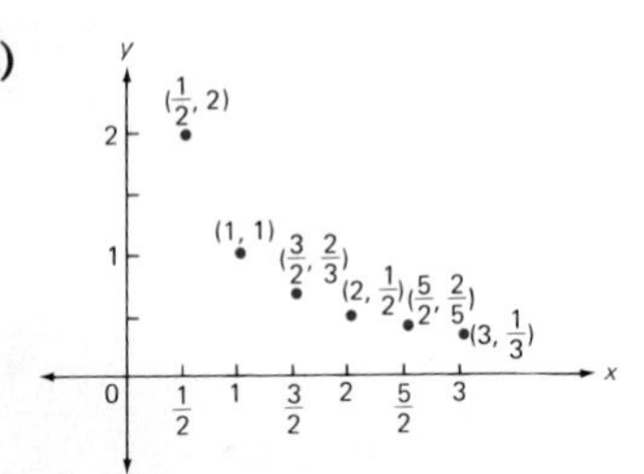

(b)

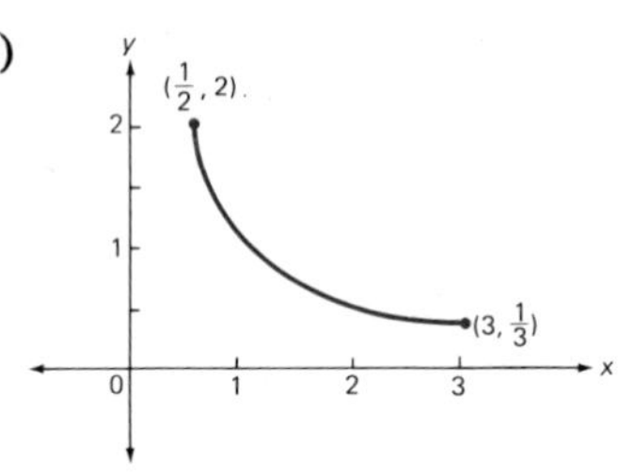

25

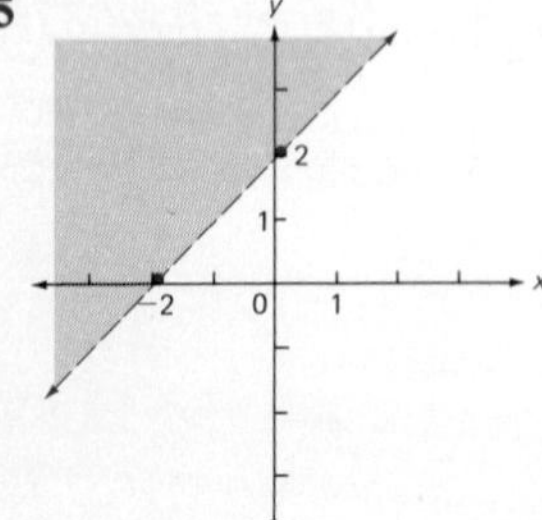

27

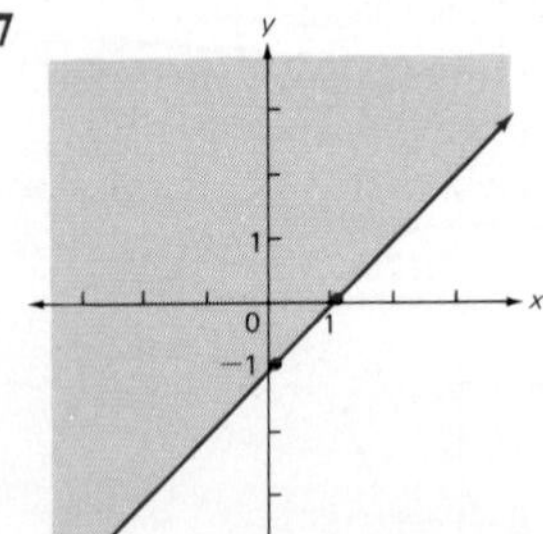

29

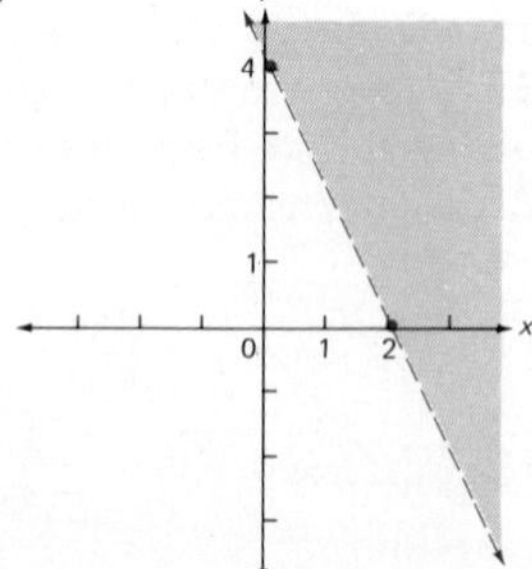

3.3 SLOPE (*Page 88*)

1 Each quotient gives the same slope, $-\frac{1}{2}$. **(a)** $\dfrac{3-1}{-2-2}$; **(b)** $\dfrac{2-0}{0-4}$; **(c)** $\dfrac{1-0}{2-4}$; **(d)** $\dfrac{3-(-1)}{-2-6}$; **(e)** $\dfrac{2-(-1)}{0-6}$; **(f)** $\dfrac{1-(-1)}{2-6}$

3 $\frac{1}{9}$ **5** 0 **7** $-\frac{17}{28}$ **9** $\dfrac{2}{\sqrt{3}+1} = \sqrt{3} - 1$

11

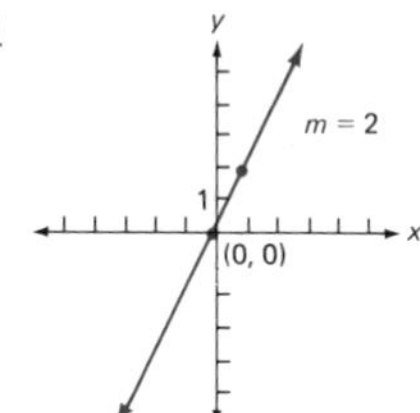

13

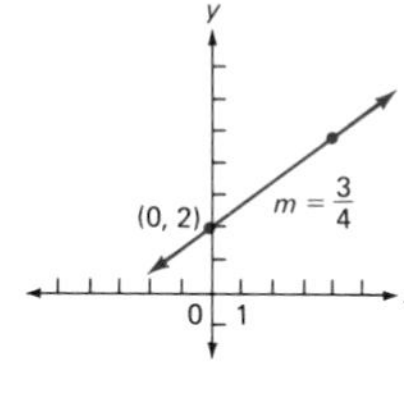

15

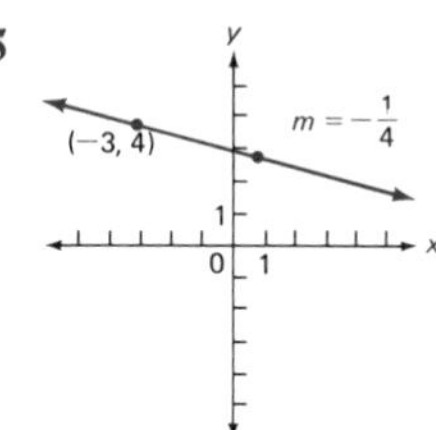

17

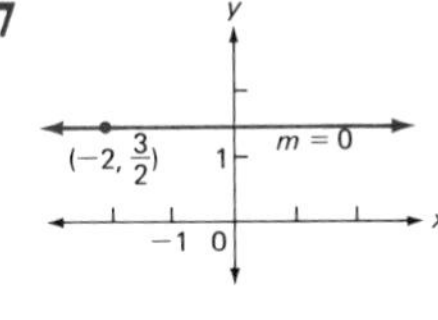

19

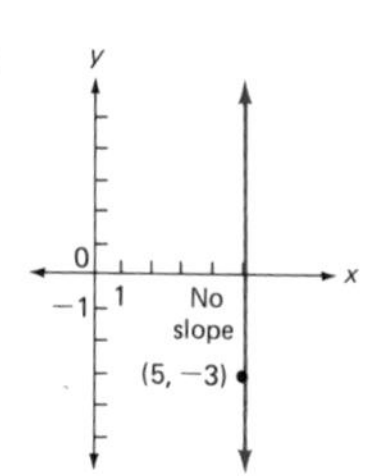

21

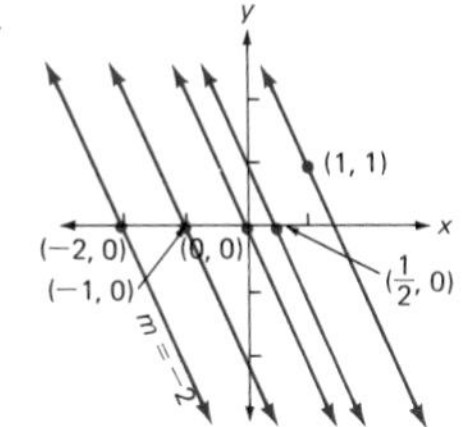

23

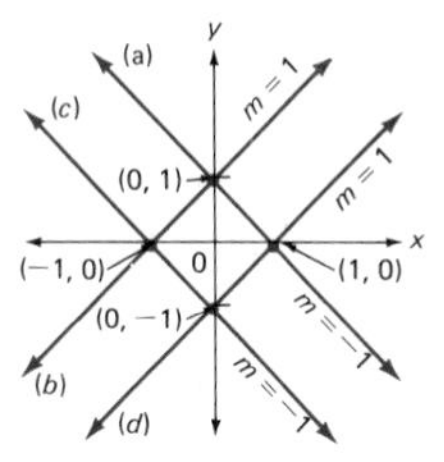

25

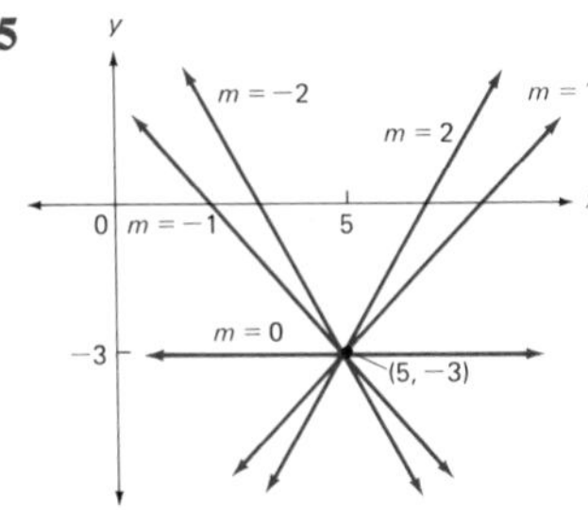

27 Both have the same slope of $-\frac{3}{2}$.

29 The slopes of PQ and RS are each $-\frac{5}{12}$; the slopes of PS and QR are each $\frac{12}{5}$. Thus the sides of the figure are perpendicular to each other and form right angles. Also, the diagonals are perpendicular because the slope of PR is $\frac{7}{17}$ and the slope of QS is the negative reciprocal, $-\frac{17}{7}$.

31 Horizontal lines have slope 0, which does not have a reciprocal. Also, vertical lines have no slope, and therefore a slope comparison cannot be made.

33 $\frac{49}{2}$

35 (a) Slope $\ell_1 = m_1 = \dfrac{DA}{CD} = \dfrac{DA}{1} = DA$; (b) slope $\ell_2 = m_2 = \dfrac{DB}{CD} = \dfrac{DB}{1} = DB$; $m_2 < 0$;

(c) $\dfrac{DA}{CD} = \dfrac{CD}{BD}$ or $\dfrac{m_1}{1} = \dfrac{1}{-m_2}$, since $BD = -DB = -m_2$.

3.4 LINEAR FUNCTIONS (*Page 93*)

1 $y = 2x + 3$ **3** $y = x + 1$ **5** $y = 5$ **7** $y = \frac{1}{2}x + 3$ **9** $y = \frac{1}{4}x - 2$ **11** $y - 3 = x - 2$
13 $y - 3 = 4(x + 2)$ **15** $y - 5 = 0$ **17** $y - 1 = \frac{1}{2}(x - 2)$ **19** $y = 5x$ **21** $y + \sqrt{2} = 10(x - \sqrt{2})$
23 $y = -3x + 4$; $m = -3$; $k = 4$ **25** $y = 2x - \frac{1}{3}$; $m = 2$; $k = -\frac{1}{3}$ **27** $y = \frac{5}{3}$; $m = 0$; $k = \frac{5}{3}$
29 $y = \frac{4}{3}x - \frac{7}{3}$; $m = \frac{4}{3}$; $k = -\frac{7}{3}$ **31** $y = \frac{1}{2}x - 2$; $m = \frac{1}{2}$; $k = -2$ **33** $y = -x + 5$ **35** $y = x - 3$
37 $y = -2x - 15$ **39** $y = 27$ **41** $x = 5, y = -7$ **43** $y = -\frac{2}{3}x - \frac{1}{3}$ **45** $y - 7 = -\frac{1}{3}(x - 4)$

47 $y - 1 = -\frac{1}{2}(x + 5)$

49

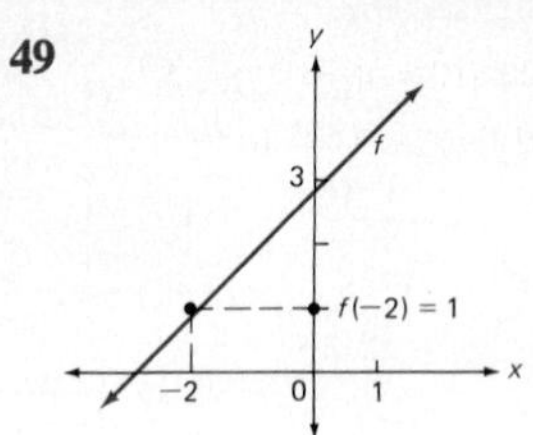

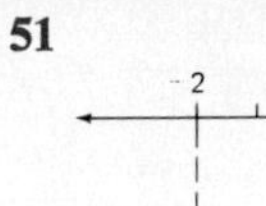

51

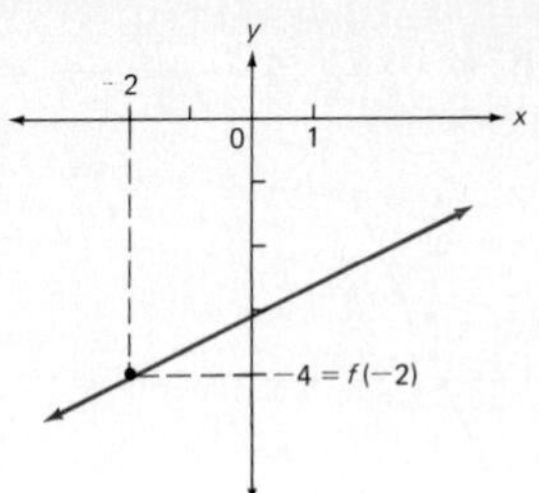

53

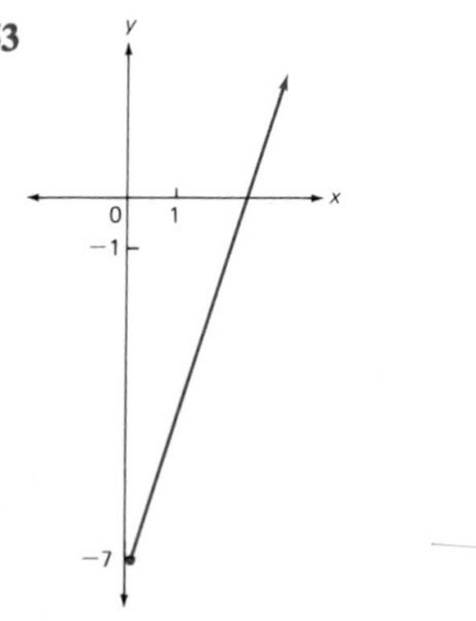

55

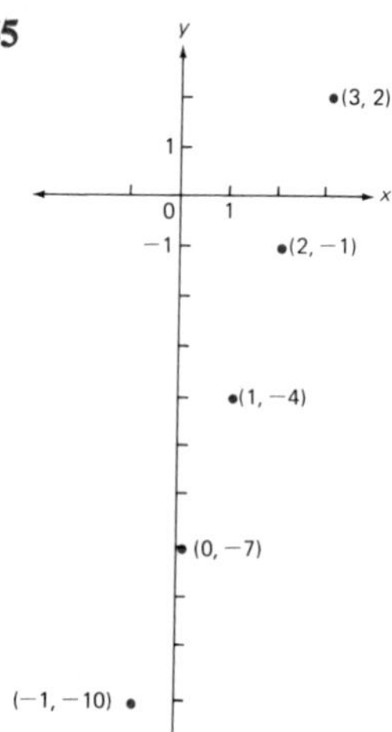

57 $x = -3$; not a function 59 $y = x + 2$; $-2 \leq x$; $0 \leq y$

61 $y = -1$ for $-1 \leq x < 0$; $y = 3$ for $0 \leq x < 3$; $y = 1$ for $3 \leq x \leq 5$

63 **(a)** If $c = 0$ then (0, 0) fits the equation and the line would then pass through the origin.

(b) $\frac{c}{b}$ = y-intercept; $\frac{c}{a}$ = x-intercept.

(c) $ax + by = c$; $\frac{ax}{c} + \frac{by}{c} = 1$; $\frac{x}{\left(\frac{c}{a}\right)} + \frac{y}{\left(\frac{c}{b}\right)} = 1$; $\frac{x}{q} + \frac{y}{p} = 1$

(d) $\frac{x}{\frac{3}{2}} + \frac{y}{-5} = 1$ or $10x - 3y = 15$

(e) $m = \frac{0 - (-5)}{\frac{3}{2} - (0)} = \frac{5}{\frac{3}{2}} = \frac{10}{3}$; $y = \frac{10}{3}x - 5$ 65 $y - 3 = -\frac{5}{7}(x + 2)$

3.5 SOME SPECIAL FUNCTIONS (*Page 99*)

1 99 3 −100 5 −4 7 1

9 Domain: all reals.
Range: $y \geq 0$.

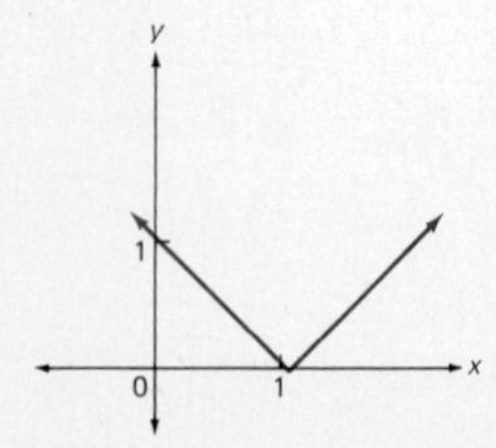

11 Domain: all reals.
Range: $y \geq 0$.

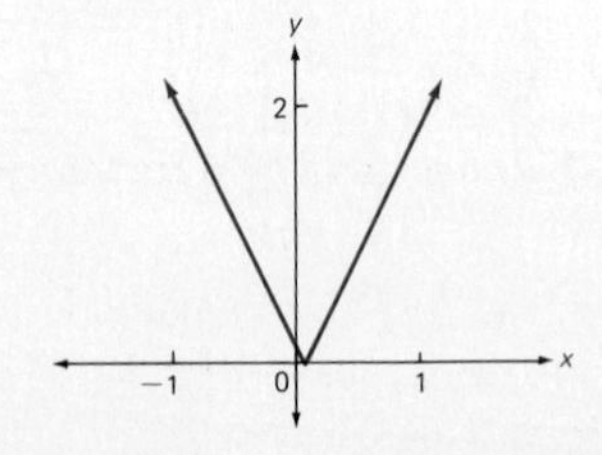

13 Domain: all reals.
Range: $y \geq 0$.

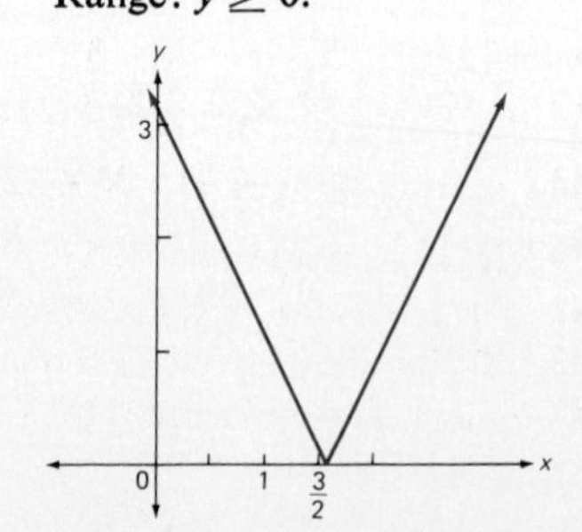

15 Domain: $x \geq -1$.
Range: $y \leq 3$.

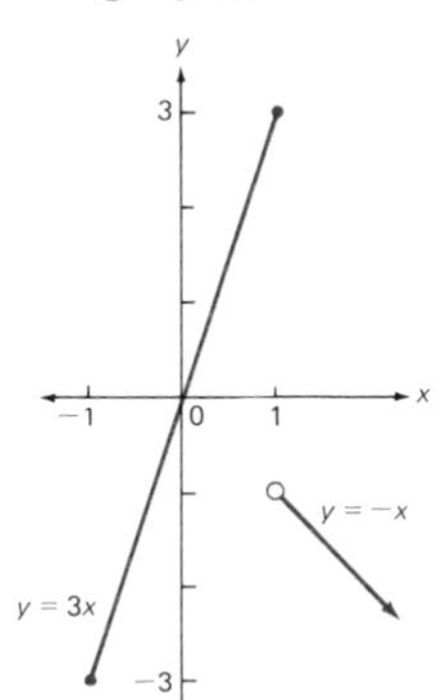

17 Domain: $-2 < x \leq 3$.
Range: $-2 < y \leq 4$.

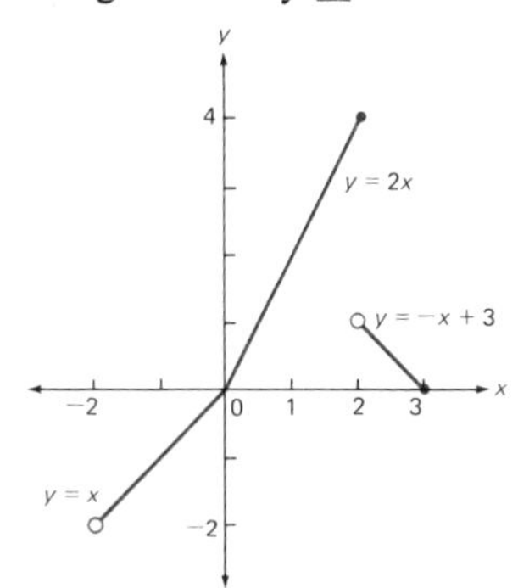

19

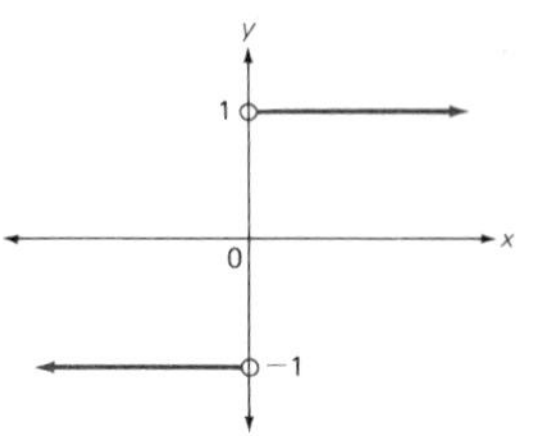

21

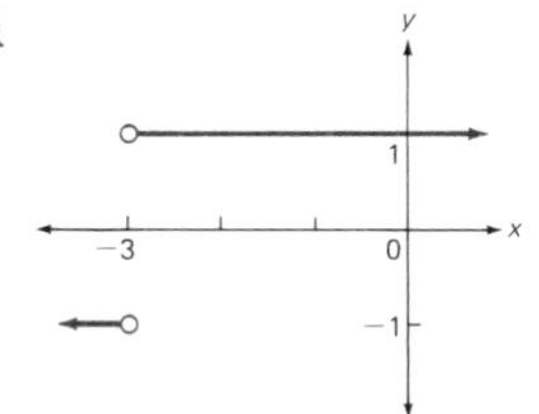

23

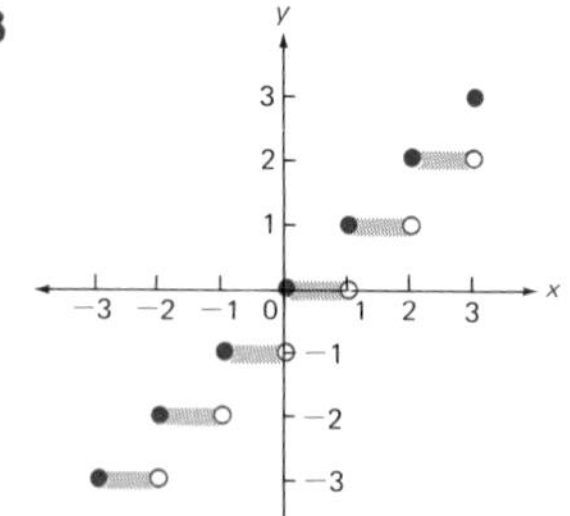

25

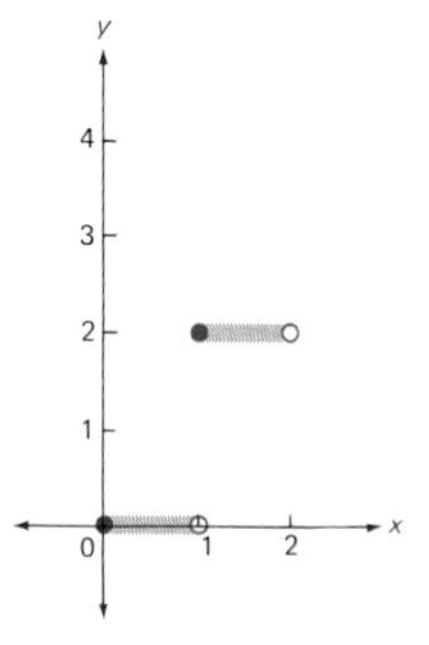

27

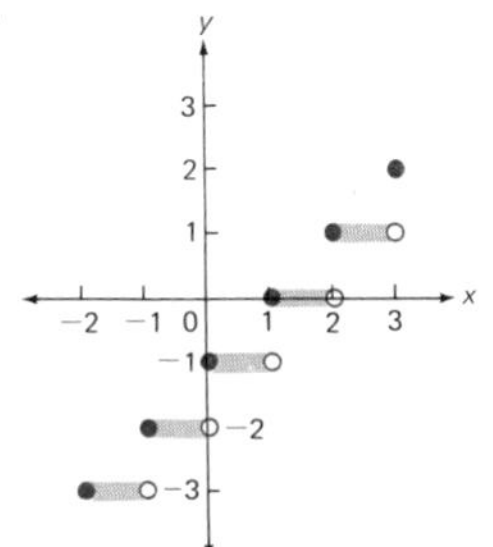

29

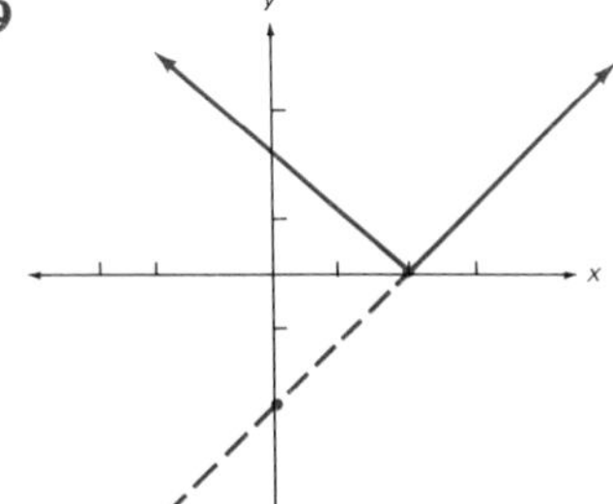

3.6 SYSTEMS OF LINEAR EQUATIONS (*Page 103*)

1 $(-2, -6)$ **3** $(4, -10)$ **5** $(-4, -1)$ **7** $(-26, -62)$ **9** $(-16, -38)$ **11** $(\frac{21}{88}, -\frac{9}{22})$ **13** $(5, 1)$ **15** $(8, 5)$
17 $(3, -11)$ **19** $(1, 0)$ **21** $(1, \frac{1}{2})$ **23** $(1, 0)$ **25** $(\frac{17}{18}, \frac{7}{18})$ **27** $(\frac{7}{2}, 4)$ **29** $(1, -2, 3)$ **31** $(2, -1, -1)$
33 $\left(\frac{ce - bf}{ae - bd}, \frac{af - cd}{ae - bd}\right)$ **35** $(-20, -20)$ **37** 40 field goals; 16 free throws.
39 \$1700 room and board; \$3600 tuition.
41 Speed of plane = 410 miles per hour; wind velocity = 10 miles per hour.
43 180 quarts of 74¢ oil; 220 quarts of 94¢ oil. **45** \$2800 at 8%; \$3200 at $7\frac{1}{2}$%.
47 6 miles by car; 72 miles by train. **49** $a = 4$, $b = -7$

3.7 CLASSIFICATION OF LINEAR SYSTEMS (*Page 109*)

1 Consistent. **3** Inconsistent. **5** Dependent. **7** Consistent. **9** Inconsistent. **11** Consistent. **13** Dependent. **15** Inconsistent. **17** Parallel.

19

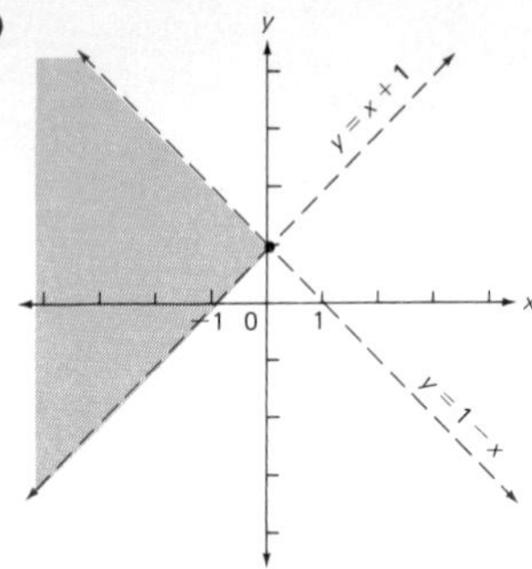

21

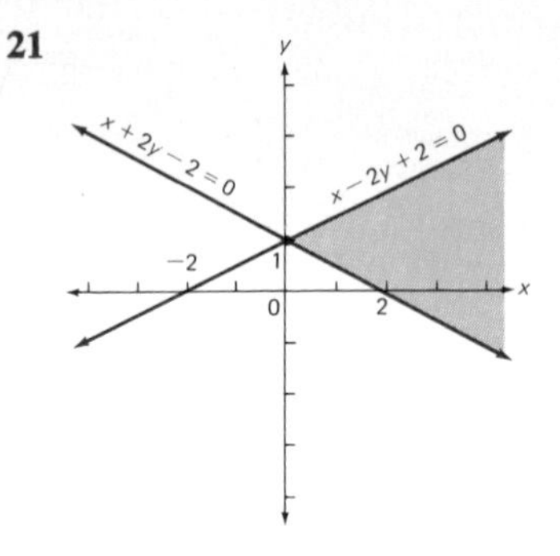

23

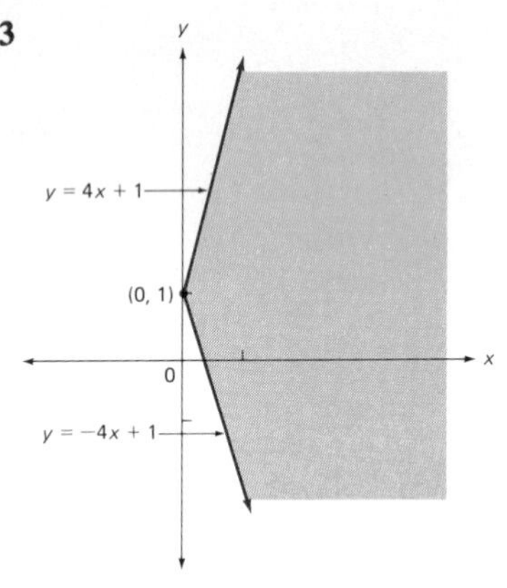

25

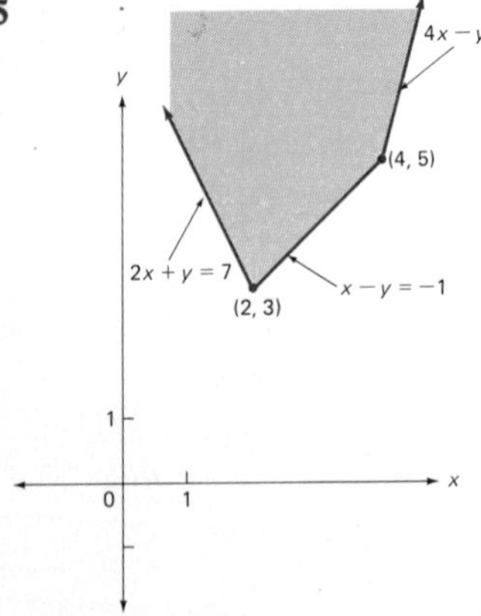

27

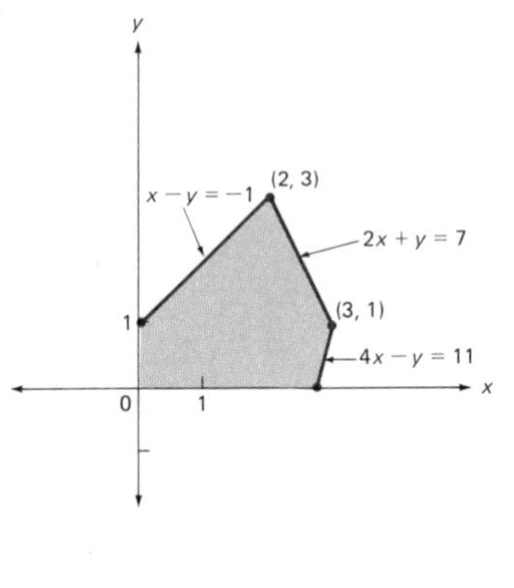

29 An infinite number of answers.

31 Yes, the clerk was wrong. If there were common unit prices, say x = cost per orange and y = cost per tangerine, then

$$6x + 12y = 234$$
$$2x + 4y = 77$$

which is an inconsistent system. (*Note:* The smaller bag is a better buy since 2 (oranges) + 4 (tangerines) = $\frac{1}{3}$(6 oranges + 12 tangerines) = $\frac{1}{3}$(2.34) = 0.78, and 77¢ is cheaper.)

3.8 EXTRACTING FUNCTIONS FROM GEOMETRIC FIGURES (*Page 113*)

1 $h(x) = \frac{5}{2}x$ **3** $w(h) = \frac{20}{9}h$ **5** $s(x) = \frac{1}{3}x$ **7** $A(x) = \frac{5}{2}x(4 - x)$ **9** $V(x) = x(50 - 2x)^2$
11 $A(x) = \pi r^2 + 2r(200 - \pi r)$ **13** $V(x) = \frac{4}{3}x(30 - x^2)$ **15** $\frac{7}{5}x(10 - 2x)$

SAMPLE TEST QUESTIONS FOR CHAPTER 3 (*Page 119*)

1 $x \leq -1$ or $x \geq 1$ **2** **(a)** $\frac{3}{2 + x}$; **(b)** $\frac{3}{2} + \frac{3}{x} = \frac{3x + 6}{2x}$ **3** $-\frac{2}{3(3 + h)}$

4

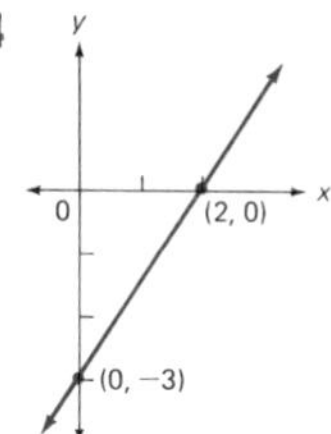

5

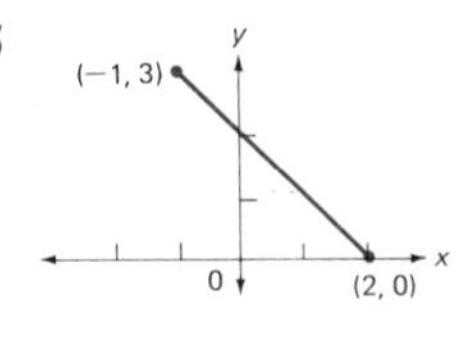

6 $-\frac{7}{3}$ 7 $y - 3 = -\frac{3}{4}(x + 2)$ 8 $y = -\frac{9}{5}x + \frac{2}{5}$ 9 $\frac{3}{4}; \frac{1}{2}$

10 $y - 8 = \frac{5}{2}(x - 2)$

11 Domain: all x.
Range: all y.

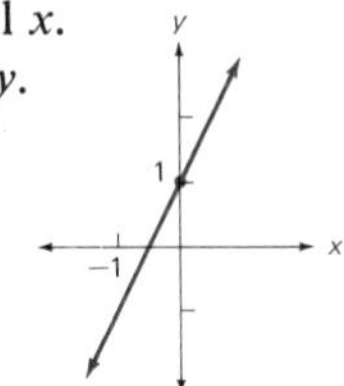

12

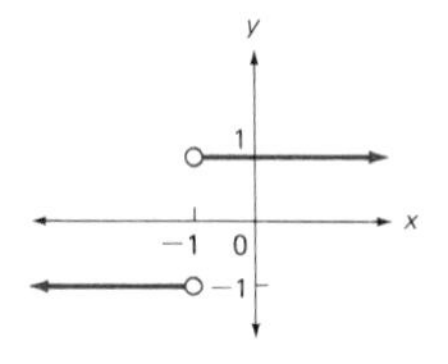

13 $x = 3, y = -1$ 14 $x = -1, y = 2, z = 5$ 15 25 at 20¢; 15 at 37¢ 16 Dependent. 17 Consistent; (3, −5).

18 Inconsistent. 19

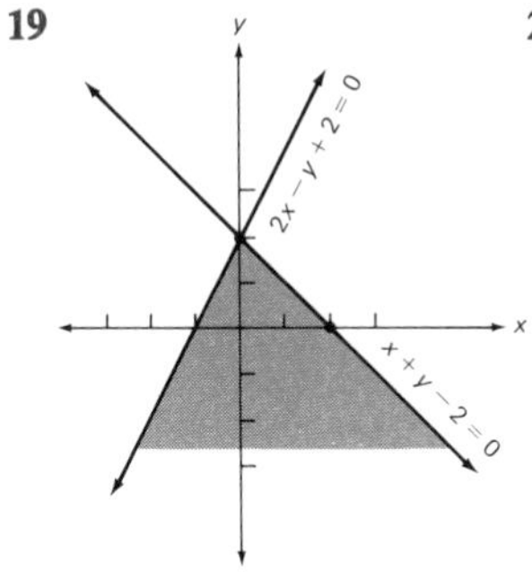

20 $V(x) = x(40 - 2x)^2$

CHAPTER 4: QUADRATIC FUNCTIONS AND EQUATIONS

4.1 GRAPHING QUADRATIC FUNCTIONS (*Page 121*)

1

3

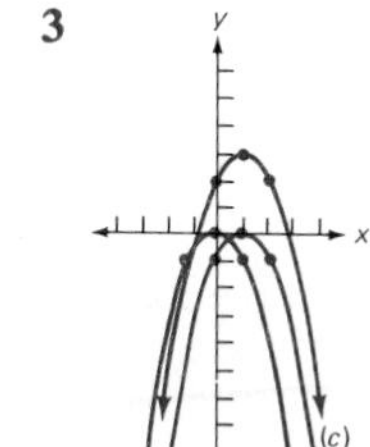

(a) (3, 5)

5

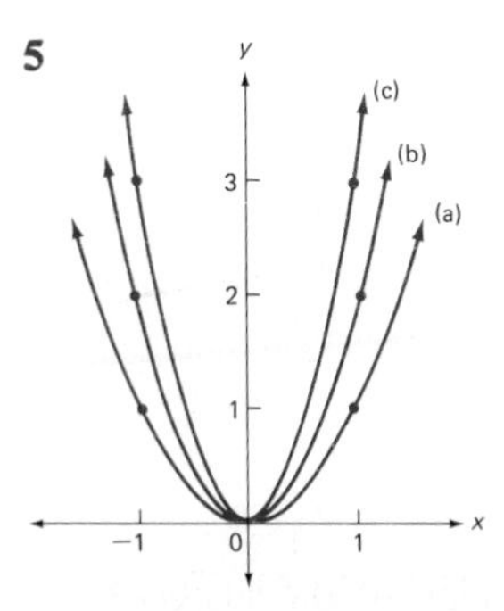

7

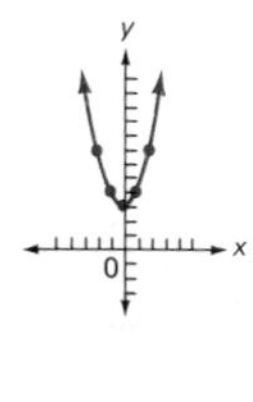

9

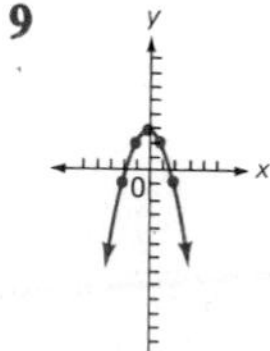

11

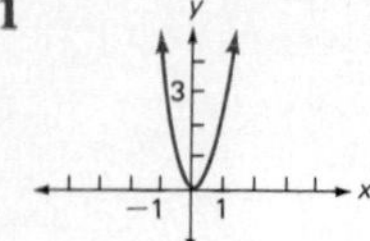

13

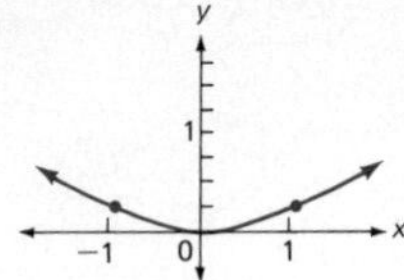

15

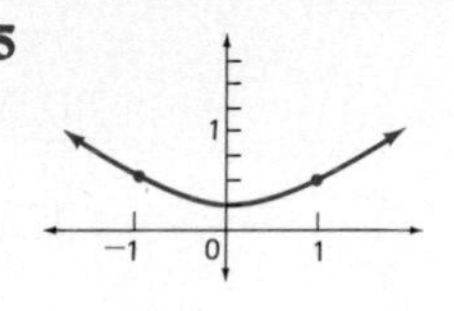

17

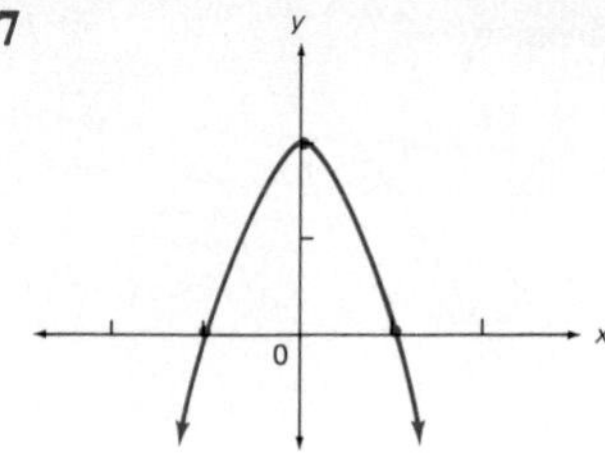

19 Decreasing on $(-\infty, 1]$; increasing on $[1, \infty)$; concave up.

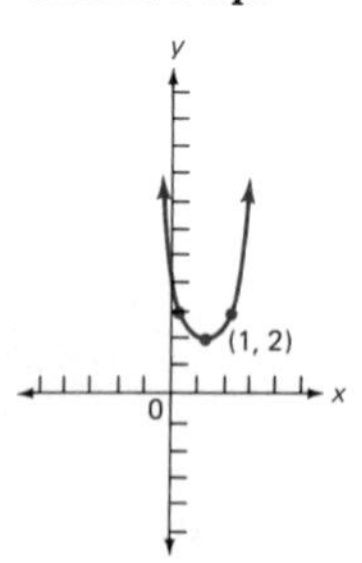

21 Increasing on $(-\infty, -1]$; decreasing on $[-1, \infty)$; concave down.

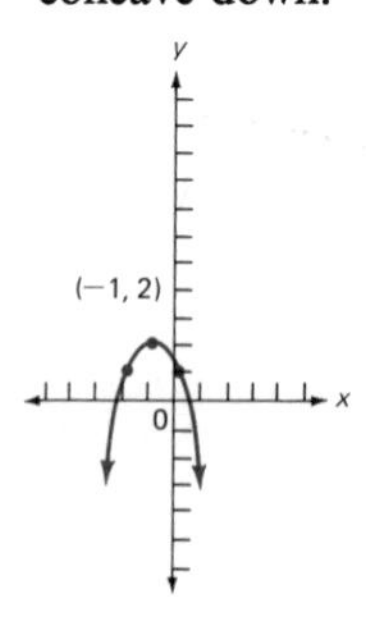

23 Decreasing on $(-\infty, 3]$; increasing on $[3, \infty)$; concave up.

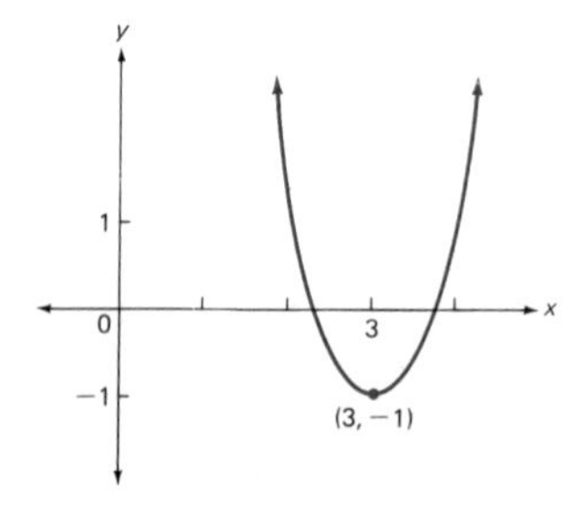

25 **(a)** (3, 5); **(b)** $x = 3$; **(c)** Set of real numbers; **(d)** $y \geq 5$

27 **(a)** (3, 5); **(b)** $x = 3$; **(c)** Set of real numbers; **(d)** $y \leq 5$

29 **(a)** $(-1, -3)$; **(b)** $x = -1$; **(c)** Set of real numbers; **(d)** $y \geq -3$

31

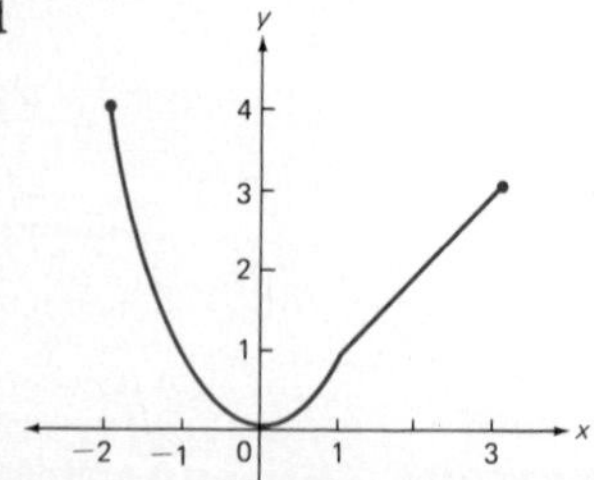

33 Not a function because for $x > 0$ there are two corresponding y-values.

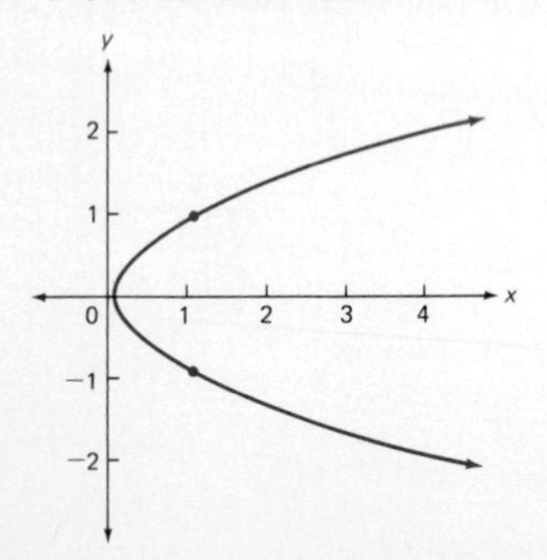

35 The graph of $x^2 - 4 > 0$ consists of all points on the number line where $x < -2$ or where $x > 2$. On the coordinate plane, the graph of $y = x^2 - 4$ is positive (above the x-axis) for $x < -2$ or $x > 2$.

37 $a = -2$ 39 $k = 3$ 41 43

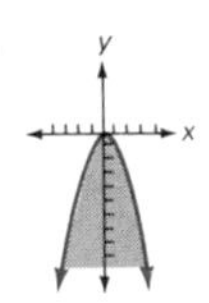

45

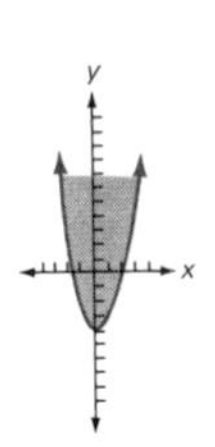

4.2 COMPLETING THE SQUARE (*Page 127*)

1 $(x+1)^2 - 6$ 3 $-(x+3)^2 + 11$ 5 $-(x-\frac{3}{2})^2 - \frac{7}{4}$ 7 $(x+\frac{5}{2})^2 - \frac{33}{4}$ 9 $2(x+1)^2 - 5$
11 $3(x+1)^2 - 8$ 13 $(x-\frac{1}{4})^2 + \frac{15}{16}$ 15 $\frac{3}{4}(x-\frac{2}{3})^2 - \frac{2}{3}$ 17 $-5(x+\frac{1}{5})^2 + 1$ 19 $(x-\frac{1}{2})^2 - 3$
21 $h = -\dfrac{b}{2a}$; $k = \dfrac{4ac-b^2}{4a}$ 23 $y = (x-2)^2 + 3$; $(2, 3)$; $x = 2$; $y = 7$
25 $y = (x-3)^2 - 4$; $(3, -4)$; $x = 3$; $y = 5$ 27 $y = 2(x-1)^2 - 6$; $(1, -6)$; $x = 1$; $y = -4$

29

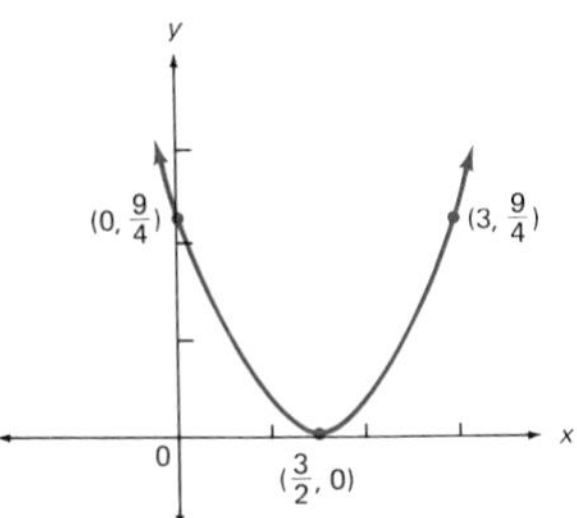

31

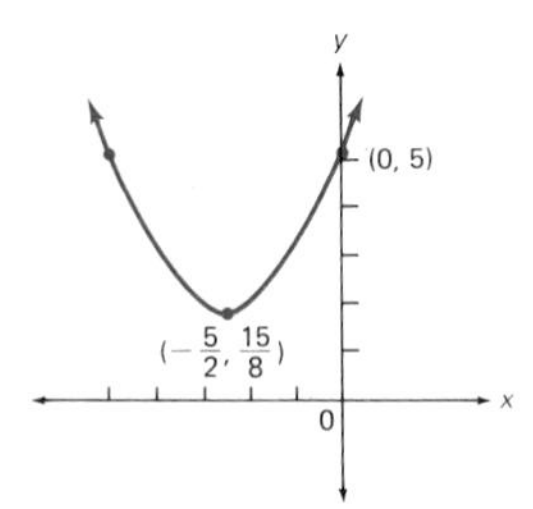

33

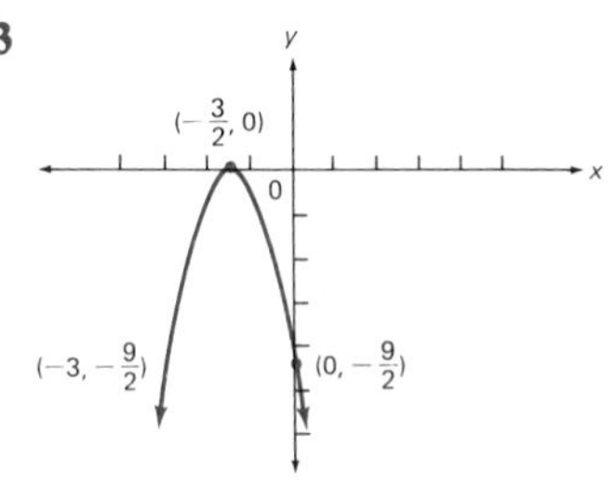

35

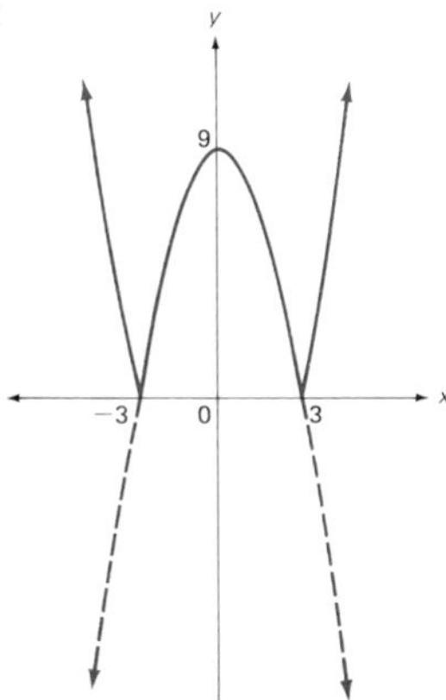

37

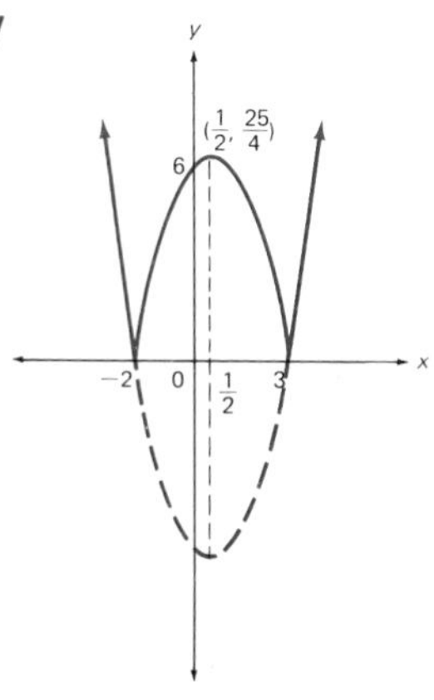

39

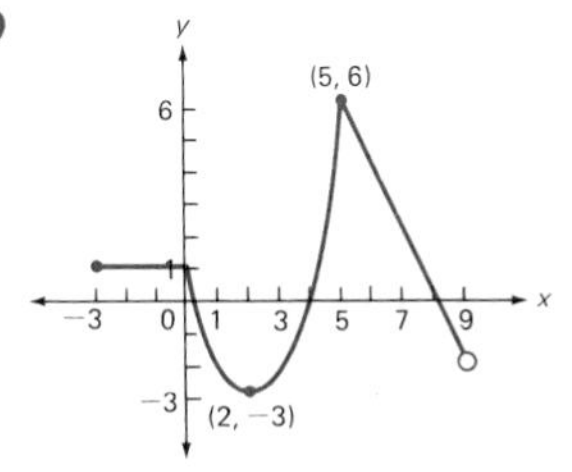

41 $a > 0$; $k = 0$. 43 $a > 0$; $k = 2$. 45 $y = -2x^2 + 3x + 7$

4.3 THE QUADRATIC FORMULA (*Page 132*)

1 $x = -2$; $x = 5$ 3 $x = 3$ 5 $x = -2$; $x = -\frac{1}{3}$ 7 $x = 1 - \sqrt{5}$; $x = 1 + \sqrt{5}$

9 $x = \frac{3 - \sqrt{17}}{4}$; $x = \frac{3 + \sqrt{17}}{4}$ **11** 3; $-\frac{1}{2}$ **13** None. **15** $\frac{3}{2}$; 2 **17** None. **19** 0; (b) **21** 36; (c) **23** 13; (d)
25 41; (d) **27** -7; (a) **29** Once; **(a)** $(\frac{1}{3}, 0)$; **(b)** 1; **(c)** $\frac{1}{3}$ **31** $-1 < x < 3$ **33** $x \leq -5$ or $x \geq 2$
35 No solution.

37

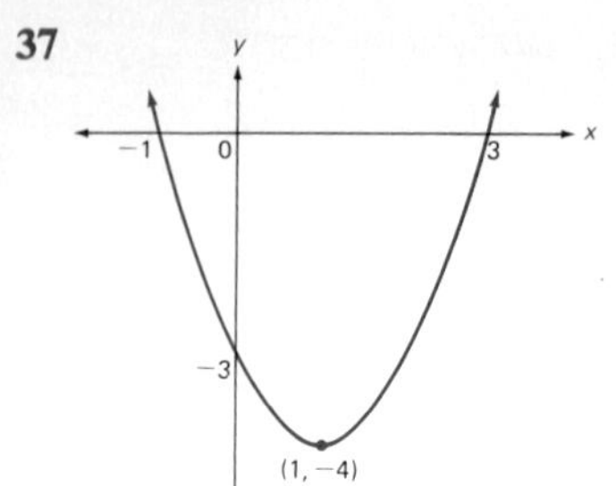

39

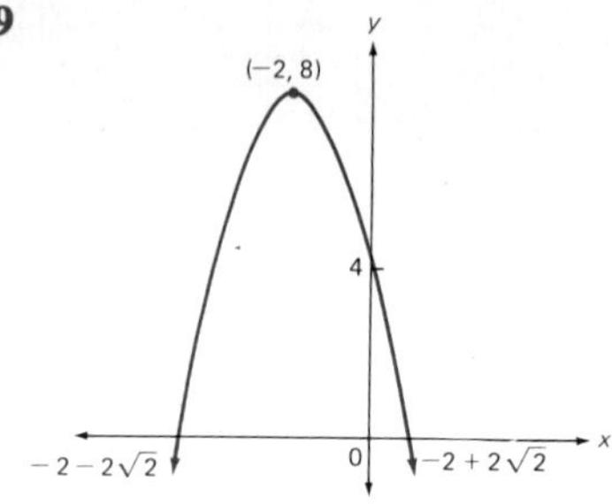

41

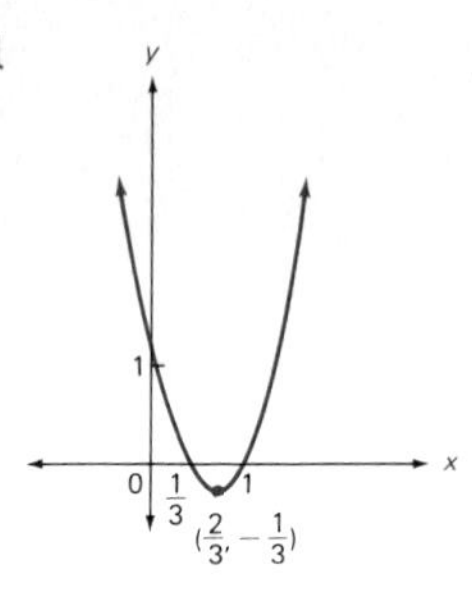

43

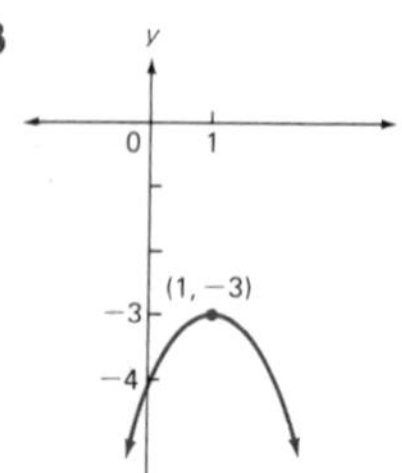

45 -6 or 6 **47** $\pm 2\sqrt{7}$ **49** $k > -4$ **51** $k > -\frac{1}{4}$ **53** $a < -\frac{1}{4}$ **55** **(a)** $-\frac{b}{a}$; **(b)** $\frac{c}{a}$

57 sum $= -\frac{5}{6}$, product $= -\frac{2}{3}$ **59** sum $= \frac{26}{3}$, product $= \frac{35}{3}$ **61** sum $= -3$, product $= -\frac{9}{2}$ **63** $x = -3$; $x = -2$; $x = 2$

4.4 APPLICATIONS OF QUADRATIC FUNCTIONS *(Page 138)*

1

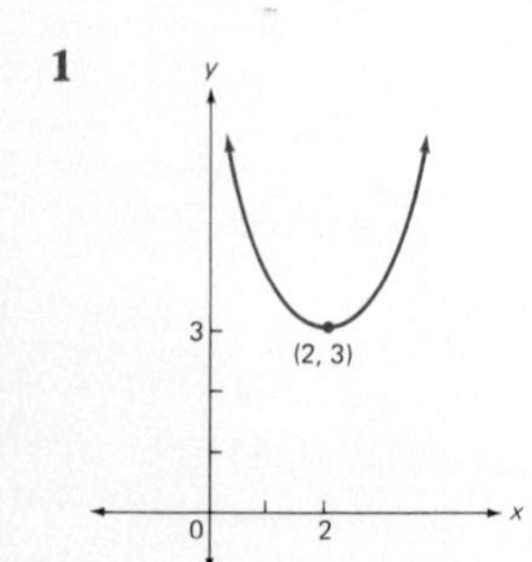

3

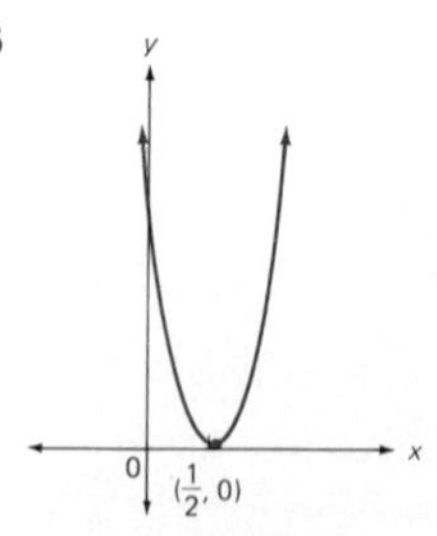

5 Maximum = 7 at $x = 5$.

7 Minimum = 36 at $x = 2$. **9** Minimum = 0 at $x = \frac{7}{2}$. **11** Maximum = $\frac{1}{9}$ at $x = -\frac{1}{3}$. **13** 60; $600 **15** 6, 6
17 No; the product $y = x(n - x) = -x^2 + nx$ gets arbitrarily small as x becomes very large or very small since the graph of this quadratic opens downward.
19 10 feet by 10 feet. **21** 256 feet; 4 seconds. **23** 2, 6 **25** $9.75 **27** (1, 2)
29 Maximum or minimum $= \frac{4ac - b^2}{4a}$ at $x = -\frac{b}{2a}$. **31** (4, 7); 7 is maximum.

33 $(\frac{5}{2}, -\frac{25}{4})$; $-\frac{25}{4}$ is minimum. **35** $(-5, 750)$; 750 is minimum. **37** 7, 9 **39** $-\frac{3}{2}$ or 1 **41** 8 centimeters. **43** 12 inches.

4.5 CIRCLES AND THEIR EQUATIONS (*Page 142*)

1

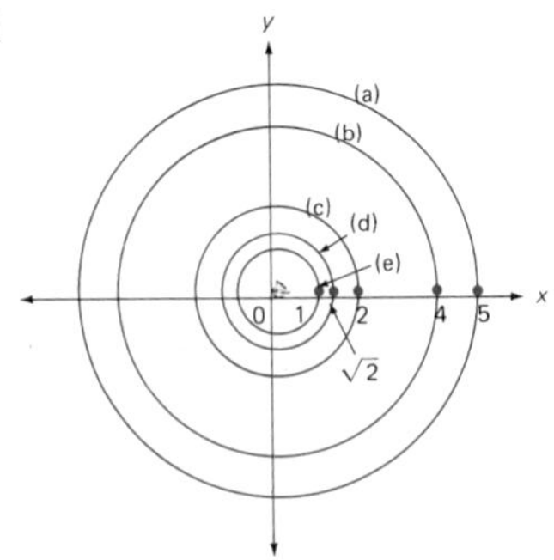

3 $(x-2)^2 + (y-5)^2 = 1$; $C(2, 5)$, $r = 1$

5 $(x-4)^2 + (y-0)^2 = 2$; $C(4, 0)$, $r = \sqrt{2}$ **7** $(x-10)^2 + (y+10)^2 = 100$; $C(10, -10)$, $r = 10$
9 $(x + \frac{3}{4})^2 + (y-1)^2 = 9$; $C(-\frac{3}{4}, 1)$, $r = 3$ **11** $(x-2)^2 + y^2 = 4$ **13** $(x+3)^2 + (y-3)^2 = 7$

15

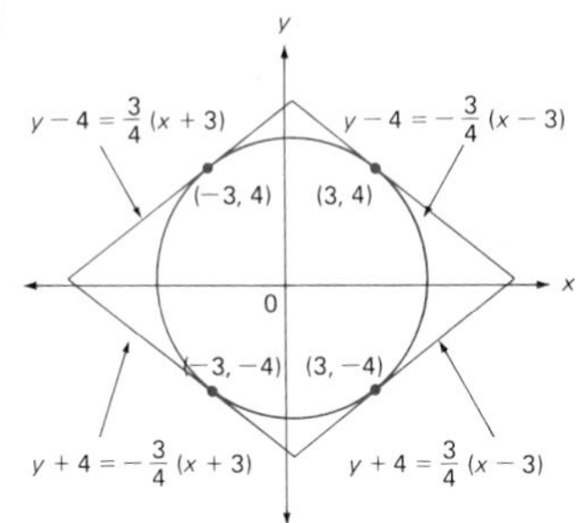

17 $x = 2$; $x = -2$

19

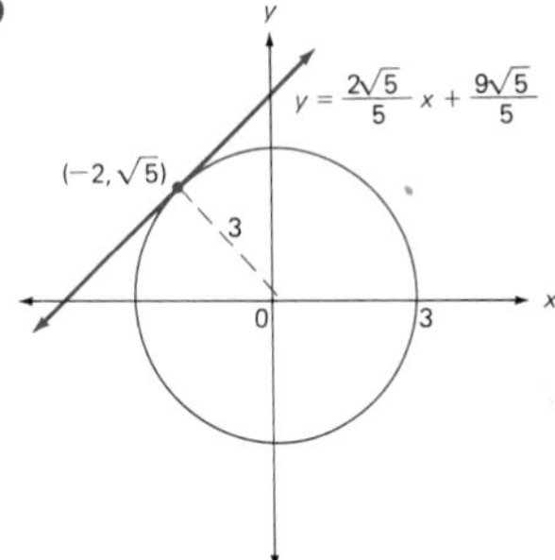

21

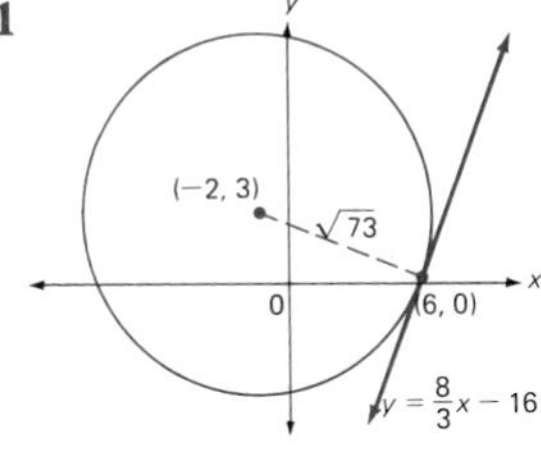

23

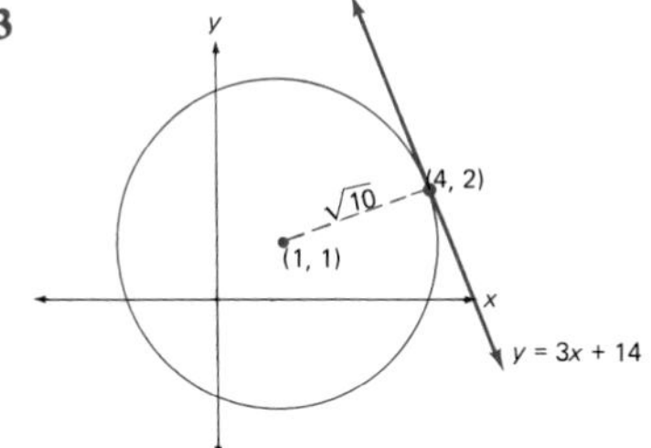

25 $(\frac{1}{2}, -2)$ **27** $(-\frac{5}{2}, \frac{1}{2})$

29 $y = -\frac{1}{2}x + 10$ **31** $5x + 12y = 26$ **33** (**a**) $y(x) = \sqrt{25 + 24x - x^2} - 5$; (**b**) $y(7) = 7$

4.6 SOLVING NONLINEAR SYSTEMS (*Page 149*)

1

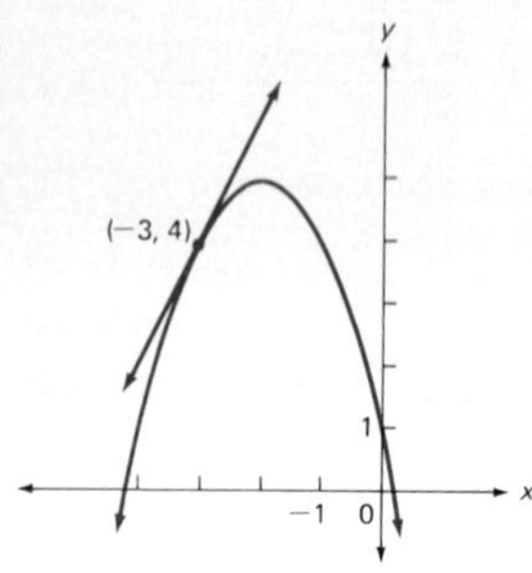

3

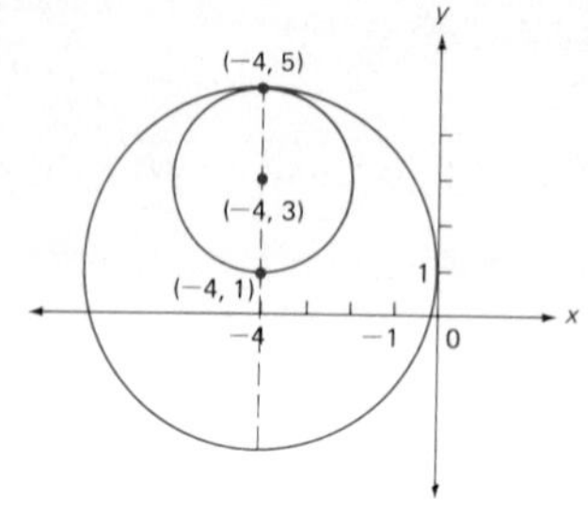

5

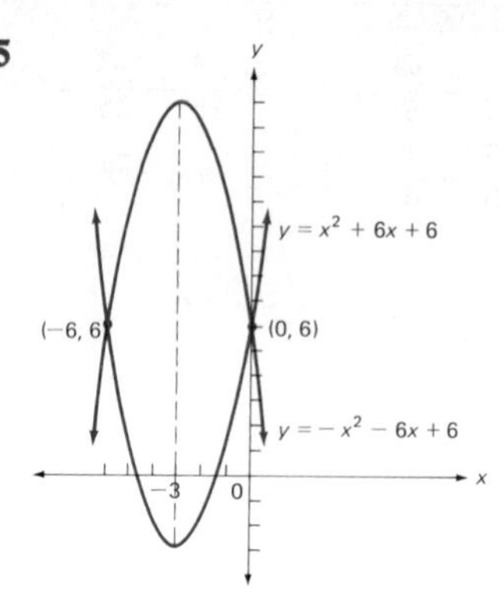

7 (0, 1) **9** No solutions. **11** No solutions. **13** (−1, −1); (−2, −2). **15** (0, 0); (2, 0). **17** (0, 0); (1, 1).
19 (3, −3); $(3 + \sqrt{3}, -2)$; $(3 - \sqrt{3}, -2)$.

4.7 THE ELLIPSE AND THE HYPERBOLA (*Page 152*)

1 Ellipse; center (0, 0), vertices (±5, 0), foci (±3, 0).
3 Hyperbola; center (0, 0), vertices (±6, 0), foci $(\pm\sqrt{61}, 0)$, asymptotes $y = \pm\frac{5}{6}x$.
5 Ellipse; center (0, 0), vertices (0, ±5), foci $(0, \pm\sqrt{21})$.
7 Ellipse; center (0, 1), vertices (0, −1) and (0, 3), foci $(0, 1 - \sqrt{3})$ and $(0, 1 + \sqrt{3})$.

9 Ellipse: **11** Hyperbola: **13** Ellipse:

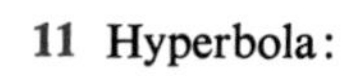

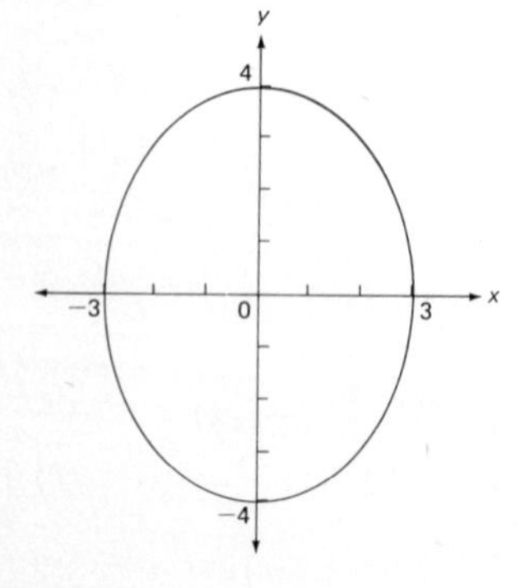

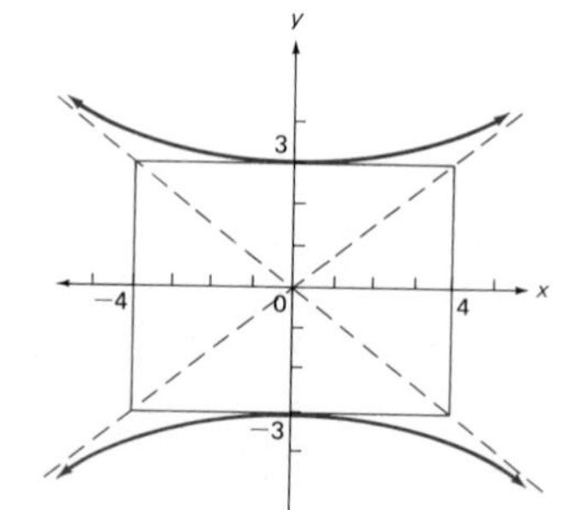

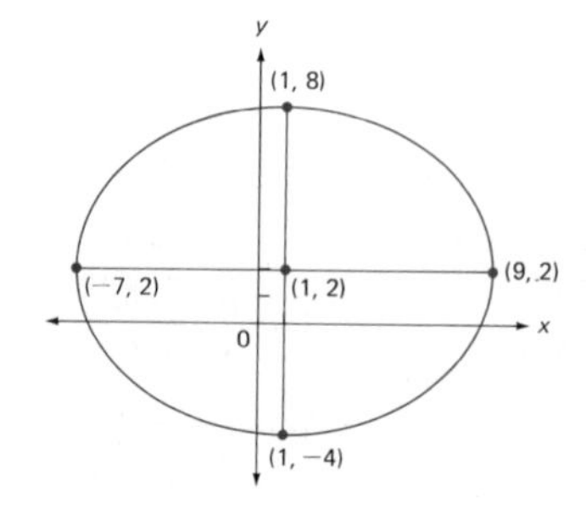

15 Hyperbola:

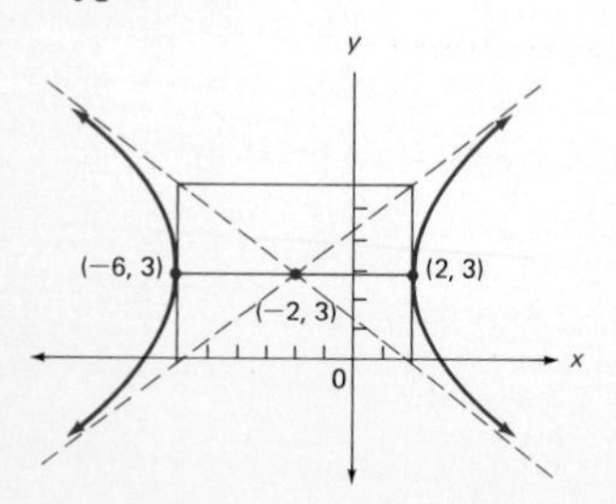

17 $(x - 1)^2 + (y + 2)^2 = 4$; circle; center (1, −2); $r = 2$.

19 $\frac{(x + 1)^2}{4} + \frac{y^2}{1} = 1$; ellipse; center (−1, 0); vertices (−3, 0) and (1, 0); foci $(-1 - \sqrt{3}, 0)$ and $(-1 + \sqrt{3}, 0)$.

21 $\dfrac{(x+1)^2}{16} - \dfrac{(y-3)^2}{9} = 1$; hyperbola; center $(-1, 3)$; vertices $(-5, 3)$ and $(3, 3)$; foci $(-6, 3)$ and $(4, 3)$.

23 $\dfrac{x^2}{25} + \dfrac{y^2}{9} = 1$ 25 $\dfrac{x^2}{16} + \dfrac{y^2}{25} = 1$ 27 $\dfrac{(x-2)^2}{16} + \dfrac{(y+3)^2}{36} = 1$ 29 $\dfrac{x^2}{16} - \dfrac{y^2}{20} = 1$ 31 $\dfrac{(y-3)^2}{9} - \dfrac{(x+2)^2}{7} = 1$

33 $\dfrac{x^2}{16} - \dfrac{y^2}{4} = 1$ 35 **(a)** $\sqrt{x^2 + (y-p)^2}$; **(b)** $y + p$. **(c)** $\sqrt{x^2 + (y-p)^2} = y + p$

$$x^2 + (y-p)^2 = (y+p)^2$$
$$x^2 = 4py$$

37 $x^2 = -12y$

39 $(0, -1)$; $y = 1$ 41 $(x-2)^2 = 8(y+5)$; $y = \frac{1}{8}x^2 - \frac{1}{2}x - \frac{9}{2}$

4.8 COMPLEX NUMBERS (*Page 160*)

1 True. 3 True. 5 False. 7 $5 + 2i$ 9 $-5 + 0i$ 11 $4i$ 13 $12i$ 15 $\frac{3}{4}i$ 17 $-\sqrt{5}\,i$ 19 -27 21 $-\sqrt{6}$
23 $15i$ 25 $12i$ 27 $5i\sqrt{2}$ 29 $(3 - \sqrt{3})i$ 31 $10 + 7i$ 33 $5 - 3i$ 35 $10 + 2i$ 37 $-10 + 6i$ 39 $13i$
41 $23 + 14i$ 43 $5 - 3i$ 45 $\frac{13}{5} + \frac{1}{5}i$ 47 $\frac{4}{5} - \frac{3}{5}i$

49 $\overline{z + w} = \overline{(-6 + 8i) + (\frac{1}{2} + \frac{1}{2}i)} = \overline{-\frac{11}{2} + \frac{17}{2}i} = -\frac{11}{2} - \frac{17}{2}i$

$\bar{z} + \bar{w} = \overline{-6 + 8i} + \overline{\frac{1}{2} + \frac{1}{2}i} = (-6 - 8i) + (\frac{1}{2} - \frac{1}{2}i) = -\frac{11}{2} - \frac{17}{2}i$

51 $\overline{zw} = \overline{(-6 + 8i)(\frac{1}{2} + \frac{1}{2}i)} = \overline{-7 + i} = -7 - i$; $\bar{z}\bar{w} = \overline{(-6 + 8i)}\,\overline{(\frac{1}{2} + \frac{1}{2}i)} = (-6 - 8i)(\frac{1}{2} - \frac{1}{2}i) = -7 - i$

53 $-3i$ 55 $-2i$ 57 4 59 $\frac{3}{13} - \frac{2}{13}i$

61 $\sqrt{(-4)(-9)} = \sqrt{36} = 6$; $\sqrt{-4}\sqrt{-9} = (2i)(3i) = 6i^2 = -6$

63 **(a)** $(x + 2i)(x - 2i)$; **(b)** $3(x + 3i)(x - 3i)$; **(c)** $(x + i\sqrt{2})(x - i\sqrt{2})$; **(d)** $5(x + i\sqrt{3})(x - i\sqrt{3})$

65 $-3 \pm 4i$ 67 $-\dfrac{1}{4} \pm \dfrac{\sqrt{39}}{4}i$ 69 $3 \pm \sqrt{3}\,i$ 71 $\dfrac{ac + bd}{c^2 + d^2} + \dfrac{bc - ad}{c^2 + d^2}i$

73 $[(3 + i)(3 - i)](4 + 3i) = (9 - i^2)(4 + 3i) = 10(4 + 3i) = 40 + 30i$

$(3 + i)[(3 - i)(4 + 3i)] = (3 + i)(15 + 5i) = 40 + 30i$

75 $4 - 3i$ 77 $\frac{53}{13} - \frac{21}{13}i$ 79 $\pm 3, \pm 3i$ 81 $\pm 3, \pm \dfrac{1}{2i}$

83 $\overline{z + w} = \overline{(a + bi) + (c + di)} = \overline{(a + c) + (b + d)i} = (a + c) - (b + d)i$

$\bar{z} + \bar{w} = \overline{a + bi} + \overline{c + di} = (a - bi) + (c - di) = (a + c) - (b + d)i$

85 $\overline{\left(\dfrac{z}{w}\right)} = \overline{\left(\dfrac{a + bi}{c + di}\right)} = \overline{\dfrac{ac + bd}{c^2 + d^2} + \dfrac{bc - ad}{c^2 + d^2}i} = \dfrac{ac + bd}{c^2 + d^2} - \dfrac{bc - ad}{c^2 + d^2}i$

$\dfrac{\bar{z}}{\bar{w}} = \dfrac{\overline{a + bi}}{\overline{c + di}} = \dfrac{a - bi}{c - di} = \dfrac{a - bi}{c - di} \cdot \dfrac{c + di}{c + di} = \dfrac{ac + bd}{c^2 + d^2} - \dfrac{bc - ad}{c^2 + d^2}i$

SAMPLE TEST QUESTIONS FOR CHAPTER 4 (*Page 169*)

1

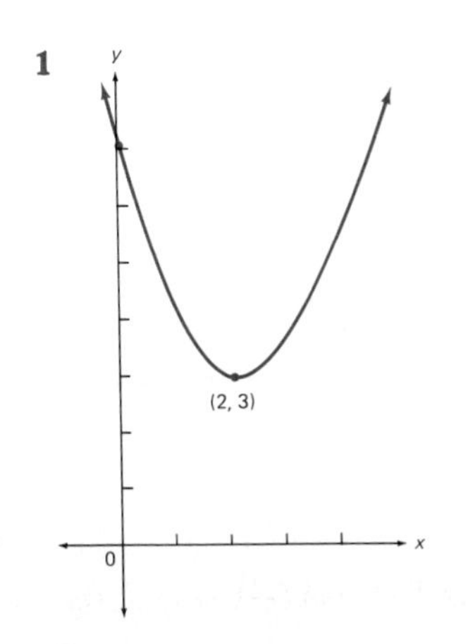

2

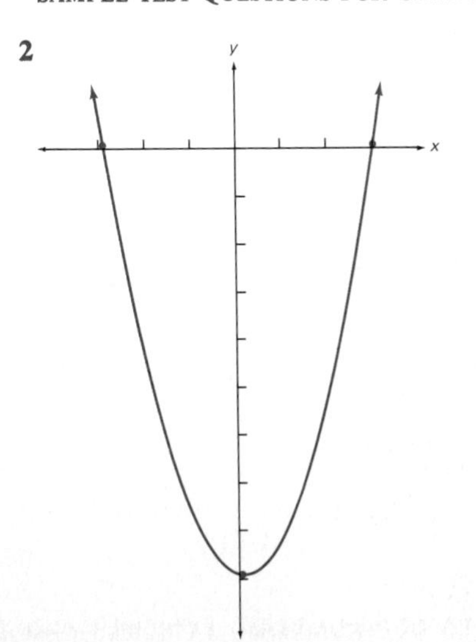

3 **(a)** $y = -5(x-2)^2 + 19$; **(b)** (2, 19); **(c)** $x = 2$; **(d)** domain: all real numbers; range: all $y \le 19$.
4 $x = -\frac{1}{3}, x = 3$ **5** $2 - \sqrt{11}, 2 + \sqrt{11}$ **6** 5; two irrational numbers. **7** 169; two rational numbers.
8 Maximum = 20 at $x = -6$. **9** 6 feet. **10** 64 feet. **11** **(a)**

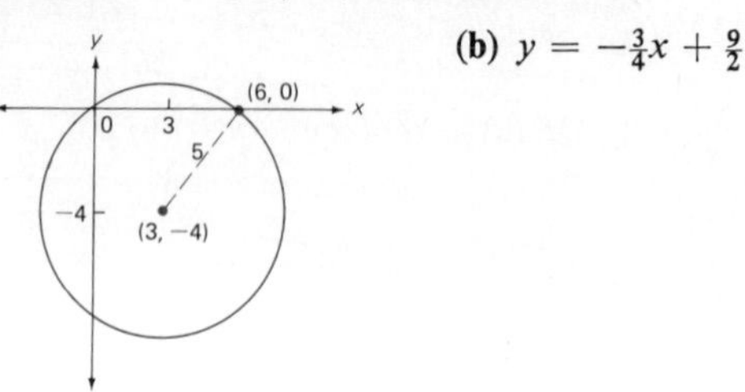

(b) $y = -\frac{3}{4}x + \frac{9}{2}$

12 $C(-\frac{1}{2}, 7); r = 5$ **13** $\sqrt{106}$ **14** $(-1, -1)$; $\sqrt{20}$; $(x+1)^2 + (y+1)^2 = 20$
15 (0, 2); (4, 2); $(2 + \sqrt{3}, 1)$; $(2 - \sqrt{3}, 1)$
16 $\frac{(x+2)^2}{1} - \frac{(y-3)^2}{4} = 1$; a hyperbola with center at $(-2, 3)$, and vertices at $(-3, 3)$, and $(-1, 3)$.
17 $\frac{x^2}{16} + \frac{y^2}{9} = 1$ **18** $\frac{x^2}{36} - \frac{y^2}{28} = 1$ **19** $1 - 6i$ **20** $43 + 23i$ **21** $-\frac{13}{41} + \frac{47}{41}i$ **22** $\frac{3}{25} + \frac{4}{25}i$ **23** $x = \frac{5}{4} \pm \frac{\sqrt{7}}{4}i$

CHAPTER 5: POLYNOMIAL AND RATIONAL FUNCTIONS

5.1 HINTS FOR GRAPHING (*Page 171*)

1

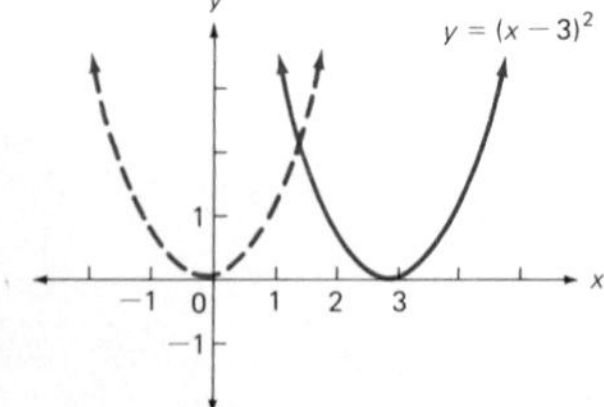

3

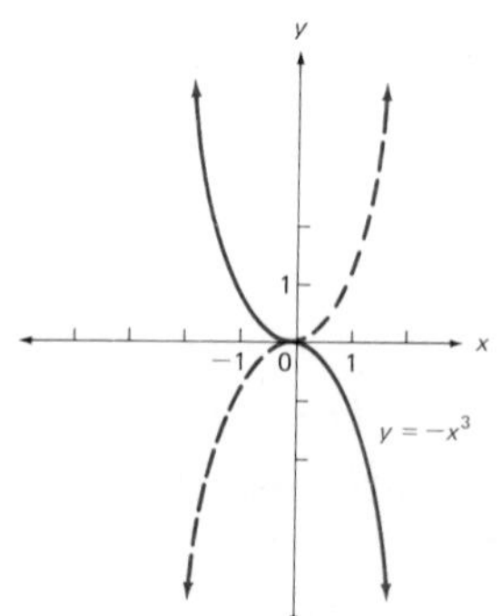

5

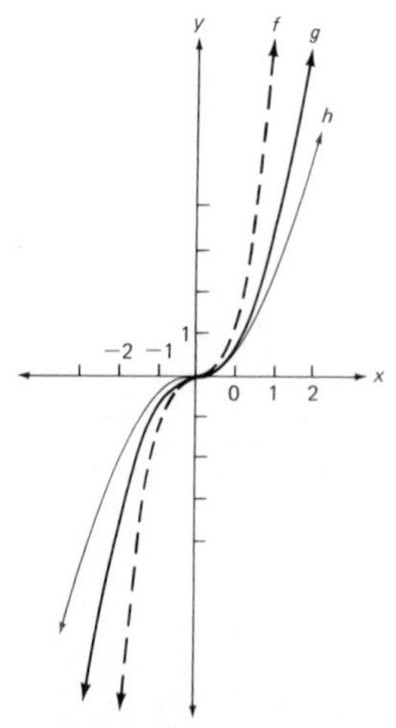

7

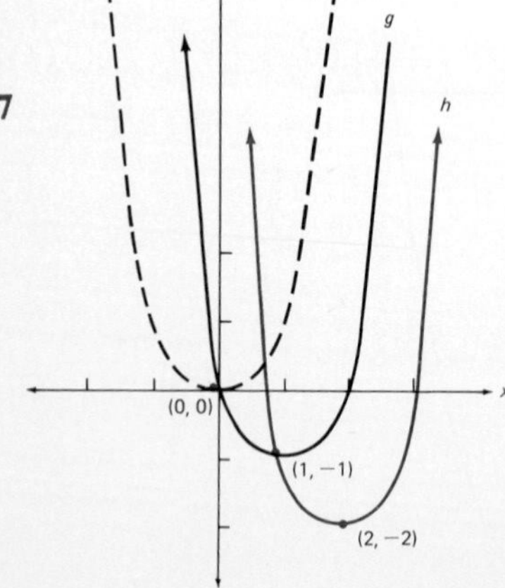

9

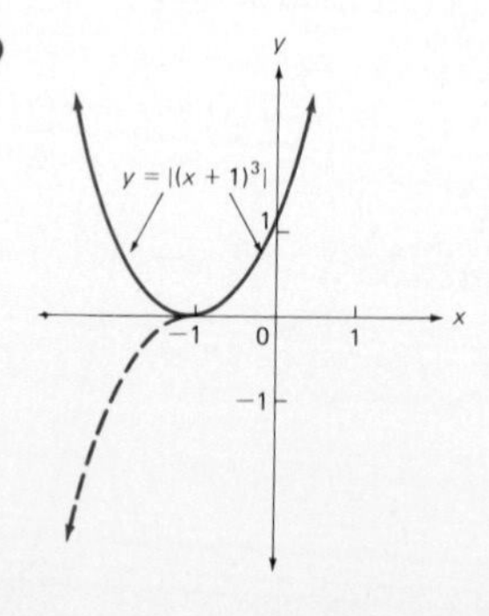

11

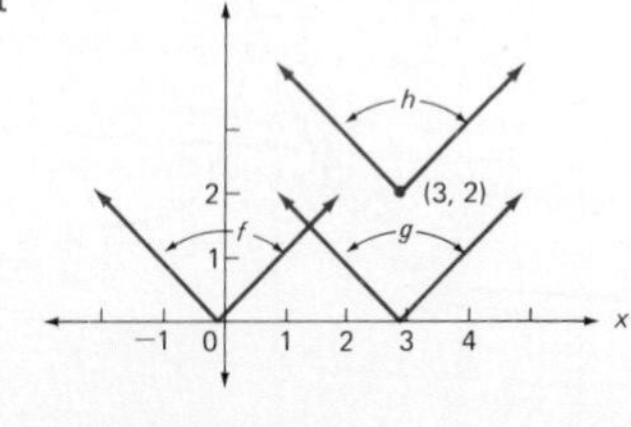

13 $y = (x - 3)^4 + 2$ **15** $y = |x + \frac{3}{4}|$ **17**

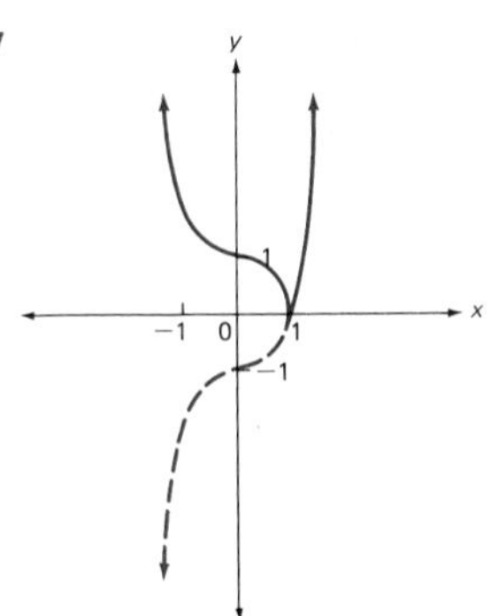

19

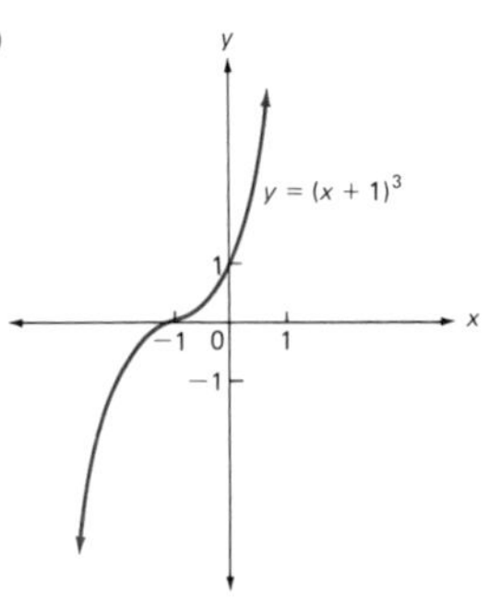

21

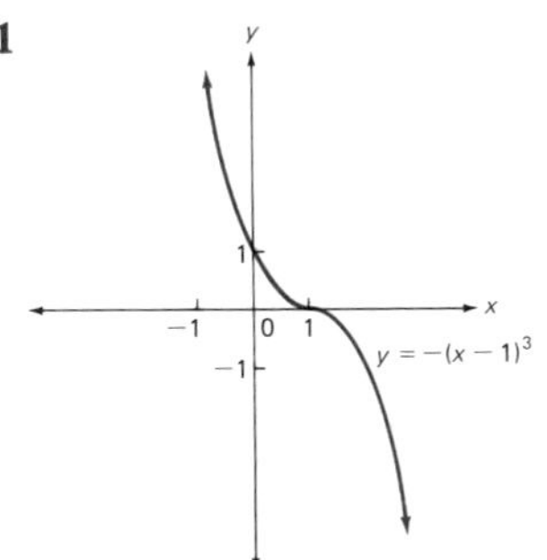

23 $x^2 + 3x + 9$ **25** $4 + 6h + 4h^2 + h^3$

5.2 GRAPHING SOME SPECIAL RATIONAL FUNCTIONS (*Page 176*)

1

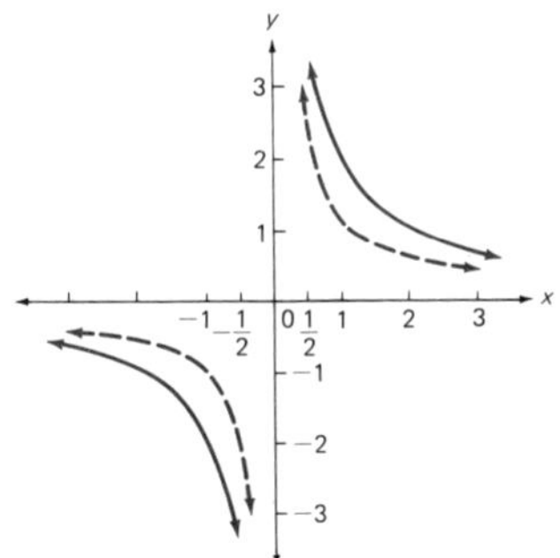

3

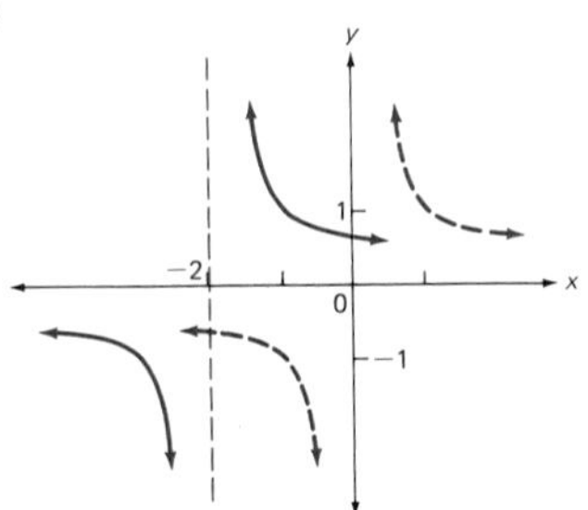

5

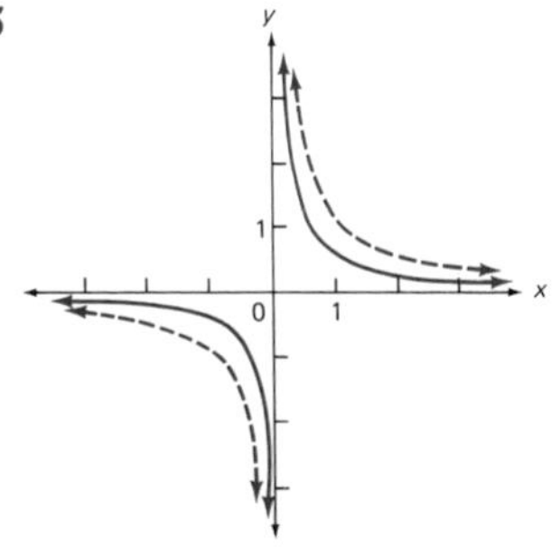

7

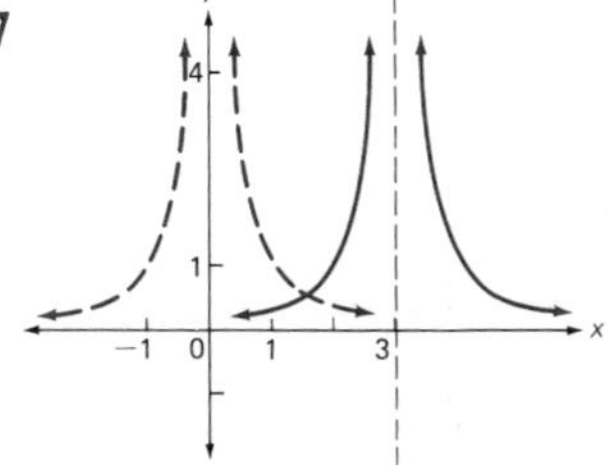

9 Asymptotes: $x = -4$, $y = -2$.

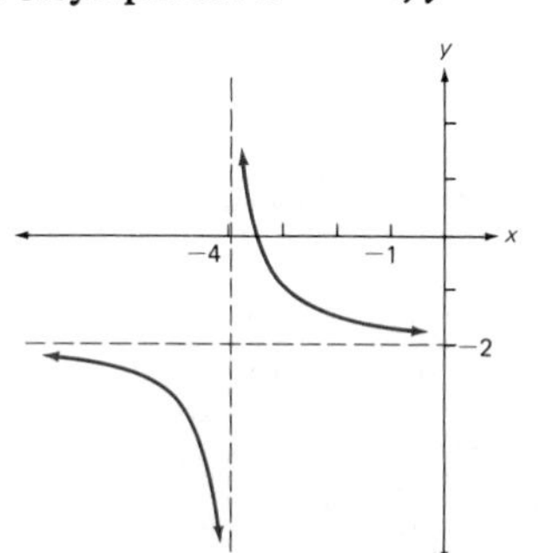

11 Asymptotes: $x = 2$, $y = 0$.

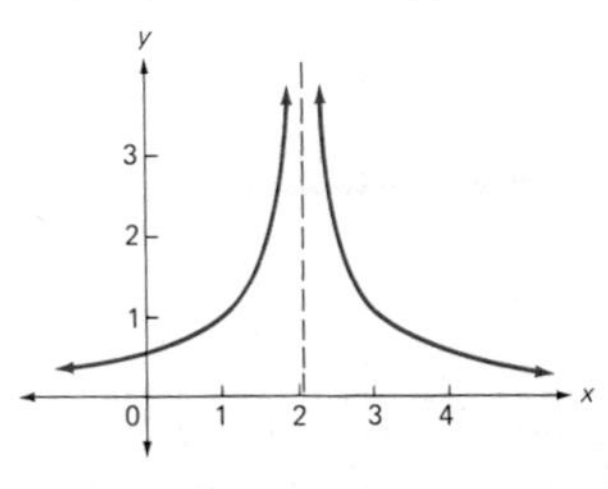

13 Asymptotes: $y = 0$; $x = 0$.

15 Asymptotes: $y = 0$; $x = 1$.

17

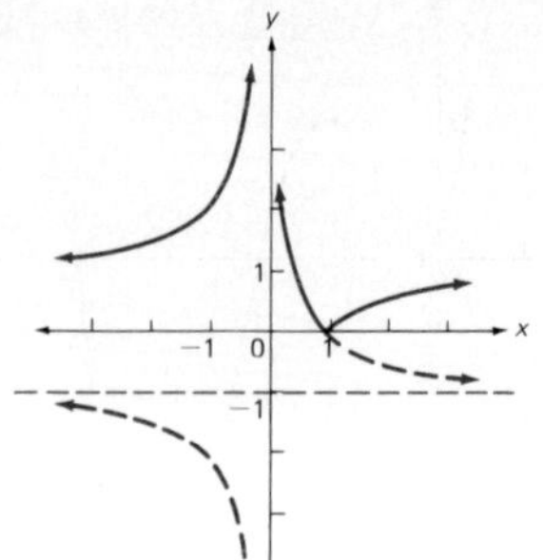

19 $-\dfrac{1}{3x}$

5.3 POLYNOMIAL AND RATIONAL FUNCTIONS (*Page 180*)

1 $f(x) < 0$ on both $(-\infty, 1)$ and $(2, 3)$; $f(x) > 0$ on both $(1, 2)$ and $(3, \infty)$.

3 $f(x) < 0$ on $(-\infty, -4)$ and $(\frac{1}{3}, 2)$; $f(x) > 0$ on $(-4, 0)$, $(0, \frac{1}{3})$, and $(2, \infty)$. **5** $x < -1$ or $x > 4$

7 $x < -1$ or $\frac{1}{5} < x < 10$

9

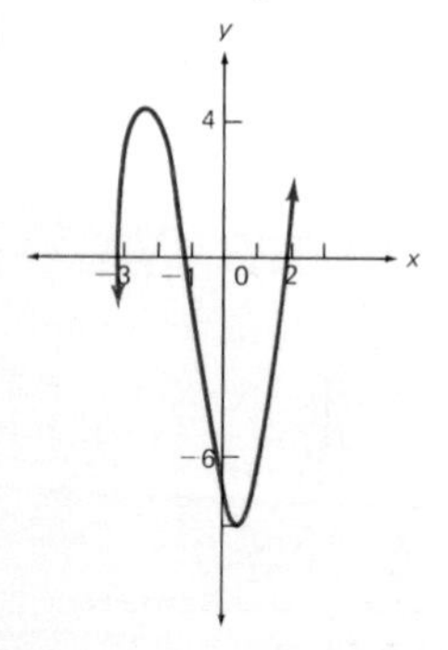

11

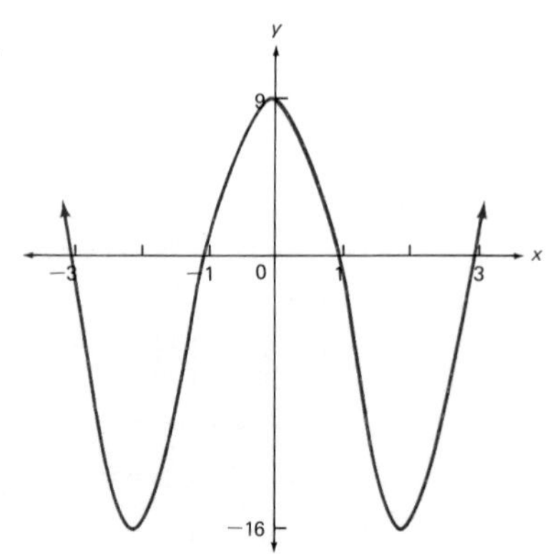

13

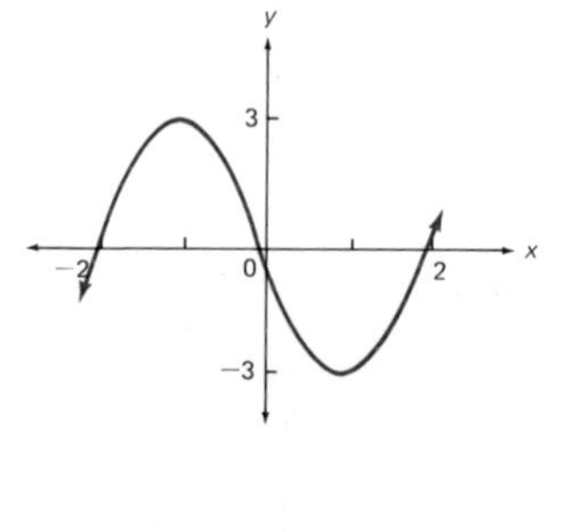

15

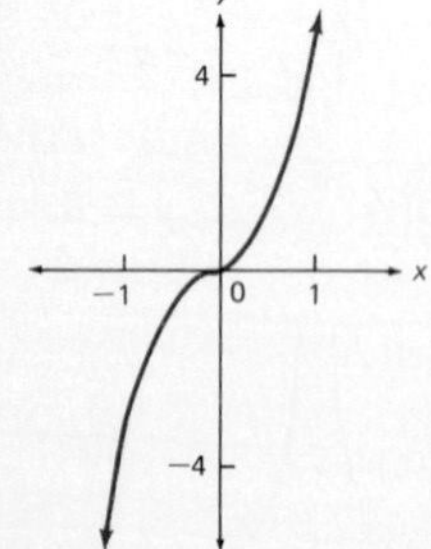

17

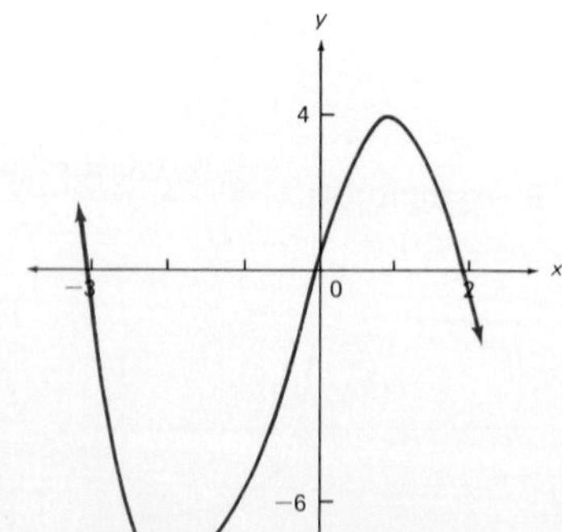

19

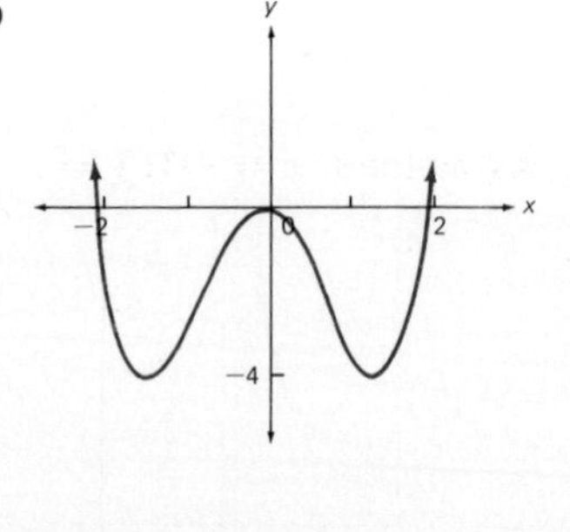

21

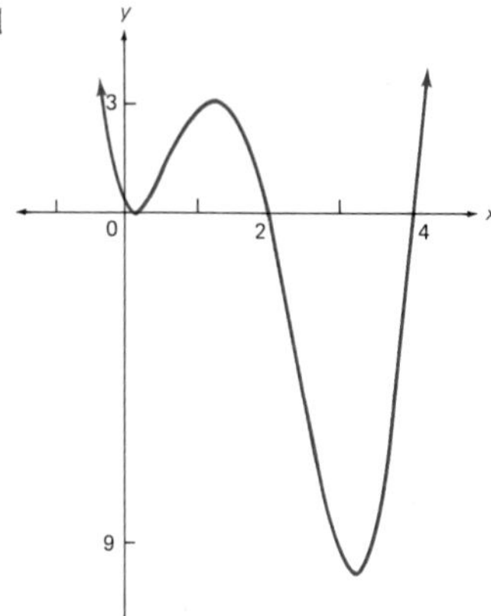

23 Asymptotes: $x = -2$; $x = 1$; $y = 0$.

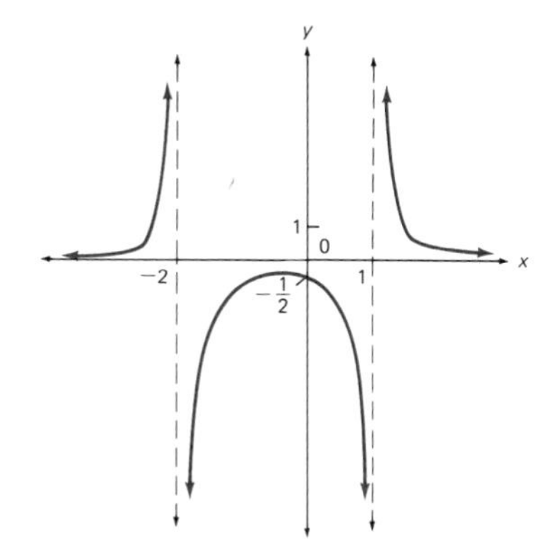

25 Asymptotes: $x = -2$; $x = 2$; $y = 0$.

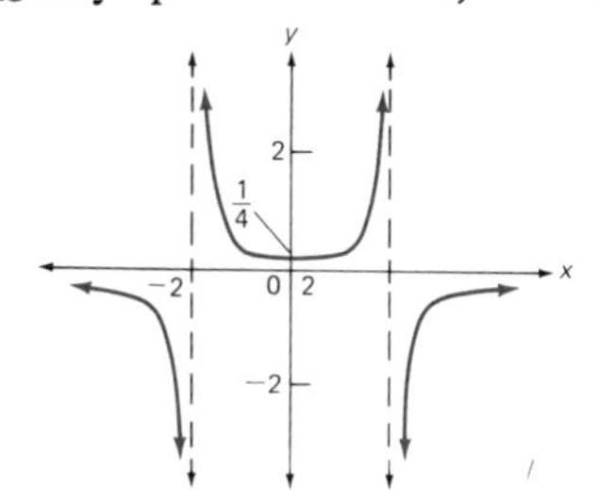

27 Asymptotes: $x = -1$; $x = 1$; $y = 0$.

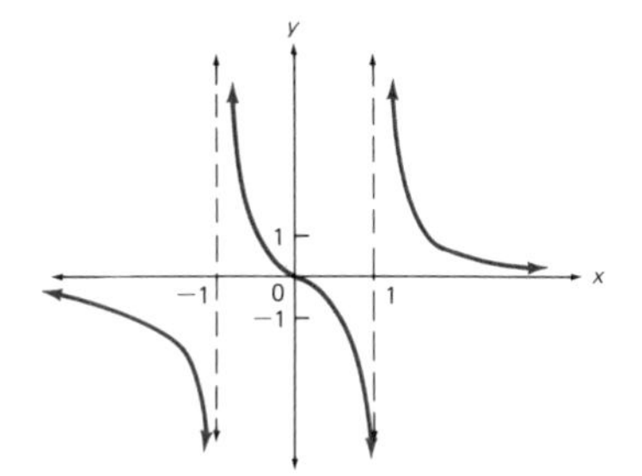

29 Asymptote: $y = 0$.

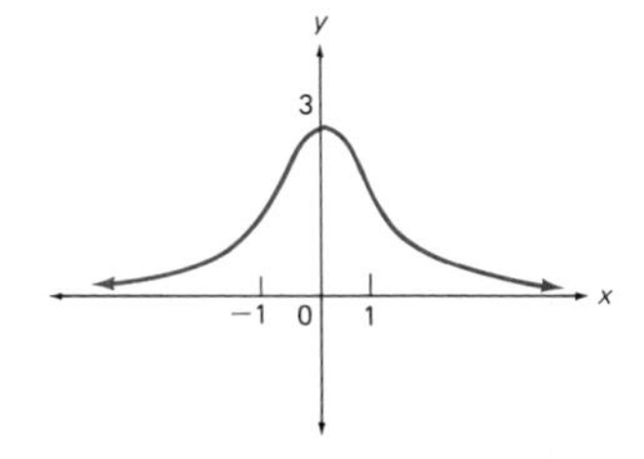

31 Asymptotes: $x = 2$; $y = 1$.

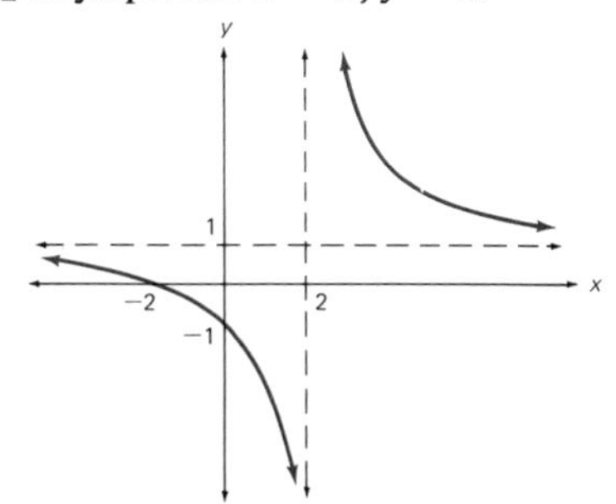

33 Asymptotes: $x = -3$; $y = 1$.

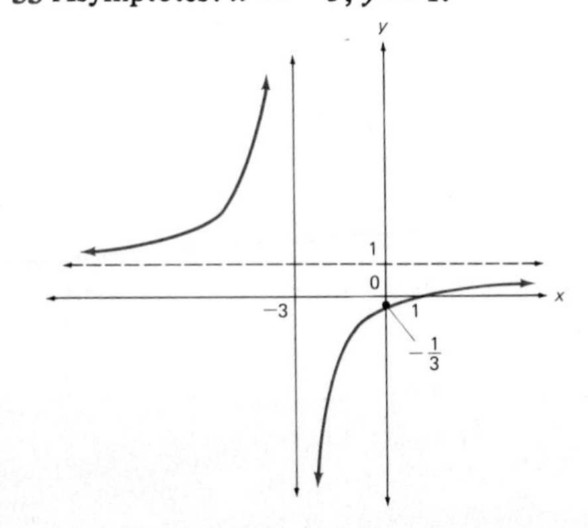

35 Asymptotes: $x = -4$; $x = 3$; $y = 1$.

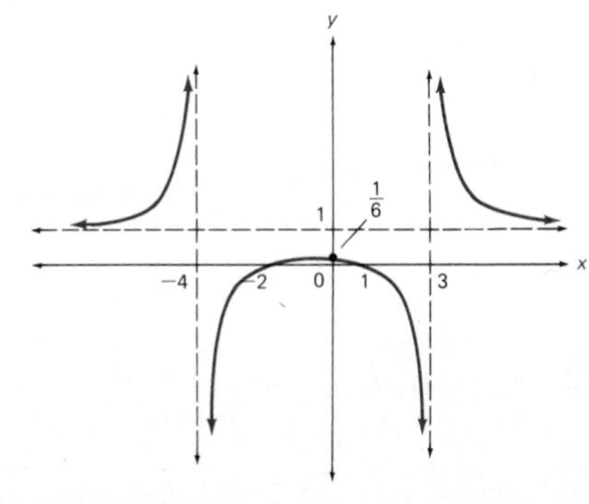

37 $g(x) = 6(x-2)(x-1)$; $g(x) > 0$ on both $(-\infty, 1)$ and $(2, \infty)$; $g(x) < 0$ on $(1, 2)$.

39 $g(x) = -\dfrac{x+4}{x^3}$; $g(x) > 0$ on $(-4, 0)$; $g(x) < 0$ on both $(-\infty, -4)$ and $(0, \infty)$.

41 $g(x) = \dfrac{14(x-5)}{(x+2)^3}$; $g(x) > 0$ on both $(-\infty, -2)$ and $(5, \infty)$; $g(x) < 0$ on $(-2, 5)$.

5.4 EQUATIONS AND INEQUALITIES WITH FRACTIONS (*Page 184*)

1 $x = 20$ 3 $x = 4$ 5 $x = \frac{17}{4}$ 7 $x = 2$ 9 $x = 12$ 11 No solutions. 13 $x = -\frac{5}{2}$; $x = 2$ 15 $x = 1$; $x = 2$
17 $x = 20$ 19 $x = -3$; $x = 3$ 21 $x = -\frac{1}{2}$; $x = 3$ 23 $x = 5$; $x = \frac{10}{7}$ 25 $x = 7$; $x = -5$ 27 $m = \dfrac{2gK}{v^2}$

29 $r_1 = \dfrac{S}{\pi s} - r_2$ 31 $s = a + (n-1)d$ 33 $m = \dfrac{fp}{p-f}$ 35 $x \leq 30$ 37 $x < -1$ 39 $x > 10$ 41 $-3 < x < 0$
43 3 45 $1\frac{5}{7}$ hours. 47 $\frac{2}{9}, \frac{4}{9}$ 49 60 feet. 51 19 53 $\frac{2}{3}$

55 $\dfrac{a}{b} = \dfrac{c}{d}$ implies $ad = bc$. Then, $ad + bd = bc + bd$; $d(a+b) = b(c+d)$; therefore $\dfrac{a+b}{b} = \dfrac{c+d}{d}$.

57 110 paperbacks and 55 hardcover copies.

59 Use $\dfrac{4(4) + 4(3) + 3(3) + 1(2) + x(3)}{15} = 3.4$ to get $x = 4$; grade must be A. 61

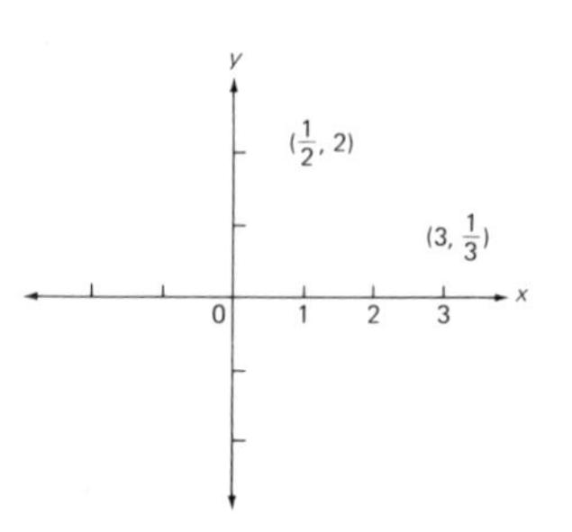

5.5 VARIATION (*Page 190*)

1 $P = 4s$; $k = 4$ 3 $A = 5l$; $k = 5$ 5 $z = kxy^3$ 7 $z = \dfrac{kx}{y^3}$ 9 $w = \dfrac{kx^2}{yz}$ 11 $\frac{1}{2}$ 13 $\frac{1}{5}$ 15 $\frac{2}{5}$ 17 $\frac{243}{8}$

19 27 feet. 21 4 kilograms. 23 75 kilograms. 25 784 feet. 27 $\dfrac{32\pi}{3}$ cubic inches. 29 5.12 ohms; 512 ohms.

31 **(a)** $s(y) = ky(\frac{25}{4} - y^2)$; **(b)** $s(x) = \frac{1}{2}kx^2\sqrt{25 - 4x^2}$; **(c)** $s\left(\dfrac{2.5}{\sqrt{2}}\right) = 5.52k$

5.6 SYNTHETIC DIVISION (*Page 195*)

1 $x^2 + x - 2$; $r = 0$ 3 $2x^2 - x - 2$; $r = 9$ 5 $x^2 + 7x + 7$; $r = 22$ 7 $x^3 - 5x^2 + 17x - 36$; $r = 73$
9 $2x^3 + 2x^2 - x + 3$; $r = 1$ 11 $x^2 + 3x + 9$; $r = 0$ 13 $x^2 - 3x + 9$; $r = 0$ 15 $x^3 + 2x^2 + 4x + 8$; $r = 0$
17 $x^3 - 2x^2 + 4x - 8$; $r = 32$ 19 $x^3 + \frac{1}{2}x^2 + \frac{5}{6}x + \frac{7}{12}$; $r = \frac{47}{60}$ 21 $x + 2$; $r = 0$
23 $N = Q \cdot D + R$; to obtain the dividend, multiply the quotient by the divisor and add the remainder.

5.7 THE REMAINDER AND FACTOR THEOREMS (*Page 199*)

1 8 3 59 5 0 7 1 9 67 11 -4

For Exercises 13–23, use synthetic division to obtain $p(c) = 0$, showing $x - c$ to be a factor of $p(x)$. The remaining factors of $p(x)$ are obtained by factoring the quotient obtained in the synthetic division.

13 $(x+1)(x+2)(x+3)$ 15 $(x-2)(x+3)(x+4)$ 17 $-(x+2)(x-3)(x+1)$
19 $(x-5)(3x+4)(2x-1)$ 21 $(x+2)^2(x+1)(x-1)$ 23 $x^2(x+3)^2(x+1)(x-1)$ 25 -36 27 2

1 2; 3; -7 **3** $-\frac{1}{2}$; 4 **5** -1 **7** -3; $-\sqrt{2}$; $\sqrt{2}$ **9** -2; 3; $-i$; i **11** 2; 3; $\frac{1}{2}+\frac{3}{2}i$; $\frac{1}{2}-\frac{3}{2}i$
13 -3; 3; $-2-\sqrt{2}$; $-2+\sqrt{2}$ **15** $-\frac{1}{3}$; 2; $\frac{5}{2}$ **17** $(x-2)(x-4)(x+3)$ **19** $(x+2)(x+4)(x-6)$
21 $(-2, -27)$; $(2, 1)$; $(3, 8)$ **23** $(-3, -161)$; $(5, 335)$; $(\frac{1}{3}, \frac{253}{27})$
25 By the rational root theorem, the only possible rational roots are: ±1, $\pm\frac{1}{2}$, ±2, ±4, ±8. Using synthetic division, we see that none of these are roots.

1 $\frac{1}{x+1}+\frac{1}{x-1}$ **3** $\frac{2}{x-3}-\frac{1}{x+2}$ **5** $\frac{6}{x-4}+\frac{3}{x-2}-\frac{4}{x+1}$ **7** $\frac{3}{x-2}+\frac{3}{(x-2)^2}$ **9** $\frac{6}{5x+2}-\frac{3}{3x-4}$
11 $\frac{1}{x-2}+\frac{3}{(x-2)^2}-\frac{2}{(x-2)^3}$ **13** $x+\frac{1}{x-1}-\frac{1}{x+1}$ **15** $3x^2+1+\frac{1}{2x-1}-\frac{3}{(2x-1)^2}$
17 $10x-5+\frac{6}{x-3}+\frac{14}{x+2}$ **19** $x^2-1-\frac{1}{x-1}+\frac{2}{(x-1)^2}+\frac{6}{(x-1)^3}$

1 No asymptotes.

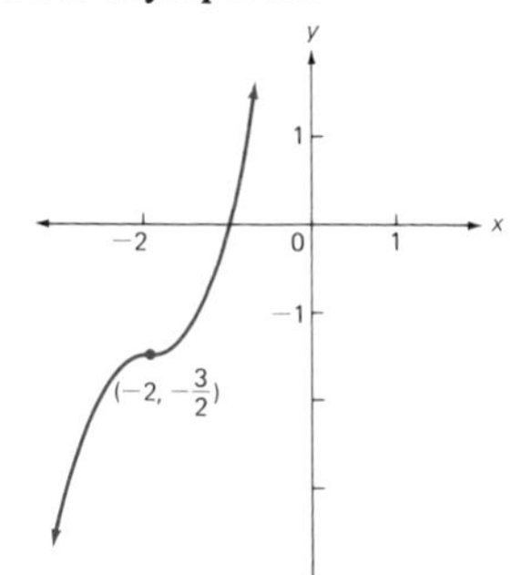

2 No asymptotes.

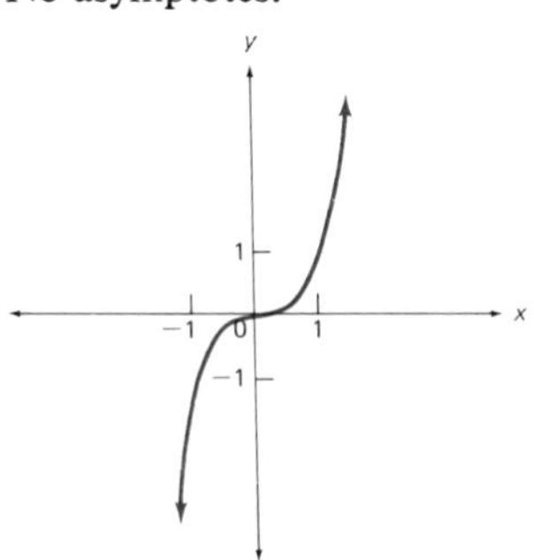

3 Asymptotes: $x = 2$, $y = 0$.

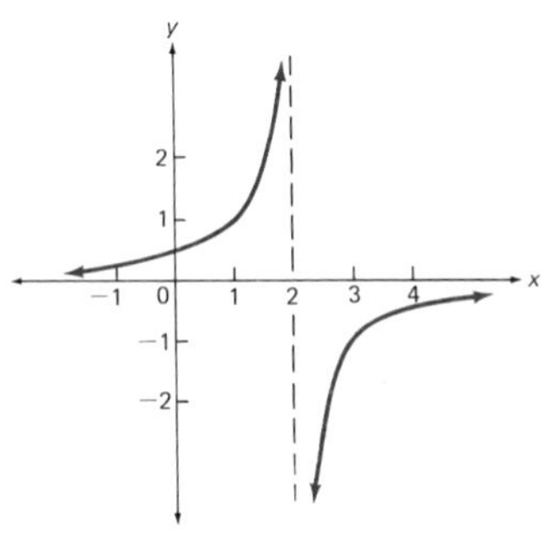

4 Asymptotes: $x = -1$, $x = 2$, $y = 0$.

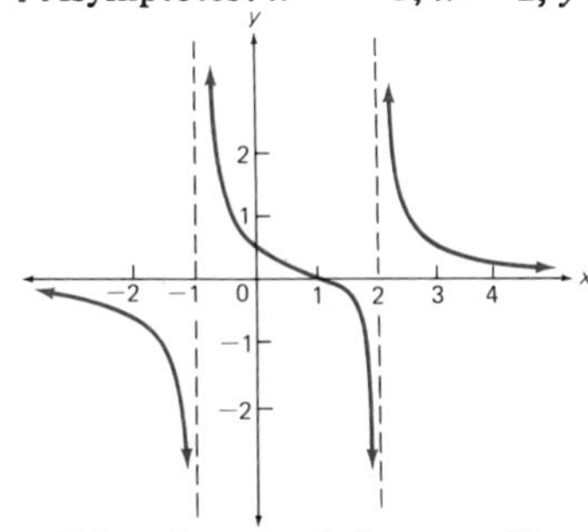

5

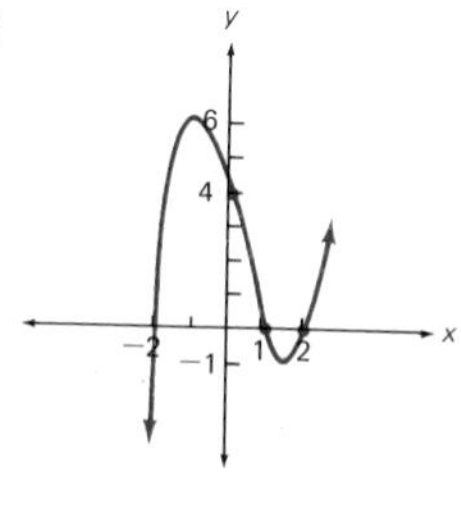

6 $f(x) < 0$ on both $(-\infty, -3)$ and $(0, 2)$;
$f(x) > 0$ on both $(-3, 0)$ and $(2, \infty)$.

7 $-\frac{3}{2}$; 2 **8** $x < -\frac{14}{3}$ **9** $5\frac{5}{7}$ feet. **10** 90 minutes. **11** $z = 2$ **12** $V = 288\pi$ cubic inches.

13
$$\begin{array}{r|rrrrrr} -3 & 2 & +5 & +0 & -1 & -21 & +7 \\ & & -6 & +3 & -9 & +30 & -27 \\ \hline & 2 & -1 & +3 & -10 & +9 & \vert\ -20 \end{array}$$
quotient: $2x^4 - x^3 + 3x^2 - 10x + 9$
remainder: -20

14 **(a)** When $p(x)$ is divided by $x - \frac{1}{3}$, the remainder is $\frac{2}{3}$. Then, by the remainder theorem, we have $p(\frac{1}{3}) = \frac{2}{3}$.
(b) Since $p(\frac{1}{3}) \neq 0$, the factor theorem says that $x - \frac{1}{3}$ is not a factor of $p(x)$.

15 $\begin{array}{r|l} 2 & 1 - 4 + 7 - 12 + 12 \\ & + 2 - 4 + 6 - 12 \\ \hline & 1 - 2 + 3 - 6 \,|\, + 0 = r \end{array}$

Since $r = 0$, $x - 2$ is a factor of $p(x)$, and we get

$$p(x) = (x - 2)(x^3 - 2x^2 + 3x - 6) = (x - 2)[x^2(x - 2) + 3(x - 2)] = (x - 2)(x^2 + 3)(x - 2)$$
$$= (x - 2)^2(x^2 + 3)$$

16 $(x + 3)^2(x^2 - x + 1)$ **17** $(2, 8); (3, 27); (-5, -125)$ **18** $\dfrac{2}{x + 5} - \dfrac{1}{x - 5}$

CHAPTER 6: RADICAL FUNCTIONS AND EQUATIONS

6.1 GRAPHING SOME SPECIAL RADICAL FUNCTIONS (*Page 215*)

1

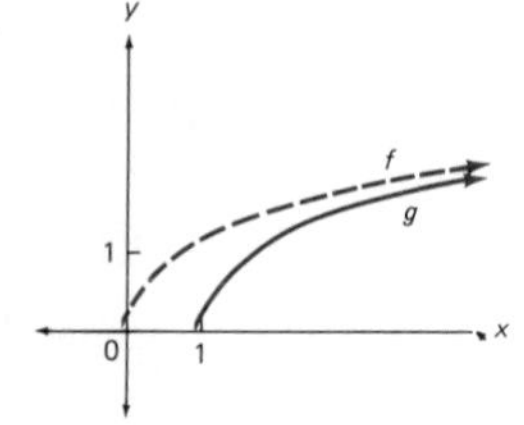

3

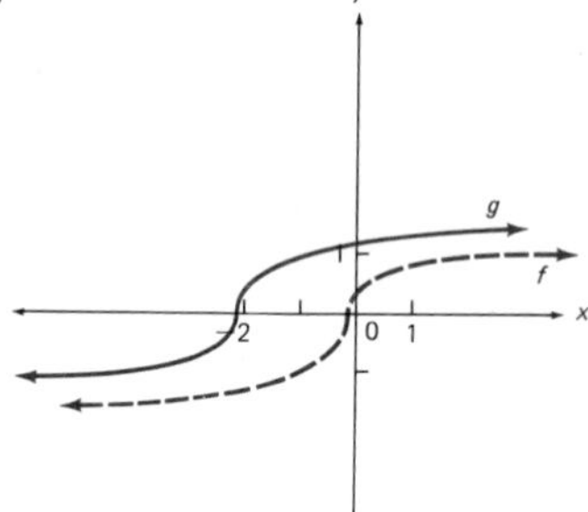

5 Domain: $x \geq -2$.

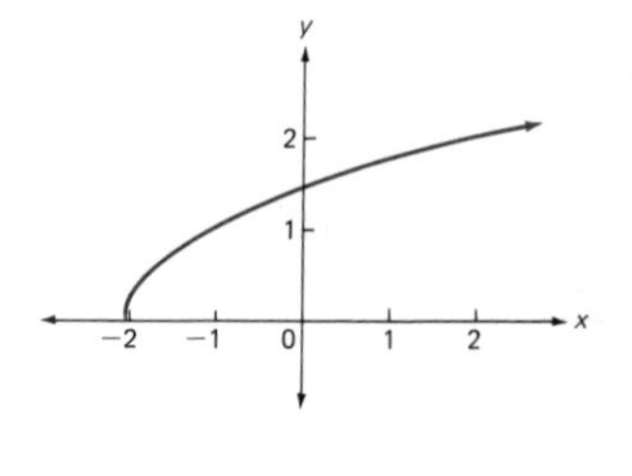

7 Domain: $x \geq 3$.

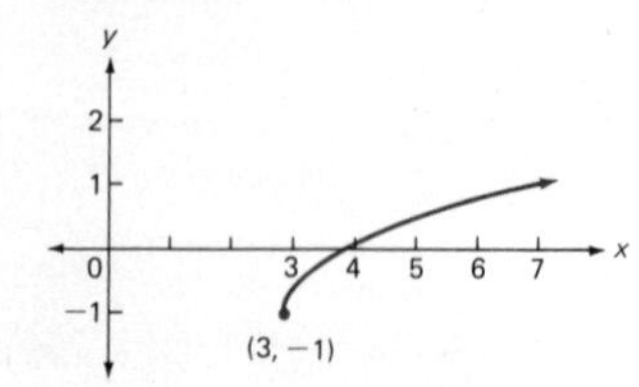

9 Domain: $x \leq 0$.

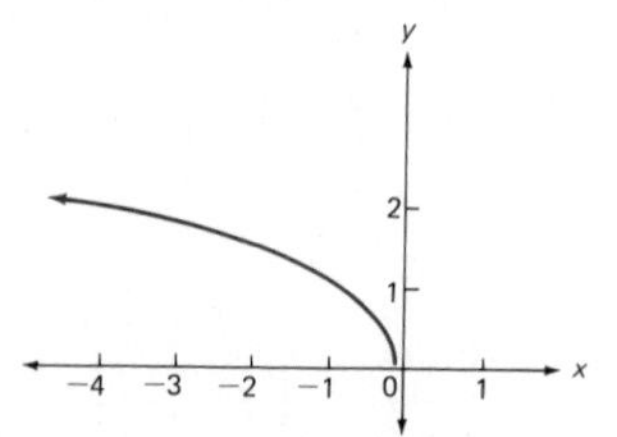

11 Domain: all real x.

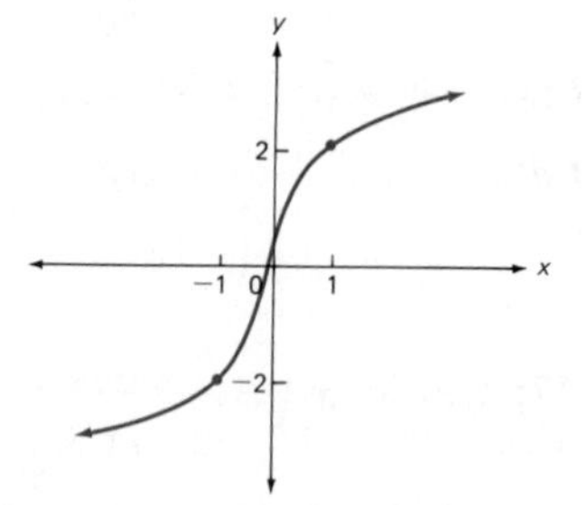

13 Domain: all real x.

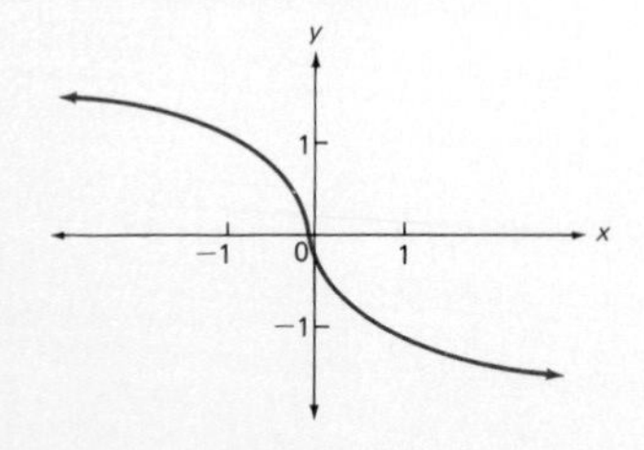

15 Domain: $x > 0$.
Asymptotes: $x = 0$, $y = -1$.

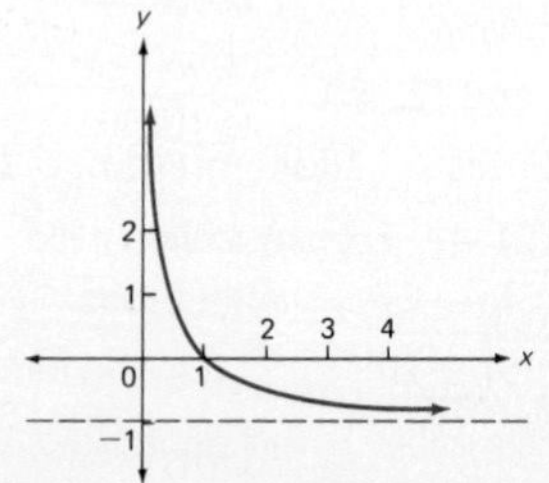

17 (a) $f(-x) = \dfrac{1}{\sqrt[3]{-x}} = \dfrac{1}{-\sqrt[3]{x}} = -\dfrac{1}{\sqrt[3]{x}} = -f(x)$

(b) All $x \neq 0$.

(c)

x	$\frac{1}{27}$	$\frac{1}{8}$	1	8
y	3	2	1	$\frac{1}{2}$

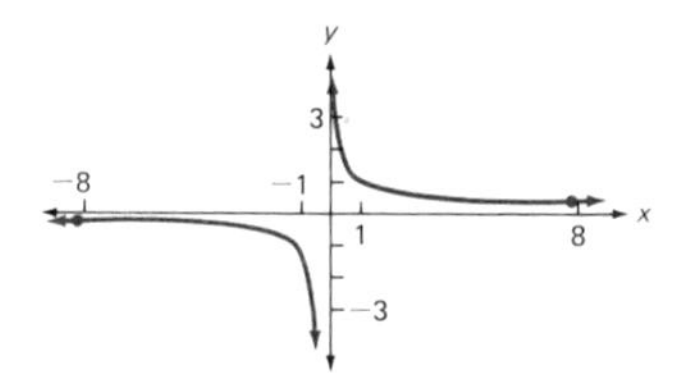

(d) $x = 0$; $y = 0$

19 $y = \sqrt[4]{x}$ is equivalent to $y^4 = x$ for $x \geq 0$.

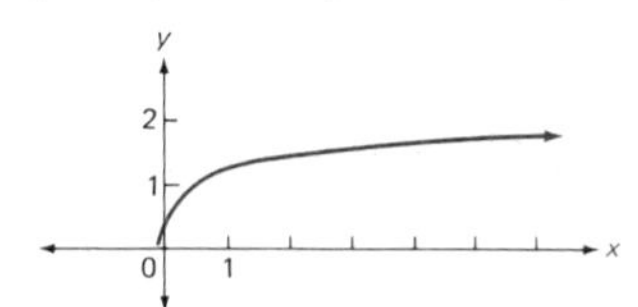

21

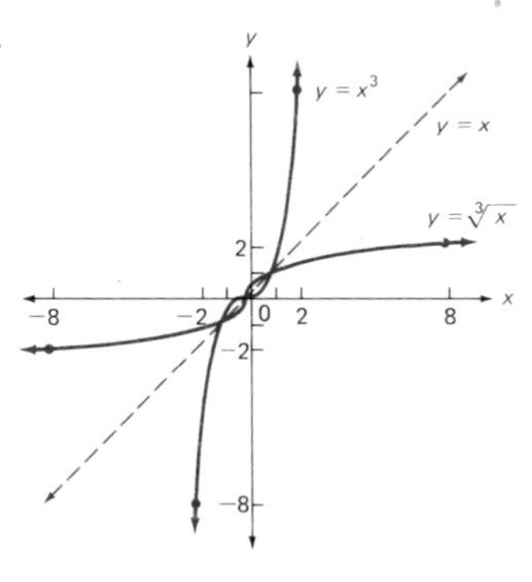

23

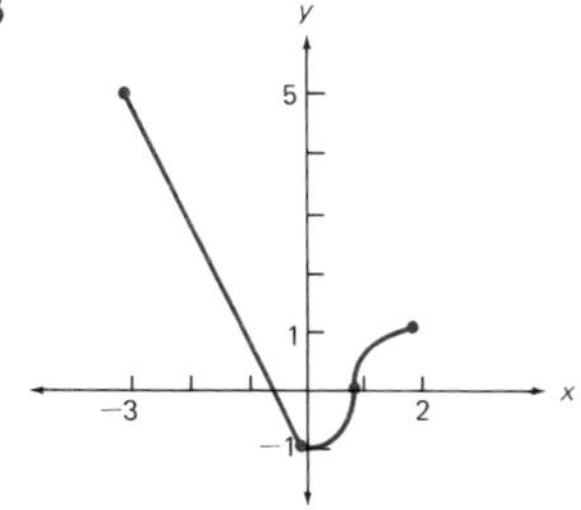

25 $\dfrac{\sqrt{4+h}-2}{h} = \dfrac{(\sqrt{4+h}-2)(\sqrt{4+h}+2)}{h(\sqrt{4+h}+2)} = \dfrac{4+h-4}{h(\sqrt{4+h}+2)} = \dfrac{1}{\sqrt{4+h}+2}$

27 (a) $d(x) = \sqrt{25+x^2} + \sqrt{100+(20-x)^2}$; (b) $t(x) = \dfrac{\sqrt{25+x^2}}{12} + \dfrac{\sqrt{100+(20-x)^2}}{10}$

29 $d(x) = \sqrt{x^2+(6-3x)^2}$

6.2 RADICAL EQUATIONS (*Page 220*)

1 17; domain: $x \geq 1$. **3** $-2, 0$; domain: $x \leq -2$ or $x \geq 0$. **5** $-1, 6$; domain: $x \leq -1$ or $x \geq 6$. **7** $x = 10$ **9** $x = 5$ **11** $x = \pm 10$ **13** No solution. **15** $x = 2$ **17** $x = \frac{2}{3}$ **19** $x = \frac{5}{16}$ **21** $x = 9$ **23** $x = 4$ **25** $x = 4$

27 $x = 0, x = 8$ **29** $x = \frac{27}{8}$ **31**

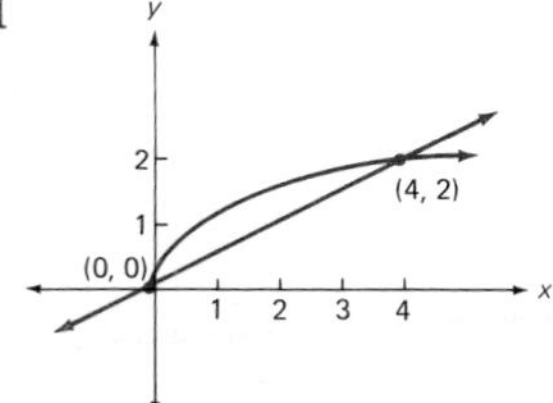

33 $x = \sqrt{16-y^2}$ **35** $f = -\dfrac{\sqrt{y}}{\sqrt{x}}$

37 (9, 3) **39** $S = \dfrac{10}{h} + 2\sqrt{5\pi h}$

6.3 DETERMINING THE SIGNS OF RADICAL FUNCTIONS (*Page 225*)

1 **(a)** $x \geq 2$; **(b)** $f(x) > 0$ for all $x > 2$; **(c)** 2. **3** **(a)** All real x; **(b)** $f(x) > 0$ for all $x \neq 2$; **(c)** 2.
5 **(a)** All $x > 0$; **(b)** $f(x) > 0$ for $x > 0$; **(c)** none.
7 **(a)** All $x \neq 0$; **(b)** $f(x) < 0$ on both $(-\infty, -4)$, $(0, 2)$; $f(x) > 0$ on both $(-4, 0)$ $(2, \infty)$; **(c)** -4, 2.
9 **(a)** $-3 \leq x \leq 3$, $x \neq 0$; **(b)** $f(x) < 0$ on $(-3, 0)$; $f(x) > 0$ on $(0, 3)$; **(c)** -3, 3.
11 $x^{1/2} + (x-4)\frac{1}{2}x^{-1/2} = x^{1/2} + \frac{x-4}{2x^{1/2}} = \frac{2x + x - 4}{2\sqrt{x}} = \frac{3x-4}{2\sqrt{x}}$
13 $\frac{1}{2}(4-x^2)^{-1/2}(-2x) = \frac{-2x}{2(4-x^2)^{1/2}} = -\frac{x}{\sqrt{4-x^2}}$
15 $\frac{x}{3(x-1)^{2/3}} + (x-1)^{1/3} = \frac{x + 3(x-1)}{3(x-1)^{2/3}} = \frac{4x-3}{3(\sqrt[3]{x-1})^2}$
17 $\frac{x}{(5x-6)^{4/5}} + (5x-6)^{1/5} = \frac{x + (5x-6)}{(5x-6)^{4/5}} = \frac{6(x-1)}{(5x-6)^{4/5}}$
19 $f(x) = \frac{x-1}{\sqrt{x}}$; $f(x) < 0$ on $(0, 1)$, $f(x) > 0$ on $(1, \infty)$.
21 $f(x) = \frac{3(x-3)}{2\sqrt{x}}$; $f(x) < 0$ on $(0, 3)$, $f(x) > 0$ on $(3, \infty)$. **23** $f(x) = \frac{x(5x-8)}{2\sqrt{x-2}}$; $f(x) > 0$ on $(2, \infty)$.

6.4 COMBINING FUNCTIONS (*Page 228*)

1 **(a)** $-1, 5, 4$; **(b)** $5x - 1$, all reals; **(c)** 4. **3** **(a)** $-2, \frac{7}{2}, -7$; **(b)** $6x^2 - 5x - 6$, all reals; **(c)** -7.
5 **(a)** 2, 1; **(b)** $6x + 1$, all reals; **(c)** 1. **7** **(a)** $x^2 + \sqrt{x}$; $x \geq 0$; **(b)** $x^{3/2}$; $x > 0$; **(c)** x; $x \geq 0$.
9 **(a)** $x^3 - 1 + \frac{1}{x}$; all $x \neq 0$; **(b)** $x^4 - x$; all $x \neq 0$; **(c)** $\frac{1}{x^3} - 1$; all $x \neq 0$.
11 **(a)** $x^2 + 6x + 8 + \sqrt{x-2}$; $x \geq 2$; **(b)** $\frac{x^2 + 6x + 8}{\sqrt{x-2}}$; $x > 2$; **(c)** $x + 6 + 6\sqrt{x-2}$; $x \geq 2$
13 **(a)** $6x - 6$, all reals; **(b)** $-8x^2 + 22x - 5$, all reals; **(c)** $-8x + 19$, all reals.
15 **(a)** $\frac{1}{2x} - 2x^2 + 1$, all $x \neq 0$; **(b)** $x - \frac{1}{2x}$, all $x \neq 0$; **(c)** $\frac{1}{4x^2 - 2}$, all $x \neq \pm\frac{1}{\sqrt{2}}$.
17 $(f \circ g)(x) = (x-1)^2$; $(g \circ f)(x) = x^2 - 1$ **19** $(f \circ g)(x) = \frac{x+3}{3-x}$; $(g \circ f)(x) = 4 - \frac{6}{x}$
21 $(f \circ g)(x) = x^2$; $(g \circ f)x = x^2 + 2x$ **23** $(f \circ g)(x) = 2$; $(g \circ f)(x) = 4$
25 **(a)** $\frac{1}{2\sqrt[3]{x} - 1}$; **(b)** $\frac{2}{\sqrt[3]{x}} - 1$; **(c)** $\frac{1}{\sqrt[3]{2x-1}}$ **27** $(f \circ f)(x) = x$; $(f \circ f \circ f)(x) = \frac{1}{x}$ **29** $g(x) = \sqrt[3]{x}$
31 $x^2 - 2x$

6.5 DECOMPOSITION OF COMPOSITE FUNCTIONS (*Page 234*)

(Other answers are possible for Exercises 1–23.)

1 $g(x) = 3x + 1$; $f(x) = x^2$ **3** $g(x) = 1 - 4x$; $f(x) = \sqrt{x}$ **5** $g(x) = \frac{x+1}{x-1}$; $f(x) = x^2$
7 $g(x) = 3x^2 - 1$; $f(x) = x^{-3}$ **9** $g(x) = \frac{x}{x-1}$; $f(x) = \sqrt{x}$ **11** $g(x) = (x^2 - x - 1)^3$; $f(x) = \sqrt{x}$
13 $g(x) = 4 - x^2$; $f(x) = \frac{2}{\sqrt{x}}$ **15** $f(x) = 2x + 1$; $g(x) = x^{1/2}$; $h(x) = x^3$
17 $f(x) = \frac{x}{x+1}$; $g(x) = x^5$; $h(x) = x^{1/2}$ **19** $f(x) = x^2 - 9$; $g(x) = x^2$; $h(x) = x^{1/3}$
21 $f(x) = x^2 - 4x + 7$; $g(x) = x^3$; $h(x) = -\sqrt{x}$ **23** $f(x) = 2x - 11$; $g(x) = 1 + \sqrt{x}$; $h(x) = x^2$
25 $f(x) = (x+1)^2$

1 Domain: all reals.

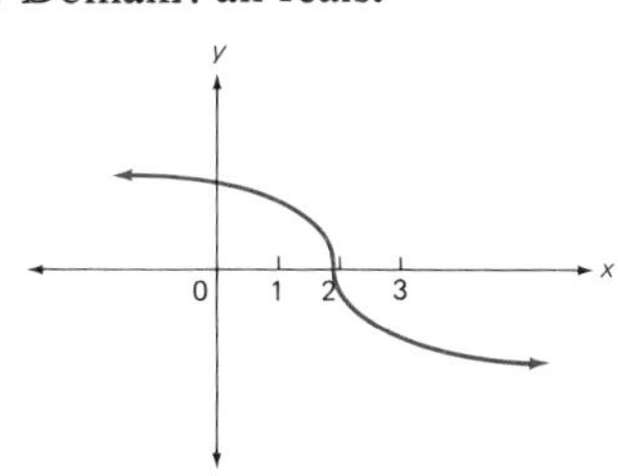

2 Domain: $x > 0$.
Asymptotes: $y = 2$, $x = 0$.

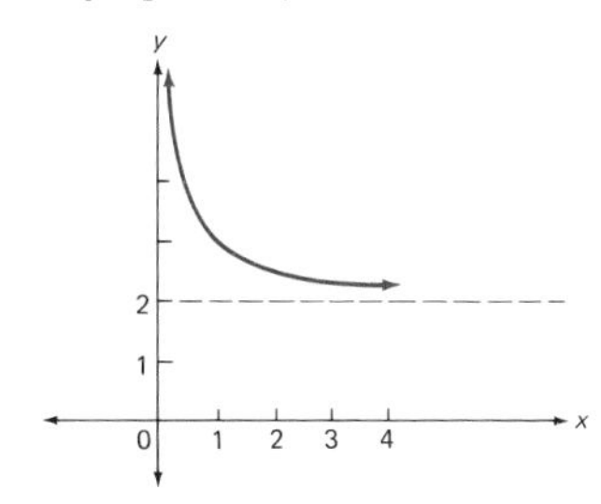

3 $\frac{2}{9}$ **4** $-\frac{64}{27}$; $\frac{1}{8}$. **5** $x = 16$

6 $(x+1)^{1/2}(2x) + (x+1)^{-1/2}\left(\frac{x^2}{2}\right) = 2x(x+1)^{1/2} + \frac{x^2}{2(x+1)^{1/2}} = \frac{4x(x+1)+x^2}{2\sqrt{x+1}} = \frac{5x^2+4x}{2\sqrt{x+1}} = \frac{x(5x+4)}{2\sqrt{x+1}}$

7 $f(x) < 0$ on $(2, 4)$; $f(x) > 0$ on $(-\infty, 2)$ and $(4, \infty)$. **8** $f(x) < 0$ on $(0, 4)$; $f(x) > 0$ on $(4, \infty)$.

9 $(f-g)(x) = \frac{1}{x^2-1} - \sqrt{x+2}$; all $x \geq -2$ and $x \neq \pm 1$. $\frac{f}{g}(x) = \frac{1}{(x^2-1)\sqrt{x+2}}$; all $x > -2$ and $x \neq \pm 1$.

10 $(f \circ g)(x) = \frac{1}{1-x}$; all $x > 0$ and $x \neq 1$. $(g \circ f)(x) = \frac{1}{\sqrt{1-x^2}}$; $-1 < x < 1$.

(Other answers are possible for Exercises 11 and 12.)

11 $g(x) = x - 2$; $f(x) = x^{2/3}$ **12** $h(x) = 2x - 1$; $g(x) = x^{3/2}$; $f(x) = \frac{1}{x}$ **13** (d)

14 All real numbers; $x = -3$, $x = 0$. **15** $(2, 2)$, $(10, 6)$

CHAPTER 7: EXPONENTIAL AND LOGARITHMIC FUNCTIONS

7.1 INVERSE FUNCTIONS (*Page 239*)

1 Not one-to-one. **3** One-to-one. **5** One-to-one.

7 $(f \circ g)(x) = \frac{1}{3}(3x + 9) - 3 = x$
$(g \circ f)(x) = 3(\frac{1}{3}x - 3) + 9 = x$

9 $(f \circ g)(x) = (\sqrt[3]{x} - 1 + 1)^3 = x$
$(g \circ f)(x) = \sqrt[3]{(x+1)^3} - 1 = x$

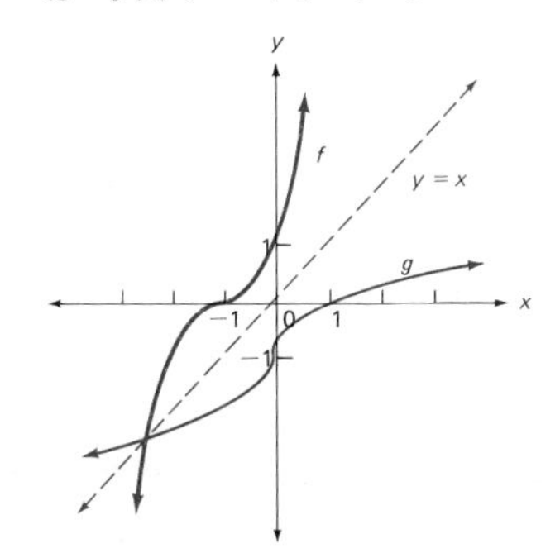

11 $(f \circ g)(x) = \frac{1}{\frac{1}{x} + 1 - 1} = x$

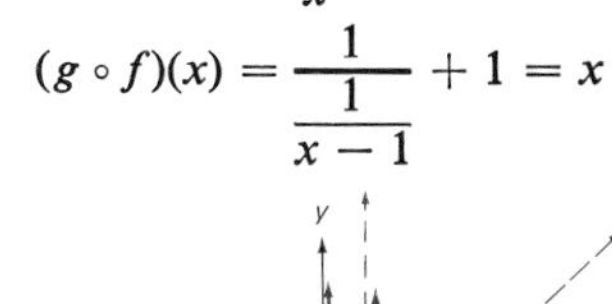

$(g \circ f)(x) = \frac{1}{\frac{1}{x-1}} + 1 = x$

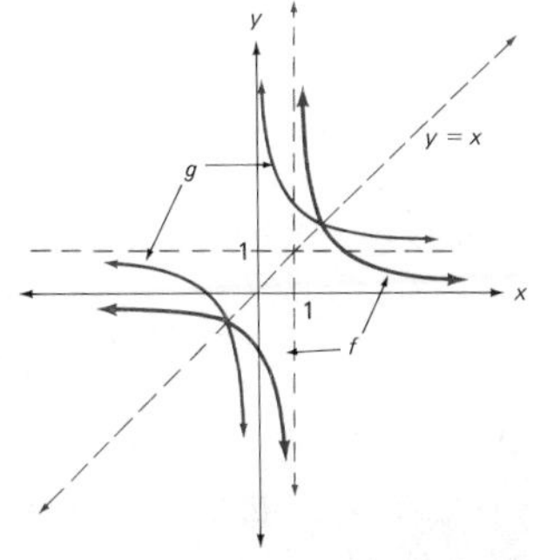

13 $g(x) = \sqrt[3]{x} + 5$ 15 $g(x) = \frac{3}{2}x + \frac{3}{2}$ 17 $g(x) = \sqrt[5]{x} + 1$ 19 $g(x) = x^{5/3}$ 21 $g(x) = 2 + \frac{2}{x}$

23 $g(x) = \frac{1}{\sqrt[5]{x}}$ 25 $(f \circ f)(x) = f(f(x)) = f\left(\frac{1}{x}\right) = \frac{1}{\frac{1}{x}} = x$

27 $(f \circ f)(x) = f(f(x)) = f\left(\frac{x}{x-1}\right) = \frac{\frac{x}{x-1}}{\frac{x}{x-1} - 1} = \frac{x}{x - (x-1)} = x$

29 $g(x) = \sqrt{x} - 1$

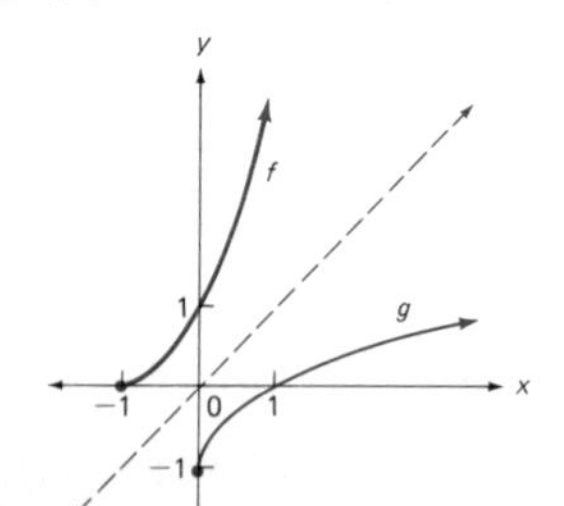

31 $g(x) = \frac{1}{x^2}$

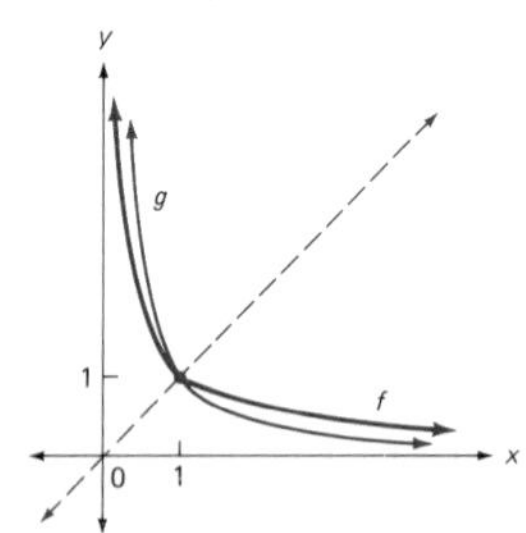

33 $g(x) = \sqrt{x + 4}$

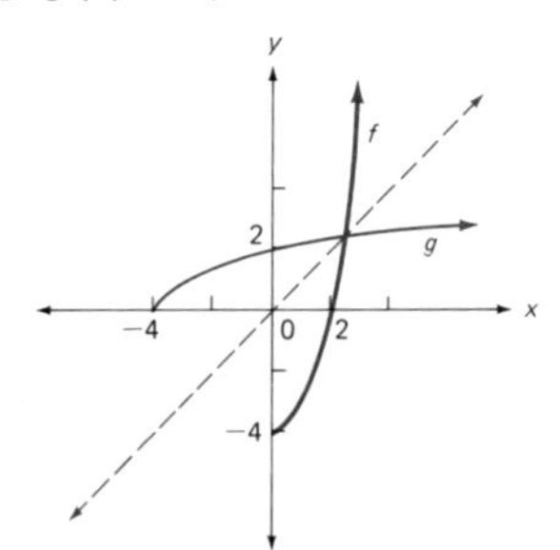

35 $y = mx + k$, where $m = -1$.

7.2 EXPONENTIAL FUNCTIONS (*Page 244*)

1

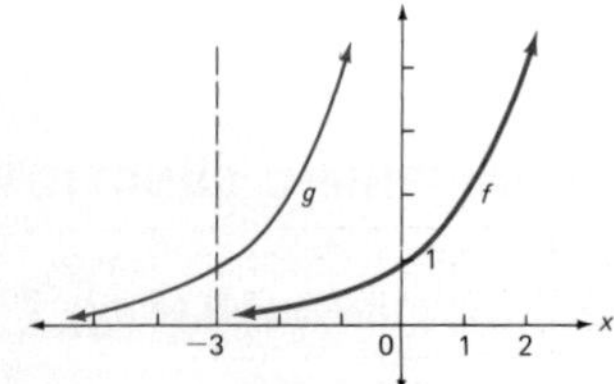

3

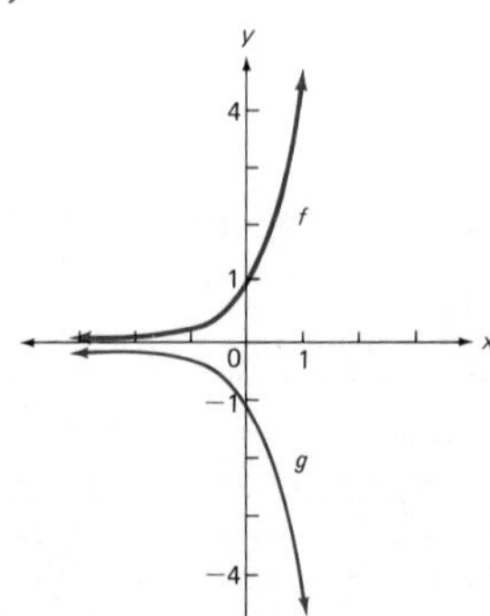

5

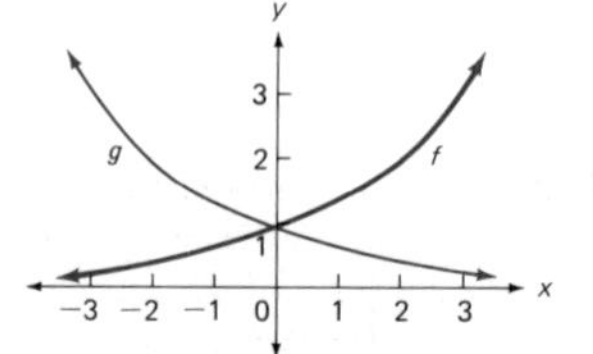

7

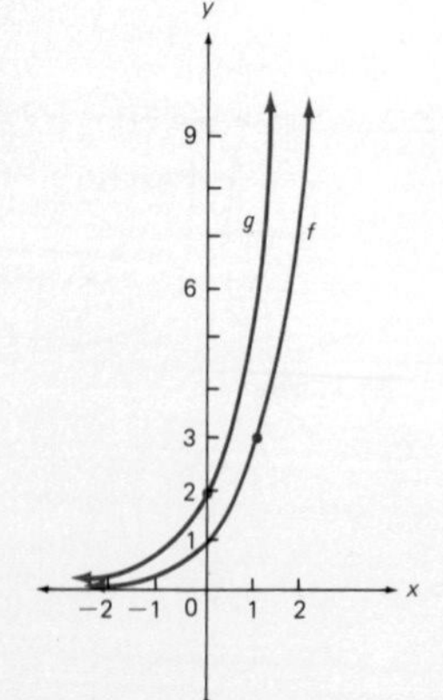

9

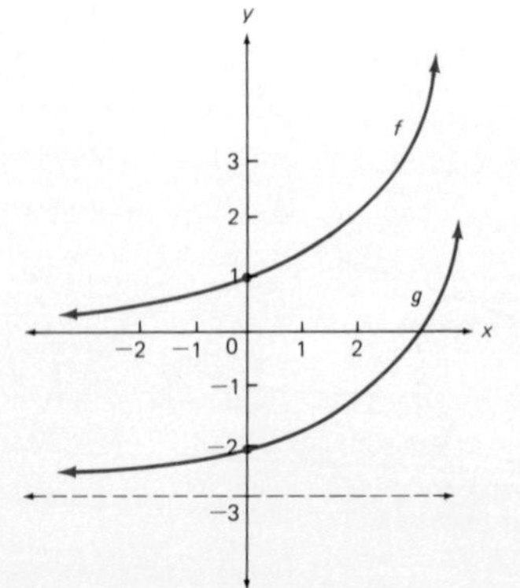

11

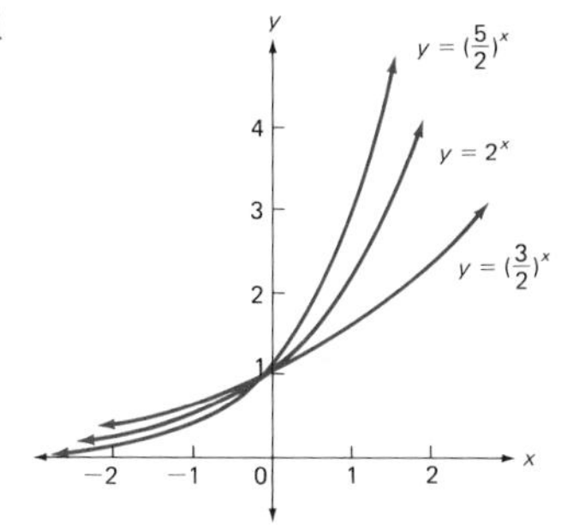

13

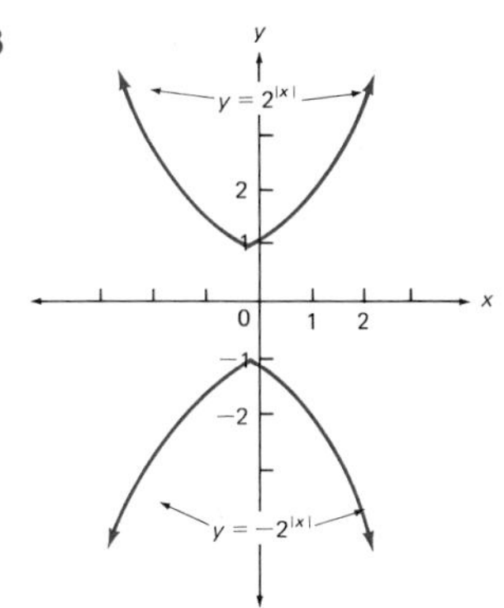

15 6 **17** ± 3 **19** 1 **21** $-2; 1$ **23** -5 **25** $\frac{1}{2}$ **27** $\frac{3}{2}$ **29** $-\frac{1}{2}$ **31** $\frac{2}{3}$ **33**

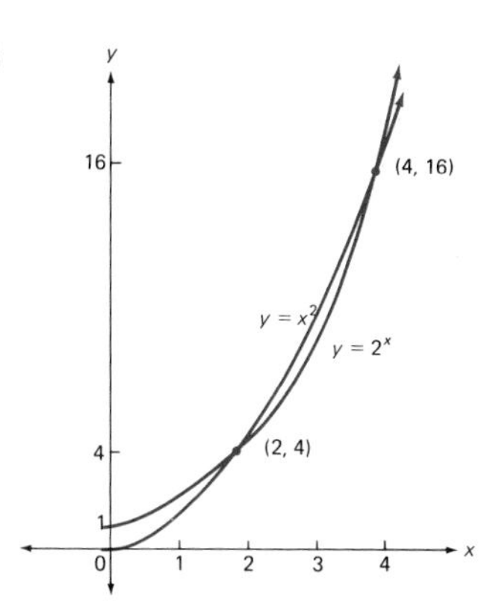

35 (a) $\sqrt{2} = 1.414213\ldots$

x	1.4	1.41	1.414	1.4142	1.41421
3^x	4.6555	4.7070	4.7277	4.7287	4.7288

(b) $\sqrt{3} = 1.732050\ldots$

x	1.7	1.73	1.732	1.7320	1.73205
3^x	6.4730	6.6899	6.7046	6.7046	6.7050

(c) $\sqrt{5} = 2.236068\ldots$

x	2.2	2.23	2.236	2.2360	2.23606
2^x	4.5948	4.6913	4.7109	4.7109	4.71109

(d) $\pi = 3.141592\ldots$

x	3.1	3.14	3.141	3.1415	3.14159
4^x	73.5167	77.7085	77.8163	77.8702	77.8799

37 $x = \frac{1}{2}$

7.3 LOGARITHMIC FUNCTIONS (*Page 249*)

1 $g(x) = \log_4 x$

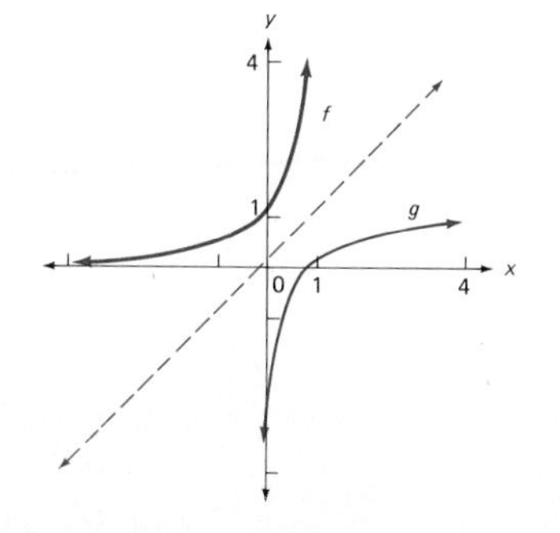

3 $g(x) = \log_{1/3} x$

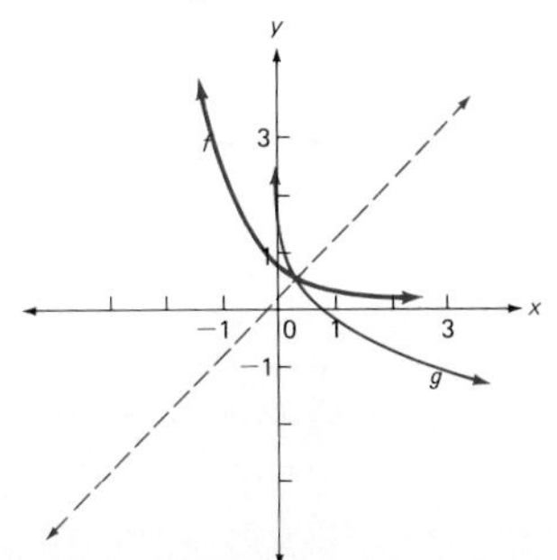

5 Shift 2 units left; $x > -2$; $x = -2$. 7 Shift 2 units upward; $x > 0$; $x = 0$.

9

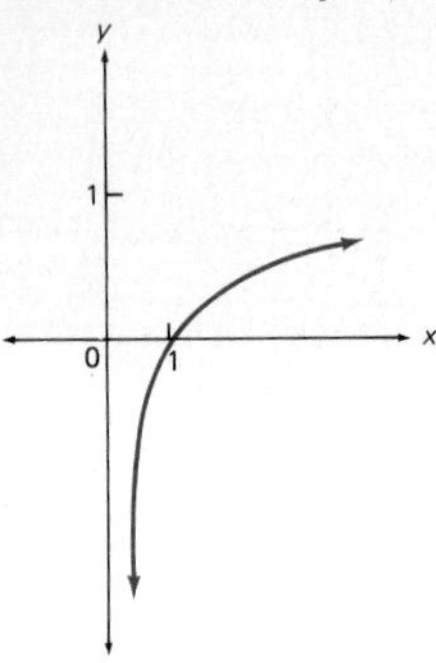

11

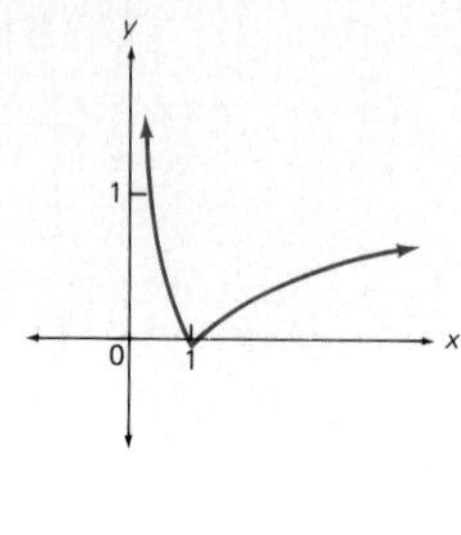

13

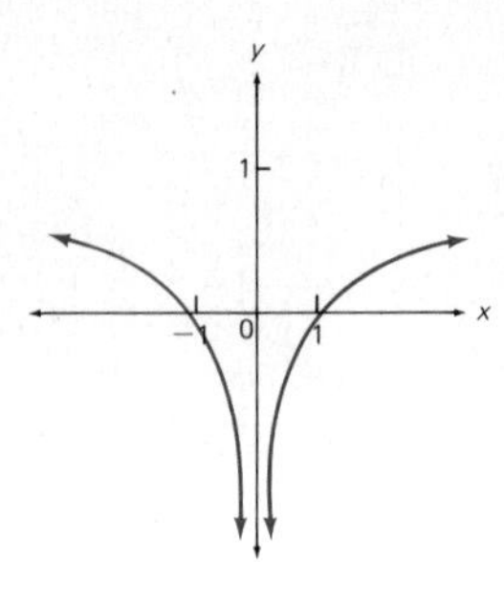

15 $\log_2 256 = 8$ 17 $\log_{1/3} 3 = -1$ 19 $\log_{17} 1 = 0$ 21 $10^{-4} = 0.0001$ 23 $(\sqrt{2})^2 = 2$ 25 $12^{-3} = \frac{1}{1728}$
27 4 29 −3 31 $\frac{1}{216}$ 33 5 35 4 37 $\frac{1}{3}$ 39 $\frac{2}{3}$ 41 9 43 −2 45 $-\frac{1}{3}$ 47 9 49 −1
51 $g(x) = 3^x - 3$, $(f \circ g)(x) = \log_3(3^x - 3 + 3) = \log_3 3^x = x$; $(g \circ f)(x) = 3^{\log_3(x+3)} - 3 = (x + 3) - 3 = x$

7.4 THE LAWS OF LOGARITHMS (*Page 261*)

1 $\log_b 3 + \log_b x - \log_b(x + 1)$ 3 $\frac{1}{2}\log_b(x^2 - 1) - \log_b x = \frac{1}{2}\log_b(x + 1) + \frac{1}{2}\log_b(x - 1) - \log_b x$
5 $-2\log_b x$ 7 $\log_b \frac{x + 1}{x + 2}$ 9 $\log_b \sqrt{\frac{x^2 - 1}{x^2 + 1}}$ 11 $\log_b \frac{x^3}{2(x + 5)}$

13 $\log_b 27 + \log_b 3 = \log_b 81$ (Law 1)
$\log_b 243 - \log_b 3 = \log_b 81$ (Law 2)

15 $-2\log_b \frac{4}{9} = \log_b(\frac{4}{9})^{-2}$ (Law 3)
$= \log_b \frac{81}{16}$

17 (a) 0.6020; (b) 0.9030; (c) −0.3010 19 (a) 1.6811; (b) −0.1761; (c) 2.0970
21 (a) 0.2330; (b) 1.9515; (c) 1.4771 23 20 25 $\frac{1}{20}$ 27 17 29 8 31 2 33 7 35 1.01 37 5
39 Let $r = \log_b M$ and $s = \log_b N$. Then $b^r = M$ and $b^s = N$. Divide:

$$\frac{M}{N} = \frac{b^r}{b^s} = b^{r-s}$$

Convert to log form and substitute:

$$\log_b \frac{M}{N} = r - s = \log_b M - \log_b N$$

41 −1; 0. 43 8
45 $B^{\log_B N} = N$

$\log_b B^{\log_B N} = \log_b N$ (take log base b)

$(\log_B N)(\log_b B) = \log_b N$ (Law 3)

$\log_B N = \frac{\log_b N}{\log_b B}$ (divide by $\log_b B$)

7.5 THE BASE *e* (*Page 257*)

1

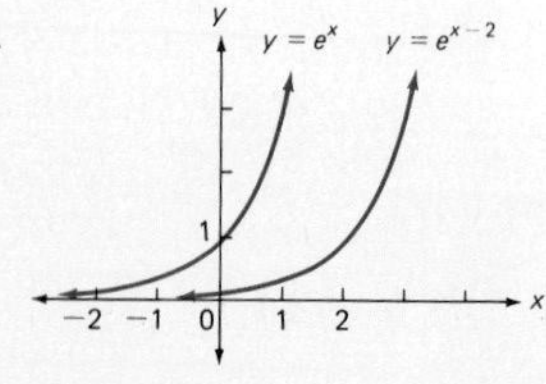

3

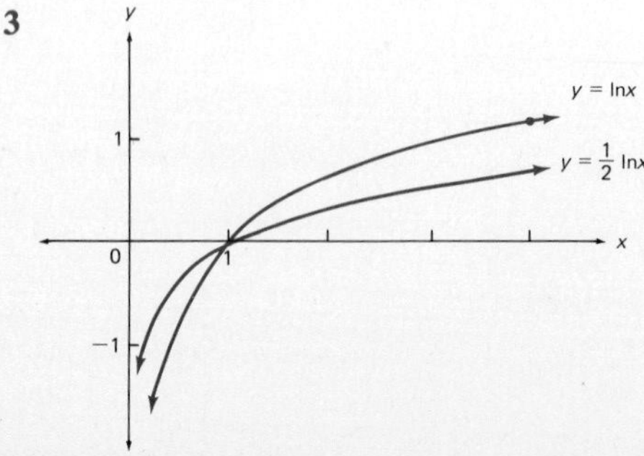

5

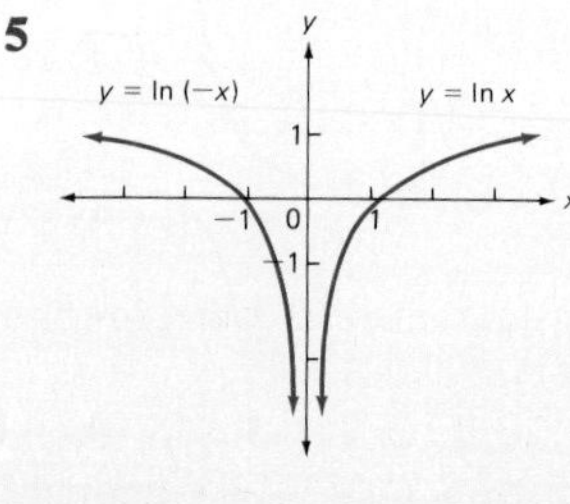

7

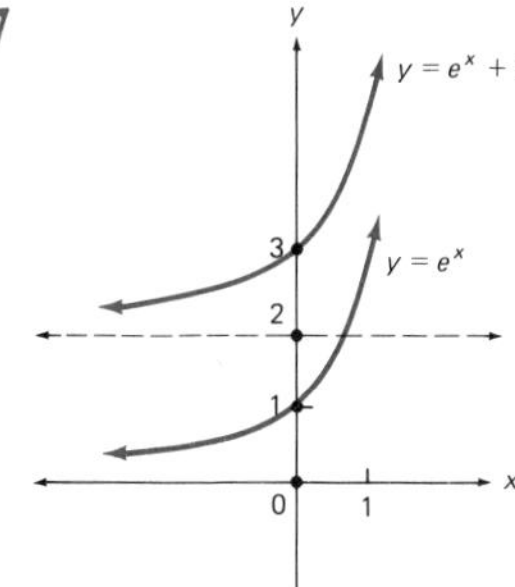

9

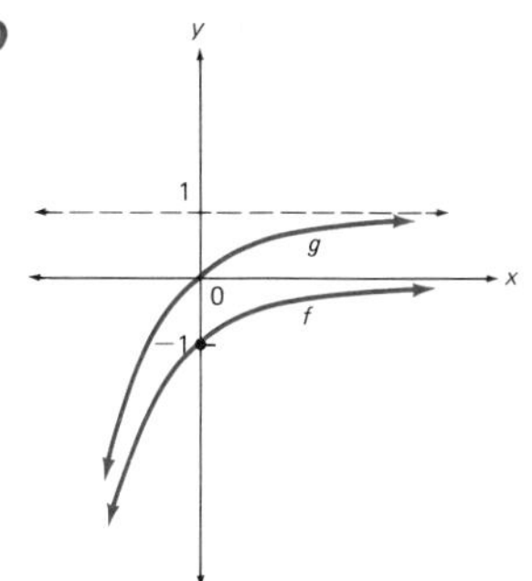

11

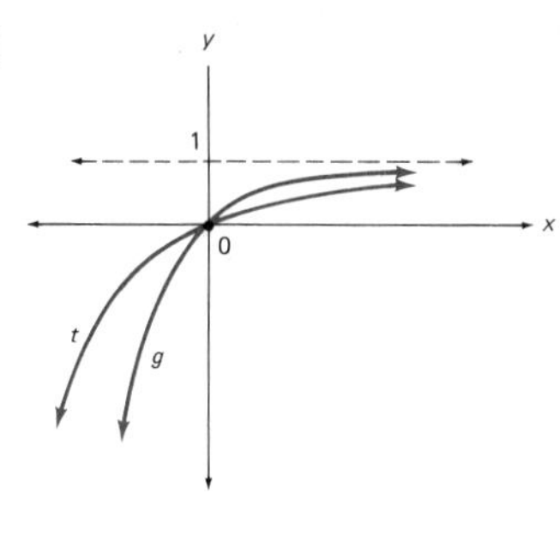

13 Since $f(x) = 1 + \ln x$, shift one unit upward. **15** Since $f(x) = \frac{1}{2}\ln x$, multiply the ordinates by $\frac{1}{2}$.
17 Since $f(x) = \ln(x - 1)$, shift one unit to the right. **19** $x > -2$ **21** $x > \frac{1}{2}$ **23** All $x > 1$ except for $x = 2$.
25 $\ln 5 + \ln x - \ln(x^2 - 4) = \ln 5 + \ln x - \ln(x + 2) - \ln(x - 2)$
27 $\ln(x - 1) + 2\ln(x + 3) - \frac{1}{2}\ln(x^2 + 2)$ **29** $\frac{3}{2}\ln x + \frac{1}{2}\ln(x + 1)$ **31** $\ln \sqrt{x}(x^2 + 5)$ **33** $\ln(x^2 - 1)^3$

35 $\ln \dfrac{\sqrt{x}}{(x - 1)^2\sqrt[3]{x^2 + 1}}$ **37** $\sqrt{x}$ **39** $\dfrac{1}{x^2}$ **41** 1 **43** $x = -100 \ln 27$ **45** $x = \frac{1}{3}$ **47** $x = 0$ **49** $x = 8$

51 $x = 3$ **53** $x = 8$ **55** $x = \dfrac{e^2}{1 + e^2}$

57

n	2	10	100	500	1000	5000	10000
$\left(1 + \frac{1}{n}\right)^n$	2.2500	2.5937	2.7048	2.7156	2.7169	2.7181	2.7181

(Other answers are possible for Exercises 59–67.)
59 Let $g(x) = -x^2 + x$ and $f(x) = e^x$. Then $(f \circ g)(x) = f(g(x)) = f(-x^2 + x) = e^{-x^2+x} = h(x)$.
61 Let $g(x) = \dfrac{x}{x + 1}$ and $f(x) = \ln x$. Then $(f \circ g)(x) = f(g(x)) = f\left(\dfrac{x}{x + 1}\right) = \ln \dfrac{x}{x + 1} = h(x)$.
63 Let $g(x) = \ln x$ and $f(x) = \sqrt[3]{x}$. Then $(f \circ g)(x) = f(g(x)) = f(\ln x) = \sqrt[3]{\ln x} = h(x)$.
65 Let $h(x) = 3x - 1$, $g(x) = x^2$, and $f(x) = e^x$. Then

$$(f \circ g \circ h)(x) = f(g(h(x))) = f(g(3x - 1)) = f((3x - 1)^2) = e^{(3x-1)^2} = F(x)$$

67 Let $h(x) = e^x + 1$, $g(x) = \sqrt{x}$, and $f(x) = \ln x$. Then

$$(f \circ g \circ h)(x) = f(g(h(x))) = f(g(e^x + 1)) = f(\sqrt{e^x + 1}) = \ln \sqrt{e^x + 1} = F(x)$$

69 $f(x) > 0$ on the interval $(-\infty, \frac{1}{2})$; $f(x) < 0$ on the interval $(\frac{1}{2}, \infty)$.
71 $f(x) > 0$ on the interval $\left(\dfrac{1}{e}, \infty\right)$; $f(x) < 0$ on the interval $\left(0, \dfrac{1}{e}\right)$.

73 $$\ln\left(\frac{x}{4} - \frac{\sqrt{x^2 - 4}}{4}\right) = \ln\left(\frac{x - \sqrt{x^2 - 4}}{4}\right) = \ln\left(\frac{x - \sqrt{x^2 - 4}}{4} \cdot \frac{x + \sqrt{x^2 - 4}}{x + \sqrt{x^2 - 4}}\right)$$
$$= \ln\left(\frac{x^2 - (x^2 - 4)}{4(x + \sqrt{x^2 - 4})}\right) = \ln\left(\frac{1}{x + \sqrt{x^2 - 4}}\right) = -\ln(x + \sqrt{x^2 - 4})$$

75 $x = \ln(y + \sqrt{y^2 + 1})$

7.6 EXPONENTIAL GROWTH AND DECAY (*Page 264*)

(These answers were found using Tables II and III. Using a calculator may result in slightly different answers in some cases.)
1 2010 **3** 27 **5** $\frac{1}{2}\ln 100$ **7** $\frac{1}{4}\ln\frac{1}{3}$ **9** 667,000 **11** 1.83 days. **13** 17.31 years.
15 **(a)** $\frac{1}{5}\ln\frac{4}{5}$; **(b)** 6.4 grams; **(c)** 15.5 years. **17** 93.2 seconds. **19** 18.47 years. **21** 4200 years. **23** 114,000 years.

FIC NOTATION (*Page 268*)

10^3 **3** 9.2×10^{-1} **5** 7.583×10^6 **7** 2.5×10^1 **9** 5.55×10^{-7} **11** 2.024×10^2 **13** 78,900
0 **17** 0.174 **19** 17.4 **21** 0.0906 **23** 10^2 **25** 10^0 **27** 10^{17} **29** 2000 **31** 400 **33** 0.02 **35** 3125
200 **39** 0.000008 **41** 0.0125 **43** 500 seconds.

.8 COMMON LOGARITHMS (*Page 270*)

(These answers were found using Table IV. If a calculator is used for the common logarithms, then some of these answers will be slightly different.)

1 43,000,000 **3** 2770 **5** 0.000125 **7** 1.22 **9** 0.00887 **11** 6.58 **13** \$1.21 per gallon. **15** 2.27 years.
17 \$4050
19 For $1 \leq x < 10$ we get $\log 1 \leq \log x < \log 10$ because $f(x) = \log x$ is an increasing function. Substituting $0 = \log 1$ and $1 = \log 10$ into the preceding inequality gives $0 \leq \log x < 1$.
21 **(a)** 2.825; **(b)** 9.273; **(c)** 1.378; **(d)** 5.489; **(e)** 4.495; **(f)** 7.497; **(g)** 1.021; **(h)** 9.508; **(i)** 2.263.

SAMPLE TEST QUESTIONS FOR CHAPTER 7 (*Page 277*)

1 **(i)** a; **(ii)** c; **(iii)** e; **(iv)** h; **(v)** d; **(vi)** b. **2** Each range value corresponds to exactly one domain value.
3 $g(x) = (x+1)^3$; $(f \circ g)(x) = f(g(x)) = f((x+1)^3) = \sqrt[3]{(x+1)^3} - 1 = x$ **4** **(a)** $\frac{1}{2}$; **(b)** 9 **5** **(a)** $\frac{2}{3}$; **(b)** -2.
6 All real values x; $y = -4$. **7** All $x > -4$; $x = -4$. **8** 5 **9**

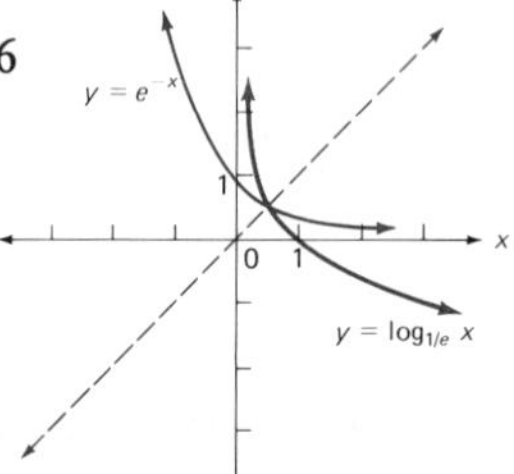

10 $3 \ln x - \ln(x+1) - \frac{1}{2}\ln(x^2+2)$ **11** $\dfrac{\ln 5}{2 \ln 4}$ **12** $25 \ln 2$ **13** 6
14 1,330,000 **15** $x = \dfrac{e-2}{e+2}$

CHAPTER 8: MATRICES, DETERMINANTS, AND LINEAR SYSTEMS

8.1 SOLVING LINEAR SYSTEMS USING MATRICES (*Page 279*)

1 $(-4, -1)$ **3** $(0, 9)$ **5** $(\frac{1}{8}, \frac{1}{2})$ **7** $(1, 0)$ **9** $(2 - \frac{3}{2}c, c)$ for all c. **11** No solutions. **13** $(5, 2, -1)$ **15** $(3, 4, 2)$
17 No solutions. **19** $(\frac{3}{11} + \frac{5}{11}c, -\frac{4}{11} + \frac{8}{11}c, c)$ for all c. **21** $(-1, 3, 4, -2)$ **23** No solutions.

8.2 MATRIX ALGEBRA (*Page 285*)

1 $\begin{bmatrix} 6 & 1 \\ 0 & 8 \end{bmatrix}$ **3** Does not exist. **5** $\begin{bmatrix} 3 & -1 & 4 \\ -7 & 4 & 2 \\ -2 & -3 & 5 \end{bmatrix}$ **7** $[-7]$ **9** $\begin{bmatrix} 2 & 6 \\ 9 & 10 \end{bmatrix}$ **11** $\begin{bmatrix} 7 \\ -11 \end{bmatrix}$ **13** $\begin{bmatrix} 21 & -11 & 2 \\ 17 & -17 & 19 \end{bmatrix}$

15 Does not exist. **17** $\begin{bmatrix} 2 & -5 \\ 4 & 3 \\ 4 & 2 \end{bmatrix}$ **19** $\begin{bmatrix} 1 & 2 & -15 & 4 \\ 27 & -6 & -15 & 8 \end{bmatrix}$ **21** $\begin{bmatrix} 0 \\ 0 \\ 0 \end{bmatrix}$ **23** H

25 $B + C = \begin{bmatrix} 0+3 & 6+7 \\ 5-1 & -4+(-2) \end{bmatrix} = \begin{bmatrix} 3 & 13 \\ 4 & -6 \end{bmatrix}$; $C + B = \begin{bmatrix} 3+0 & 7+6 \\ -1+5 & -2+(-4) \end{bmatrix} = \begin{bmatrix} 3 & 13 \\ 4 & -6 \end{bmatrix}$

27 $A + (B + C) = A + \begin{bmatrix} 3 & 13 \\ 4 & -6 \end{bmatrix} = \begin{bmatrix} 5 & 14 \\ 0 & -5 \end{bmatrix}$; $(A + B) + C = \begin{bmatrix} 2 & 7 \\ 1 & -3 \end{bmatrix} + C = \begin{bmatrix} 5 & 14 \\ 0 & -5 \end{bmatrix}$

29 $(A + B)C = \begin{bmatrix} 2 & 7 \\ 1 & -3 \end{bmatrix} C = \begin{bmatrix} -1 & 0 \\ 6 & 13 \end{bmatrix}$; $AC + BC = \begin{bmatrix} 5 & 12 \\ -13 & -30 \end{bmatrix} + \begin{bmatrix} -6 & -12 \\ 19 & 43 \end{bmatrix} = \begin{bmatrix} -1 & 0 \\ 6 & 13 \end{bmatrix}$

31 $(AB)C = \begin{bmatrix} -5 & 4 & 2 \\ 3 & 21 & 33 \\ 8 & 4 & 12 \end{bmatrix} C = \begin{bmatrix} -31 \\ 6 \\ 44 \end{bmatrix}$; $A(BC) = A\begin{bmatrix} -9 \\ 11 \end{bmatrix} = \begin{bmatrix} -31 \\ 6 \\ 44 \end{bmatrix}$

33 (a) $\begin{bmatrix} d & e & f \\ a & b & c \\ g & h & i \end{bmatrix}$ (b) $F = \begin{bmatrix} 0 & 0 & 1 \\ 0 & 1 & 0 \\ 1 & 0 & 0 \end{bmatrix}$

35 (a) $x = -2, y = -3$ (b) If there were such x, y, then $\begin{bmatrix} 2 & 2 \\ x & x \end{bmatrix} = \begin{bmatrix} 0 & 0 \\ 0 & 0 \end{bmatrix}$, which cannot be true since $2 \neq 0$.

37 (a) $\begin{bmatrix} x^2 & 0 & 0 \\ 0 & y^2 & 0 \\ 0 & 0 & z^2 \end{bmatrix}$ (b) $\begin{bmatrix} x^3 & 0 & 0 \\ 0 & y^3 & 0 \\ 0 & 0 & z^3 \end{bmatrix}$ (c) $\begin{bmatrix} x^n & 0 & 0 \\ 0 & y^n & 0 \\ 0 & 0 & z^n \end{bmatrix}$

8.3 INVERSES (*Page 292*)

1 $\begin{bmatrix} 0 & \frac{1}{2} \\ -1 & 2 \end{bmatrix}$ 3 Does not exist. 5 $\begin{bmatrix} -\frac{4}{7} & -\frac{5}{7} \\ -\frac{3}{7} & -\frac{2}{7} \end{bmatrix}$ 7 $\begin{bmatrix} -1 & 2 \\ 2 & -1 \end{bmatrix}$ 9 $\begin{bmatrix} 0 & \frac{1}{b} \\ \frac{1}{a} & 0 \end{bmatrix}$ 11 $A^{-1} = A$

13 $\begin{bmatrix} -\frac{1}{2} & \frac{3}{4} & \frac{1}{4} \\ -1 & \frac{1}{2} & \frac{1}{2} \\ \frac{3}{4} & -\frac{3}{8} & -\frac{1}{8} \end{bmatrix}$ 15 Does not exist. 17 $\begin{bmatrix} 1 & 0 & 2 \\ 2 & -1 & 3 \\ 4 & 1 & 8 \end{bmatrix}$ 19 $\begin{bmatrix} -5 & 0 & \frac{5}{2} & 3 \\ 2 & 0 & -\frac{1}{2} & -1 \\ -\frac{3}{2} & \frac{1}{2} & \frac{3}{4} & 1 \\ 2 & 0 & -1 & -1 \end{bmatrix}$ 21 $\begin{aligned} 3x + y &= 9 \\ 2x - 2y &= 14 \end{aligned}$

23 (6, 1) 25 (18, −21) 27 (−8, −8, 6) 29 $(-\frac{1}{2}, -\frac{1}{2}, -2)$ 31 (13, 7, −6) 33 (14, −4, 2, −4)

35 $(AB)^{-1} = \begin{bmatrix} 1 & 2 \\ -4 & 8 \end{bmatrix}^{-1} = \begin{bmatrix} \frac{1}{2} & -\frac{1}{8} \\ \frac{1}{4} & \frac{1}{16} \end{bmatrix}$; $B^{-1}A^{-1} = \begin{bmatrix} 1 & 1 \\ \frac{1}{2} & 0 \end{bmatrix}\begin{bmatrix} \frac{1}{2} & \frac{1}{8} \\ 0 & -\frac{1}{4} \end{bmatrix} = \begin{bmatrix} \frac{1}{2} & -\frac{1}{8} \\ \frac{1}{4} & \frac{1}{16} \end{bmatrix}$

37 (a) $\begin{aligned} AB &= I \\ C(AB) &= CI \\ (CA)B &= C \\ IB &= C \\ B &= C \end{aligned}$ (b) The inverse of a matrix is unique.

8.4 INTRODUCTION TO DETERMINANTS AND CRAMER'S RULE (*Page 299*)

1 17 3 94 5 −60 7 0 9 (−1, 2) 11 (3, 2) 13 (1, −1) 15 (12, 24) 17 (5, 5) 19 (−5, −7) 21 $(\frac{6}{5}, -\frac{3}{10})$

23 Inconsistent. 25 Dependent. 27 Inconsistent. 29 6 31 $\begin{vmatrix} a_1 & b_1 \\ a_2 & b_2 \end{vmatrix} = a_1b_2 - a_2b_1 = a_1b_2 - b_1a_2 = \begin{vmatrix} a_1 & a_2 \\ b_1 & b_2 \end{vmatrix}$

33 Let $b_1 = ka_1$, $b_2 = ka_2$; then $\begin{vmatrix} a_1 & b_1 \\ a_2 & b_2 \end{vmatrix} = \begin{vmatrix} a_1 & ka_1 \\ a_2 & ka_2 \end{vmatrix} = \begin{vmatrix} a_1 & a_2 \\ ka_1 & ka_2 \end{vmatrix}$ by Exercise 31

$= 0$ by Exercise 32

35 $\begin{vmatrix} 27 & 3 \\ 105 & -75 \end{vmatrix} = 3\begin{vmatrix} 9 & 1 \\ 105 & -75 \end{vmatrix} = 45\begin{vmatrix} 9 & 1 \\ 7 & -5 \end{vmatrix} = -2340$

$\begin{vmatrix} 27 & 3 \\ 105 & -75 \end{vmatrix} = 3\begin{vmatrix} 9 & 1 \\ 105 & -75 \end{vmatrix} = 9\begin{vmatrix} 3 & 1 \\ 35 & -75 \end{vmatrix} = 45\begin{vmatrix} 3 & 1 \\ 7 & -15 \end{vmatrix} = -2340$

37 $\begin{vmatrix} a_1 + kb_1 & b_1 \\ a_2 + kb_2 & b_2 \end{vmatrix} = \begin{vmatrix} a_1 & b_1 \\ a_2 & b_2 \end{vmatrix} + \begin{vmatrix} kb_1 & b_1 \\ kb_2 & b_2 \end{vmatrix}$ (By Exercise 36)

$= \begin{vmatrix} a_1 & b_1 \\ a_2 & b_2 \end{vmatrix} + 0$ (By Exercise 33)

$= \begin{vmatrix} a_1 & b_1 \\ a_2 & b_2 \end{vmatrix}$

39 (Sample solution)

$$\begin{vmatrix} 12 & -42 \\ -6 & 27 \end{vmatrix} \underset{\text{Exercise 34}}{=} 6\begin{vmatrix} 2 & -42 \\ -1 & 27 \end{vmatrix} \underset{\text{Exercise 34}}{=} 12\begin{vmatrix} 1 & -21 \\ -1 & 27 \end{vmatrix} \underset{\text{Exercise 34}}{=} 36\begin{vmatrix} 1 & -7 \\ -1 & 9 \end{vmatrix} \underset{\text{Exercise 37}}{=} 36\begin{vmatrix} 1 & 0 \\ -1 & 2 \end{vmatrix} = 36(2) = 72$$

8.5 THIRD-ORDER DETERMINANTS (*Page 305*)

1 35 **3** -45 **5** 4

7 $\begin{vmatrix} a_1 & b_1 & c_1 \\ a_2 & b_2 & c_2 \\ a_3 & b_3 & c_3 \end{vmatrix} = a_1b_2c_3 + a_2b_3c_1 + a_3b_1c_2 - a_1b_3c_2 - a_2b_1c_3 - a_3b_2c_1 = \begin{vmatrix} a_1 & a_2 & a_3 \\ b_1 & b_2 & b_3 \\ c_1 & c_2 & c_3 \end{vmatrix}$

9 $\begin{vmatrix} ka_1 & b_1 & c_1 \\ ka_2 & b_2 & c_2 \\ ka_3 & b_3 & c_3 \end{vmatrix} = ka_1b_2c_3 + kb_1c_2a_3 + kc_1a_2b_3 - ka_3b_2c_1 - kb_3c_2a_1 - kc_3a_2b_1$

$= k(a_1b_2c_3 + b_1c_2a_3 + c_1a_2b_3 - a_3b_2c_1 - b_3c_2a_1 - c_3a_2b_1)$

$= k\begin{vmatrix} a_1 & b_1 & c_1 \\ a_2 & b_2 & c_2 \\ a_3 & b_3 & c_3 \end{vmatrix}$

11 0, since third column is -2 times first column. **13** $80 = 20(4)$ **15** $(-1, 0, 2)$ **17** $(-1, 7, 2)$ **19** $(\frac{2}{3}, 1, \frac{1}{2})$
21 -144

8.6 LINEAR PROGRAMMING (*Page 309*)

1 (a)

y
(0, 2)
x + 3y = 6
x
(0, 0)
(6, 0)

(b) (0, 0), (6, 0), (0, 2) **(c)** Maximum = 6; minimum = 0 **(d)** Maximum = 36; minimum = 0.
(e) Maximum = 18; minimum = 0.

3 (a)

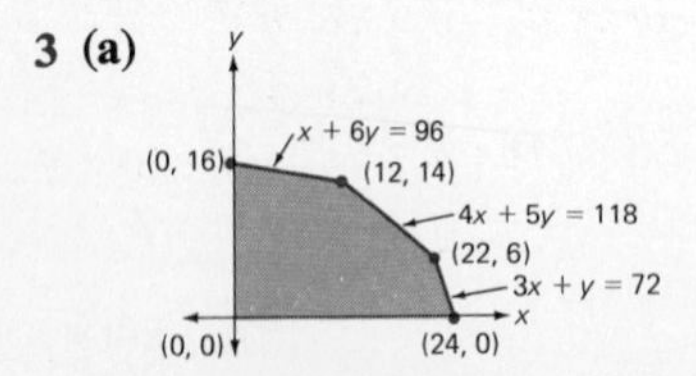

(b) (0, 0); (24, 0); (22, 6); (12, 14); (0, 16) **(c)** Maximum = 28; minimum = 0
(d) Maximum = 212; minimum = 0 **(e)** Maximum = 150; minimum = 0

5 (a)

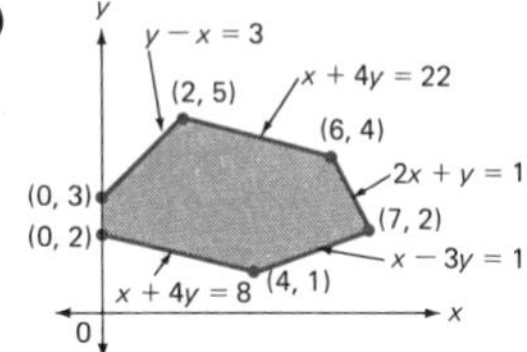

(b) (0, 2), (4, 1), (7, 2), (6, 4), (2, 5), (0, 3) **(c)** Maximum = 10; minimum = 2
(d) Maximum = 76; minimum = 20 **(e)** Maximum = 49; minimum = 17

7

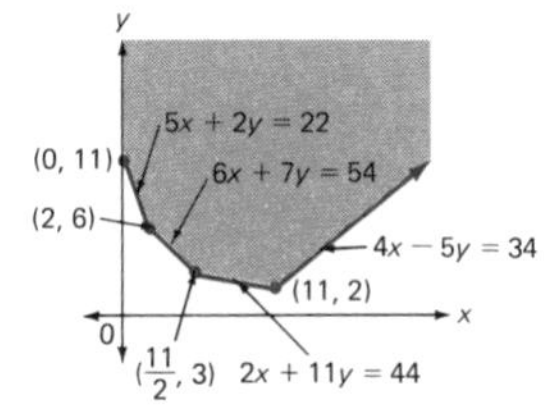

(a) No maximum value; minimum = 8 **(b)** No maximum value; minimum = 63
(c) No maximum value; minimum = 10

9 (a) $x \geq 0$, $y \geq 0$ because a negative number of either model is not possible; $\frac{3}{2}x + y$ is the amount of time that machine M_1 works per day, and $\frac{3}{2}x + y \leq 12$ says that M_1 works at most 12 hours daily. The remaining inequalities are the constraints for machines M_2 and M_3; the explanations are similar, as for M_1.

(b) $p = 5x + 8y$ **(c)**

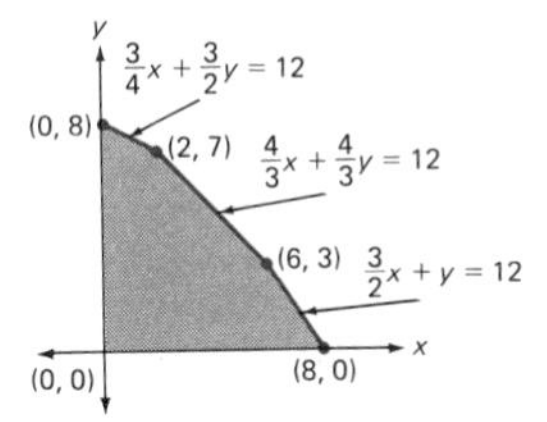

(d) Maximum = \$66 when $x = 2$ and $y = 7$. **(e)** Maximum = \$64 when $x = 8$ and $y = 0$.

SAMPLE TEST QUESTIONS FOR CHAPTER 8 (*Page 318*)

1 $(-3, 2)$ **2** No solutions. **3** $(4, -1, -6)$ **4** $(-\frac{2}{3}c + \frac{2}{3}, -\frac{4}{3}c + \frac{7}{3}, c)$ for all c

5 $\begin{bmatrix} 2 & 5 \\ 4 & -3 \end{bmatrix}\begin{bmatrix} x \\ y \end{bmatrix} = \begin{bmatrix} 4 \\ -18 \end{bmatrix}$ **6** $\begin{bmatrix} 2 & -1 & 2 \\ 1 & 4 & -3 \\ -4 & 2 & -3 \end{bmatrix}\begin{bmatrix} x \\ y \\ z \end{bmatrix} = \begin{bmatrix} -3 \\ 18 \\ 0 \end{bmatrix}$ **7** $\begin{bmatrix} 1 & -1 & 0 \\ 6 & 1 & 4 \\ -6 & -3 & 3 \end{bmatrix}$ **8** Does not exist.

9 $\begin{bmatrix} 25 & -14 \\ -3 & 23 \end{bmatrix}$ **10** $\begin{bmatrix} -1 & 0 & 31 \\ -10 & -4 & 104 \end{bmatrix}$ **11** -8 **12** $\begin{bmatrix} -11 & -11 \\ -18 & -18 \end{bmatrix}$ **13** Does not exist. **14** $\begin{bmatrix} 27 & 0 \\ 13 & 1 \end{bmatrix}$

15 $\begin{bmatrix} -2 & 2 & -4 \\ 2 & -2 & 4 \\ 4 & -4 & 8 \end{bmatrix}$ **16** 4 by 3 **17** Does not exist. **18** 51 **19** -41 **20** $\begin{bmatrix} -8 \end{bmatrix}$ **21** $\begin{bmatrix} \frac{1}{4} & \frac{1}{8} \\ \frac{1}{4} & -\frac{3}{8} \end{bmatrix}$ **22** $(4, -3)$

23 $\begin{bmatrix} 0 & -\frac{2}{3} & \frac{1}{3} \\ 0 & 1 & 0 \\ -\frac{1}{2} & -\frac{1}{3} & \frac{1}{6} \end{bmatrix}$ **24** $(-1, 1, -5)$ **25** $(1, -\frac{11}{2})$ **26** $(1, -2, 3)$

27

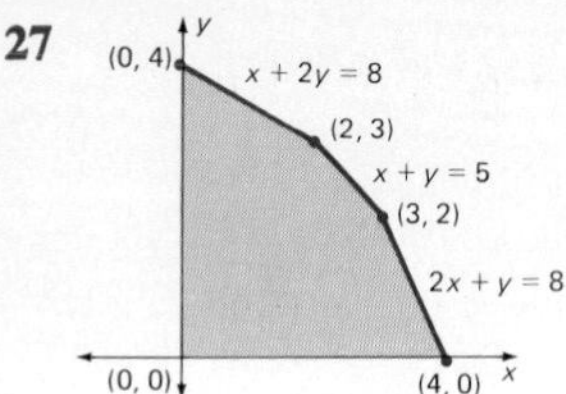

28 13

CHAPTER 9: SEQUENCES AND SERIES

9.1 SEQUENCES (*Page 321*)

1 $s_1 = 1, s_2 = 3, s_3 = 5, s_4 = 7, s_5 = 9$

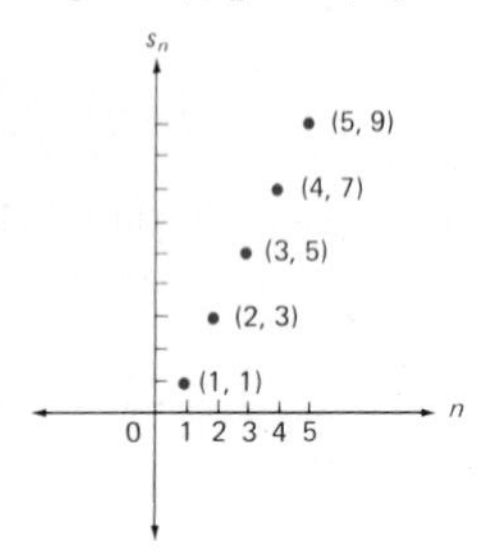

3 $s_1 = -1, s_2 = 1, s_3 = -1, s_4 = 1, s_5 = -1$

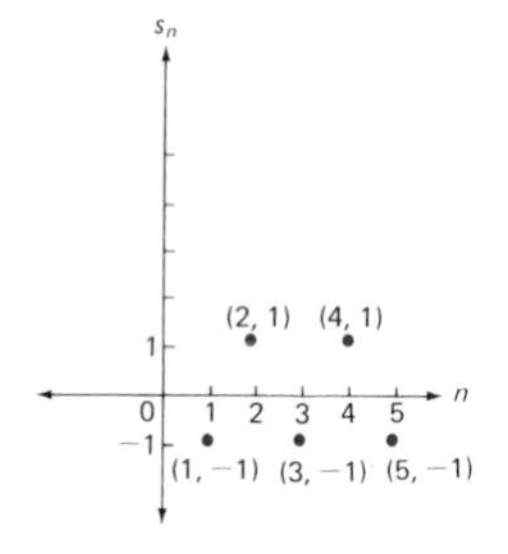

5 $s_1 = -4, s_2 = 2, s_3 = -1, s_4 = \frac{1}{2}, s_5 = -\frac{1}{4}$

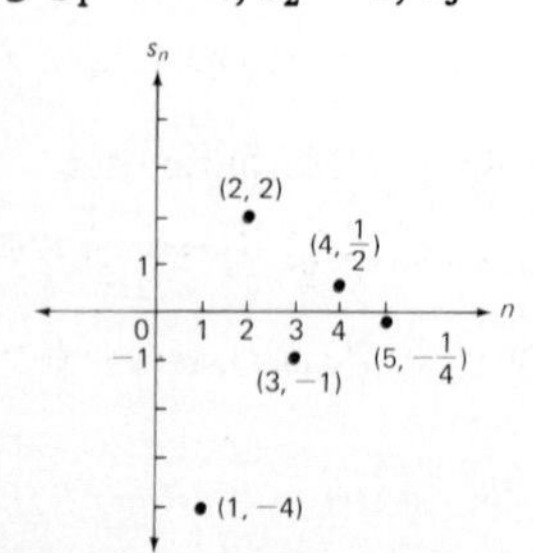

7 $-1, 4, -9, 16$

9 $\frac{3}{10}, \frac{3}{100}, \frac{3}{1000}, \frac{3}{10,000}$ **11** $\frac{3}{100}, \frac{3}{10,000}, \frac{3}{1,000,000}, \frac{3}{100,000,000}$
13 $\frac{1}{2}, \frac{1}{6}, \frac{1}{12}, \frac{1}{20}$ **15** 64, 36, 16, 4 **17** $-2, 1, 4, 7$ **19** $0, \frac{1}{3}, \frac{1}{2}, \frac{3}{5}$ **21** $1, \frac{3}{2}, \frac{16}{9}, \frac{125}{64}$ **23** $-2, -\frac{3}{2}, -\frac{9}{8}, -\frac{27}{32}$
25 $\frac{1}{2}, \frac{1}{2}, \frac{3}{8}, \frac{1}{4}$ **27** $-\frac{3}{2}, -\frac{5}{6}, -\frac{7}{12}, -\frac{9}{20}$ **29** 4, 4, 4, 4 **31** 122 **33** 0.000003 **35** 12 **37** -1331 **39** 4, 6, 8, 10
41 5, 10, 15, 20, 25; $s_n = 5n$ **43** $-5, 25, -125, 625, -3125$; $s_n = (-5)^n$ **45** 15, 21, 28 **47** $1, \frac{9}{5}, 3, \frac{81}{17}, \frac{81}{11}$
49 $\frac{3}{2}, \frac{15}{8}, \frac{35}{16}, \frac{315}{128}$

9.2 SUMS OF FINITE SEQUENCES (*Page 325*)

1 45 **3** 55 **5** 0.33333 **7** 510 **9** 60 **11** 254 **13** $\frac{381}{64}$ **15** 105 **17** 40 **19** $\frac{13}{8}$ **21** 0 **23** 57 **25** 1111.111
27 60 **29** $\frac{35}{12}$ **31** 0.010101 **33** **(a)** 4, 9, 16, 25, 36; **(b)** n^2.

35 $s_n = s_1 + s_{n-1} = s_1 + (s_1 + s_{n-2}) = \cdots = \overbrace{s_1 + s_1 + s_1 + \cdots + s_1}^{n \text{ terms}} = 2 + 2 + 2 + \cdots + 2 = 2n$

37 $\sum_{k=1}^{n} s_k + \sum_{k=1}^{n} t_k = (s_1 + s_2 + \cdots + s_n) + (t_1 + t_2 + \cdots + t_n) = (s_1 + t_1) + (s_2 + t_2) + \cdots + (s_n + t_n)$
$= \sum_{k=1}^{n} (s_k + t_k)$

39 $\sum_{k=1}^{n} (s_k + c) = (s_1 + c) + (s_2 + c) + \cdots + (s_n + c) = (s_1 + s_2 + \cdots + s_n) + (c + c + \cdots + c)$
$= \left(\sum_{k=1}^{n} s_k\right) + nc$

9.3 ARITHMETIC SEQUENCES AND SERIES (*Page 328*)

1 5, 7, 9; $2n - 1$; 400 **3** $-10, -16, -22$; $-6n + 8$; -1100 **5** $\frac{17}{2}, 9, \frac{19}{2}$; $\frac{1}{2}n + 7$; 245 **7** $-\frac{4}{5}, -\frac{7}{5}, -2$; $-\frac{3}{5}n + 1$; -106 **9** 150, 200, 250; $50n$; 10,500 **11** 30, 50, 70; $20n - 30$; 3600 **13** 455 **15** $\frac{63}{4}$ **17** 15,150 **19** 94,850 **21** $\frac{173{,}350}{7}$ **23** 567,500 **25** 5824 **27** 15,300 **29** (a) 10,000; (b) n^2 **31** -36 **33** 4620 **35** $\frac{3577}{4}$ **37** $\frac{5}{2}n(n + 1)$ **39** $-\frac{9}{4}$ **41** $\frac{228}{15} = \frac{76}{5}$ **43** $1183 **45** 930 **47** $a = -7$; $d = -13$ **49** $u = \frac{16}{3}$, $v = \frac{23}{3}$ **51** 9080 **53** $-16{,}200$ **55** 520

9.4 GEOMETRIC SEQUENCES AND SERIES (*Page 333*)

1 16, 32, 64; 2^n **3** 27, 81, 243; 3^{n-1} **5** $\frac{1}{9}, -\frac{1}{27}, \frac{1}{81}$; $-3(-\frac{1}{3})^{n-1}$ **7** $-125, -625, -3125$; -5^{n-1} **9** $-\frac{16}{9}, -\frac{32}{27}, -\frac{64}{81}$; $-6(\frac{2}{3})^{n-1}$ **11** 1000, 100,000, 10,000,000; $\frac{1}{1000}(100)^{n-1}$ **13** 126 **15** $-\frac{1330}{81}$ **17** 1024 **19** 3 **21** 1023 **23** $2^n - 1$ **25** $\frac{121}{27}$ **27** $\frac{211}{54}$ **29** $-\frac{85}{64}$ **31** $-\frac{5}{21}$ **33** 512,000; $1000(2^{n-1})$ **35** $3117.79 **37** (a) $\$800(1.11)^n$; (b) $1350 **39** $703.99

9.5 INFINITE GEOMETRIC SERIES (*Page 339*)

1 4 **3** $\frac{125}{4}$ **5** $\frac{2}{3}$ **7** $\frac{10}{9}$ **9** $-\frac{16}{7}$ **11** $\frac{3}{2}$ **13** $\frac{1}{6}$ **15** 4 **17** $\frac{20}{9}$ **19** No finite sum. **21** $\frac{10}{3}$ **23** No finite sum. **25** 30 **27** $\frac{20}{11}$ **29** 4 **31** $\frac{4}{21}$ **33** $\frac{7}{9}$ **35** $\frac{13}{99}$ **37** $\frac{13}{990}$ **39** 1 **41** (a) $\frac{4}{3}, \frac{4}{9}, \frac{4}{27}, \ldots, \frac{4}{3^n}, \ldots$; (b) $\sum_{n=1}^{\infty} \frac{4}{3^n} = \frac{\frac{4}{3}}{1 - \frac{1}{3}} = 2$. **43** 8 hours. **45** The time for the last $\frac{1}{2}$ mile would have to be $\sum_{n=1}^{\infty} \frac{2}{5}(\frac{10}{9})^{n-1}$, which is not a finite sum, since $\frac{10}{9} > 1$.

9.6 MATHEMATICAL INDUCTION (*Page 345*)

For these exercises, $\mathcal{S}_n$ represents the given statement where n is an integer ≥ 1 ($n \geq 2$ when appropriate). The second part of each proof begins with the hypothesis $\mathcal{S}_k$, where k is an arbitrary positive integer.

1 Since $1 = \frac{1(1 + 1)}{2}$, $\mathcal{S}_1$ is true. Assume $\mathcal{S}_k$ and add $k + 1$ to obtain

$$1 + 2 + 3 + \cdots + k + (k + 1) = \frac{k(k + 1)}{2} + (k + 1) = \frac{k^2 + 3k + 2}{2} = \frac{(k + 1)(k + 2)}{2} = \frac{(k + 1)[(k + 1) + 1]}{2}$$

Therefore, $\mathcal{S}_{k+1}$ holds. Since $\mathcal{S}_1$ is true and $\mathcal{S}_k$ implies $\mathcal{S}_{k+1}$, the principle of mathematical induction makes $\mathcal{S}_n$ true for all integers $n \geq 1$. *Note:* The preceding sentence is an appropriate final statement for the remaining proofs. For the sake of brevity, however, it will not be repeated.

3 Since $\sum_{i=1}^{1} 3i = 3 = \frac{3(1 + 1)}{2}$, $\mathcal{S}_1$ is true. Assume $\mathcal{S}_k$ and add $3(k + 1)$ to obtain the following. $\sum_{i=1}^{k+1} 3i = \left(\sum_{i=1}^{k} 3i\right) + 3(k + 1) = \frac{3k(k + 1)}{2} + 3(k + 1)$; $\sum_{i=1}^{k+1} 3i = \frac{3(k^2 + 3k + 2)}{2} = \frac{3(k + 1)(k + 2)}{2} = \frac{3(k + 1)[(k + 1) + 1]}{2}$.
Therefore, $\mathcal{S}_{k+1}$ holds.

5 Since $\frac{5}{3} = \frac{1(11-1)}{6}$, $\mathcal{S}_1$ is true. Assume $\mathcal{S}_k$ and add $-\frac{1}{3}(k+1)+2$ to obtain $\frac{5}{3} + \frac{4}{3} + 1 + \cdots +$ $\left(-\frac{1}{3}k+2\right) + \left[-\frac{1}{3}(k+1)+2\right] = \frac{k(11-k)}{6} + \left[-\frac{1}{3}(k+1)+2\right] = \frac{10+9k-k^2}{6} = \frac{(k+1)(10-k)}{6}$ $= \frac{(k+1)[11-(k+1)]}{6}$. Therefore, $\mathcal{S}_{k+1}$ holds.

7 Since $\frac{1}{1 \cdot 2} = \frac{1}{1+1}$, $\mathcal{S}_1$ is true. Assume $\mathcal{S}_k$ and add $\frac{1}{(k+1)[(k+1)+1]}$ to obtain $\frac{1}{1 \cdot 2} + \frac{1}{2 \cdot 3} + \cdots + \frac{1}{k(k+1)}$ $+ \frac{1}{(k+1)(k+2)} = \frac{k}{k+1} + \frac{1}{(k+1)(k+2)} = \frac{k^2+2k+1}{(k+1)(k+2)} = \frac{(k+1)^2}{(k+1)(k+2)} = \frac{k+1}{k+2} = \frac{k+1}{(k+1)+1}$

Therefore, $\mathcal{S}_{k+1}$ holds.

9 $\mathcal{S}_1$ is true since $-2 = -1 - (1^2)$. Assume $\mathcal{S}_k$ and add $-2(k+1)$ to obtain

$$\begin{aligned} -2-4-6-\cdots-2k-2(k+1) &= -k-k^2-2(k+1) = -(k+1)-(k^2+2k+1) \\ &= -(k+1)-(k+1)^2 \end{aligned}$$

Therefore, $\mathcal{S}_{k+1}$ holds.

11 $\mathcal{S}_1$ is true since $1 = \frac{5}{3}[1-(\frac{2}{5})^1]$. Assume $\mathcal{S}_k$ and add $(\frac{2}{5})^k$ to obtain

$$\begin{aligned} 1 + \tfrac{2}{5} + \tfrac{4}{25} + \cdots + (\tfrac{2}{5})^{k-1} + (\tfrac{2}{5})^k &= \tfrac{5}{3}[1-(\tfrac{2}{5})^k] + (\tfrac{2}{5})^k \\ &= \tfrac{5}{3}[1-(\tfrac{2}{5})^k + \tfrac{3}{5}(\tfrac{2}{5})^k] \\ &= \tfrac{5}{3}[1-(\tfrac{2}{5})^k(1-\tfrac{3}{5})] \\ &= \tfrac{5}{3}[1-(\tfrac{2}{5})^k(\tfrac{2}{5})] \\ &= \tfrac{5}{3}[1-(\tfrac{2}{5})^{k+1}] \end{aligned}$$

Therefore, $\mathcal{S}_{k+1}$ holds.

13 $\mathcal{S}_1$ is true since $1^3 = 1 = \frac{1^2(1+1)^2}{4}$. Assume $\mathcal{S}_k$ and add $(k+1)^3$ to obtain

$$\begin{aligned} 1^3+2^3+3^3+\cdots+k^3+(k+1)^3 &= \frac{k^2(k+1)^2}{4} + (k+1)^3 \\ &= \frac{k^2(k+1)^2+4(k+1)^3}{4} \\ &= \frac{(k+1)^2[k^2+4k+4]}{4} \\ &= \frac{(k+1)^2[k+2]^2}{4} \\ &= \frac{(k+1)^2[(k+1)+1]^2}{4} \end{aligned}$$

Therefore, $\mathcal{S}_{k+1}$ holds.

15 $\mathcal{S}_1$ is true since $\sum_{i=1}^{1} ar^{i-1} = a = \frac{a(1-r^1)}{1-r}$. Assume $\mathcal{S}_k$ and add $ar^{(k+1)-1}$ to obtain

$$\sum_{i=1}^{k} ar^{i-1} + ar^k = \frac{a(1-r^k)}{1-r} + ar^k$$

$$\begin{aligned} \sum_{i=1}^{k+1} ar^{i-1} &= \frac{a(1-r^k)}{1-r} + ar^k \\ &= \frac{a(1-r^k)+ar^k-ar^{k+1}}{1-r} \\ &= \frac{a-ar^{k+1}}{1-r} \\ &= \frac{a(1-r^{k+1})}{1-r} \end{aligned}$$

Therefore, $\mathcal{S}_{k+1}$ holds.

17 Since $a^1 < 1$, $\mathcal{S}_1$ is true. Assume $\mathcal{S}_k$ and multiply by a to obtain $a \cdot a^k < a \cdot 1$ or $a^{k+1} < a$. But $a < 1$; hence $a^{k+1} < 1$ and, therefore, $\mathcal{S}_{k+1}$ holds.

19 $\mathcal{S}_1$ is true since $(ab)^1 = a^1b^1$. Assume that $(ab)^k = a^kb^k$ and multiply by ab to obtain

$$\begin{aligned}(ab)^k(ab) &= (a^kb^k)ab\\(ab)^{k+1} &= (a^ka)(b^kb)\\(ab)^{k+1} &= a^{k+1}b^{k+1}\end{aligned}$$

Therefore, $\mathcal{S}_{k+1}$ holds.

21 $\mathcal{S}_1$ is true since $|a_0 + a_1| \le |a_0| + |a_1|$. Assume $\mathcal{S}_k$. Then

$$\begin{aligned}|a_0 + a_1 + \cdots + a_k + a_{k+1}| &= |(a_0 + a_1 + \cdots + a_k) + a_{k+1}|\\&\le |a_0 + a_1 + \cdots + a_k| + |a_{k+1}| \quad \text{(by } \mathcal{S}_1)\\&\le (|a_0| + |a_1| + \cdots + |a_k|) + |a_{k+1}| \quad \text{(by } \mathcal{S}_k)\\&= |a_0| + |a_1| + \cdots + |a_{k+1}|\end{aligned}$$

Therefore, $\mathcal{S}_{k+1}$ holds.

23 **(a)** Since $\dfrac{a^2 - b^2}{a - b} = a + b = a^{2-1} + b^{2-1}$, $\mathcal{S}_2$ is true. Assume $\mathcal{S}_k$. Then

$$\begin{aligned}\frac{a^{k+1} - b^{k+1}}{a - b} &= \frac{a^ka - b^kb}{a - b}\\&= \frac{a^ka - b^ka + b^ka - b^kb}{a - b}\\&= \frac{a(a^k - b^k) + b^k(a - b)}{a - b} = \frac{a(a^k - b^k)}{a - b} + b^k\\&= a[a^{k-1} + a^{k-2}b + \cdots + ab^{k-2} + b^{k-1}] + b^k \quad \text{(by } \mathcal{S}_k)\\&= a^k + a^{k-1}b + \cdots + a^2b^{k-1} + ab^{k-1} + b^k\end{aligned}$$

Therefore, $\mathcal{S}_{k+1}$ holds.

(b) Since $\dfrac{a^n - b^n}{a - b} = a^{n-1} + a^{n-2}b + \cdots + ab^{n-2} + b^{n-1}$, multiplying by $a - b$ gives

$$a^n - b^n = (a - b)(a^{n-1} + a^{n-2}b + \cdots + ab^{n-2} + b^{n-1})$$

25 **(a)** 1; 4; 9; 16; 25; **(b)** n^2

(c) $\mathcal{S}_1$ is true since $1 = 1^2$. Assume $\mathcal{S}_k$ and add $k + (k + 1)$ to obtain

$$\begin{aligned}1 + 2 + 3 + \cdots + (k - 1) + k + (k + 1) + k + (k - 1) + \cdots + 3 + 2 + 1 &= k^2 + k + (k + 1)\\&= k^2 + 2k + 1\\&= (k + 1)^2\end{aligned}$$

Therefore, $\mathcal{S}_{k+1}$ holds.

SAMPLE TEST QUESTIONS FOR CHAPTER 9 (*Page 351*)

1 $1, -1, -1, -\frac{8}{7}$ **2** $-\frac{25}{11}$ **3** 21 **4** 35 **5** $12, -3, \frac{3}{4}$ **6** $-768(-\frac{1}{4})^{n-1}$ **7** $\displaystyle\sum_{k=1}^{4} 8\left(\frac{1}{2}\right)^k = \frac{4\left(1 - \frac{1}{2^4}\right)}{1 - \frac{1}{2}} = \frac{15}{2}$

8 15,554 **9** 18 **10** No finite sum since $r = \frac{3}{2} > 1$. **11** $\frac{6}{115}$ **12** $\frac{4}{11}$ **13** $652.60 **14** 36 feet.

15 For $n = 1$, we have $5 = \dfrac{5 \cdot 1(1 + 1)}{2}$. If $5 + 10 + \cdots + 5k = \dfrac{5k(k + 1)}{2}$, then $5 + 10 + \cdots + 5k + 5(k + 1)$

$$= \frac{5k(k + 1)}{2} + 5(k + 1) = \frac{5k(k + 1) + 10(k + 1)}{2} = \frac{5(k + 1)(k + 2)}{2} = \frac{5(k + 1)[(k + 1) + 1]}{2}$$

Thus the statement holds for the $k + 1$ case, and by the principle of mathematical induction the statement is true for all $n \ge 1$.

16 For $n = 1$, $\dfrac{1}{1 \cdot 3} = \dfrac{1}{2 \cdot 1 + 1}$. For $n = k$, assume that

$$\frac{1}{1 \cdot 3} + \frac{1}{3 \cdot 5} + \cdots + \frac{1}{(2k - 1)(2k + 1)} = \frac{k}{2k + 1}$$

Add $\frac{1}{(2k+1)(2k+3)}$ to get

$$\frac{1}{1\cdot 3}+\frac{1}{3\cdot 5}+\cdots+\frac{1}{(2k-1)(2k+1)}+\frac{1}{(2k+1)(2k+3)}$$
$$=\frac{k}{2k+1}+\frac{1}{(2k+1)(2k+3)}$$
$$=\frac{k(2k+3)+1}{(2k+1)(2k+3)}$$
$$=\frac{2k^2+3k+1}{(2k+1)(2k+3)}$$
$$=\frac{(2k+1)(k+1)}{(2k+1)(2k+3)}$$
$$=\frac{k+1}{2(k+1)+1}$$

Thus the statement holds for $n = k + 1$, and by the principle of mathematical induction the statement is true for all $n \geq 1$.

CHAPTER 10: PERMUTATIONS, COMBINATIONS, AND PROBABILITY

10.1 PERMUTATIONS (*Page 353*)

1 7 **3** 66 **5** 120 **7** 4 **9** 15 **11** 60 **13** 12 **15** 6 **17** 36 **19** 504 **21** 224 **23** 224 **25** 120
27 362,880; 40,320 **29** 720 **31** 468,000; 676,000 **33** **(a)** 10 **(b)** 4
35 36;

(1, 1) (2, 1) (3, 1) (4, 1) (5, 1) (6, 1)
(1, 2) (2, 2) (3, 2) (4, 2) (5, 2) (6, 2)
(1, 3) (2, 3) (3, 3) (4, 3) (5, 3) (6, 3)
(1, 4) (2, 4) (3, 4) (4, 4) (5, 4) (6, 4)
(1, 5) (2, 5) (3, 5) (4, 5) (5, 5) (6, 5)
(1, 6) (2, 6) (3, 6) (4, 6) (5, 6) (6, 6)

10.2 COMBINATIONS (*Page 358*)

1 10 **3** 1 **5** 1 **7** 4060 **9** **(a)** 4845 **(b)** 116,280 **11** 20 **13** 3003 **15** 435 **17** 9900 **19** **(a)** n **(b)** $\frac{n(n-1)}{2}$
21 70 **23** **(a)** 2^2 **(b)** 2^3 **(c)** 2^4 **(d)** 2^5 **25** $4\binom{13}{5} = 5148$ **27** $13\binom{4}{2}12\binom{4}{3} = 3744$
29 **(a)** 205,931,880 **(b)** 147,094,200 **31** 256
33
$$\binom{n}{r-1}+\binom{n}{r}=\frac{n!}{(r-1)!\,(n-(r-1))!}+\frac{n!}{r!\,(n-r)!}$$
$$=\frac{rn!}{r(r-1)!\,(n-(r-1))!}+\frac{(n-(r-1))n!}{r!\,(n-(r-1))(n-r)!}$$
$$=\frac{rn!}{r!\,(n-r+1)!}+\frac{(n-r+1)n!}{r!\,(n-r+1)!}$$
$$=\frac{(r+n-r+1)n!}{r!\,(n-r+1)!}$$
$$=\frac{(n+1)n!}{r!\,(n-r+1)!}=\frac{(n+1)!}{r!\,(n-r+1)!}=\binom{n+1}{r}$$

10.3 THE BINOMIAL EXPANSION (*Page 392*)

1 $x^5 + 5x^4 + 10x^3 + 10x^2 + 5x + 1$ **3** $x^7 + 7x^6 + 21x^5 + 35x^4 + 35x^3 + 21x^2 + 7x + 1$

5 $a^4 - 4a^3b + 6a^2b^2 - 4ab^3 + b^4$ **7** $243x^5 - 405x^4y + 270x^3y^2 - 90x^2y^3 + 15xy^4 - y^5$
9 $a^{10} + 5a^8 + 10a^6 + 10a^4 + 5a^2 + 1$
11 $1 - 10h + 45h^2 - 120h^3 + 210h^4 - 252h^5 + 210h^6 - 120h^7 + 45h^8 - 10h^9 + h^{10}$ **13** $3 + 3h + h^2$
15 $3c^2 + 3ch + h^2$ **17** $2(5x^4 + 10x^3h + 10x^2h^2 + 5xh^3 + h^4)$
19 $(1 + 1)^{10} = 1 + 10 + 45 + 120 + 210 + 252 + 210 + 120 + 45 + 10 + 1 = 1024$
21 $c^{20} + 20c^{19}h + 190c^{18}h^2 + 1140c^{17}h^3 + 4845c^{16}h^4 + \cdots + 4845c^4h^{16} + 1140c^3h^{17} + 190c^2h^{18} + 20ch^{19} + h^{20}$
23 The nth row of the triangle contains the coefficients in the expansion of $(a + b)^n$.
25

1 7 21 35 35 21 7 1
1 8 28 56 70 56 28 8 1
1 9 36 84 126 126 84 36 9 1
1 10 45 120 210 252 210 120 45 10 1

27 $x^{10} - 10x^9h + 45x^8h^2 - 120x^7h^3 + 210x^6h^4 - 252x^5h^5 + 210x^4h^6 - 120x^3h^7 + 45x^2h^8 - 10xh^9 + h^{10}$

29 $8064a^5b^5$ **31** $35x^{-5/2}$ **33** $48{,}384x^5y^3$ **35** $-\frac{15309}{8}a^5b^5$

10.4 PROBABILITY (*Page 368*)

1 $\frac{1}{8}$ **3** $\frac{7}{8}$ **5** $\frac{1}{2}$ **7** $\frac{1}{6}$ **9** $\frac{1}{6}$ **11** $\frac{1}{9}$ **13** $\frac{1}{2}$ **15** $\frac{25}{102}$ **17** 0 **19** $\frac{1}{2}$ **21** $\frac{1}{2704}$ **23** $\frac{144}{169}$ **25** $\frac{1}{169}$ **27** $\frac{28}{143}$ **29** $\frac{70}{143}$

31 (a) $\frac{1}{1024}$ (b) $\frac{1}{1024}$ (c) $\frac{7}{128}$ **33** (a) $\frac{48}{\binom{52}{5}} = 0.0000185$ (b) $\frac{13 \cdot 48}{\binom{52}{5}} = 0.0002401$ (c) $\frac{4\binom{13}{5}}{\binom{52}{5}} = 0.0019808$

35 \$5 **37** \$10 **39** 1 to 12; 12 to 1 **41** 7 to 2

SAMPLE TEST QUESTIONS FOR CHAPTER 10 (*Page 374*)

1 (a) 720 (b) 120 **2** 120 **3** (a) 900 (b) 648 **4** 455 **5** 2970 **6** 360 **7** (a) 32 (b) $\frac{1}{32}$ (c) $\frac{1}{32}$ **8** $\frac{2}{3}$

9 (a) $\frac{1}{36}$ (b) $\frac{35}{36}$ (c) $\frac{1}{6}$ **10** (a) $\frac{2}{169}$ (b) $\frac{2}{221}$ **11** $\frac{\binom{12}{4} \cdot 40}{\binom{52}{5}} = 0.0076184$ **12** $\frac{\binom{20}{4} \cdot 15}{\binom{35}{5}} = 0.2238689$ **13** (a) $\frac{1}{16}$ (b) $\frac{1}{16}$

14 $\frac{2}{11}$ **15** $\frac{9}{40}$ **16** $\frac{39}{80}$ **17** $x^5 - 10x^4y + 40x^3y^2 - 80x^2y^3 + 80xy^4 - 32y^5$ **18** $\frac{1}{1024a^{10}} - \frac{5b}{256a^9} + \frac{45b^2}{256a^8} - \frac{15b^3}{16a^7}$

19 $\binom{11}{6}(3a)^5b^6 = 112{,}266a^5b^6$ **20** $\binom{16}{8}(2x)^8(-y)^8 = 3{,}294{,}720x^8y^8$

INDEX

A

B

C

D

E

F

G

H

I

L

M

N

O

S

T

U

V

W

X

Y

Z

GEOMETRIC FORMULAS

Triangle

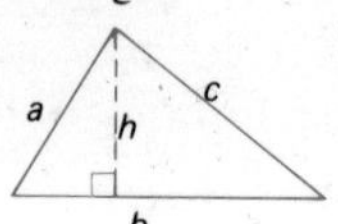

Area $= \frac{1}{2}bh$

Perimeter $= a + b + c$

Right Triangle

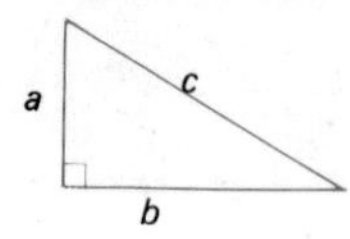

Pythagorean theorem

$a^2 + b^2 = c^2$

Similar Triangles

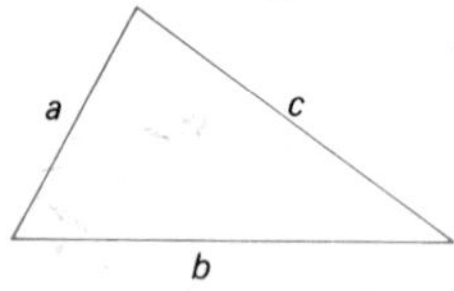

$\frac{a}{a'} = \frac{b}{b'} = \frac{c}{c'}$

a′ c′ b′

Special Triangles

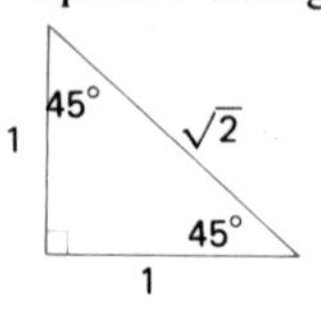

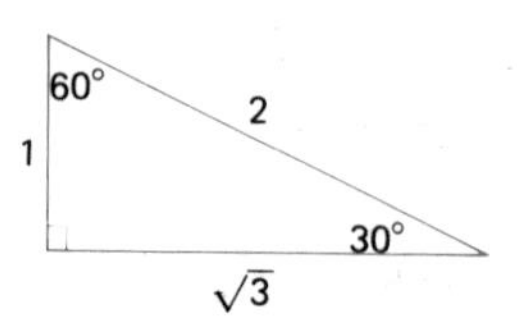

Square

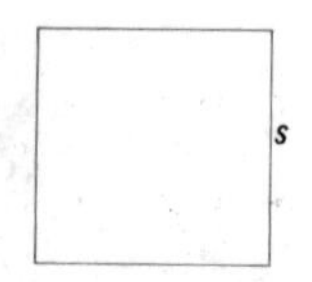

Area $= s^2$

Perimeter $= 4s$

Rectangle

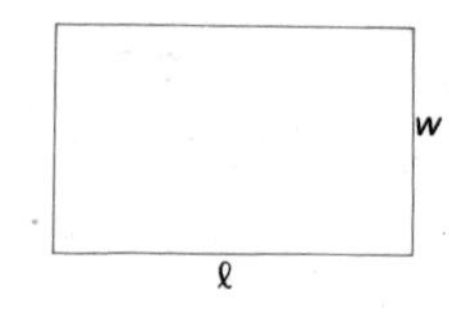

Area $= \ell w$

Perimeter $= 2\ell + 2w$

Parallelogram

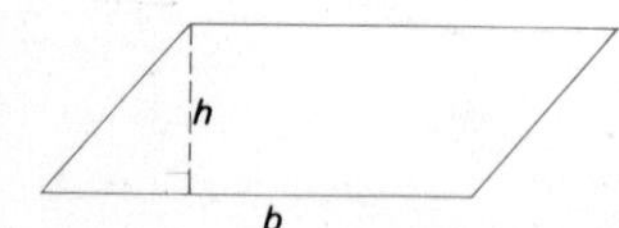

Area $= bh$

Trapezoid

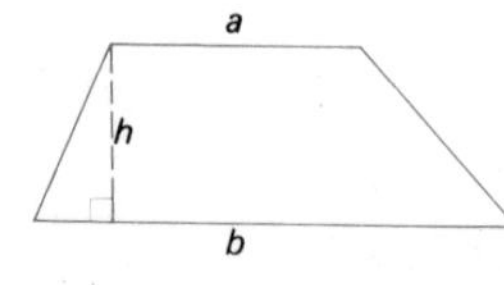

Area $= \frac{1}{2}h(a + b)$

Circle

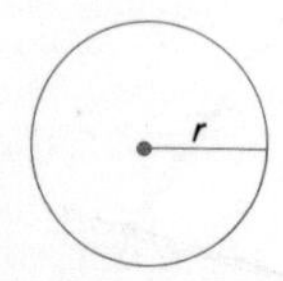

Area $= \pi r^2$

Circumference $= 2\pi r$

Circular Sector

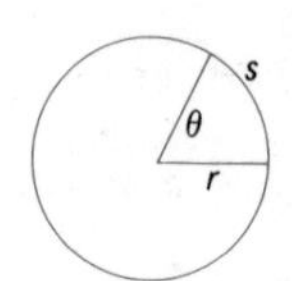

Area $= \frac{1}{2} r^2 \theta$

Arc length $s = r\theta$

Sphere

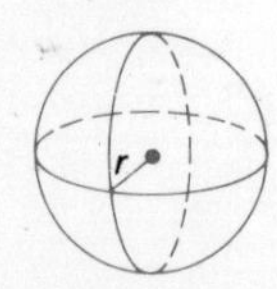

Volume $= \frac{4}{3}\pi r^3$

Surface area $= 4\pi r^2$

Right Circular Cylinder

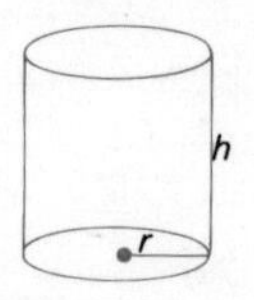

Volume $= \pi r^2 h$

Lateral surface area $= 2\pi rh$

Right Circular Cone

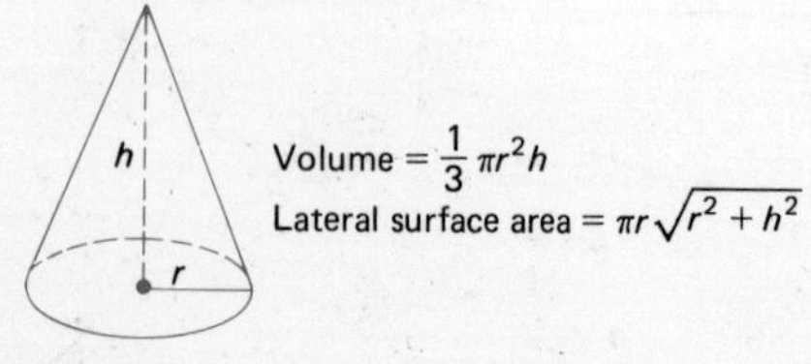

Volume $= \frac{1}{3}\pi r^2 h$

Lateral surface area $= \pi r\sqrt{r^2 + h^2}$

Rectangular Solid

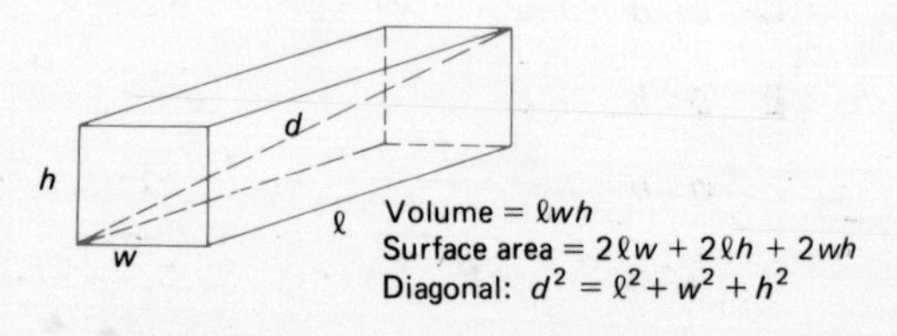

Volume $= \ell wh$

Surface area $= 2\ell w + 2\ell h + 2wh$

Diagonal: $d^2 = \ell^2 + w^2 + h^2$